Nanotechnology in the Life Sciences

Series Editor

Ram Prasad
Department of Botany
Mahatma Gandhi Central University
Motihari, Bihar, India

Nano and biotechnology are two of the 21st century's most promising technologies. Nanotechnology is demarcated as the design, development, and application of materials and devices whose least functional make up is on a nanometer scale (1 to 100 nm). Meanwhile, biotechnology deals with metabolic and other physiological developments of biological subjects including microorganisms. These microbial processes have opened up new opportunities to explore novel applications, for example, the biosynthesis of metal nanomaterials, with the implication that these two technologies (i.e., thus nanobiotechnology) can play a vital role in developing and executing many valuable tools in the study of life. Nanotechnology is very diverse, ranging from extensions of conventional device physics to completely new approaches based upon molecular self-assembly, from developing new materials with dimensions on the nanoscale, to investigating whether we can directly control matters on/in the atomic scale level. This idea entails its application to diverse fields of science such as plant biology, organic chemistry, agriculture, the food industry, and more.

Nanobiotechnology offers a wide range of uses in medicine, agriculture, and the environment. Many diseases that do not have cures today may be cured by nanotechnology in the future. Use of nanotechnology in medical therapeutics needs adequate evaluation of its risk and safety factors. Scientists who are against the use of nanotechnology also agree that advancement in nanotechnology should continue because this field promises great benefits, but testing should be carried out to ensure its safety in people. It is possible that nanomedicine in the future will play a crucial role in the treatment of human and plant diseases, and also in the enhancement of normal human physiology and plant systems, respectively. If everything proceeds as expected, nanobiotechnology will, one day, become an inevitable part of our everyday life and will help save many lives.

More information about this series at http://www.springer.com/series/15921

Devarajan Thangadurai
Jeyabalan Sangeetha • Ram Prasad
Editors

Bioprospecting Algae for Nanosized Materials

Springer

Editors
Devarajan Thangadurai
Department of Botany
Karnatak University
Dharwad, Karnataka, India

Jeyabalan Sangeetha
Department of Environmental Science
Central University of Kerala
Kasaragod, Kerala, India

Ram Prasad
Department of Botany
Mahatma Gandhi Central University
Motihari, Bihar, India

ISSN 2523-8027 ISSN 2523-8035 (electronic)
Nanotechnology in the Life Sciences
ISBN 978-3-030-81556-1 ISBN 978-3-030-81557-8 (eBook)
https://doi.org/10.1007/978-3-030-81557-8

This Springer imprint is published by the registered company Springer Nature Switzerland AG
The registered company address is: Gewerbestrasse 11, 6330 Cham, Switzerland

Preface

Nanotechnology is an emerging interdisciplinary field that holds many application strategies in various sectors. Nanoparticles are the smallest products of nanotechnology, size ranging from 1 to 100 nanometers. These nanoparticles are produced through three major ways, such as natural, accidental, and engineered nanoparticles. Based on the components and sources of nanoparticles, these are further categorized into natural organic nanoparticles, anthropogenic organic nanoparticles, natural inorganic nanoparticles, anthropogenic inorganic nanoparticles, and DNA- and peptide-based nanoparticles. Among all the categorized nanoparticles, anthropogenic inorganic nanoparticles are playing a major role in the research and significance of industrial applications in many fields as those materials are known for their unique characteristics like magnetic and optical polarizability and conductivity. The wide usage of inorganic nanoparticles is often associated with many adverse effects on the environment and living organisms including human beings. The adverse effects of inorganic nanoparticles urged researchers to think about the production of nanoparticles that are ecofriendly, nontoxic, easy to operate, energy efficient, use renewable resources, and cost effective. Green nanotechnology is being explored to synthesize inorganic nanoparticles by involving more ecofriendly and energy efficient methodologies which may have minimal or negligible adverse effects without causing any damages to the environment and human health. Biological systems including microorganisms and plants have been used under green nanotechnology processes.

Fascinatingly, research works on nanotechnology aspects are focusing on the biosynthesis of inorganic nanoparticles with biological sources. Algae are one of the potential biological systems for the reduction of metal ions for the synthesis of bionanoparticles through economical, high yielding, ecofriendly, speedy, and energy efficient processes. Many algal species from the classes of Chlorophyceae, Phaeophyceae, Rhodophyceae, and Cyanophyceae, including diatoms and euglenoids, are also involved in the biosynthesis process of inorganic nanoparticles using silver, gold, and palladium metal ions. Moreover, the algal species are more physiologically expedient to handle than other biological sources and have vast biodiversity in addition to synthesis of the nanoparticles at low temperature, zero or less toxicity, and zero or negligible adverse effect on the environment. Microalgae are used in the different ways to synthesis bionanomaterials at either molecular or cellular level scale. Compared to other microorganisms, algae also equally contribute to the biosynthesis of nanoparticles. Algal nanoparticles have been studied and proved as one of the promising materials to be considered in various sectors, including medical, environmental, agricultural, and industrial sectors. Microalgal-mediated metallic nanoparticles are explored for the combined advantages of treatment of wastewater by use as a growth medium and for the production of algal biomass which in turn portray them as a powerful nanofactory. Algae are commonly known as the bionanofactories among other biological materials, as both living as well as dead dried algae are used for the biosynthesis of bionanoparticles. Algal-mediated nanoparticles are playing vital role in various sectors like medical, food, cosmetics, and environment. Algal nanoparticles are possessing high potential on anticancer and antimicrobial as well as it is being used in the process of imaging, drug delivery, gene therapy and biosensors.

This book has five major parts that provide a clear understanding of the chapters covered under the broad title. Part I elaborates on the merits and limitations of the algal nanotechnology. Part II discusses the different processes involved in the biosynthesis of algal-mediated nanoparticles and characterization of algal nanoparticles. Part III discusses the production of nanomaterials and nanoproducts from algae. The authors have discussed elaborately the importance and merits of algae in biosynthesis of nanoparticles. Two chapters in this part selectively discuss seaweeds and Pelagic Sargassum as the progressing sources in the field of nanotechnology. Part IV is occupied by chapters mainly dealing with the medical applications of algal nanomaterials. The authors have reviewed well the potential application of algal nanomaterials in effective encapsulation techniques, general disease therapy, cancer therapy, drug delivery, and tissue engineering. The final part, Part V, discusses industrial and environmental applications of algal-mediated nanomaterials. In this part, the authors have given an overview on the application of algal nanomaterials for the production of nanocellulose for sustainable plastic production, wastewater treatment, bio-absorption, and remediation of environmental pollutants.

Prominent experts in the field of algal nanotechnology have been selected as contributing authors for this book. The editors appreciatively acknowledge their acceptance of our invitation and convey their willingness to contribute to this valuable book. The editors of this book are also very much grateful to Mr. Eric Stannard, Senior Editor, Springer Nature, for his industrious, valuable, and prudent support throughout this publication.

This book will serve as an essential source on the synthesis and various applications of algal-mediated nanoparticles for the researchers and professionals in the fields of botany, phycology, environmental biology, therapeutics, and nanobiology. This book is suitable for graduate students, young researchers, scientists, academicians, and industrial professionals who are working in the field of algal nanomaterials.

Dharwad, Karnataka, India — Devarajan Thangadurai
Kasaragod, Kerala, India — Jeyabalan Sangeetha
Motihari, Bihar, India — Ram Prasad

Contents

Contributors

Bilal Haider Abbasi Department of Biotechnology, Quaid-i-Azam University, Islamabad, Pakistan

Mrinal Kanti Adak Department of Chemistry, Sidho-Kanho-Birsha University, Purulia, West Bengal, India

Department of Chemical Sciences, Indian Institute of Science Education and Research, Mohali, Punjab, India

Iffat Zareen Ahmad Department of Bioengineering, Integral University, Lucknow, Uttar Pradesh, India

Bakrudeen Ali Ahmed Department of Biochemistry and Biotechnology, Centre for Research and Development, PRIST Deemed University, Vallam, Thanjavur, Tamil Nadu, India

Institute of Biological Sciences, Faculty of Science, University of Malaya, Kuala Lumpur, Malaysia

Sumaira Anjum Department of Biotechnology, Kinnaird College for Women, Lahore, Pakistan

Vishal Aparadh PG Department of Botany, Shri Pancham Khemraj Mahavidyalay, Sawantwadi, Maharashtra, India

Rathindra Mohan Banik School of Biochemical Engineering, Indian Institute of Technology (Banaras Hindu University), Varanasi, Uttar Pradesh, India

Luísa Barreira Centre for Marine Sciences, University of Algarve, Faro, Portugal

Mahesh P. Bhat Centre for Nano and Material Sciences, Jain University, Jain Global Campus, Bengaluru, Karnataka, India

Georgina Calderón-Domínguez Instituto Politécnico Nacional, Escuela Nacional de Ciencias Biológicas, Departamento de Ingeniería Bioquímica, Ciudad de México, Mexico

Héctor Uriel Castañeda-Serna Instituto Politécnico Nacional, Escuela Nacional de Ciencias Biológicas, Departamento de Ingeniería Bioquímica, Ciudad de México, Mexico

Rimsha Chaudhary Department of Biotechnology, Kinnaird College for Women, Lahore, Pakistan

Jorge Alberto Vieira Costa Laboratory of Biochemical Engineering, College of Chemistry and Food Engineering, Federal University of Rio Grande, Rio Grande, Brazil

Ajit Das Department of Chemistry, Sidho-Kanho-Birsha University, Purulia, West Bengal, India

Department of Chemistry, Balarampur College, Balarampur, West Bengal, India

Ana Cláudia Araujo de Almeida Laboratory of Microbiology and Biochemistry, College of Chemistry and Food Engineering, Federal University of Rio Grande, Rio Grande, Brazil

Ma. De la Paz Salgado Cruz Instituto Politécnico Nacional, Escuela Nacional de Ciencias Biológicas, Departamento de Ingeniería Bioquímica, Ciudad de México, Mexico

Consejo Nacional de Ciencia y Tecnología (CONACYT), Ciudad de México, Mexico

Etiele Greque de Morais Centre for Marine Sciences, University of Algarve, Faro, Portugal

Michele Greque de Morais Laboratory of Microbiology and Biochemistry, College of Chemistry and Food Engineering, Federal University of Rio Grande, Rio Grande, Brazil

Nivas Desai PG Department of Botany, Shri Pancham Khemraj Mahavidyalay, Sawantwadi, Maharashtra, India

Uttam Dethe PG Department of Botany, Shri Pancham Khemraj Mahavidyalay, Sawantwadi, Maharashtra, India

Daiane Dias Laboratório de Eletroespectro Analítica (LEEA), Escola de Química e Alimentos, Universidade Federal do Rio Grande, Rio Grande, RS, Brazil

Sonam Dwivedi Natural Products Laboratory, Department of Bioengineering and Biosciences, Integral University, Lucknow, Uttar Pradesh, India

Reynold Ramón Farrera-Rebollo Instituto Politécnico Nacional, Escuela Nacional de Ciencias Biológicas, Departamento de Ingeniería Bioquímica, Ciudad de México, Mexico

Dattatraya Gaikwad Department of Botany, Shivaji University, Kolhapur, India

Arturo García-Bórquez Instituto Politécnico Nacional, Escuela Superior de Fisico Matemáticas, Ciudad de México, Mexico

Ana Grenha Centre for Marine Sciences, University of Algarve, Faro, Portugal

Christophe Hano Laboratoire de Biologie des Ligneux et des Grandes Cultures (LBLGC), INRAE USC1328, Université d'Orléans, Chartres, France

Mônika Grazielle Heinemann Laboratório de Eletroespectro Analítica (LEEA), Escola de Química e Alimentos, Universidade Federal do Rio Grande, Rio Grande, RS, Brazil

Ho-Young Jung Department of Environment and Energy Engineering, Chonnam National University, Gwangju, Republic of Korea

Amna Komal Khan Department of Biotechnology, Kinnaird College for Women, Lahore, Pakistan

Mahaveer D. Kurkuri Centre for Nano and Material Sciences, Jain University, Jain Global Campus, Bengaluru, Karnataka, India

Sanjaya Lama Analytical and Circular Chemistry, Institute for Materials Research IMO-IMOMEC, University of Hasselt, Diepenbeek, Belgium

Juliana Botelho Moreira Laboratory of Microbiology and Biochemistry, College of Chemistry and Food Engineering, Federal University of Rio Grande, Rio Grande, Brazil

Prakashkumar Nallasamy Bionanomaterials Research Lab, Department of Nanoscience and Technology, Alagappa University, Karaikudi, Tamil Nadu, India

Suganthy Natarajan Bionanomaterials Research Lab, Department of Nanoscience and Technology, Alagappa University, Karaikudi, Tamil Nadu, India

Khadija Nawaz Department of Biotechnology, Kinnaird College for Women, Lahore, Pakistan

Michaela Pappa Analytical and Circular Chemistry, Institute for Materials Research IMO-IMOMEC, University of Hasselt, Diepenbeek, Belgium

Umesh Pawar PG Department of Botany, Shri Pancham Khemraj Mahavidyalay, Sawantwadi, Maharashtra, India

Jorge F. Pontes Centre for Marine Sciences, University of Algarve, Faro, Portugal

Arunkumar Ramu Department of Biochemistry and Biotechnology, Centre for Research and Development, PRIST Deemed University, Vallam, Thanjavur, Tamil Nadu, India

Pieter Samyn Analytical and Circular Chemistry, Institute for Materials Research IMO-IMOMEC, University of Hasselt, Diepenbeek, Belgium

Laura M. Sanchez Materiales Compuestos Termoplásticos (CoMP), Instituto de Investigaciones de Ciencia y Tecnología de Materiales (INTEMA) CONICET-UNMdP), Mar del Plata, Argentina

Shailendra Singh Shera Department of Biotechnology, Faculty of Engineering and Technology, Rama University, Kanpur, Uttar Pradesh, India

Heena Tabassum Department of Bioengineering, Integral University, Lucknow, Uttar Pradesh, India

U. T. Uthappa Centre for Nano and Material Sciences, Jain University, Jain Global Campus, Bengaluru, Karnataka, India

Department of Environment and Energy Engineering, Chonnam National University, Gwangju, Republic of Korea

Dries Vandamme Analytical and Circular Chemistry, Institute for Materials Research IMO-IMOMEC, University of Hasselt, Diepenbeek, Belgium

João Varela Centre for Marine Sciences, University of Algarve, Faro, Portugal

Aysun Yucetepe Department of Food Engineering, Faculty of Engineering, Aksaray University, Aksaray, Turkey

About the Editors

Devarajan Thangadurai is associate professor at Karnatak University, Dharwad, Karnataka, India. He received his PhD in botany from Sri Krishnadevaraya University in South India as CSIR Senior Research Fellow with funding from the Ministry of Science and Technology, Government of India. He served as a postdoctoral fellow at the University of Madeira, Portugal; University of Delhi, India; and ICAR National Research Centre for Banana, India. He is the recipient of the Best Young Scientist Award with a Gold Medal from Acharya Nagarjuna University, India, and the VGST-SMYSR Young Scientist Award of the Government of Karnataka, India. He has authored and edited more than 25 books with publishers of national/international repute. He has also visited 24 countries in Africa, Middle East, Asia, and Europe for academic visits, scientific meetings, and international collaborations.

Jeyabalan Sangeetha is assistant professor in the Central University of Kerala at Kasaragod, Kerala, India. She earned her BSc in microbiology and PhD in environmental science from Bharathidasan University, Tiruchirappalli, Tamil Nadu, India. She holds an MSc in environmental science from Bharathiar University, Coimbatore, Tamil Nadu, India. Sangeetha is the recipient of Tamil Nadu Government Scholarship and Rajiv Gandhi National Fellowship of University Grants Commission, Government of India, for her doctoral studies. She served as Dr. D.S. Kothari Postdoctoral Fellow and UGC Postdoctoral Fellow at Karnatak University, Dharwad, Karnataka, India, during 2012–2016 with funding from University Grants Commission, Government of India, New Delhi. She has published more than 80 research articles, book chapters, and books with international publishers including Springer Nature, Switzerland; Taylor & Francis, UK; and IK International, India. Her research interests are in the fields of environmental toxicology, environmental microbiology, environmental biotechnology, and environmental nanotechnology.

Ram Prasad is associate professor in the Department of Botany, Mahatma Gandhi Central University, Motihari, Bihar, India. His research interest includes applied microbiology, plant-microbe-interactions, sustainable agriculture, and nanobiotechnology. Dr. Prasad has more than 200 publications to his credit, including research papers, review articles and book chapters, and six patents issued or pending, and he has edited or authored several books. Dr. Prasad has 12 years of teaching experience, and he has been awarded the Young Scientist Award (2007), Prof. J.S. Datta Munshi Gold Medal (2009), FSAB Award (2010), American Cancer Society UICC International Fellowship for Beginning Investigators (2014), Outstanding Scientist Award (2015), BRICPL Science Investigator Award (2017), and Research Excellence Award (2018). He has been serving as editorial board member for *BMC Microbiology, BMC Biotechnology, IET Nanobiotechnology, Journal of Nanomaterials, Current Microbiology, Annals of Microbiology, Archives of Microbiology, Archives of Phytopathology and Plant Protection, Journal of Renewable Materials, Journal of Agriculture and Food Research*, and book series editor of Nanotechnology in the Life Sciences by Springer Nature. Previously, Dr. Prasad served as assistant professor, Amity University, Uttar Pradesh, India; visiting assistant professor at Johns Hopkins University, USA; and research associate professor at Sun Yat-sen University, Guangzhou, China.

Part I
Introduction

Chapter 1
Algal Nanotechnology: Scope and Limitations

Prakashkumar Nallasamy and Suganthy Natarajan

Contents

1.1 Introduction

Nanotechnology focuses on synthesis, arrangement and technical applications of nanosized materials (nano – 10^{-9} m or billionth of a metre). Nanoscale materials exhibit modified properties and behaviours with respect to size, shape and dimensional aspects, when compared to all other micro- and macro-scaled materials thereby exhibiting unique physical, chemical and mechanical properties. Relatively, the above-mentioned modifications in properties and behaviours of nanoparticles enhance its application in various fields of science and technology like medicine, agriculture, biotechnology, wastewater treatment, energy, textile, optical, electrical,

P. Nallasamy · S. Natarajan (✉)
Bionanomaterials Research Lab, Department of Nanoscience and Technology, Alagappa University, Karaikudi, Tamil Nadu, India
e-mail: suganthyn@alagappauniversity.ac.in

D. Thangadurai et al. (eds.), *Bioprospecting Algae for Nanosized Materials*, Nanotechnology in the Life Sciences,
https://doi.org/10.1007/978-3-030-81557-8_1

sensors, cosmetics, etc. (Prasad 2014; Prasad et al. 2014, 2017, 2018). Till date understanding the fundamentals of nanostructure was based on the synthesis of nanomaterials using top-down and bottom-up approaches (Cao 2010; Kajbafvala et al. 2013). In general, top-down and bottom-up approaches for the synthesis of nanoscale materials involve physical, chemical and biological methods. Physical method of nanoparticle synthesis focuses on abrading, evaporating, melting or condensing the material sources using high energy emissions, mechanical forces, electrical and thermal energies. In the case of chemical methods, techniques like sol-gel, polyol, hydrothermal, microemulsion, chemical vapour deposition and plasma-enhanced chemical vapour deposition were employed for synthesis of nanoscale materials. However, these physical and chemical methods were found to be costlier and pose severe environmental and health hazards. Hence, researchers have turned their focus towards less toxic and cost-effective biogenic route for effective fabrication of nanomaterials using biological entities like animals, plants, bacteria, virus, fungi, algae, etc. (Dhand et al. 2015; Prasad 2014, 2016, 2017, 2019a, b, c, d; Prasad et al. 2016, 2018; Srivastava et al. 2021). These essential technical findings reveal that the potential on reducing, capping and stabilizing agents for the green synthesis of nanosized materials was achieved with several kinds of bioactive compounds present in most of the biological origin (Rai et al. 2013). The first report on phycosynthesis of nanoparticles was synthesis of Au nanoparticles using the marine algae *Sargassum wightii* (Singaravelu et al. 2007). Sharma et al. (2016) reported synthesis of nanoparticles using algae belonging to the Phaeophyceae, Rhodophyceae and Chlorophyceae and the mechanism behind the phycosynthesis.

Algae are crucial photosynthetic organisms mostly available in different water resources environment such as freshwater, marine water, brackish water and some moist rocks or soil (Lee 2018). The archaeological witness collected from the archaeological site of southern Chile states that several thousand years ago the source of algae was utilized by human beings as food and medicinal supplement (Dillehay et al. 2008). Furthermore, the diverse categories of green algae (Chlorophyta), red algae (Rhodophyta) and brown algae (Phaeophyceae) can be employed for the production of food, pharmaceutics, essential biological supplements, cosmetics, energy and other environmental safety applications (Khan et al. 2018). The biogenic synthesis of nanoparticles using algae was widely termed as phycosynthesis of nanoparticles (Khalafi et al. 2019).

In general, the Greek word "phykos" means algae or seaweed (Lee 2018); hence, the algae-mediated synthesis of nanoparticles was termed as phyco (phykos)-synthesis of nanoparticles. Accordingly, the last decades were observed with synthesis and applications of different types of organic, inorganic and composite nanomaterials using various micro- and macro-algal sources collected from diverse biological environment (LewisOscar et al. 2016). Consequently, the phycosynthesis of nanoparticles was typically noted with some specific advantages including frequently beneficial, simple storage, economically cheap and minimal consumption of various source of energy and time with lower toxic or risk factors (Aziz et al. 2014, 2015). Phycosynthesis of nanoparticles significantly overcomes the drawbacks of harmful effects caused by the radiations, thermal and electrical impulses of physical method or otherwise the toxic effects caused by the chemical substrates as a part of reducing and stabilizing agents (Sharma et al. 2016). Therefore, the present chapter focuses on phycosynthesis of nanoparticles and their applications in various divisions of science and technology.

1.2 Steps Involved in Phycosynthesis

The biological methods adapted for the synthesis of nanoparticles show technical variations when compared to all other physical and chemical methods. The process on phycosynthesis of nanoparticles has to be initiated with the collection of different classes of algae from their dwelling bodies mostly from marine water (LewisOscar et al. 2016). The algae-mediated synthesis of nanoparticles can be carried out with three major steps as detailed in Fig. 1.1.

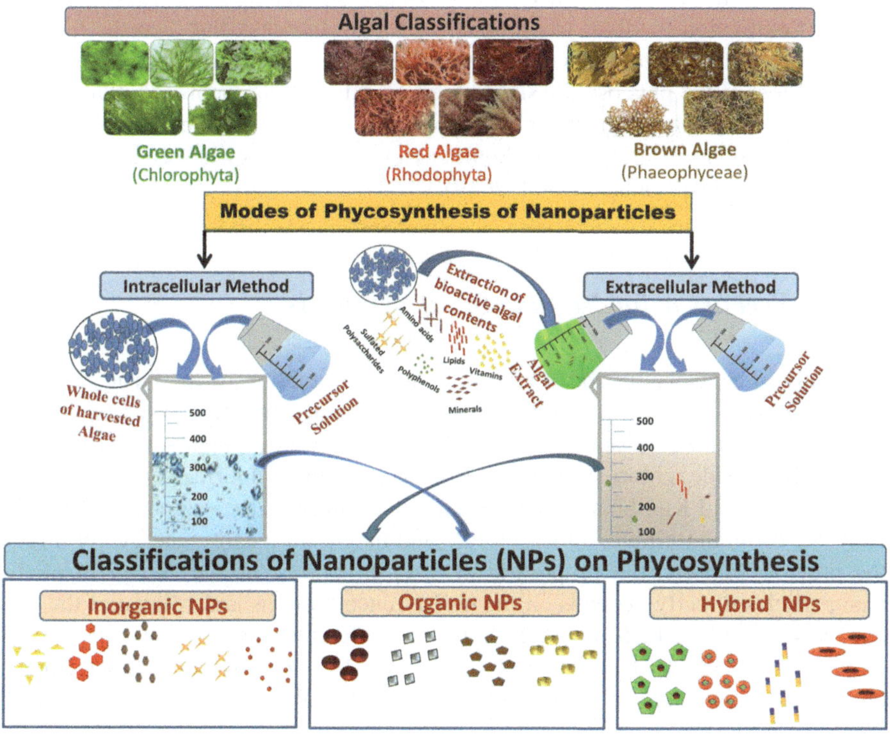

Fig. 1.1 Graphical representation of synthesis of various nanoparticles using different class of algal sources

1.2.1 Preparation of Algal Extract

The bioactive elements present in the algal resources act as reducing and stabilizing agents for the synthesis of nanoparticles. Particularly, the seaweeds were naturally enriched by the process of photosynthesis to produce essential bioactive compounds such as amino acids, fatty acids, lipids, phenols, proteins, polysaccharides, phlorotannins and pigment accessories (Salehi et al. 2019). Table 1.1 reveals the bioactive compounds present in algal sources involved in nanoparticle synthesis and their descriptions throughout the previous reports. After the collection of seaweeds, it was washed several times with fresh tap water and soaked in distilled water at room temperature to remove the contaminant factors such as dirt, sand and the epiphytes followed by the drying of algae at the appropriate temperature. In addition to the above detailed washing and drying, the algal sources were exposed to cutting or graining techniques to get the yield of powdered samples before the extraction process and in some exceptional cases the whole algae was used for extraction. The extraction process from algae was mostly done with deionized water or organic solvents at different temperatures either by boiling or soaking the algal content to excrete the bioactive compounds from the origin of algal cells. Moreover, another mode of extraction involves freeze drying of algal biomass and then the biomass was subjected to different types of chromatographic and membrane filtration techniques. Most of the phycosynthesis is based on extracellular approach, that is, synthesis of nanoparticles using bioactive compounds from aqueous and organic extract of algae (Abboud et al. 2014; Dahoumane et al. 2016; LewisOscar et al. 2016; Mashjoor et al. 2018; Khalafi et al. 2019).

1.2.2 Precursor for Nanomaterial Synthesis

Generally, precursors for the synthesis of nanoparticles were the basic chemical substance available in various solid, liquid and gaseous state based on the nature of synthesis of nanoscale materials. Organic and inorganic precursor reagents are mostly their oxide, chloride, nitrate, carbonate, sulphate, fluorides and hydroxides form (Kumar and Kumbhat 2016), such as silver nitrate, zinc sulphate, copper (II) chloride, cadmium nitrate, auric chloride and magnesium carbonate. According to the previous reports, the proportion of size, structure, surface and yield of the nanoparticles directly depends on the molar concentrations and volume of the precursor solutions, which interact with the different classes of algal extracts (Dahoumane et al. 2014a, b). Table 1.2 illustrates the list of various ionic precursors and their molar concentrations adapted for the algae-mediated synthesis of nanoparticles.

Table 1.1 Major bioactive compounds extracted from algal sources involved in nanoparticle synthesis

Bioactive compounds detected in algae	Available classifications of the compound in algae	References
Polysaccharides	Galactans	Venkatesan et al. (2016)
	Fucoidan	
	Laminarin	
	Alginates	
	Carrageenan	Necas and Bartosikova (2013)
	Agar	Holdt and Kraan (2011)
Proteins	Variations in its low and higher content	
Pigments	Chlorophylls	Salehi et al. (2019)
	Carotenoids	
	Phycobiliproteins	
Polyphenols	Phenolic acids	Machu et al. (2015)
	Flavonoids	
	Isoflavones	
	Cinnamic acid	
	Benzoic acid	
	Quercetin	
	Lignans	
Amino acids	Aspartic acid	Punnett and Derrenbacker (1966)
	Alanine	
	Cysteine	
	Glycine	
	Glutamic acid	
	Histidine	
	Lysine	
	Leucine	
	Methionine	
	Threonine	
	Tyrosine	
	Tryptophan Valine	

(continued)

Table 1.1 (continued)

Bioactive compounds detected in algae	Available classifications of the compound in algae	References
Vitamins	Retinol	Wells et al. (2017)
	Thiamine	
	Riboflavin	
	Niacin	
	Pantothenic acid	
	Pyridoxine	
	Biotin	
	Cyanocobalamin	
	Ascorbic acid	
	Tocopherol	
Plant growth hormones	Auxins	Chojnacka (2012)
	Cytokinins	
	Gibberellins	
Minerals	Potassium	
	Magnesium	
	Calcium	
	Some other traceable minerals	
Lipids	Phospholipids	Kumari et al. (2013)
	Glycolipids	
Fatty acids	Monounsaturated fatty acids (MUFAs)	Rohit and Venkata Mohan (2018)
	Polyunsaturated fatty acids (PUFAs)	

1.2.3 Incubation of Algal Extracts with Precursors of Nanomaterials

The incubation period involves mixing of algal extract with ionic precursors of nanoparticles in a particular ratio under constant stirring/ agitation or without using any kind of physical/mechanical enforces (Shah et al. 2015; Fatima et al. 2020). Incubation is carried out under suitable pH, temperature and illumination to produce the nanoparticles under controlled condition (Dahoumane et al. 2016). The incubation period was carried out till colorimetric change on the reaction medium was observed as nanomaterials exhibit colour changes with respect to the size (Rahman et al. 2019). The solvent and all other physiochemical conditions were scrutinized based on the properties and behaviours of the algal extract and precursor sources taken for the synthesis of nanoscale materials. In the case of synthesis with whole-algal cell instead of the algal extract, algal cell is inoculated into the precursor media for the synthesis of the nanoparticles (Hulkoti and Taranath 2014; Li et al. 2015).

Table 1.2 Ionic precursors and their specific molar concentrations used in phycosynthesis of nanoparticles

Nanoparticles (NPs)	Precursors	Molar concentration	References
Ag NPs	$AgNO_3$	1.0 mM	Barwal et al. (2011)
		10 mM	Mora-Godínez et al. (2020)
Au NPs	$AuCl_3$	1.0 mM	Rajeshkumar et al. (2013)
	$HAuCl_4$	1.0 mM	Rajathi et al. (2012)
ZnO NPs	$Zn(NO_3)_2.6H_2O$	–	Francavilla et al. (2014)
	$Zn(Ac)_2.2H_2O$	2.0 mM	Azizi et al. (2014)
	$Zn(CH_3COO)_2$	–	Khalafi et al. (2019)
CuO NPs	$CuSO_4$	1.0 mM	Abboud et al. (2014)
Cds NPs	$Cd(NO_3)_2$	1.8 mM	Mandal et al. (2016)
MgO NPs	$Mg\ (NO_3)_2.6H_2O$	–	Pugazhendhi et al. (2018)
Pd NPs	$PdCl_2$	1.0 mM	Momeni and Nabipour (2015)
	Pd $(OAc)_2$	1.0 mM	Arya et al. (2020)
Pt NPs	H_2PtCl_6	1.0 mM	Arya et al. (2020)
Fe_3O_4 NPs	$FeCl_2.4H_2O$	1.0 mM	Mashjoor et al. (2018)
	$FeCl_3.6H_2O$	2.0 mM	
TiO_2 NPs	$C_{12}H_{28}O_4Ti$	0.5 M	Sharma et al. (2018)
ZrO_2 NPs	$ZrO(NO_3)_2$	5 mM	Kumaresan et al. (2018)

1.3 Mechanism Behind the Algal-Mediated Synthesis of Nanoparticles

The phycosynthesis approaches involve (i) extracellular and (ii) intracellular methods for the synthesis of inorganic nanoparticles and (iii) conversation of total mass or the extracted compounds of algae into organic nanoparticles.

1.3.1 Extracellular Synthesis

Initially, the role of extracellular synthesis incorporates the precursor solution and the aqueous phase of the bioactive compounds extracted from algal sources such as peptides, proteins, enzymes, pigments, antioxidants, polysaccharides and reducing sugars as tabulated in Table 1.1. Bioactive compounds tend to reduce and precipitate the ions of precursor source towards the synthesis of nanoparticles, and also the carboxyl, hydroxyl and amino-functional additives of the algal phytochemicals play a major role in reducing, capping and stabilizing agents for the synthesis of nanoparticles.

1.3.2 Intracellular Synthesis

Intracellular synthesis of nanoparticles confers the reduction in particle size into several nanometres as a result of respiration, photosynthesis, nitrogen fixation and other metabolic activities of the algal cells infused within the precursor medium. In intracellular process, the precursor material interacts with algal cells via electrostatic force of interaction between the positively charged precursor ions and the negatively charged phospholipids present in the cell membrane of the algal cells respectively (Hulkoti and Taranath 2014; Vijayan et al. 2016). Intracellular methods for nanomaterial synthesis using live algal cells have certain drawbacks such as (i) algal cells were unable to retain their metabolic activity for a long duration and (ii) the algal cells breakdown due to stressful surrounding medium (Dahoumane et al. 2016; Rahman et al. 2019). Hence, conventional approach using whole entities and extracted biocompounds of algae was utilized for synthesis of carbon-based and polymeric nanoparticles (Bhatia 2016; Ramanan et al. 2016; Venkatesan et al. 2016).

1.4 Types of Algal-Mediated Nanoparticles

Generally, nanomaterials were classified into three major classes: (i) inorganic nanomaterials – metals or their oxides (Pandey and Dahiya 2016); (ii) organic nanoparticles – mostly polymer-based and lipid-based nanoparticles (Dong and Hu 2013) and (iii) hybrid nanomaterials (Ananikov 2019) – a combination of different kinds of nanoparticles like alloys or composites and they are extensively subjected with combination of two or more organic and/or inorganic nanosubstrates, that is, combination of inorganic to inorganic, organic to organic and organic to inorganic nanosubstrates. Organic, inorganic and hybrid nanomaterials synthesized using algal technology have a wide array of application in various fields of bioimaging, drug delivery system, wastewater treatment, dye degradation, antimicrobial, anticancer, antiproliferative, antiapoptotic and biosensor, which are clearly tabulated in Table 1.3.

1.4.1 Algae-Based Inorganic Nanoparticles

The overall mechanism behind the algae-based intracellular and extracellular synthesis of metal and metal oxide nanoparticles has been discussed in the previous section. First report on phycosynthesis of nanoparticles was synthesis of Au nanoparticles using extract of marine algae *Sargassum wightii* with aqueous solution of $HAuCl_4$ as precursor, which yielded uniformly dispersed Au nanoparticles (Singaravelu et al. 2007). Later, different kinds of metal and metal oxide

Table 1.3 Classification and applications of phycosynthesized nanoparticles

Type of nanoparticles	Algae/source derived from algae	Nanoparticles	Shape	Size (nm)	Applications	References
Organic nanoparticles	*Pectinodesmus* sp.	Carbon nanodots	Spherical	67	Anticancer effect on breast cancer (HCC1954) and colorectal cancer (HCT116) cell lines	Amjad et al. (2019)
	Eutrophic algal blooms	Fluorescent carbon dots	Quasi-spherical	8.5 ± 5.6	Bioimaging of MCF-7 breast cancer cells	Ramanan et al. (2016)
	Microalgae	Carbon quantum dots	Spherical	1 < 8	Bioimaging of MCF-7 breast cancer cells	Guo et al. (2017)
	Sodium alginate	Alginate NPs	Spherical	~200	–	Machado et al. (2012)
	Sodium alginate	Alginate nanoparticles		50–100	Loading of pesticide (imidacloprid) into sodium alginate nanoparticles	Kumar et al. (2014)
	Fucoidan	Acetylated fucoidan NPs	Spherical	137.52	Doxorubicin-loaded fucoidan nanoparticles for the treatment of HCT-116 and HCT-8 cancer cells	Lee et al. (2013)
	Fucoidan	Protamine/ fucoidan NPs	Spherical	180	Doxorubicin-loaded protamine/fucoidan nanoparticles for the treatment of MDA-MB-231 breast cancer cells	Lu et al. (2017)
	Chitosan	Chitosan NPs	–	30–50	Chitosan nanoparticles for controlled release of bovine lactoferrin (bLF) to prevent carbendazim-induced toxicity	Sunkireddy et al. (2016)

(continued)

Table 1.3 (continued)

Type of nanoparticles	Algae/source derived from algae	Nanoparticles	Shape	Size (nm)	Applications	References
Inorganic nanoparticles	*Sargassum wightii*	Au NPs	–	11	–	Singaravelu et al. (2007)
	Stoechospermum marginatum	Au NPs	Spherical, hexagonal and triangle	62.5	Antibacterial studies on certain pathogenic bacterial strains	Rajathi et al. (2012)
	Chlorella	Au NPs	Spherical	~14.65 ± 3.95	Antioxidant and catalytic applications	Zayadi and Bakar (2020)
	Turbinaria conoides	Au NPs	Spherical	6–10	–	Rajeshkumar et al. (2013)
	Gelidium corneum	Ag NPs	Angular	20–40	Antibacterial and antibiofilm applications	Özturk et al. (2020)
	Chlorella pyrenoidosa	Ag NPs	–	>20	Antibacterial and photocatalytic application	Aziz et al. (2015)

Type of nanoparticles	Algae/source derived from algae	Nanoparticles	Shape	Size (nm)	Applications	References
	Portieria hornemannii	Ag NPs	Spherical	35–50	Antibacterial effect on fish pathogens	Fatima et al. (2020)
	Chlorella	ZnO NPs	Hexagonal	20	Photocatalytic degradation of dibenzothiophene (DBT)	Khalafi et al. (2019)
	Kappaphycus alvarezii	ZnO NPs	Rod	62–74	–	Pachiyappan et al. (2020)
	Sargassum muticum	ZnO NPs	Hexagonal	42	–	Azizi et al. (2014)
	Sargassum wightii	MgO NPs	Flower	68.06	Antibacterial, antifungal and photocatalytic evaluations	Pugazhendhi et al. (2018)
	Turbinaria ornata	MgO NPs	Flakes	–	Antibacterial effect on *Mycobacterium tuberculosis*	Govindaraju et al. (2020)
	Bifurcaria bifurcata	Cu NPs	Spherical	20.66	Antibacterial effect on *Enterobacter aerogenes* and *Staphylococcus aureus* strains	Abboud et al. (2014)
	Arthrospira platensis	Cds NPs	Spherical	8.4	Photocatalytic degradation of malachite green (MG) dye	Mandal et al. (2016)
	Ulva flexuosa	Fe_3O_4 NPs	Cubo-spherical	12.3	Antibacterial and antifungal applications	Mashjoor et al. (2018)
	Sargassum bovinum	Pd NPs	Octahedral	5–10	Electrocatalytic hydrogen peroxide sensor	Momeni and Nabipour (2015)
	Botryococcus braunii	Pd and Pt NPs	Cubical, spherical and triangular	4.89 and 86.96	Antimicrobial and antioxidant effects of Pd and Pt NPs	Arya et al. (2020)
	Sargassum wightii	ZrO_2 NPs	Spherical	5	Antibacterial effect on gram-positive and gram-negative strains	Kumaresan et al. (2018)

(continued)

Table 1.3 (continued)

Type of nanoparticles	Algae/source derived from algae	Nanoparticles	Shape	Size (nm)	Applications	References
Hybrid nanoparticles	*Sargassum muticum*	HA/ZnO nanocomposite	Spherical	20–25	Antibiofilm and anticorrosion applications	Sadek et al. (2019)
	Algal powder	Trimetallic nanocomposite (AlBc@La/Cu/Zr/TNC)	Irregular	30–80	Adsorption/photocatalytic degradation of malachite green (MG) dye	Sharma et al. (2019)
	Chondracanthus chamissoi	Biopolymer nanocomposite	Irregular	–	–	Rodriguez et al. (2016)
	Schizochytrium sp.	Chelating nanocomposite	Granule- like structure	68.16 ± 0.50	Evaluation of calcium nutritional supplement	Lin et al. (2015)
	Sargassum muticum	HA/ZnO nanocomposite	Hexagonal wurtzite	10.2 ± 1.5	Anticancer potential on PANC-1, CaOV-3, COLO205 and HL-60 cancer cells	Namvar et al. (2018)
	Turbinaria ornata	ZnO-MgO nanocomposite	Grains and clusters	25	Photocatalytic degradation of rhodamine B dye	Vijai Anand et al. (2019)
	Kappaphycus alvarezii	ZnO-MgO nanocomposite	Flakes	57–68	–	Pachiyappan et al. (2020)
	Chitosan and alginate	Chitosan-alginate NPS	Spherical	155	Potential against human lymphatic filarial parasite	Ali et al. (2013)
	Chitosan and fucoidan	Chitosan-fucoidan NPs	Spherical	173	pH-sensitive chitosan-fucoidan NPs for oral drug delivery of curcumin	Huang and Lam (2011)

nanoparticles such as Au, Ag, Pd, Pt, CdS, ZnO, Fe_3O_4, CuO and MgO were synthesized through kinetic incubation of appropriate ionic precursors in a specific molar concentrations (Table 1.2) with the whole cells (intracellular approach) or aqueous extract (extracellular approach) of various algal species. Algal species widely used for phycosynthesis includes *Chlamydomonas reinhardtii* (Barwal et al. 2011), *Sargassum plagiophyllum* (Dhanalakshmi et al. 2012), *Chlorella vulgaris* (Eroglu et al. 2013), *Stoechospermum marginatum* (Rajathi et al. 2012), *Sargassum muticum* (Azizi et al. 2014), *Turbinaria conoides* (Rajeshkumar et al. 2013), *Bifurcaria bifurcata* (Abboud et al. 2014), *Codium capitatum* (Kannan et al. 2013), *Gracilaria gracilis* (Francavilla et al. 2014), *Padina tetrastromatica* (Princy and Gopinath 2015) and other classifications of algal sources (Sharma et al. 2016).

1.4.2 Algae-Based Organic Nanoparticles

In contrast to the algae-mediated inorganic nanoparticles, the organic nanoparticles were synthesized by reducing the size of the precursor materials, mostly the total mass or the extracted bioactive compounds of the algal entities. Carbon-based and polymeric nanoparticles were synthesized using different kinds of algal species collected from the diverse aqueous ecological system. Basically, the algae-based carbon nanoparticles were prepared by treating the algal precursors with the appropriate pH, temperature and pressure under the controlled medium, whereas the algal precursors are the well-dried and fine-granulated powder of the algal cells treated with various physiochemical and mechanical circumstances for producing the carbon-based nanomaterials. Hence, carbon-mediated nanomaterials were obtained as carbon nanodots or the carbon quantum dots from several kinds of algal spores (Ramanan et al. 2016; Guo et al. 2017; Amjad et al. 2019).

In most cases, the polymer materials from the algal secretions or from the cell wall of algae were utilized for the synthesis of algae-based polymeric nanoparticles. Polymers present in the algae are mostly polysaccharides that include galactans, carrageenan, fucoidan, laminarin, ulvan, alginates, agar and chitin. Especially, algal polysaccharides mostly vary from all other polysaccharide sources derived from plants, animals and other earthbound resources because the algal polysaccharides are sulphate-enriched anionic compounds in nature. Certain classifications of the algal polysaccharides have been utilized as a precursor source for the synthesis of polymeric nanoparticles as these polysaccharides are smoothly soluble due to the presence of hydrophilic groups such as the carboxyl, hydroxyl and the sulphated groups on their surface (Bhatia 2016; Venkatesan et al. 2016). In addition, algal polysaccharides were converted into nanoparticles through different methods including cross-linking method (ionic gelation of the substrate) (Hudson and Margaritis 2014), emulsification techniques (Xing et al. 2010; Machado et al. 2012), polyelectrolyte complexation (interaction of oppositely charged contents) (Motwani et al. 2008; Arora et al. 2011), precipitation (Gazori et al. 2009), coacervation (produces the droplet substrates) (da Silva et al. 2012) and coprecipitation methods

(Zhao et al. 2012). Algal polysaccharide-based nanoparticles produced by the above-mentioned approaches were observed to be less toxic, biodegradable and biocompatible to the living mammalian system, and hence, these algal polysaccharide-based nanoparticles have been widely used in the field of biomedicine (Bhatia 2016; Venkatesan et al. 2016).

1.4.3 Algae-Based Hybrid Nanoparticles

Hybrid nanomaterials were mostly fabricated as organic and/or inorganic composite materials, in which two or more phases of nanoscale materials have to be combined for the fabrication of nanocomposites (Ananikov 2019). Algae-based synthesis of nanocomposites involves fabrication of two or more organic substrates like biopolymers, lipids, carbon sources and other kinds of biomolecules and/or the inorganic substrates like metals and metal oxides (Namvar et al. 2018). Fabrication of algae-based hybrid nanomaterials based on the appropriate physiochemical and mechanical parameters involves couple of methods like (i) incubation of two or more ionic concentrations of inorganic precursor solutions with the extracellular (extract with bioactive elements) or the intracellular (whole cells in powdered or cultured state) algal sources (Namvar et al. 2018; Sharma et al. 2018; Vijai Anand et al. 2019; Sadek et al. 2019; Pachiyappan et al. 2020) and (ii) treatment of two or more organic polymers or the carbon-based/other biomolecules derived from the algal resources with the same kind of organic substrates and/or the other kind of inorganic substrates using the intracellular/extracellular algal medium or the desired physiochemical medium incorporated with suitable chemical substituent as a reducing, stabilizing and capping agents (Motwani et al. 2008; Xing et al. 2010; Lin et al. 2015; Rodriguez et al. 2016; Zia et al. 2017; Collado González et al. 2020).

1.5 Conclusion

This chapter illustrated the fundamentals of green chemistry involved in the synthesis of nanomaterials using a different kind of algal sources. It describes the reasonable naming factor for the term phyco (phykos)-synthesis towards the synthesis of nanoparticles and, at the same time, the potential of several bioactive compounds on the synthesis of nanomaterials was well explored. Moreover, the intracellular and extracellular mechanical kinetics involved in algal nanotechnology was elaborated by the standard review of previous reports in the last decades. Several classifications of algae-based inorganic, organic and hybrid nanoparticles were clearly explored so far. Algae-based nanomaterials exhibit wide array of applications in the field of energy, electronics, cosmetics, constructions, agriculture, biotechnology, clinical diagnosis, nanocarriers as drug delivery system for anticancer, antibacterial, antifungal, antitumor and other therapies. This chapter also ensures that the algae-based nanoparticles act as promising factors for various futuristic applications through their enhanced properties and behaviours in the scale of nanometry.

Acknowledgements Dr. N. Suganthy thanks Rashtriya Uchchatar Shiksha Abhiyan (RUSA) for the financial support through RUSA Phase 2.0 grant (No. F. 24-51/2014-U, Policy (TN Multi-Gen), Department of Education, Government of India (dated 09.10.2018)). Prakashkumar Nallasamy thanks Department of Science and Technology (DST), New Delhi, under Promotion of University Research and Scientific Excellence (PURSE) for the financial support in general and infrastructure facilities sponsored under the PURSE 2nd Phase programme (Order No. SR/PURSE Phase 2/38 (G)) as Project Fellow.

References

Abboud Y, Saffaj T, Chagraoui A, El Bouari A, Brouzi K, Tanane O, Ihssane B (2014) Biosynthesis, characterization and antimicrobial activity of copper oxide nanoparticles (CONPs) produced using brown alga extract (*Bifurcaria bifurcata*). Appl Nanosci 4:571–576. https://doi.org/10.1007/s13204-013-0233-x

Ali M, Afzal M, Verma M, Misra-Bhattacharya S, Ahmad FJ, Dinda AK (2013) Improved antifilarial activity of ivermectin in chitosan-alginate nanoparticles against human lymphatic filarial parasite, *Brugia malayi*. Parasitol Res 112:2933–2943. https://doi.org/10.1007/s00436-013-3466-4

Amjad M, Iqbal M, Faisal A, Junjua AM, Hussain I, Hussain SZ, Ghramh HA, Khan KA, Janjua HA (2019) Hydrothermal synthesis of carbon nanodots from bovine gelatin and PHM3 microalgae strain for anticancer and bioimaging applications. Nanoscale Adv 1:2924–2936. https://doi.org/10.1039/c9na00164f

Ananikov VP (2019) Organic – inorganic hybrid nanomaterials. Nanomaterials (Basel) 9(9):1197. https://doi.org/10.3390/nano9091197

Arora S, Gupta S, Narang RK, Budhiraja RD (2011) Amoxicillin loaded chitosan-alginate polyelectrolyte complex nanoparticles as mucopenetrating delivery system for *H. pylori*. Sci Pharm 79:673–694. https://doi.org/10.3797/scipharm.1011-05

Arya A, Gupta K, Chundawat TS (2020) In vitro antimicrobial and antioxidant activity of biogenically synthesized palladium and platinum nanoparticles using *Botryococcus braunii*. Turkish J Pharm Sci 17:299–306. https://doi.org/10.4274/tjps.galenos.2019.94103

Aziz N, Fatma T, Varma A, Prasad R (2014) Biogenic synthesis of silver nanoparticles using *Scenedesmus abundans* and evaluation of their antibacterial activity. Journal of Nanoparticles, Article ID 689419, 2014. https://doi.org/10.1155/2014/689419

Aziz N, Faraz M, Pandey R, Sakir M, Fatma T, Varma A, Barman I, Prasad R (2015) Facile algae-derived route to biogenic silver nanoparticles: synthesis, antibacterial, and photocatalytic properties. Langmuir 31:11605–11612. https://doi.org/10.1021/acs.langmuir.5b03081

Azizi S, Ahmad MB, Namvar F, Mohamad R (2014) Green biosynthesis and characterization of zinc oxide nanoparticles using brown marine macroalga *Sargassum muticum* aqueous extract. Mater Lett 116:275–277. https://doi.org/10.1016/j.matlet.2013.11.038

Barwal I, Ranjan P, Kateriya S, Yadav SC (2011) Cellular oxido-reductive proteins of *Chlamydomonas reinhardtii* control the biosynthesis of silver nanoparticles. J Nanobiotechnol 9:1–12. https://doi.org/10.1186/1477-3155-9-56

Bhatia S (2016) Marine polysaccharides based nano-materials and its applications. In: Natural polymer drug delivery systems. Springer International Publishing, Cham, pp 185–225. https://doi.org/10.1007/978-3-319-41129-3

Cao G (2010) Nanostructures and nanomaterials - synthesis, properties and applications. Imperial College Press, London. https://doi.org/10.1142/9781860945960

Chojnacka K (2012) Biologically active compounds in seaweed extracts - the prospects for the application. Open Conf Proc J 3:20–28. https://doi.org/10.2174/1876326x01203020020

Collado González M, Cristina Ferreri M, Freitas AR, Santos AC, Ferreira NR, Carissimi G, Sequeira JAD, Guillermo Díaz Baños F, Villora G, Veiga F, Ribeiro A (2020) Complex polysaccharide based nanocomposites for oral insulin delivery. Mar Drugs 18:1–18. https://doi.org/10.3390/md18010055

Dahoumane SA, Wijesekera K, Filipe CDM, Brennan JD (2014a) Stoichiometrically controlled production of bimetallic gold-silver alloy colloids using micro-alga cultures. J Colloid Interface Sci 416:67–72. https://doi.org/10.1016/j.jcis.2013.10.048

Dahoumane SA, Yéprémian C, Djédiat C, Couté A, Fiévet F, Coradin T, Brayner R (2014b) A global approach of the mechanism involved in the biosynthesis of gold colloids using micro-algae. J Nanopart Res 16:1–12. https://doi.org/10.1007/s11051-014-2607-8

Dahoumane SA, Mechouet M, Alvarez FJ, Agathos SN, Jeffryes C (2016) Microalgae: an outstanding tool in nanotechnology. Rev Bionatura 1:196–201. https://doi.org/10.21931/RB/2016.01.04.7

da Silva LC, Garcia T, Mori M, Sandri G, Bonferoni MC, Finotelli PV, Cinelli LP, Caramella C, Cabral LM (2012) Preparation and characterization of polysaccharide-based nanoparticles with anticoagulant activity. Int J Nanomedicine 7:2975–2986. https://doi.org/10.2147/IJN.S31632

Dhanalakshmi PK, Riyazulla A, Rekha RP, Thangaraju SN (2012) Synthesis of silver nanoparticles using green and brown seaweeds. Phykos 42:39–45

Dhand C, Dwivedi N, Loh XJ, Jie Ying AN, Verma NK, Beuerman RW, Lakshminarayanan R, Ramakrishna S (2015) Methods and strategies for the synthesis of diverse nanoparticles and their applications: a comprehensive overview. RSC Adv 5:105003–105037. https://doi.org/10.1039/c5ra19388e

Dillehay TD, Ramírez C, Pino M, Collins MB, Rossen J, Pino-Navarro JD (2008) Monte Verde: seaweed, food, medicine, and the peopling of South America. Science 320(5877):784–786. https://doi.org/10.1126/science.1156533

Dong H, Hu W (2013) Organic nanomaterials. In: Vajtai R (ed) Springer handbook of nanomaterials, pp 905–940. https://doi.org/10.1007/978-3-642-20595-8_25

Eroglu E, Chen X, Bradshaw M, Agarwal V, Zou J, Stewart SG, Duan X, Lamb RN, Smith SM, Raston CL, Iyer KS (2013) Biogenic production of palladium nanocrystals using microalgae and their immobilization on chitosan nanofibers for catalytic applications. RSC Adv 3:1009–1012. https://doi.org/10.1039/c2ra22402j

Fatima R, Priya M, Indurthi L, Radhakrishnan V, Sudhakaran R (2020) Biosynthesis of silver nanoparticles using red algae *Portieria hornemannii* and its antibacterial activity against fish pathogens. Microb Pathog 138:103780. https://doi.org/10.1016/j.micpath.2019.103780

Francavilla M, Pineda A, Romero AA, Colmenares JC, Vargas C, Monteleone M, Luque R (2014) Efficient and simple reactive milling preparation of photocatalytically active porous ZnO nanostructures using biomass derived polysaccharides. Green Chem 16:2876–2885. https://doi.org/10.1039/c3gc42554a

Gazori T, Khoshayand MR, Azizi E, Yazdizade P, Nomani A, Haririan I (2009) Evaluation of alginate/chitosan nanoparticles as antisense delivery vector: formulation, optimization and in vitro characterization. Carbohydr Polym 77:599–606. https://doi.org/10.1016/j.carbpol.2009.02.019

Princy K.F, Gopinath A (2015) Eco-friendly synthesis and characterization of silver nanoparticles using marine macroalga *Padina tetrastromatica*. Int J Sci Res 4:1050–1054

Govindaraju K, Anand KV, Anbarasu S, Theerthagiri J, Revathy S, Krupakar P, Durai G, Kannan M, Subramanian KS (2020) Seaweed (*Turbinaria ornata*)-assisted green synthesis of magnesium hydroxide [$Mg(OH)_2$] nanomaterials and their anti-mycobacterial activity. Mater Chem Phys 239:2–5. https://doi.org/10.1016/j.matchemphys.2019.122007

Guo LP, Zhang Y, Li WC (2017) Sustainable microalgae for the simultaneous synthesis of carbon quantum dots for cellular imaging and porous carbon for CO_2 capture. J Colloid Interface Sci 493:257–264. https://doi.org/10.1016/j.jcis.2017.01.003

Holdt SL, Kraan S (2011) Bioactive compounds in seaweed: functional food applications and legislation. J Appl Phycol 23:543–597. https://doi.org/10.1007/s10811-010-9632-5

Huang YC, Lam UI (2011) Chitosan/fucoidan pH sensitive nanoparticles for oral delivery system. J Chin Chem Soc 58:779–785. https://doi.org/10.1002/jccs.201190121

Hudson D, Margaritis A (2014) Biopolymer nanoparticle production for controlled release of biopharmaceuticals. Crit Rev Biotechnol 34:161–179. https://doi.org/10.3109/07388551.2012.743503

Hulkoti NI, Taranath TC (2014) Biosynthesis of nanoparticles using microbes – a review. Colloids Surf B Biointerfaces 121:474–483. https://doi.org/10.1016/j.colsurfb.2014.05.027

Kajbafvala A, Li M, Bahmanpour H, Maneshian MH, Kauffmann A (2013) Nano/microstructured materials: rapid, low-cost, and eco-friendly synthesis methods. J Nanomater 2013:1–3. https://doi.org/10.1155/2013/530170

Kannan RRR, Stirk WA, Van Staden J (2013) Synthesis of silver nanoparticles using the seaweed *Codium capitatum* P.C. Silva (Chlorophyceae). S Afr J Bot 86:1–4. https://doi.org/10.1016/j.sajb.2013.01.003

Khalafi T, Buazar F, Ghanemi K (2019) Phycosynthesis and enhanced photocatalytic activity of zinc oxide nanoparticles toward organosulfur pollutants. Sci Rep 9:1–10. https://doi.org/10.1038/s41598-019-43368-3

Khan MI, Shin JH, Kim JD (2018) The promising future of microalgae: current status, challenges, and optimization of a sustainable and renewable industry for biofuels, feed, and other products. Microb Cell Factories 17:1–21. https://doi.org/10.1186/s12934-018-0879-x

Kumar N, Kumbhat S (2016) Nanomaterials: general synthetic approaches. Essent Nanosci Nanotechnol 2:29–76. https://doi.org/10.1002/9781119096122.ch2

Kumari P, Kumar M, Reddy CR, Jha B (2013) Algal lipids, fatty acids and sterols. In Functional Ingredients from Algae for Foods and Nutraceuticals; Domínguez, H., Ed.; Woodhead Publishing: Cambridge, UK, pp. 87–134

Kumar S, Bhanjana G, Sharma A, Sidhu MC, Dilbaghi N (2014) Synthesis, characterization and on field evaluation of pesticide loaded sodium alginate nanoparticles. Carbohydr Polym 101:1061–1067. https://doi.org/10.1016/j.carbpol.2013.10.025

Kumaresan M, Vijai Anand K, Govindaraju K, Tamilselvan S, Ganesh Kumar V (2018) Seaweed *Sargassum wightii* mediated preparation of zirconia (ZrO_2) nanoparticles and their antibacterial activity against gram positive and gram negative bacteria. Microb Pathog 124:311–315. https://doi.org/10.1016/j.micpath.2018.08.060

Lee R (2018) Phycology. Cambridge University Press. https://doi.org/10.1017/9781316407219

Lee KW, Jeong D, Na K (2013) Doxorubicin loading fucoidan acetate nanoparticles for immune and chemotherapy in cancer treatment. Carbohydr Polym 94:850–856. https://doi.org/10.1016/j.carbpol.2013.02.018

LewisOscar F, Vismaya S, Arunkumar M, Thajuddin N, Dhanasekaran D, Nithya C (2016) Algal nanoparticles: synthesis and biotechnological potentials. Algae – Org Imminent Biotechnol 7:157–182. https://doi.org/10.5772/62909

Li Y, Tang X, Song W, Zhu L, Liu X, Yan X, Jin C, Ren Q (2015) Biosynthesis of silver nanoparticles using *Euglena gracilis*, *Euglena intermedia* and their extract. IET Nanobiotechnol 9:19–26. https://doi.org/10.1049/iet-nbt.2013.0062

Lin J, Cai X, Tang M, Wang S (2015) Preparation and evaluation of the chelating nanocomposite fabricated with marine algae *schizochytrium* sp. protein hydrolysate and calcium. J Agric Food Chem 63:9704–9714. https://doi.org/10.1021/acs.jafc.5b04001

Lu KY, Li R, Hsu CH, Lin CW, Chou SC, Tsai ML, Mi FL (2017) Development of a new type of multifunctional fucoidan-based nanoparticles for anticancer drug delivery. Carbohydr Polym 165:410–420. https://doi.org/10.1016/j.carbpol.2017.02.065

MacHado AHE, Lundberg D, Ribeiro AJ, Veiga FJ, Lindman B, Miguel MG, Olsson U (2012) Preparation of calcium alginate nanoparticles using water-in-oil (W/O) nanoemulsions. Langmuir 28:4131–4141. https://doi.org/10.1021/la204944j

Machu L, Misurcova L, Ambrozova JV, Orsavova J, Mlcek J, Sochor J, Jurikova T (2015) Phenolic content and antioxidant capacity in algal food products. Molecules 20:1118–1133. https://doi.org/10.3390/molecules20011118

Mandal RP, Sekh S, Sarkar NS, Chattopadhyay D, De S (2016) Algae mediated synthesis of cadmium sulphide nanoparticles and their application in bioremediation. Mater Res Express 3. https://doi.org/10.1088/2053-1591/3/5/055007

Mashjoor S, Yousefzadi M, Zolgharnain H, Kamrani E, Alishahi M (2018) Organic and inorganic nano-Fe_3O_4: alga *Ulva flexuosa*-based synthesis, antimicrobial effects and acute toxicity to briny water rotifer *Brachionus rotundiformis*. Environ Pollut 237:50–64. https://doi.org/10.1016/j.envpol.2018.02.036

Momeni S, Nabipour I (2015) A simple green synthesis of palladium nanoparticles with *Sargassum* alga and their electrocatalytic activities towards hydrogen peroxide. Appl Biochem Biotechnol 176:1937–1949. https://doi.org/10.1007/s12010-015-1690-3

Mora-Godínez S, Abril-Martínez F, Pacheco A (2020) Green synthesis of silver nanoparticles using microalgae acclimated to high CO_2. Mater Today Proc. https://doi.org/10.1016/j.matpr.2020.04.761

Motwani SK, Chopra S, Talegaonkar S, Kohli K, Ahmad FJ, Khar RK (2008) Chitosan-sodium alginate nanoparticles as submicroscopic reservoirs for ocular delivery: formulation, optimisation and in vitro characterisation. Eur J Pharm Biopharm 68:513–525. https://doi.org/10.1016/j.ejpb.2007.09.009

Namvar F, Azizi S, Rahman HS, Mohamad R, Rasedee A, Soltani M, Rahim RA (2018) Green synthesis, characterization, and anticancer activity of hyaluronan/zinc oxide nanocomposite. Mater Lett 216:58–62. https://doi.org/10.1016/j.matlet.2017.12.081

Necas J, Bartosikova L (2013) Carrageenan: a review. Vet Med 58(4):187–205. https://doi.org/10.17221/6758-VETMED

Özturk BY, Gürsu BY, Dağ İ (2020) Antibiofilm and antimicrobial activities of green synthesized silver nanoparticles using marine red algae *Gelidium corneum*. Process Biochem 89:208–219. https://doi.org/10.1016/j.procbio.2019.10.027

Pachiyappan J, Gnanasundaram N, Rao GL, Gnanasundaram N, Rao GL (2020) Preparation and characterization of ZnO, MgO and ZnO–MgO hybrid nanomaterials using green chemistry approach. Results in Mater 7:1–6. https://doi.org/10.1016/j.rinma.2020.100104

Pandey P, Dahiya M (2016) A brief review on inorganic nanoparticles. J Crit Rev 3(3):18–26

Prasad R (2014) Synthesis of silver nanoparticles in photosynthetic plants. Journal of Nanoparticles, Article ID 963961, 2014. https://doi.org/10.1155/2014/963961

Prasad R (2016) Advances and Applications through Fungal Nanobiotechnology. Springer, International Publishing Switzerland (ISBN: 978-3-319-42989-2).

Prasad R (2017) Fungal Nanotechnology: Applications in Agriculture, Industry, and Medicine. Springer Nature Singapore Pte Ltd. (ISBN 978-3-319-68423-9).

Prasad R (2019a) Microbial Nanobionics: Basic Research and Applications. Springer International Publishing (ISBN 978-3-030-16534-5) https://www.springer.com/gp/book/9783030165338

Prasad R (2019b) Microbial Nanobionics: State of Art. Springer International Publishing (ISBN 978-3-030-16383-9) https://www.springer.com/gp/book/9783030163822

Prasad R (2019c) Plant Nanobionics: Advances in the Understanding of Nanomaterials Research and Applications. Springer International Publishing (ISBN 978-3-030-12495-3) https://www.springer.com/gp/book/9783030124953

Prasad R (2019d) Plant Nanobionics: Approaches in Nanoparticles Biosynthesis and Toxicity. Springer International Publishing (ISBN 978-3-030-16379-2) https://www.springer.com/gp/book/9783030163785

Prasad R, Kumar V, Prasad KS (2014) Nanotechnology in sustainable agriculture: present concerns and future aspects. Afr J Biotechnol 13(6):705–713

Prasad R, Pandey R, Barman I (2016) Engineering tailored nanoparticles with microbes: quo vadis. WIREs Nanomed Nanobiotechnol 8:316–330. https://doi.org/10.1002/wnan.1363

Prasad R, Bhattacharyya A, Nguyen QD (2017) Nanotechnology in sustainable agriculture: Recent developments, challenges, and perspectives. Front Microbiol 8:1014. https://doi.org/10.3389/fmicb.2017.01014

Prasad R, Jha A, Prasad K (2018) Exploring the Realms of Nature for Nanosynthesis. Springer International Publishing (ISBN 978-3-319-99570-0) https://www.springer.com/978-3-319-99570-0

Pugazhendhi A, Prabhu R, Muruganantham K, Shanmuganathan R, Natarajan S, Prabhu R, Muruganantham K, Shanmuganathan R, Natarajan S (2018) Anticancer, antimicrobial and photocatalytic activities of green synthesized magnesium oxide nanoparticles (MgO NPs) using aqueous extract of *Sargassum wightii*. J Photochem Photobiol B 190:86–97. https://doi.org/10.1016/j.jphotobiol.2018.11.014

Punnett T, Derrenbacker EC (1966) The amino acid composition of algal cell walls. J Gen Microbiol 44:105–114. https://doi.org/10.1099/00221287-44-1-105

Rahman A, Kumar S, Bafana A, Dahoumane SA, Jeffryes C (2019) Biosynthetic conversion of Ag^+ to highly stable Ag^0 nanoparticles by wild type and cell wall deficient strains of *Chlamydomonas reinhardtii*. Molecules 24. https://doi.org/10.3390/molecules24010098

Rai M, Ingle A, Gupta I, Birla S, Yadav A, Abd-Elsalam K (2013) Potential role of biological systems in formation of nanoparticles: mechanism of synthesis and biomedical applications. Curr Nanosci 9:576–587. https://doi.org/10.2174/15734137113099990092

Rajathi FAA, Parthiban C, Ganesh Kumar V, Anantharaman P (2012) Biosynthesis of antibacterial gold nanoparticles using brown alga, *Stoechospermum marginatum* (Kützing). Spectrochim Acta – Part A Mol Biomol Spectrosc 99:166–173. https://doi.org/10.1016/j.saa.2012.08.081

Rajeshkumar S, Malarkodi C, Gnanajobitha G, Paulkumar K, Vanaja M, Kannan C, Annadurai G (2013) Seaweed-mediated synthesis of gold nanoparticles using *Turbinaria conoides* and its characterization. J Nanostruct Chem 3:1–7. https://doi.org/10.1186/2193-8865-3-44

Ramanan V, Thiyagarajan SK, Raji K, Suresh R, Sekar R, Ramamurthy P (2016) Outright green synthesis of fluorescent carbon dots from eutrophic algal blooms for in vitro imaging. ACS Sustain Chem Eng 4:4724–4731. https://doi.org/10.1021/acssuschemeng.6b00935

Rodriguez SA, Weese E, Nakamatsu J, Torres F (2016) Development of biopolymer nanocomposites based on polysaccharides obtained from red algae *Chondracanthus chamissoi* reinforced with Chitin Whiskers and Montmorillonite. Polym – Plast Technol Eng 55:1557–1564. https://doi.org/10.1080/03602559.2016.1163583

Rohit MV, Venkata Mohan S (2018) Quantum yield and fatty acid profile variations with nutritional mode during microalgae cultivation. Front Bioeng Biotechnol 6:1–12. https://doi.org/10.3389/fbioe.2018.00111

Sadek RF, Farrag HA, Abdelsalam SM, Keiralla ZMH, Raafat AI, Araby E (2019) A powerful nanocomposite polymer prepared from metal oxide nanoparticles synthesized via brown algae as anti-corrosion and anti-biofilm. Front Mater 6:1–17. https://doi.org/10.3389/fmats.2019.00140

Salehi B, Sharifi-Rad J, Seca AML, Pinto DCGA, Michalak I, Trincone A, Mishra AP, Nigam M, Zam W, Martins N (2019) Current trends on seaweeds: looking at chemical composition, phytopharmacology, and cosmetic applications. Molecules 24(22):1–50. https://doi.org/10.3390/molecules24224182

Shah M, Fawcett D, Sharma S, Tripathy SK, Poinern GEJ (2015) Green synthesis of metallic nanoparticles via biological entities. Materials (Basel) 8(11):7278–7308. https://doi.org/10.3390/ma8115377

Sharma A, Sharma S, Sharma K, Chetri SPK, Vashishtha A, Singh P, Kumar R, Rathi B, Agrawal V (2016) Algae as crucial organisms in advancing nanotechnology: a systematic review. J Appl Phycol 28:1759–1774. https://doi.org/10.1007/s10811-015-0715-1

Sharma M, Behl K, Nigam S, Joshi M (2018) TiO_2-GO nanocomposite for photocatalysis and environmental applications: a green synthesis approach. Vacuum 156:434–439. https://doi.org/10.1016/j.vacuum.2018.08.009

Sharma G, Bhogal S, Gupta VK, Agarwal S, Kumar A, et al (2019) Algal biochar reinforced trimetallic nanocomposite as adsorptional/photocatalyst for remediation of malachite green from aqueous medium. J Mol Liq 275:499–509. https://doi.org/10.1016/j.molliq.2018.11.070

Singaravelu G, Arockiamary JS, Kumar VG, Govindaraju K (2007) A novel extracellular synthesis of monodisperse gold nanoparticles using marine alga, *Sargassum wightii* Greville. Colloids Surf B: Biointerfaces 57:97–101. https://doi.org/10.1016/j.colsurfb.2007.01.010

Srivastava S, Usmani Z, Atanasov AG, Singh VK, Singh NP, Abdel-Azeem AM, Prasad R, Gupta G, Sharma M, Bhargava A (2021) Biological nanofactories: Using living forms for metal nanoparticle synthesis. Mini-Reviews in Medicinal Chemistry 21(2):245–265

Sunkireddy P, Kanwar RK, Ram J, Kanwar JR (2016) Ultra-small algal chitosan ocular nanoparticles with iron-binding milk protein prevents the toxic effects of carbendazim pesticide. Nanomedicine 11:495–511. https://doi.org/10.2217/nnm.15.205

Venkatesan J, Anil S, Kim SK, Shim MS (2016) Seaweed polysaccharide-based nanoparticles: preparation and applications for drug delivery. Polymers (Basel) 8:1–25. https://doi.org/10.3390/polym8020030

Vijai Anand K, Aravind Kumar J, Keerthana K, Deb P, Tamilselvan S, Theerthagiri J, Rajeswari V, Sekaran SMS, Govindaraju K (2019) Photocatalytic degradation of rhodamine B dye using biogenic hybrid ZnO-MgO nanocomposites under visible light. Chem Sel 4:5178–5184. https://doi.org/10.1002/slct.201900213

Vijayan SR, Santhiyagu P, Ramasamy R, Arivalagan P, Kumar G, Ethiraj K, Ramaswamy BR (2016) Seaweeds: a resource for marine bionanotechnology. Enzym Microb Technol 95:45–57. https://doi.org/10.1016/j.enzmictec.2016.06.009

Wells ML, Potin P, Craigie JS, Raven JA, Merchant SS, Helliwell KE, Smith AG, Camire ME, Brawley SH (2017) Algae as nutritional and functional food sources: revisiting our understanding. J Appl Phycol 29:949–982. https://doi.org/10.1007/s10811-016-0974-5

Xing J, Deng L, Dong A (2010) Chitosan/alginate nanoparticles stabilized by poloxamer for the controlled release of 5-fluorouracil. J Appl Polym Sci 117(4):2354–2359. https://doi.org/10.1002/app.32083

Zayadi RA, Bakar FA (2020) Comparative study on stability, antioxidant and catalytic activities of bio-stabilized colloidal gold nanoparticles using microalgae and cyanobacteria. J Environ Chem Eng 8(4):1–8. https://doi.org/10.1016/j.jece.2020.103843

Zhao D, Liu CJ, Zhuo RX, Cheng SX (2012) Alginate/$CaCO_3$ hybrid nanoparticles for efficient codelivery of antitumor gene and drug. Mol Pharm 9:2887–2893. https://doi.org/10.1021/mp3002123

Zia F, Sobhani H, Mohammadi M, Athar M, Afzal M, Sultana T, Shchipunov Y (2017) Alginate-based hybrid nanocomposite materials. In: Algae based polymers, blends, and composites chemistry, biotechnology and materials science, vol 17. Elsevier, pp 603–648. https://doi.org/10.1016/B978-0-12-812360-7.00017-3

Part II
Synthesis and Properties of Algal Nanomaterials

Chapter 2
Algal Nanoparticles: Synthesis and Characterization

Shailendra Singh Shera and Rathindra Mohan Banik

Contents

S. S. Shera
Department of Biotechnology, Faculty of Engineering and Technology, Rama University Uttar Pradesh, Kanpur, India

R. M. Banik (✉)
School of Biochemical Engineering, Indian Institute of Technology (Banaras Hindu University), Varanasi, India

D. Thangadurai et al. (eds.), *Bioprospecting Algae for Nanosized Materials*, Nanotechnology in the Life Sciences,
https://doi.org/10.1007/978-3-030-81557-8_2

2.1 Introduction

Nanotechnology is a multidisciplinary field including science, engineering, and technology, which encompasses the understanding of the fundamental physics, chemistry, biology, and technology of nanometer-scale objects. The word "nano" has been derived from Greek word "nanos" meaning dwarf and is used to describe a billionth of a meter, or 10^{-9} of a meter. It is concerned with manipulation of matter at an atomic, molecular, and supramolecular scale, which is about 1–100 nanometer (Abdelghany et al. 2018). For comparison purpose, a red blood cell is approximately 7000 nm wide and a water molecule is almost 0.3 nm. Nanotechnology gave us an ability to see, control, and manipulate atoms and molecules, which has enabled us to manipulate the properties of object at nanoscale at will suiting our requirement. The extraordinarily different physical and chemical properties of nanoparticles to behave differently at nanoscale compared to their bulk (macrosized) counterparts and its potential to upgrade, enhance, and develop futuristic technologies have expanded its application in every field imaginable. Today nanotechnology is everywhere from simple household items such as cosmetics, paints, clothes, sunglasses, soaps, and toiletries, to biomedical such as chemotherapy, controlled drug delivery, and wound healing, to energy storage such as high-performance solar cells and energy storage devices, and in space explorations in the form of nanosensors, bionano robot-laced spacesuits to radiation shielding (Prasad et al. 2018, 2019, 2020).

The technological advancement in applied science is directly related to advancement in material science and engineering. Nanotechnology is increasingly used to fabricate novel nanomaterials such as nanoparticles, tubes, rods, and fibers. Nanomaterials can be synthesized from minerals and nearly from any chemical substance and they differ with respect to composition, primary particle size, shape, surface coatings, and strength of particle bonds. Given the potential and burgeoning application of nanoparticles in any field imaginable, there is a focus to search for cost-effective, nontoxic, and environment friendly process for nanoparticle synthesis. Before delving deeper, let us understand what nanoparticles actually are? Nanoparticle is defined as ultrafine microscopic particle of matter with at least one dimension less than 100 nm. The European Commission criterion states that the particle size of at least half of the particles in the number size distribution must measure 100 nm or below to be termed as nanoparticle. Now the question arises, what makes nanoparticles so exciting? The answer to this question lies in nanodimension. It has been proven that at atomic scale, material properties are entirely different than at the macrolevel. At nanoscale, materials have high surface area-to-volume ratio and particle surface atoms dominate the particle performance. This feature enables nanoparticles to possess unexpected optical, physical, and chemical properties, as they are small enough to confine their electrons and produce quantum effects.

Natural processes such as volcanic eruptions, storms, ocean tides, and photochemical processes are nature's way of creating and releasing nanoparticles in environment. The nanoparticle synthesized by natural process is called as natural nanoparticles. Calcium carbonates ($CaCO_3$), alumina (Al_2O_3), silicate (SiO_4^{4-}), sulfur (S), manganese oxide (MnO), silver (Ag), gold (Au), and viruses are some examples of natural nanoparticle synthesized by various natural processes (Griffin et al. 2018). However, natural nanoparticles have uncontrolled growth, irregular shape, and variable size due to variable climatic natural process which is beyond the control of human. This inherent problem with natural synthesis process has shifted the attention of researchers and scientist toward manmade or laboratory-synthesized nanoparticles. The advantages of laboratories-synthesized nanoparticles are precisely controlled sizes, shapes, and compositions. In laboratories, nanoparticles can be easily synthesized by physical, chemical, and biological methods (termed as green synthesis). Some of the commonly employed nanoparticle synthesis methods in laboratories are organic reduction with or without stabilizing agents, thermal decomposition in organic solvents, chemical reduction and photoreduction in reverse micelles, laser ablation, gamma irradiation, and electron irradiation (Ranjani and Sakthivel 2013). Figure 2.1 shows the various types of nanoparticles and their most commonly used synthesis methods.

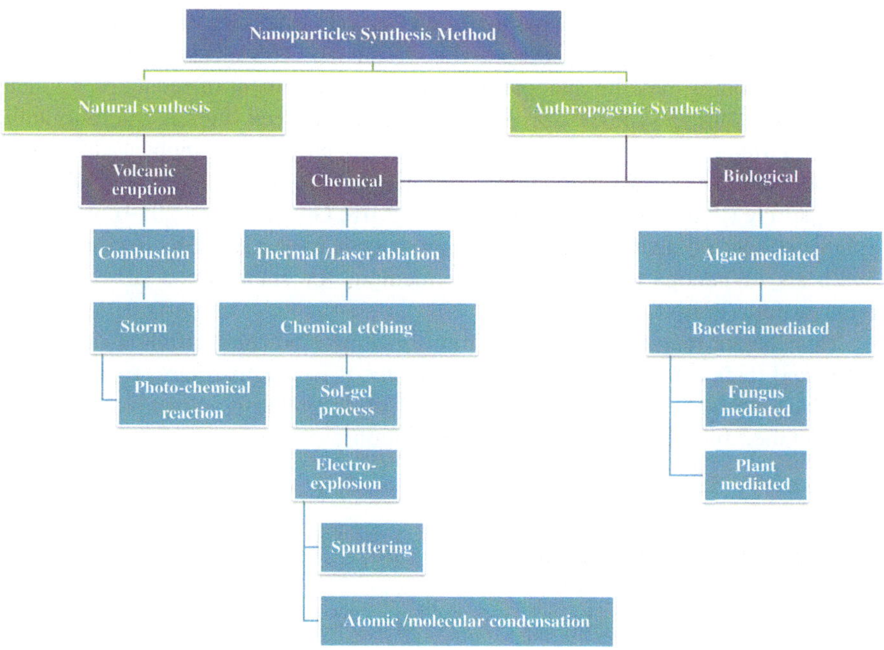

Fig. 2.1 Various methods used for synthesis of nanoparticles

Chemical reduction method is based on the principle of reduction of metal salts in various solvents and reducing agents and is essentially based on reduction in sizes from macroscale to nanoscale in sequential steps. Chemical method of synthesis includes chemical reduction, condensation, sol-gel method precipitation, and laser pyrolysis (Khan et al. 2016; Sandhu et al. 2019). Physical method of synthesis includes diffusion, irradiation, thermal decomposition, arc-discharge radiation-assisted, thermal decomposition, laser ablation, sonochemical, photochemical, and polyaniline synthesis, ball milling, and electrochemical techniques. Nanoparticles synthesized through physical and chemical methods have well-defined morphology, size, shape, and composition, and the process is well optimized without any deviation in process parameters for synthesis. It means that chemical and physical process allows greater control during synthesis of nanoparticle. However, chemical processes are being shunned and alternative route of synthesis is being explored due to the use of toxic chemicals and solvents, contamination from precursor chemicals, energy intensiveness, expensiveness, and generation of toxic byproducts (Thakkar et al. 2010). Biological or biogenic methods provide a nontoxic and eco-friendly route for synthesis of nanoparticles. This method of nanoparticle synthesis is termed as green synthesis due to use of biological origin material or extract for nanoparticle synthesis (Prasad 2014; Prasad et al. 2016, 2018; Sarma et al. 2021; Srivastava et al. 2021) (Fig. 2.1).

Green synthesis is being touted as an eco-friendly method against chemical method of nanoparticle synthesis. Now the question arises, what is green synthesis of nanoparticle? Green synthesis refers to the methods/approach/process utilizing biological components and nontoxic chemicals, requires minimum energy input, and avoids harmful byproduct for synthesis of nanoparticles through regulation, control, cleanup, and remediation process. The guiding principles for green synthesis are prevention/minimization of waste, reduction in derivatives/pollution, and the use of safer (or nontoxic) solvent/auxiliaries as well as renewable feedstock (Singh et al. 2018). With respect to nanoparticle, green synthesis methods are best suited for production of metallic nanoparticles due to requirement of reduction and capping agents during synthesis process. Biological sources such as microorganism (bacteria, fungi, actinomycetes, yeast, and blue green algae (cyanobacteria)), algae, and plant sources (fruit, root, flower, and leaves) are used for green synthesis of nanoparticle (Prasad et al. 2016; Singh et al. 2018; Gour and Jain 2019). Whole cell (plant or bacteria), their components, or extract can be used as reducing medium for synthesis of nanoparticle and capping medium for stabilization of synthesized nanoparticle (Prasad 2019a, b, c, d).

This chapter mainly focuses on synthesis of metallic nanoparticle using various algae and their extracts, characterization methods of nanoparticles, mechanism of synthesis, process parameters and factors affecting synthesis of metallic nanoparticle, and their biomedical and clinical applications.

2.2 Historical Background

Since the discovery of fire, human civilization is synthesizing and using nanoparticles unknowingly or knowingly in various forms. More than 4000 years ago, ancient Egyptian synthesized PbS nanoparticle using chemical reduction process for hair dyes.

Around third century BC, Egyptians prepared a blue pigment named Egyptian Blue by using sintered mixture of nanometer-sized glass and quartz, which were mainly used for decorative purposes (Jeevanandam et al. 2018). By the Bronze Age, incidental copper nanoparticle was unknowingly in common use by human civilization. The earliest evidence of nanotechnology can be traced back to 600 BC to India. During those times, India was known as global exporter of wootz steel that contained traces of carbon nanotubes and cementite nanowires in its microstructure. The beginning of metallic nanoparticle can be regarded to be started from fourteenth and thirteenth century BC when Egyptians and Mesopotamians started making glass using metals (Jeevanandam et al. 2018). Soluble gold nanoparticles were reported to be used by Egyptian and Chinese civilization in fifth or fourth BC. Egyptian used soluble gold nanoparticle for detection and treatment of syphilis and used it as a natural immunity booster. Nanoparticles have extraordinary optical properties and generate colored aesthetic visuals. In ninth century BC, artisan in Mesopotamia used nanoparticle to create glazed pottery. Colloidal nanoparticles of gold and silver were used for coloring ceramics to give unique appearance. The most famous example is Lycurgus Cup that was manufactured in the fourth century AD. This cup changes color depending on the incidence of light with respect to the cup. The cup turns ruby red when light is shone through it internally and it looks green in reflected light. The extraordinary optical properties were attributed to nanosized (70 nm) metal crystals of silver and gold in molar ration 14:1 (Horikoshi and Serpone 2013). Through times nanoparticles/nanocolloids were propagated as curative for various diseases, for example, heart and venereal diseases, dysentery, epilepsy, and tumors. In 1676, the German chemist, Johann Kunckels, while working with gold nanocolloids, indicated that extraordinary optical properties of nanoparticle are attributed to size. Finally, in 1857, Michael Faraday famously proved the extraordinary properties of metal nanoparticle in his paper "Experimental relations of gold (and other metals) to light." Faraday synthesized gold nanoparticle by chemical reduction methods. The synthesized nanoparticle was deep red solutions of colloidal gold by reduction of an aqueous solution of chloroaurate ($AuCl^{4-}$) by phosphorus in CS2 (a two-phase system) (Prathna et al. 2010; Horikoshi and Serpone 2013).

The foundation stone of nanotechnology in twentieth century was laid down by physicist Richard Feynman in 1959 during a talk entitled "There's Plenty of Room at the Bottom." A decade later from Richard Feynman's famous talk, Professor Norio Taniguchi coined the term "nanotechnology." The work in nanotechnology and nanoparticle accelerated with the development of scanning tunneling microscope in 1981, which could "see" individual atoms. Coincidentally, in 1981, K. Eric Drexler of Space Systems Laboratory, Massachusetts Institute of Technology, published the first paper in nanotechnology. Since then thousands of scientific papers have been published on various aspects of nanoparticles such as synthesis, properties, modifications, and applications.

2.2.1 History of Green Synthesis of Nanoparticles

The green synthesis of nanoparticle follows the principles of green chemistry championed by Paul Anastas and John C. Warner in their book published in 1998 (Anastas and Warner 1998). Green synthesis of nanoparticle includes six design principles of green chemistry such as waste prevention, less hazardous synthesis, use of safer solvent and auxiliaries, design for energy efficiency, use of renewable feedstock, and reduced derivatives. These principles enable scientists and researchers engaged into nanoparticle synthesis to find creative and innovative ways to reduce waste, conserve energy, and discover replacements for hazardous substances, which could benefit economy, society, and planet. A research article elaborating the gold nanoparticle synthesis procedure through bioreduction of $AuCl_4^-$ ions by the fungus, *Verticillium* sp., could be considered as the first report of green synthesis of nanoparticles (Mukherjee et al. 2001). Later in year 2003, plant extract was also reported to carry out bioreduction of metal salts for synthesis of metallic nanoparticle. Shankar et al. (2003) demonstrated the extracellular production of differently sized gold nanoparticle using geranium (*Pelargonium graveolens*) leaves. The authors identified terpenoids as the reducing and capping agent mediating reduction in chloroaurate ions for nanoparticle synthesis (Shankar et al. 2003). Brayner et al. (2005) published a report for synthesis of magnetic nanoparticle using alginate polysaccharide isolated from brown marine algae of the family Phaeophyceae, which can be considered as the first report for synthesis of nanoparticle using algal component. Singaravelu et al. (2007) utilized algal extract of *Sargassum wightii* Greville for synthesis of monodispersed gold nanoparticles. Xie et al. (2007) synthesized silver nanoplate using the extract of unicellular green alga *Chlorella vulgaris* at room temperature. They identified that proteins in algal extract were responsible for bioreduction and capping process during silver nanoplate synthesis. Algal extract protein performs dual function of Ag ion reduction and shape-controlled synthesis of nanosilver. Hydroxyl groups in Tyr residues and carboxyl groups in Asp and/or Glu residues were the most active functional groups for Ag ion reduction and for directing the anisotropic growth of Ag nanoplates. After 2010, the increased emphasis on eco- friendly methods and minimization of waste and toxic chemicals has led to proliferation of research in green synthesis of nanoparticles using biological sources such as plants, algae, and biomass (Aziz et al. 2014, 2015).

2.3 Types of Nanoparticle

Nanoparticles in general are highly complex entity and is are basically composed of three layers, that is, (a) the surface layer, which may be functionalized with a variety of small molecules, metal ions, surfactants, and polymers; (b) the shell layer, which is a chemically different material from the core in all aspects, and (c) the core, which is essentially the central portion of the nanoparticles (Khan et al. 2019). Nanoparticles can be broadly classified into two types: (i) organic nanoparticles and (ii) inorganic nanoparticle. Organic nanoparticles (NP) are solid particles consisting of synthetic or natural organic molecules of carbon. The main groups of organic nanoparticles are liposomes, micelles, protein/peptide, and dendrimers- based nanoparticles, polymersomes, polymer conjugates, capsules, and polymeric NPs. The properties of organic nanoparticle are dependent upon size, shape, and composition. Though organic nanoparticles have advantages of nontoxicity, biocompatibility, and biodegradability, these are often prone to degradation at high temperature and by intracellular enzymes present in cells. Some examples of organic NPs include poly-L-lysine, quaternary ammonium compounds, cationic quaternary polyelectrolytes, N-halamine compounds, and chitosan. Inorganic nanoparticles are composed of molecules or elements other than carbon. Inorganic nanoparticles can be classified into three types: (i) metallic nanoparticles, (ii) metal oxide nanoparticles, and (iii) nonmetallic nanoparticle. Metallic nanoparticles ranges from 1 to 100 nm in size and are made up of pure metals (e.g., gold, platinum, silver, titanium, zinc, cerium, iron, and thallium) or their compounds (e.g., oxides, hydroxides, sulfides, phosphates, fluorides, and chlorides) (Zaragoza et al. 2020). These nanoparticles can be synthesized by reduction and capping metal salts through chemical reduction methods or biological methods.

Nonmetallic nanoparticles are prepared from elements that are nonmetals. These nonmetallic nanoparticles are composed of various nonmetal elements belonging to various groups in periodic table such as group IV (carbon and silicon) and group V (phosphorous) elements. Group V elements usually combine with other metals and nonmetals to create doped nanoparticles. The phosphorous in form of phosphate combines to create ceramic nanoparticles. Similarly, phosphate combines with indium and gallium to create gallium phosphate (GaP) and indium phosphate (InP) semiconductor nanoparticles.

2.3.1 Organic Nanoparticles

The infinite possibility of design and modification as per the specific applications has attracted immense attention for organic nanoparticles. Organic nanoparticles can be easily synthesized just by varying the process parameters and give greater flexibility toward molecular design to create nanoparticle with unique electronic and optical properties. Due to this uniqueness, organic nanoparticles have a wide range of applications ranging from electronic to photonic, conducting materials to sensors, medicine to biotechnology, and so on. Organic nanoparticles are highly suitable for medical applications due to their biocompatibility and biodegradability. Their main field of application is in bioimaging and drug delivery purposes.

2.3.2 Metallic Nanoparticles

Nanomaterials can be classified as zero dimension, one dimension, two dimensions, and three dimensions. This classification is based on a number of dimension of materials, which are measured in nanoscale (no dimension is larger than 100 nm). Nanoparticles are an example of zero-dimension nonmaterial because none of its dimension is greater than 100 nm. Metal nanoparticles due to their unique surface plasmon resonance and optical properties have aroused the interest of scientific community worldwide. The existence of metallic nanoparticles was firstly proved by Michael Faraday in 1857 and the extraordinary color was explained by Mie in

1908. Metal nanoparticle and their oxides have widespread applications in electronics, optical system, energy storage, biomedical engineering and sciences, biotechnology, pharmacy, and diagnostic imaging (Prasad et al. 2020; Thangadurai et al. 2020; Maddela et al. 2021; Saglam et al. 2021). Metallic nanoparticle can be nanoimprinted on flexible polymers substrate to fabricate nanomaterials for electronics and display devices. For example, gold nanoparticles were deposited over flexible substrate polyimide to create nanoimprinted materials (Park et al. 2008).

The size-dependent optical properties of metallic nanoparticles are due to change in the optical energy band gap, which in turn influences the surface plasmon resonance of the nanomaterials. The extraordinary color of nanoparticle is due to surface plasmon resonance (SPR) effect. SPR effect exhibited by metal nanoparticle is due to interaction of conduction electron of metal nanoparticle with incident photon. These interactions are dependent upon size, shape, nature, and composition of dispersion media. Surface plasmon resonance too is dependent upon size of nanoparticle and is only observed if the particle size of the materials is less than the wavelength of the incident radiation (Raouf et al. 2017). Transition or heavy metals are preferred choice for nanoparticle synthesis on account of partially filled d-orbitals. Partially filled orbitals mean they can undergo reduction easily, which facilitates nanoparticle formation (Slavin et al. 2017). Though metallic nanoparticles can be synthesized by a number of different methods, this chapter focus will be on green synthesis of nanoparticles using algae or algal extract as reducing and capping agent.

2.4 Green Synthesis of Metallic Nanoparticles Using Algae and Their Components

Green synthesis is an eco-friendly and energy-optimized mode of nanoparticle synthesis using biological sources and their components. "Top Down" and "Bottom up" approaches are used for synthesis of metallic nanoparticles. Top-down approach relies on various methods of size reduction of bulk material to generate nanosized particles. The techniques used for size reduction of bulk materials are grinding, milling, sputtering, thermal/laser ablation, etc. In contrast, "Bottom Up" approach involves self-assembly of atoms onto each other, which gives rise to crystal planes; crystal planes further stack onto each other, resulting in the synthesis of the nanostructures or nanoparticles. Bottom-up approach is the preferred choice due to better control over process parameters such as solvent, temperature, pressure, and pH conditions (acidic, basic, or neutral). The green synthesis method of metal and metal oxide nanoparticle synthesis falls under "Bottom up approach," where metal salts are reduced in the presence of biological cells, extract, or its component to metal nanoparticle. Figure 2.2 provides a general overview of process involved in synthesis of nanoparticles using algal extract.

The heightened concern over increasing use of toxic chemicals in nanoparticle synthesis in recent years has created an urgency to shift toward waste minimization and exploring sustainable methods for synthesis of nanoparticles. Nanoparticle synthesis using algae and algal components is an eco-friendly and greener method

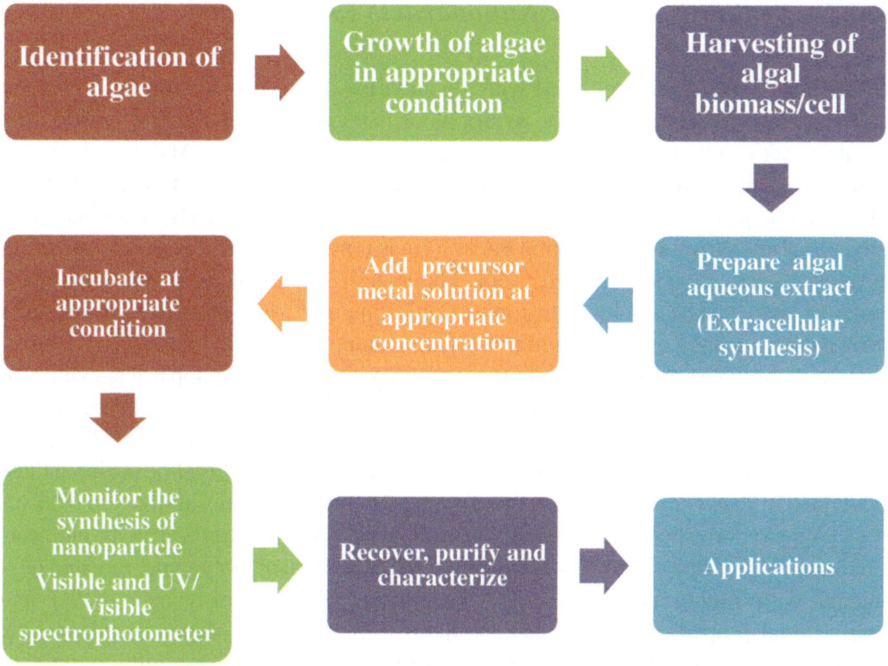

Fig. 2.2 General process flow diagram for extracellular synthesis of nanoparticles

toward sustainable process development (Aziz et al. 2014, 2015). Algae are known to hyperaccumulate heavy metal ions and possess an exceptional capability to remodel them into nanoparticles. Synthetic route for nanoparticle synthesis using biological method involves reduction of metal salts in the presence of algal biomolecules under suitable condition favoring rapid synthesis, desired size and shape, and polydispersity. The biomolecules such as amino acids, fatty acids, vitamins, minerals, phenolic compounds, aliphatic amines, bioflavonoid, and carbohydrates present in algal extract act as reducing and capping agent during nanoparticle synthesis (Raouf et al. 2017).

Algal extracts of all three categories, namely (i) Chlorophyceae, (ii) Phaeophyceae, and (iii) Rhodophyceae, are reportedly used for synthesis of various nanoparticles. Chlorophyceae (from the Greek word chloros, meaning "green") are freshwater green algae. The green colors of the pigments are due to the presence of chlorophyll a and b pigments. *Chlamydomonas*, *Scenedesmus*, *Pediastrum*, *Spirogyra*, and *Chara* are examples of Chlorophyceae. Phaeophyceae are marine brown algae containing chlorophyll a, c, carotenoids, and xanthophyll pigments. The pigments fucoxanthin mask the effect of chlorophyll giving brown appearance to algae belonging to Phaeophyceae family. *Dictyota*, *Laminaria*, and *Sargassum* are examples of algae belonging to Phaeophyceae family. The third category of algae is Rhodophyceae having red appearance due to the presence of red pigment r-phycoerythrin. *Porphyra*, *Gracilaria*, and *Gelidium* are examples of algae belonging to Rhodophyceae family.

2.4.1 Synthesis of Silver Nanoparticle Using Algae

Silver metal salts when incubated with algal extract undergo reduction from charged state to elemental state leading to synthesis of nanoparticles. Both algal powder and algal extract can be used to synthesize silver nanoparticles. Raouf-Abdel et al. (2019) synthesized silver nanoparticle using ethanolic and chloroformic algal extract of brown algae *Padina pavonia* and algal powder (Raouf-Abdel et al. 2019). The bioreduction of silver ions to elemental silver nanoparticle form is rapid in case of algal extract (2 min), whereas bioreduction using powder took 3 hours for completion. Both visual and spectrophotometry can be employed to confirm silver nanoparticle synthesis. Appearance of brown color visually confirms synthesis of silver nanoparticle, and absorbance peak in the range 400–500 nm spectrophotometrically confirms silver nanoparticle synthesis. The silver nanoparticle synthesized using chloroformic extract of *P. pavonia* exhibited absorbance at 400 nm, and for ethanolic extract, it was at 408 nm. Various bioflavonoids such as epigallocatechin, epicatechin, catechin, gallic acid, and epigallocatechin gallate contained in *P. pavonia* algal extract performed the reduction and stabilization of silver ions (Ag^{+}) to corresponding silver nanoparticle (Ag^{0}). The synthesized silver nanoparticles were polydispersed and assumed a pyramidal, spherical, polygonal, rod, and hexagonal with highly smooth edges. The surface plasmon resonance (SPR) and antibacterial properties of silver nanoparticles were reportedly dependent upon size, shape, dielectric constant of the metal, and nature of surrounding medium. Pyramidal- and polygonal-shaped nanoparticle exhibits more antibacterial activity than spherical nanoparticle solely on the basis of the presence of more active site on polygonal-shaped nanoparticle. Green alga *Sargassum plagiophyllum* extract can be used to reduce silver chloride solution to silver nanoparticle. The synthesis was rapid and it took only 24 hours for synthesis. The synthesis was confirmed by change in color from yellow to yellowish brown and UV-Vis spectra at 417 nm (Dhas et al. 2014).

Aboelfetoh et al. (2017) used green algae *Caulerpa serrulata* for synthesis of silver nanoparticle and to study the effect of various process parameters such as extract concentration, pH, temperature, contact time on the rate, size, and shape of nanoparticles. Kim et al. (2018) used extract of seaweed *Laminaria japonica* to synthesize well-dispersed spherical-/oval-shaped silver nanoparticles. Aziz et al. (2015) utilized *Chlorella pyrenoidosa* as cost-effective and eco-friendly method for biogenic synthesis of silver nanoparticles. The synthesis of silver nanoparticles by algal extract of *C. pyrenoidosa* was slow, and depending upon extract concentration and other factors, the synthesis was completed in 24–72 hours after incubation. Xie et al. (2007) described the synthesis of single crystalline silver nanoplates using extract of unicellular green alga *Chlorella vulgaris* at room temperature. The proteins contained in algal extract performed dual function of silver ion reduction and shape control. The hydroxyl groups in Tyr residues and carboxyl groups in Asp and/or Glu residues of algal protein were main site of reduction in Ag ion reduction and inducing the anisotropic growth of Ag nanocrystals. The presence of nanoplate was confirmed by appearance of three SPR band in the regions 465, 392, and 347 nm. The authors also studied the potency/reactivity of various species of algal extract in reducing silver salts. It was found that differential potency of algal extract from various species of algae is related to the presence and absence of tyrosine in adequate amount. Algal extract richer in Tyr residues or has Tyr residues that are less sterically hindered for reaction is more reactive than extract with lower tyrosine residue. Further, not only tyrosine content, which acted as electron donor, but also His and Cys residues in the proteins which could complex well with the Ag ions (reaction inhibitors) also govern the reduction kinetics of silver nanoplate synthesis (Xie et al. 2007). Priyadarshini et al. (2014) used microwave-mediated synthesis of silver nanoparticle using algal extract of *Gracilaria edulis*. The formation of silver nanoparticle was confirmed by change in color to brown color arising due to excitation of SPR and appearance of peak at 430 nm. Prasad et al. (2013) synthesized silver nanoparticle using algal extract of brown marine algae *Cystophora moniliformis*. The algal extract contained various functionalized and cyclized farnesyl acetone derivatives as well as related metabolites including the tricyclic terpene and lactone that helped in reduction of silver salt to silver nanoparticles. The nanoparticle formed was spherical and polydispersed. The size of nanoparticle was dependent upon temperature, and with increase in temperature the size of nanoparticle also got increased.

Not only macroalgae but silver nanoparticles can also be synthesized by microalgae. Microalgae offers certain advantages such as accelerated nucleation and crystal growth due to the presence of negative charge on the surface of the cell and large-scale synthesis at a very low cost. Microalgae such as *Anabaena doliolum* has been used for synthesis of silver nanoparticles (Singh et al. 2014). Jena et al. (2013) synthesized silver nanoparticle using microalgae *Chlorococcum humicola* which contains various class of biomolecules such as steam volatile acids, water-soluble phenolic compounds, lipophilic substances, proteins, and polysaccharides that acts as capping and reducing agent for synthesis of silver nanoparticle. Hamouda et al. (2019) reported synthesis of silver nanoparticle using aqueous extract of cyanobacteria, *Oscillatoria limnetica. O. limnetica* extract plays a dual part in both reducing and stabilizing *Oscillatoria*-silver nanoparticles (O-AgNPs). The synthesized silver nanoparticle exhibited characteristic absorbance maxima in the range 320–580 nm. It was believed that active biomolecules of *O. limnetica* aqueous extract such as saponins, quinines, flavonoids and terpenoids, proteins, enzymes, and metabolites like flavonoids caused bioreduction of silver ion to elemental silver in a light-induced reaction. Selvam and Sivakumar (2015) synthesized silver nanoparticle using algal extract of *Hypnea musciformis* (Wulfen) J.V. Lamouroux. The presence of peak at 440 nm confirmed the synthesis of silver NPs. The silver nanoparticle was stabilized by the (NH) C=O groups of cyclic peptides present in microalgal extract.

2.4.2 Synthesis of Gold Nanoparticles

The unique optical, physicochemical, and biological properties coupled with non-toxicity have made gold nanoparticles (AuNPs) an ideal candidate for biomedical and nonbiomedical fields alike. The size- and shape-dependent control on optical and electronic properties of gold nanoparticles give flexibility to engineer application-related AuNps. In addition, their surface chemistry is easy to modify. Similar to silver nanoparticle, gold nanoparticles can also be synthesized by several chemical methods and also through green chemistry route. Recently,

Gonzalez-Ballesteros et al. (2017) synthesized AuNPs using brown macroalgae *Cystoseira baccata.* The bioactive compounds in *C. baccata* extract act as a protective agent where the particles are embedded, keeping them separated, and avoiding aggregation and coalescence. The algal extract was able to produce spherical, stable, and polycrystalline nanoparticles with mean diameter of 8.4 ± 2.2 nm. The extract of marine algae are rich source of enriched biomolecules such as phenolic compounds, especially bromophenols, and other various secondary molecules, which have the ability to bind to metal salts (M^+) and reduce it to nanoparticle (M^0) (Raouf et al. 2017). Red marine algae *Galaxaura elongata* is a rich source of biochemicals such as andrographolide, alloaromadendrene oxide, glutamic acid, hexadecanoic acid, oleic acid, 11-eicosenoic acid, stearic acid, and gallic acid that effectively helps in synthesis of gold nanoparticles. However, among the various chemicals, epigallocatechin catechin and epicatechin gallate of the algal extract are more active as reducing agent during AuNP synthesis. Once the nanoparticle synthesis is over, it should be capped for stabilization. *G. elongata* extract is a rich source of secondary metabolite such as gallic acid; chitosan and cellulose have been used as stabilizing agents for gold metal nanoparticles against oxidation and coalescence. Li and Zhang et al. (2016) synthesized Au nanoparticles (AuNPs) using green algae *Pithophora oedogonia* extract as reducing agent. Oza et al. (2012) synthesized gold nanoparticle using aqueous extract of freshwater algae *Chlorella pyrenoidusa. C. pyrenoidusa* mediates gold nanoparticle synthesis via reductase and hydrogenase enzyme rather than phenolic compounds as electron donor. These reductases present in algal extract in combination with NADH help in extracellular reduction of Au^{+3} to Au^0 and stabilization leading to the formation of gold nanoparticles (GNPs). In another study, Rajathi et al. (2012) synthesized antibacterial gold nanoparticles using brown alga, *Stoechospermum marginatum* (Kützing), extract. The synthesis was extremely rapid and nanoparticle formation took only 10 minutes. The hydroxyl groups present in the diterpenoids of the brown seaweed were the main bioactive agent responsible for synthesis of nanoparticle at fast rate. The synthesis of nanoparticle was confirmed visually by appearance of ruby red color and UV-Vis spectrophotometer spectra. The appearance of peaks at 550 nm confirmed the formation of AuNp in solution. The extraordinary vibrant color produced by colloidal gold nanoparticle is due to interaction of gold nanoparticles with light. These interactions give rise to surface plasmon resonance phenomenon and various optical properties of gold nanoparticle. Ogi et al. (2010) reported synthesis of gold nanoparticle using *Shwanella* algal extract and whole cells. The author was able to synthesize, truncated triangular, and hexagonal nanoplates with an edge length of 100–200 nm using molecular H_2 acting as electron donor for reduction of soluble $AuCl^{4-}$ ions to insoluble Au^0. The *Shwanella* algae extract contained carbonyl group which caused formation of gold nanoplates. The yield of gold nanoplates was comprised of up to 60% of the nanoparticle population. Parial and Pal (2014) used eukaryotic algae *Rhizoclonium fontinale* for intracellular synthesis of gold nanoparticle. The formation of nanoparticle was related to metal stress response of algae biomass leading to formation of stress-related compounds and enzymes such as carotenoids, catalase, ascorbate peroxidase, and superoxide dismutase that helped in reduction of toxic ionic form of gold to nontoxic elemental form as gold nanoparticles. Further, it was reported that the formation of AuNPs was related to concentration of gold solution that was used for formation of gold nanoparticle. An increased concentration of gold solution reduced nanoparticle synthesis rate possibly due to inhibition of certain enzyme responsible for reduction of gold solution owing to cellular toxicity at high metal concentration. The optimum gold solution concentration for nanoparticle synthesis without causing cell death was found to be 10–55 mg L^{-1}.

Brown algae extract such as *Turbinaria conoides* can also be used for synthesis of gold nanoparticles. The nanoparticle synthesis process took from 1 hour to 24 hours for completion and initial yellow color changed to dark pink indicating formation of gold nanoparticle. The variations in synthesis time have been widely reported in literature. This variation in synthesis time of metal nanoparticle arises due to variability in concentration of phytochemical component in algal extract, nature of surrounding medium, and reaction time (Rajeshkumar et al. 2013). Ramakrishna et al. (2016) reported the synthesis of gold nanoparticle using marine brown algae *Turbinaria conoides* and *Sargassum tenerrimum*. There were significant differences between the rates of synthesis of nanoparticle in two algal species. *T. conoides* has higher rate of synthesis than *S. tenerrimum* due to higher fucose content in *T. conoides*. Algal pigments such as fucoxanthins, which are carotenoids rich in hydroxyl groups, and algal polysaccharides such as fucoidans – a type of polysaccharide containing considerable percentages of L-fucose and sulfate ester groups – have been believed to mediate the bioreduction of Au^{3+} present in gold solution to Au^0 nanoparticle. Further, polysaccharide presence also indicates abundant amount of hydroxyl group in algal extract. The hydroxyl group strongly interacts with nanoparticle and stabilizes the nanoparticle in solution.

Roychoudhury et al. (2016) used fractioned cellular components like intact pigment bearing thylakoids/chloroplasts, carotenoids, protein, and polysaccharides from the cyanobacterium *Anabaena sphaerica* and green alga *Chlorococcum infusionum*. Isolated thylakoid and chloroplast both were able to produce spherical nanoparticle at pH 4 intracellularly. However, extracellular synthesis was also reported but it took longer duration than intracellular synthesis. There was pH-dependent control over shape and size of gold nanoparticle; nanoparticle synthesized at pH 4 had size between 1 and 50 nm, whereas gold nanoparticle synthesized at pH 9 was about 2 and 7 nm. Suganya et al. (2015) used blue green algae *Spirulina platensis* protein to synthesize gold nanoparticles with enhanced antibacterial activity. The author reported increased antibacterial activity of biogenically synthesized nanoparticle due to the presence of a variety of antibacterial active biomolecules on surface of nanoparticle which synergistically increase microbe killing property of the nanoparticle.

2.4.3 Zinc Oxide Nanoparticle

Zinc oxide (ZnO) nanoparticle is an n-type semiconducting metal oxide which has a wide applicability in electronics, cosmetics, rubber, medicine, chemical fiber, and biomedical system (Agarwal et al. 2017; Hou et al. 2018). Azizi et al. (2014) synthesized zinc oxide nanoparticles using brown marine macroalga *Sargassum muticum* aqueous extract. The formation of zinc oxide nanoparticle was visibly confirmed by observing color change in reaction mixture from dark brown to a pale white color during the reaction, indicating the synthesis of ZnO nanoparticles. The functional group such as amino, sulfate, carboxyl, and hydroxyl groups was responsible for reduction of zinc acetate dehydrate solution to zinc oxide nanoparticle. The hydroxyl and sulfate groups of sulfated polysaccharide were implicated in the formation of zinc oxide nanoparticle. The UV-vis spectra of ZnO nanoparticle exhibited SPR absorbance at 334 nm, thereby confirming and validating the visual observation. The nanoparticles formed were agglomerated with hexagonal structures and particle size ranging from 30 nm to 57 nm with some deviations. This agglomeration is due to polarity and electrostatic attraction of ZnO nanoparticles. Priyadharshini et al. (2014) carried out experiment to synthesize zinc oxide nanoparticle (ZnO) using aqueous extract of macroalga *Gracilaria edulis.* The alga extract was able to reduce Zn^{2+} aqueous solution to Zn^0 nanoparticle. Due to formation of nanoparticle, the color of precursor solution changed from white to reddish brown for ZnO NPs. The nanoparticle was rod shape and was capped with quinines which together contributed toward increased cellular toxicity against human PC3 cancerous cell line. Zinc oxide nanoparticle being rod shaped has higher surface area for attachment to cells and can penetrate more efficiently and effect cell death. Nagarajan and Kuppusamy (2013) reported extracellular synthesis using *Sargassum myriocystum, Caulerpa peltata,* and *Hypnea valencia.* Brown alga *S. myriocystum* exhibited rapid and stable synthesis of zinc oxide nanoparticles. *S. myriocystum,* which is enriched with photochemicals such as alginic acid, ascorbic acid, protein, carbohydrates, flavonoids, tannins, mannitol, and lipids, could have acted as both reduction and stabilizing agents preventing the aggregation of zinc nanoparticles in solution.

2.4.4 Copper Oxide Nanoparticle

Nanoparticles can inflict high toxic effect to cells compared to their bulk counterparts on account of increased permeability. Copper due to its biocidal properties has been used in a number of applications such as pesticidal formulation, health-related applications, environmental cleanup, and water disinfection (Yadav et al. 2017). Bhattacharya et al. (2019) synthesized rod-shaped copper oxide (CuO) nanoparticle using *Anabaena cylindrica* extract and used it for disinfection of drinking water. The appearance of turbid brownish solution indicated the formation of Cu_2O nanoparticle. Brown alga extract *Bifurcaria bifurcata* can also be used to produce copper oxide nanoparticle. Reduction of aqueous copper (II) sulfate solution to CuO nanoparticles was carried out by α,β-unsaturated ketone group of diterpenoids present in brown algae extract. Usually, two types of nanoparticles are formed using brown algae extract: (i) cuprous oxide nanoparticle (Cu_2O), giving absorbance peak at 260 nm; and (ii) cupric oxide, with peak centered on 650 nm. Copper oxide nanoparticles can be synthesized by using *Brassica oleracea* var. *italica* algal extract (Renuga et al. 2020).

2.4.5 *Synthesis of Cadmium Sulfide Nanoparticles*

A wide applicability of cadmium sulfide nanoparticle (CdS) in photocatalysis, biological sensors, and light-emitting diodes had led to search for novel eco-friendly route for CdS nanoparticle synthesis. Microbial extract, plant, as well as algal extract were explored as nontoxic and eco-friendly method for CdS nanoparticle synthesis. Rao and Pennathura (2017) synthesized CdS nanoparticle using algal extract of microalgae *Chlamydomonas reinhardtii.* R-phycoerythrins, pigment from marine cyanobacterium, were implicated for formation of CdS nanoparticle in the 6–80 nm size range. The formation of CdS nanoparticle was confirmed by visual observation of orange color in cell-free extract and confirmed spectrophotometrically by appearance of absorption peaks at 430 nm (Rao and Pennathura 2017).

2.4.6 *Synthesis of Palladium Nanoparticle*

Arsiya et al. (2017) synthesized palladium nanoparticles using aqueous extract of *Chlorella vulgaris.* $PdCl_2$ was reduced by algal extract to Pd (II) and the formation of nanoparticle was visually confirmed by transformation of yellow color to dark brown. The formation of palladium nanoparticles was confirmed by the presence of an absorption peak between 410 and 420 nm using UV-visible spectrophotometer and the particle size was in the range 5–20 nm.

2.4.7 *Aluminum Oxide Nanoparticle*

Koopi and Buazar (2018) synthesized α-aluminum oxide nanoparticle using brown marine algae *Sargassum ilicifolium.* The algal extract functions as a bioreducing as well as a stabilizing agent. The presence of an absorption peak at 227 nm indicated that the phytochemicals present in algal extract were able to reduce aluminum ions (Al^{3+}) to Al_2O_3 NPs. The α-aluminum oxide nanoparticles were spherical and were in the size range approximately 21–42 nm. The stabilization of alumina nanoparticle was a result of hydroxyl and carbonyl groups present in algal extract.

2.4.8 *Iron Oxide Nanoparticle*

Kassas et al. (2016) synthesized iron oxide nanoparticle using algal extract of *Padina pavonica* (Linnaeus) Thivy and *Sargassum acinarium* (Linnaeus) Setchell through reduction of $FeCl_3$. The appearance of brown coloration indicated formation of iron oxide nanoparticles, and peak positions were centered at 402 nm and 415 nm. The particle size of Fe_3O_4-NPs ranged from 10 to 19.5 nm and 21.6 to 27.4 nm for *P. pavonica* and *S. acinarium*, respectively. FTIR analyses revealed that sulfated polysaccharides are the main biomolecules in the algae extracts which do dual function of reducing the $FeCl_3$ and stabilizing the phytogenic Fe_3O_4-NPs (Table 2.1).

Table 2.1 Different types of metallic nanoparticles synthesized by various algae

Algal species	Category	NPs type synthesized	NPs size (nm)	Shape	Reducing media (intra- or extracellular	Reducing agent	References
Padina pavonia	Brown algae	Silver	49.58–86.37	Spherical and polygonal	Extracellular (algal extract)	Bioflavonoids and terpenes such as epigallocatechin, gallic acid, and catechin	Raouf-Abdel et al. (2019)
Botryococcus braunii	Green algae	Silver	40–100	Cubical, spherical, and truncated triangular	Extracellular	Proteins, polysaccharides, amides, and long-chain fatty acids	Arya et al. (2018)
Sargassum plagiophyllum	Brown algae	Silver chloride	18–42	Spherical	Extracellular	C=C bond, alcoholic and carboxylic group	Dhas et al. (2014)
Caulerpa serrulata	Marine green algae	Silver	8–30	Spherical ellipsoidal, and minor amounts of silver nanocubic and nanorods	Extracellular	Caulerpenyne and/or its derivatives	Aboelfetoh et al. (2017)
Hypnea muciformis	Marine red algae	Silver	30–40	Cubical	Extracellular	Flavones, terpenoids, and polyphenols	Roni et al. (2015)
Laminaria japonica	Marine brown algae	Silver	23–31	Spherical and oval shaped	Extracellular	Carboxylic acid groups of constituents present in algal extract	Kim et al. (2018)
Chlorella vulgaris	Unicellular green algae	Silver nanoplate	44 ± 6	Circular disk and triangular	Extracellular (algal extract)	Hydroxyl groups in Tyr residues and carboxyl groups in Asp and/or Glu residues of algal proteins	Xie et al. (2007)
Gracilaria edulis	Red algae	Silver	55–99	Spherical	Extracellular (algal extract)	Alcohols/polyphenols and quinines	Priyadharshini et al. (2014)

Algal species	Category	NPs type synthesized	NPs size (nm)	Shape	Reducing media (intra- or extracellular	Reducing agent	References
Sargassum polycystum	Brown	Silver	5–7	Spherical	Extracellular (algal extract)	Various sugars, mannitol, proteins, amino acids, lipids, pigments, vitamins, and inorganic salts of primarily Na^+, Ca^{2+}, K^+, and Mg^{2+} fatty acids	Nallamuthu et al. (2012)
Cystophora moniliformis	Marine brown algae	Silver	50–100	Spherical	Extracellular (algal extract)	Tricyclic terpene, lactone, cyclized farnesyl acetone, acetogenins, and phenolics	Prasad et al. (2013)
Pterocladia capillacae		Silver	7	Spherical	Extracellular (algal extract)	Polysaccharides containing reducing sugars	Rafie and Zahran (2013)
Jania rubins		Silver	12	Spherical	Extracellular (algal extract)	Polysaccharides containing reducing sugars (glucose and galactose)	Rafie and Zahran (2013)
Ulva faciata		Silver	7	Spherical	Extracellular (algal extract)	Polysaccharides containing reducing sugars: rhammnose and xylose	Rafie and Zahran (2013)
Colpmenia sinusa		Silver	20	Spherical	Extracellular (algal extract)	Polysaccharides containing reducing sugars: alcoholic sugars (sorbitol and mannitol)	Rafie and Zahran (2013)
Pithophora oedogonia	Freshwater green algae	Silver	25–44	Cubical and hexagonal	Extracellular (aqueous algal extract)	Carbohydrate, saponins, steroid, tannins, terpenoids, and proteins	Sinha et al. (2015)
Caulerpa racemosa	Green algae	Silver	5–25	Spherical	Extracellular (aqueous algal extract)	O-H group in polyphenols or proteins/enzymes or polysaccharide and (NH)C=O groups of algal peptide	Kathiraven et al. (2015)

(continued)

Table 2.1 (continued)

Algal species	Category	NPs type synthesized	NPs size (nm)	Shape	Reducing media (intra- or extracellular	Reducing agent	References
Padina tetrastromatica	Brown algae	Silver	14	Pseudo-spherical and plate-like morphology	Extracellular	Polysaccharides and carbohydrate	Rajeshkumar et al. (2012)
Chlorococcum humicola	Green microalgae	Silver	2–16	Spherical	Extracellular (algal extract)	Proteins, lipids, and carbohydrates: free amine groups or cysteine residues in proteins and via the electrostatic attraction of negatively charged carboxylate groups in enzymes	Jena et al. (2013)
Oscillatoria limnetica	Cyanobacteria	Silver	3.30–17.97	Quasi-spherical	Extracellular	Flavonoids, phycobiliproteins, and chlorophyll	Hamouda et al. (2019)
Spyrogira insignis	Green microalgae	Silver	50	Aggregated nanoparticles	Extracellular (algal biomass)	Polyalcohols; hydroxyl group of galacturonic acid	Castro et al. (2013)
Chaetomorpha linum	Green algae	Silver	3–44	Cubical	Extracellular (algal extract)	Flavonoids and terpenoids	Kannan et al. (2013)
Enteromorpha flexuosa	Green algae	Silver	2–32	Spherical	Extracellular (algal extract)	–	Yousefzadi et al. (2014)
Hypnea musciformis	Red macroalgae	Silver	40–65	Spherical	Extracellular (algal extract)	Flavones and terpenoids	Roni et al. (2015)
Chlamydomonas reinhardtii	Green freshwater microalgae	Silver	4.3–5.7	Spherical	Extracellular (algal extract)	Reducing agent present in extracellular algal polysaccharide	Rahman et al. (2019)
Chlorella pyrenoidosa	Freshwater green microalgae	Gold	25–30	Spherical and icosahedral	Extracellular (algal extract)	Nitrate reductase	Oza et al. (2012)

Algal species	Category	NPs type synthesized	NPs size (nm)	Shape	Reducing media (intra- or extracellular	Reducing agent	References
Chondrus crispus	Red macroalgae	Gold	Polygonal: 200; spherical: 30–50	Acidic medium: triangular and hexagonal Alkaline: spherical	Extracellular (algal extract)	Free amine groups or cysteine residues in the proteins and hydroxyl functional group in alcohols	Castro et al. (2013)
Spyrogira insignis	Green microalgae	Gold	–	Mixed size	Extracellular (algal extract)	Free amine groups or cysteine residues in the proteins and hydroxyl functional group in alcohols	Castro et al. (2013)
Spirulina platensis	Blue green algae	Gold	2–8	Spherical	Extracellular (algal extract)	Carboxylate group of algal proteins	Suganya et al. (2015)
Chaetomorpha linum	Green algae	Silver	3–44	Cubical	Extracellular (algal extract)	Flavonoids and terpenoids	Kannan et al. (2013)
Enteromorpha flexuosa	Green algae	Silver	2–32	Spherical	Extracellular (algal extract)	–	Yousefzadi et al. (2014)
Hypnea musciformis	Red macroalgae	Silver	40–65	Spherical	Extracellular (algal extract)	Flavones and terpenoids	Roni et al. (2015)
Cystoseira baccata	Brown algae	Gold	8.4 ± 2.2	Spherical	Extracellular (algal extract)	Hydroxyl functional groups from polyphenols and polysaccharides	Ballesteros et al. (2017)
Galaxaura elongata	Red algae	Gold	3.85–77.13	Spherical	Extracellular (algal extract)	Andrographolide, alloaromadendrene oxide, glutamic acid, hexadecanoic acid, oleic acid, 11-eicosenoic acid, stearic acid, gallic acid, epigallocatechin, catechin, and epicatechin gallate	Raouf et al. (2017)

(continued)

Table 2.1 (continued)

Algal species	Category	NPs type synthesized	NPs size (nm)	Shape	Reducing media (intra- or extracellular	Reducing agent	References
Pithophora oedogonia	Green algae	Gold	33	Spherical	Extracellular (algal extract)	–	Li and Zhang (2016)
Chlorella pyrenoidusa	Freshwater algae	Gold	25–30	Spherical as well as anisotropic nanostructures	Extracellular (algal extract)	Nitrate reductase	Oza et al. (2012)
Stoechospermum marginatum	Brown algae	Gold	40–85	Spherical	Algal biomass	Hydroxyl group present in diterpenoids	Rajathi et al. (2012)
Gelidium amansii	Red algae	Gold	4–7	Cubical	Extracellular (algal extract)	Galactose and 3,6-anhydrogalactose	Kumar et al. (2017)
Shewanella algae	Green microalgae	Gold	9.6100–200	Triangular, hexagonal, nanoplate, and spherical	Extracellular (algal cell extract)	Carbonyl-containing aldehydic or ketonic compounds	Ogi et al. (2010)
Rhizoclonium fontinale	Green algae	Gold	16	Spherical	Intracellular	Chlorophyll, protein, and carbohydrate content	Parial and Pal (2014)
Turbinaria conoides	Brown algae	Gold	6 to 10	Spherical, triangle, and pseudo-spherical	Extracellular	Carboxylic, amine, and polyphenolic groups	Patil and Kim et al. (2018)
Turbinaria conoides	Brown algae	Gold	27–35	Spherical	Extracellular	Fucoidans	Ramakrishna et al. (2016)
Sargassum tenerrimum	Brown algae	Gold	12–57	Spherical	Extracellular	Fucoidans	Ramakrishna et al. (2016)
Chlorococcum infusionum (Schrank) Menegh	Green algae	Gold	1–50	Spherical	Extracellular (algal biomass and extracted algal cell components)	Thylakoids/chloroplasts, carotenoids, protein, and polysaccharides	Roychoudhury et al. (2016)

Algal species	Category	NPs type synthesized	NPs size (nm)	Shape	Reducing media (intra- or extracellular	Reducing agent	References
Anabaena sphaerica	Blue green algae	Gold	1–50	Spherical	Extracellular (algal biomass and extracted algal cell components)	Thylakoids/chloroplasts, carotenoids, protein, and polysaccharides	Roychoudhury et al. (2016)
Ecklonia cava	Brown algae	Gold	20–50	Spherical and triangular	Extracellular (algal extract)	Alginate and sulfated fucoidans, hydroxyl functional group in alcohols, and phenolic compounds	Venkatesan et al. (2014)
Sargassum muticum	Brown macroalgae	Zinc oxide	30–57	Hexagonal	Extracellular (aqueous algal extract)	Sulfated polysaccharides	Azizi et al. (2014)
Gracilaria edulis	Red algae	Zinc oxide	66–95	Rod shaped	Extracellular (aqueous algal extract)	Alcohols/polyphenols and quinines	Priyadharshini et al. (2014)
Sargassum myriocystum	Brown algae	ZnO	96–110	Hexagonal, triangular, rod, radial, and spherical nanoparticles	Extracellular (aqueous algal extract)	Alginic acid, ascorbic acid, protein, carbohydrates, flavonoids, tannins, mannitol, and lipids	Nagarajan and Kuppusamy (2013)
Anabaena cylindrica	Blue green algae	CuO	40–60	Rod	Extracellular (aqueous algal extract)	Flavones, terpenoids, and polysaccharides	Bhattacharya et al. (2019)
Bifurcaria bifurcata	Brown algae	CuO	5–45	Spherical	Extracellular (aqueous algal extract)	α,β-unsaturated ketone of diterpenoids	Abboud et al. (2014)

(continued)

Table 2.1 (continued)

Algal species	Category	NPs type synthesized	NPs size (nm)	Shape	Reducing media (intra- or extracellular	Reducing agent	References
Sargassum polycystum	Brown algae	CuO	–	–	Extracellular (aqueous algal extract)	Polysaccharides, tannins, flavonoids, phenolic acids, bromophenols, and carotenoids	Ramaswamy et al. (2016)
Padina pavonica	Brown algae	Iron oxide	10–19.5	Spherical	Extracellular (algal extract)	Sulfated polysaccharides	Kassas et al. (2016)
Sargassum acinarium	Brown algae	Iron oxide	21.6–27.4	Spherical	Extracellular (algal extract)	Sulfated polysaccharides	Kassas et al. (2016)
Sargassum bovinum	Brown algae	Palladium	5–10	Octahedral	Extracellular (algal extract)	Hydroxyl group of sulfated polysaccharides	Momeni and Nabipour (2015)
Chlorella vulgaris	Green microalgae	Palladium	50 to 90	Spherical	Extracellular (algal extract)	Polyol and amide	Arsiya et al. (2017)
Spirulina platensis	Blue green algae	Palladium	10–20	Spherical	Extracellular	Alcohols, carboxylic acids, and ester and ether groups, vegetable proteins and essential fatty acids	Sayadi et al. (2018), Chaudhary et al. (2020)
Dictyota indica	Brown algae	Palladium oxide	8–43	Spherical	Extracellular	–	Shargh et al. (2018)
Chlamydomonas reinhardtii	Freshwater green algae	CdS	2–7	Spherical	Extracellular (aqueous algal extract)	Oxidoreductases: 5-adenylylsulfatereductase and sulfite reductases	Rao and Pennathura (2017)

2.5 Process Parameters Affecting Synthesis of Nanoparticles by Algae

Formation of nanoparticles using algal extract is not dependent upon single factor rather dependent upon multiple factors such as concentration of bioactive molecules in algal extract, temperature, pH, reaction time, exposure time, and stirring speed. These factors in combination control the synthesis rate, size, shape, properties, and functionality of algal-synthesized nanoparticles.

2.5.1 pH

pH of the algal medium largely governs the shape and size of nanoparticles during synthesis. By optimizing the pH of algal extract solution, nanoparticle with a wide range of morphology and size can be synthesized easily. Alkaline pH tends to forms larger-sized nanoparticles though agglomeration than acidic pH (Aziz et al. 2015). Siddiqui et al. (2018) also observed agglomeration of silver nanoparticle at alkaline pH values. Aboelfetoh et al. (2017) found that silver nanoparticle synthesized using algal extract of *Caulerpa serrulata* is larger in size in acidic condition; however, it attains smaller size at alkaline pH. This behavior was attributed to availability of a large number of functional groups for binding of silver ions to effect reduction to nanoparticle with small diameter (Hamouda et al. 2019). These two studies reflected that pH does affect the size of nanoparticles and is dependent upon algal platform used for synthesizing nanoparticle. Figure 2.3 shows the effect of different pH (4 and 9) on shape and size of nanoparticles synthesized by *C. infusionum* and *A. sphaerica*.

2.5.2 Temperature

Temperature determines the morphology and yield of nanoparticle. The rate of nanoparticle synthesis increases at higher temperature due to rapid utilization of reactant, leading to formation of smaller nanoparticles with spherical morphology (Ibrahim 2015). Higher temperature increases the number of nuclei leading to formation of smaller-sized nanoparticles. However, at lower temperature, rate of growth controls the size of particles causing agglomeration of particles leading to increase in the size of nanoparticle. Similar observation was also made by Aboelfetoh et al. (2017) while synthesizing silver nanoparticles using extract of *Caulerpa serrulata*. Nagarajan and Kuppusamy (2013) also reported similar observation during zinc nanoparticle synthesis. There was significant decrease in particle size of zinc oxide nanoparticles from 76–186 nm to 20–36 nm with increase in temperature from 80 °C to 90 °C. Further, lower temperature favors polydispersed and spherical

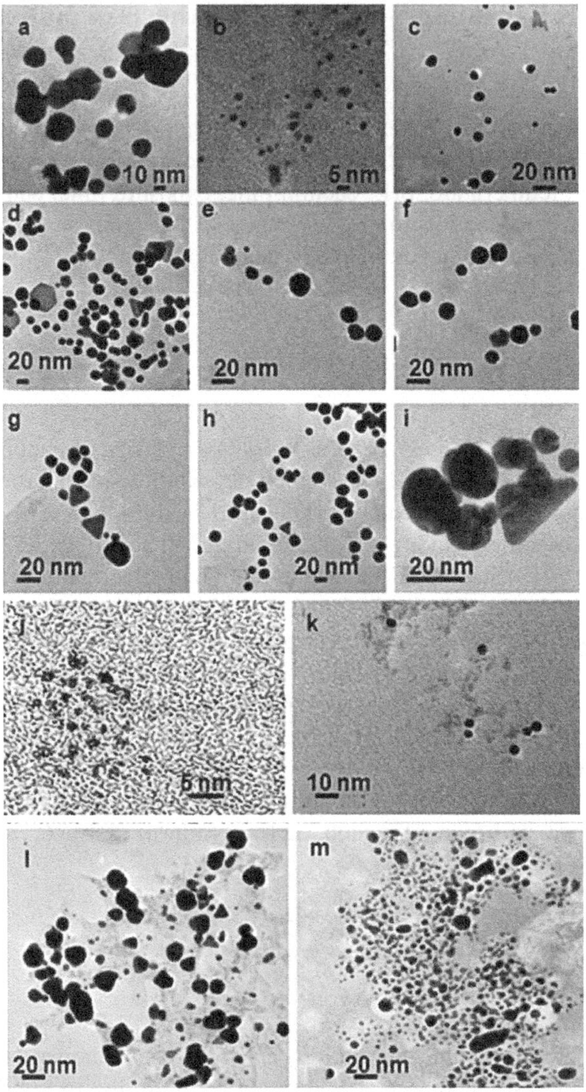

Fig. 2.3 SEM micrograph showing effect of different pH on shape and size of gold nanoparticles. SEM micrograph (a,b,c,d, e, f) at pH 4 and (g,h,i,j,k,l,m) at pH 9. (Roychudhury et al. 2016)

nanoparticle synthesis. However, Prasad et al. (2013) made contrary observation and reported that at higher temperature nanoparticle tends to aggregate, leading to formation of larger-sized nanoparticle. This indicates that there is a general acceptance among scientific community that temperature does affect nanoparticle size and shape. However, nanoparticle size (larger or smaller) at higher temperature is largely dependent upon algae species, algal extract concentration, and constituents and lastly types of nanoparticle synthesized.

2.5.3 Extract Concentration

The constituent contained in algal extract serves the dual role of bioreductant for synthesis of nanoparticle and stabilizer of synthesized nanoparticle. The concentration of the bioreducing agents in algal extract affects characteristics of nanoparticle by controlling its size and shape. High algal extract concentration increases nucleation of nanoparticles leading to faster rate of synthesis of nanoparticle. This increase in nucleation is due to increase in concentration of reducing agent contained in algal extract. However, high synthesis rate at high extract concentration decreases nanoparticle size (Kim et al. 2018). Aboelfetoh et al. (2017) reported that increase in extract concentration of *Caulerpa serrulata* from 10% to 20% reduced the silver nanoparticle size from 21–30 nm to 8–10 nm range. Further, extract concentration effected conversion of all ellipsoids, nanocube, and nanorods at 10% extract concentration to spherical shape. However, Nagarajan and Kuppusamy (2013) reported that irregular-shaped nanoparticles like hexagonal, triangular, rod, and radial zinc nanoparticles were formed at higher extract concentration as opposed to regular spherical-shaped Zinc nanoparticle at lower *S. myriocystum* extract concentration. Therefore, an optimized algal extract concentration is required for efficient production of nanoparticles because higher extract concentration may cause aggregation of nanoparticles. Bhattacharya et al. (2019) reported an optimized concentration of 5 g/100 mL algal extract of Brown algae (*Sargassum polycystum*) for synthesis of copper oxide nanoparticle (Bhattacharya et al. 2019).

2.5.4 Reaction Time and Contact Time

Reaction time refers to time required for completion of reaction or formation of nanoparticle. It generally measures how fast the nanoparticle synthesis occurs? Contact time measures the time required for completion of reaction, that is, time of exposure of metal salts with algal extract. Reaction and contact time both are largely dependent upon algal extract concentration. Algal extract with higher concentration of reductive groups tends to form nanoparticle at faster rate in short time. Raouf et al. (2019) reported synthesis of silver nanoparticle within 2 min using algal extract of *Padina pavonia,* whereas *P. pavonia* powder takes 3 hours for formation. This faster synthesis in algal extract may be due to reactive charged state of ions and solubilization of reductive group responsible for nanoparticle synthesis. Dhas et al. (2014) reported synthesis of silver nanoparticle within 24 hours using algal extract of *Sargassum plagiophyllum*. They observed that *S. plagiophyllum* extract has higher concentration of chlorine which converted the silver nitrate solution into silver chloride nanoparticles in 24 hours only. Similarly, Aboelfetoh et al. (2017) synthesized silver nanoparticles in *C. serrulata* 10% extract and studied the effect of contact time on nanoparticle formation. Prolonged contact of silver nitrate solution with *C. serrulata* extract resulted in higher formation of silver nanoparticles. Prolonged exposure gives more time for reductive groups in algal extract to react with metal salts and drives the reductive reaction to completion. It is evident that contact time for formation of nanoparticle is dependent on concentration of reductive species in algal extract, that is, extract having higher reductive species tends to form large quantity of nanoparticle at faster rate.

2.5.5 Stirring Speed

Stirring speed has positive effect on rate of synthesis and on nanoparticle size. Stirring speed controls the size and rate of synthesis of nanoparticles. Bhattacharya et al. (2019) observed that higher stirring speed reduces the dimension of copper oxide nanoparticle from 42.54 nm at 500 rpm to 3.6 nm at 900 rpm. The reduction in size was attributed to fragmentation of larger nanoparticles into finer particles at higher stirrer speed. Another aspect of higher stirring speed would be increase in nanoparticle synthesis rate due to homogeneous distribution of metal salts and bioreductant species of algal extract. Homogeneous distribution increases contact between metal salts and bioreductant species leading to faster reduction of metal salts to nanoparticle.

2.6 Mechanism of Nanoparticle Syntheses by Algae or Algal Extract

Intracellular and extracellular are the two modes for synthesis of metallic nanoparticles. Extracellular method using algal extract is the preferred choice for synthesis due to ease of handling, cost-effectiveness, and elimination of extra step of cell disruption to extract intracellular nanoparticle during downstream processing. The reduction of metal salts to metal nanoparticle using suspension of algal extract is an extracellular method of synthesis. Algal extracts are rich source of various bioactive molecules such as carbohydrates, proteins, minerals, oil, fats, and polyunsaturated fatty acids along with the soup of bioactive compounds such as antioxidants (polyphenols and tocopherols), pigments such as carotenoids (carotene and xanthophyll), chlorophylls, and phycobilins (phycocyanin and phycoerythrin), alkaloids, saponins, glycosides, steroids, tannins, polysaccharides (chitosan and cellulose), and sulfated polysaccharides (fucoxanthin) that are responsible for reducing metal salts to elemental nanoparticle metals form. Further, these bioactive compounds act as capping agent and stabilize newly formed nanoparticles (Nagarajan and Kuppusamy 2013; Li and Zhang 2016; Khanna et al. 2019). On the other hand, intracellular synthesis involves reduction of metal salts inside the cells with the help of intracellular enzymes such as nitrate reductase, NADH- and NADPH-dependent enzymes, enzymes involved in the metabolic process such as photosynthesis, respiration, and nitrogen fixation, and algal pigments such as chlorophyll (Sharma et al. 2016; Dağlıoğlu and Öztürk 2019). Intracellular biosynthesis of nanoparticles can be carried out by both live and dry biomass of algae. Figure 2.4 outlines the general procedure used to synthesize intracellular nanoparticle using algae.

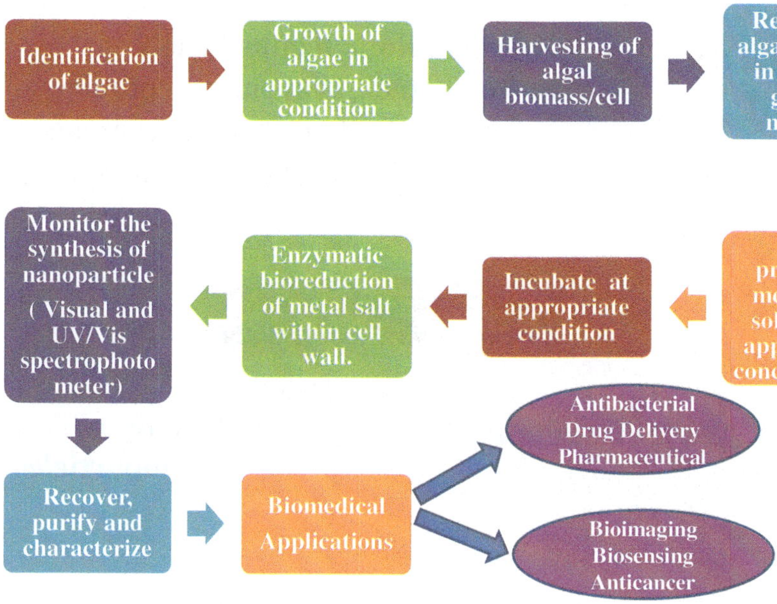

Fig. 2.4 General process flow schematic for intracellular synthesis of nanoparticles

2.6.1 Synthesis of Metallic Nanoparticles in Extracellular Mode by Algal Extract

Extracellular algal-mediated synthesis of nanoparticle involves: (i) preparation of algal extract (solvent extraction or powder), (ii) preparation of precursor metal salts, and (iii) incubation of algal extract and precursor solution together (Sharma et al. 2016). The formation of metal nanoparticle occurs through reduction of metal salts by bioactive components release by algal cells in extract medium. It has been reported that the formation of nanoparticle occurs in two steps: nucleation followed by crystal growth. Nucleation reaction is characterized by reduction of precursor metal salts (M^+) to metal nanoparticle (M^0) by constituent of algal extract. Second step, the crystal growth is characterized by complexation of metal nanoparticle (M^0) with other metal salts M^+ and forms M^0-M^+ complex followed by adsorption for nanocrystal growth as metal nanoparticle dimer. Abdel-Raouf et al. (2019) synthesized silver nanoparticle following the same mechanism of reduction of silver nitrate solution by catechin-OH (algal extract of *Padina pavonia*) followed by crystal growth (Raouf et al. 2019). Similarly, Ogi et al. (2010) adopted extracellular method for synthesis of gold nanoparticle through reduction of $HAuCl_4$ solution by algal extract solution of *Shwanella* algae. Bhattacharya et al. (2019) also reported similar mechanism for synthesis of copper oxide nanoparticle using algal extract of *Anabaena cylindrica*. The CuO nanoparticle synthesis took place in the core of extracellular polymeric moiety of the algal extract. Figure 2.5 shows the overview of mechanism involved in nanoparticle synthesis by algal extract and algal living cells with an example of silver nanoparticle synthesis (Fig. 2.5).

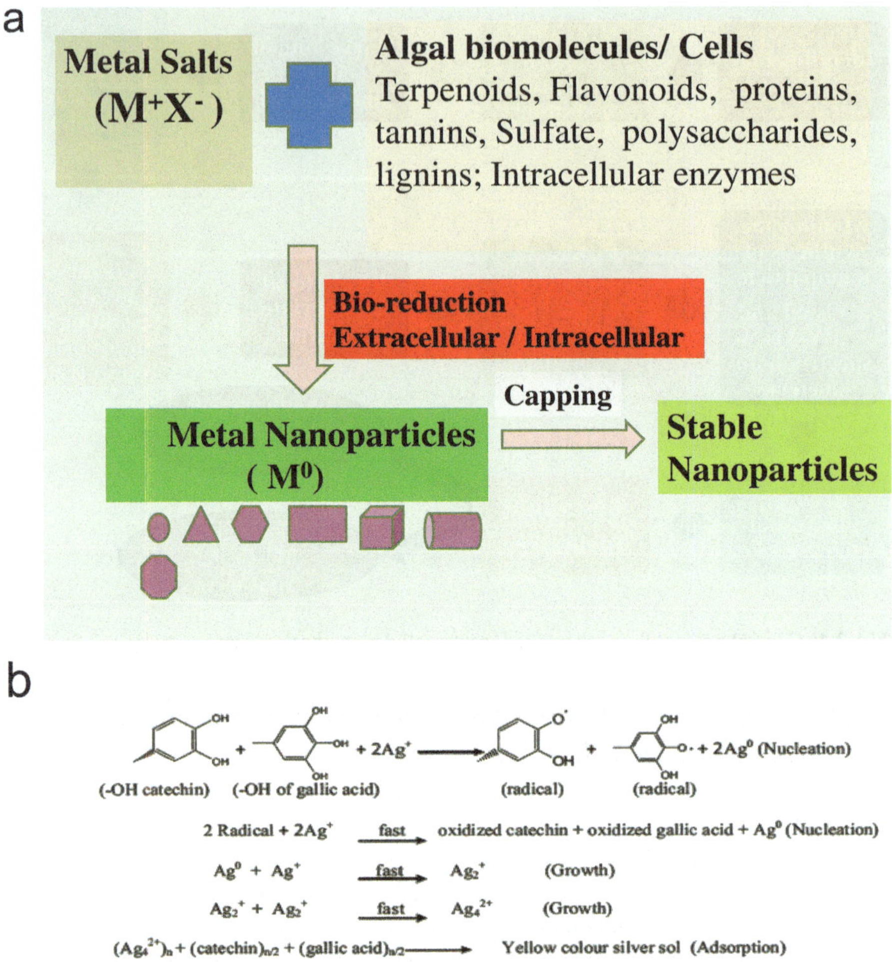

Fig. 2.5 (**a**) General mechanism leading to synthesis and stabilization of metallic nanoparticle by algal extract, (**b**) Reduction of silver salts to silver nanoparticle by algal extract of *Padina pavonia*. (Figure 5b adopted from Raouf-Abdel et al. 2019)

2.6.2 *Synthesis of Metallic Nanoparticles in Intracellular Mode by Algal Cells or Biomass*

Intracellular synthesis is the result of defense mechanism against metal-induced stress condition, diverting the metabolic machinery of algal cells to reduce toxic effects of metals. Intracellular synthesis of nanoparticles is carried out by trapping the positive charged metal ions on the surface of the cell wall or in the cytoplasm containing negative charged groups of enzymes or proteins. The trapped metal ions are reduced to form small nuclei, subsequently forming different morphological nanoparticles (Patil and Kim 2018).

Intracellular synthesis of nanoparticle relies on internalization of precursor metal salts by algal biomass. Upon internalization, metal salts are reduced to nanoparticles by intracellular enzyme present inside algal cells. The location of nanoparticle synthesis is dependent upon the types of enzymes involved in synthesis. Dağlıoğlu and Öztürk (2019) reported intracellular synthesis of silver nanoparticles using algae *Desmodesmus* sp. (Scenedesmaceae). The synthesis was localized to the compact inner region of the cells without forming aggregates. The synthesis occurred in chloroplast by chlorophyll pigments. As a result of reduction of Ag^+ to Ag^0, the photosynthetic activity and chlorophyll content decreased due to metal-induced toxicity and expenditure of energy in overcoming it. Similarly, alga *Tetraselmis kochinensis* accumulated gold nanoparticles in cell wall as well as in cytoplasm. However, concentration of gold nanoparticles was more in cell wall due to involvement of enzymes present in cell wall for bioreduction of $AuCl^{4-}$ ions to Au^0 nanoparticle. The synthesized gold nanoparticles ranged between 5 and 35 nm. Parial and Pal (2014) synthesized monodispersed gold nanoparticle using algal biomass of green alga *Rhizoclonium*. The concentration of metal stress-induced enzymes such as catalase, superoxide dismutase, and ascorbate peroxidase and carotenoid imparted protection to algal cells on incubation with auric chloride solution for a period of 3–24 hours. Later, the level of these stress enzymes reduced when the metal salts were reduced to nontoxic elemental nanoparticle form indicating possible involvement of these enzymes in nanoparticle synthesis.

2.7 Characterization of Algae-Synthesized Metallic Nanoparticles

Normally, nanoparticles are synthesized in colloidal form for ease of handling, detection, and recovery after synthesis. Nanoparticles can be isolated from colloidal solution through centrifugation at high speed or through lyophilization. Characterization of newly synthesized nanoparticles precludes any further application. Characterization refers to a description of features of nanoparticles in terms of size, shape, polydispersity, surface plasmon resonance, crystallinity, capping agents, and functional groups attached to it, and surface charge. These features govern a myriad of properties exhibited by nanoparticles. Following are the description of techniques to deduce the features outlined above.

2.7.1 Macroscopic Techniques

Macroscopic techniques with respect to nanoparticles refer to techniques that do not involve use of any sophisticated instruments. Metal nanoparticles tend to give characteristic bright color owing to surface plasmon resonance. The appearance of characteristic bright color can be used as convenient visible signature for synthesis of nanoparticles. For example, silver nanoparticles formation can be confirmed visually by appearance of brown color, and gold nanoparticle by appearance of pink or cherry red coloration. Table 2.2 lists the color shown by various nanoparticles in solution that are used to visibly confirm nanoparticle synthesis.

Table 2.2 Visible confirmatory color and UV-Visible peak position of various algae-synthesized nanoparticles

Nanoparticle types	Visible confirmatory color	Wavelength range (nm)	References
Zinc oxide	White precipitate	334, 335	Azizi et al. (2014); Agarwal et al. (2017)
Silver	Brownish black	400–470	Nallamuthu et al. (2012); Kim et al. (2018); Abdel-Raouf et al. (2019)
Silver nanoplates[a]	Pinkish Brown	465, 392, and 347	Xie et al. (2007)
Gold nanoparticle	Ruby red	530, 550, 560, and 520–525	Rajathi et al. (2012); Raouf et al. (2017); Patil and Kim (2018)
Gold nanoparticle[a]	Wine red, red brown, red orange, and orange	510–650	Dong et al. (2019)
Iron oxide	Brown	400, 402, and 415	Kassas et al. (2016)
Copper oxide	Dark brown		Arya et al. (2018)
Cadmium sulfide	Orange color	430 and 515	Rao et al. (2017)
Palladium	Dark brown	370–440	Arsiya et al. (2017)

[a]Size- and shape-dependent variation in UV-Vis peak position and color

2.7.2 *Microscopic Characterization*

Microscopic characterization includes scanning electron microscope (SEM), field emission scanning electron microscope (FESEM), transmission electron microscope (TEM), atomic force microscopy (AFM), and high-resolution transmission electron microscopy (HR-TEM). These microscopic techniques are employed to determine size, shape, and other morphological features of nanoparticles. Dhas et al. (2014) used FESEM and HR-TEM to determine size and shape of silver chloride nanoparticles. HR-TEM image revealed that nanoparticles were spherical in shape with size around 18–42 nm, whereas FESEM micrograph shows silver nanoparticles in the size range 21–48 nm. It should be understood that different microscope gives different size for the same particles due to different resolving capacity of microscope. Nagarajan and Kuppusamy (2013) used AFM microscopy to determine the size and shape of zinc oxide nanoparticle. The AFM determined size of ZnO nanoparticles was in the range 20–36 nm.

2.7.3 *Spectroscopic and Diffraction Techniques*

2.7.3.1 UV-vis Spectroscopy

A change in the absorbance or wavelength is a measure of the particle size, shape, interparticle properties, and types of nanoparticle. These changes can be ascertained using UV-vis spectrophotometer by monitoring the formation of nanoparticle in solution containing metal salts and algal extract. The metallic nanoparticle solutions are measured in the range of 190–1100 nm with each metallic nanoparticle having their characteristic wavelength range giving characteristic absorption spectra. The absorbance peaks in the specified wavelength range is the result of interaction of metallic nanoparticle with light giving rise to surface plasmon resonance phenomenon. For example, silver nanoparticle gives characteristic absorbance peaks in the range 400–470 nm (Gour and Jain 2019), and for gold nanoparticle it is 520–560 nm (Raouf et al. 2017; Khan et al. 2019). Table 2.2 lists the characteristic wavelength used to monitor synthesis of various nanoparticles using UV-Vis spectrophotometer.

The shape of absorbance peaks is frequently used to extract important data such as size, shape, aggregation, polydispersity. and the surrounding dielectric medium (Khanna et al. 2019). A high absorbance value indicates synthesis of a large number of nanoparticles. Broadening and narrowing of absorbance peaks in UV-vis spectra can reveal information about shape, size, and agglomeration status of nanoparticles. For example, smoother and broader surface plasmon research (SPR) peaks indicate spherical and polydispersed nanoparticles, peak shifting toward longer wavelength indicates the increase in size of silver nanoparticle, and an asymmetric broadening toward longer wavelength is an indication of aggregation of the particles (Prasad et al. 2013). A broad SPR band indicates aggregation of nanoparticle in the solution (Aboelfetoh et al. 2017). The shifting of SPR band toward the red or blue end depicts the size of nanoparticle. Aziz et al. (2015) reported that red shifting of SPR band from 420 nm to 460 nm indicated formation of large-sized silver nanoparticles. Further broadening of SPR peaks indicates the polydispersity of nanoparticle (Rafie and Zahran 2013). Sharp absorbance peaks are exhibited by small- and uniform-sized nanoparticles, whereas broad absorbance peaks imply a wider size distribution or aggregation of nanoparticles (Table 2.2).

2.7.3.2 FTIR Spectroscopy

Fourier transform infrared spectroscopy (FTIR) is performed in wavenumber range 4000–400 cm^{-1} with resolution of 4 cm^{-1} to extract information regarding functional groups, types of interaction, or bonding between groups. With respect to nanoparticle synthesis using algae, it is basically used to identify various functional groups present in algal solution responsible for reduction, capping, and stabilization of nanoparticles, to study surface chemistry, and to identify various functional and bioactive groups attached to the synthesized nanoparticle surface. Raouf et al. (2017) used FTIR to identify functional group in algal extract of *Galaxaura elongata*. The author was able to identify epigallocatechin catechin and epicatechin gallate of the algal extract which acted as a reducing agent for synthesis of gold nanoparticle. Rajathi et al. (2012) identified different functional groups like $-NH_2$ and CH-O-H in extract of brown algae *Stoechospermum marginatum* (Kützing) and reported that formation of gold nanoparticle was due to interplay of these groups, especially C-OH group. This group was able to attach on synthesized nanoparticle leading to its prolonged stability in colloidal solution. Normally, the presence and absence of groups on nanoparticle synthesized through algal extract are often reflective of the chemical nature of extract. -C=O-, -NH_2-, and –SH- are the most common functional groups attached to nanoparticle surface after bioreduction using algal extract solution (Jena et al. 2013; Jena et al. 2014).

2.7.3.3 XRD Analysis

Diffraction study gives an idea about crystallinity, purity, crystal size, geometry, phases, and orientation of nanoparticle (Nagarajan and Kuppusamy 2013; Khanna et al. 2019). The diffraction peaks of metal and metal oxide nanoparticles are compared with standard crystallographic databases such as JCPDS to extract out structural information. Rajathi et al. (2012) used diffraction peaks to ascertain the structure of biogenically synthesized gold nanoparticles with standard gold metal (Au^0). It also revealed that gold nanoparticle has certain crystal defect. The author argued that such crystal defects are characteristic of nanoparticle synthesized through biological methods. The presence of diffraction peaks at 38°, considered to be a facet of the face-centered cubic structure, indicated that synthesized gold nanoparticle has face-centered cubic structure (Rajathi et al. 2012). Nagarajan and Kuppusamy (2013) used width of dominant XRD peaks to determine crystallinity and size of zinc oxide nanoparticle (ZnO). The size of nanoparticle was determined to be 36 nm using Debye-Scherrer equation.

2.7.3.4 Photoluminescence Spectra

Nanoparticle of metals like gold, silver, and copper exhibits visible photoluminescence due to creation of holes in d bands, the energy of which corresponds to the energy difference between the Fermi level, and the various d bands (Smitha et al. 2008). This property can be used to ascertain the formation of metal nanoparticles in suspension. Photoluminescence spectra can be recorded using fluorescence spectrophotometer. Rajathi et al. (2012) found that suspension of gold nanoparticle exhibited photoluminescence at 550 nm. A combination of visible, UV-vis spectrophotometry, and photoluminescence spectroscopy helps in understanding finer details of nanoparticle.

2.7.4 Dynamic Light Scattering Spectroscopy and Zeta Potential Analysis

Surface charge, hydrodynamic diameter, and distribution of nanoparticles (NPs) in liquid form are measured by dynamic light scattering (DLS) spectroscopy, and particle stability is determined by zeta potential (Khanna et al. 2019). Parial and Pal (2014) determined the hydrodynamic size and polydispersity index of gold nanoparticles using dynamic light scattering method. The gold nanoparticle hydrodynamic size including hydration layer to be 156.6, 148, and 145 nm and corresponding polydispersity index of 0.432, 0.265, and 0.232, respectively, for gold nanoparticles synthesized at different pH (Parial and Pal 2014). Similarly, Rajathi et al. (2012) used light scattering method to measure size distribution profile of nanoparticles in suspension. The average mean size of silver nanoparticles comes out to <100 nm.

A stable nanoparticle system is represented by zeta potential values higher than 30 mV or lesser than -30 mV. A higher positive or negative zeta potential value of nanoparticles in suspension indicates repulsion due to interparticle electrostatic forces. These electrostatic forces help in prevention of aggregation of nanoparticle in suspension (Edison and Sethuraman 2012). Ramakrishna et al. (2016) determined the zeta potential value of gold nanoparticles synthesized by algal extract of *Turbinaria conoides* to be −26 mV and by *Sargassum tenerrimum* to be −29.1 mV. The large negative potential value suggested the presence of negatively charged moieties in the extracts that confer electrostatic stability to the nanoparticles. Similarly, Rao et al. (2017) measured stability of biogenically synthesized cadmium sulfide nanoparticle and reported zeta value of −30.7 mV indicating stability of nanoparticle.

2.8 Applications

2.8.1 General Applications of Algal-Synthesized Nanoparticle

Metallic nanoparticles owing to their small size and quantum confinement effect exhibit extraordinary absorption, photoluminescence emission, and nonlinear optical properties (Kumbhakar et al. 2014). For example, introduction of spherical aluminum nanoparticles with a 3 nm oxide layer of diameter 120 nm in a periodic array with period 360 nm on top of flat silicon solar cells can enhance the transmission of light by 7.46% over all incident light angles. This leads to an overall increase in the solar cell efficiency of 1.71%, or a relative increase of 7.23%. The extraordinary optical properties of nanoparticles are extremely useful in biosensing and sensing applications. Wang et al. (2019) used dual gold nanoparticles based on surface plasmon resonance of aptamer molecules to detect cancerous exosomes. Yuan et al. (2019) used 4-mercaptopyridine (4-MPY)-functionalized Au nanoparticles (Au NPs/4-MPY) as a signal amplification tag for fabrication of fiber optic sensor for detection of mercury toxicity in environmental samples.

Biotechnological applications require that nanoparticle size should not exceed 700 nm and it should preferably lie between 50 and 100 nm (Mody et al. 2010). The ability of nanoparticles to communicate with biomolecules or receptors of cells can be advantageously used for various biological applications such as drug delivery, gene, delivery, targeted drug delivery, bioluminescence, pharmaceutical, and bioimaging applications. Targeted drug delivery can be achieved by conjugating nanoparticles with ligands specific to cell surface receptors of targeted cells. Moreover, nanoparticles with multifunctional attachment sites can be conjugated with multiple copies of therapeutic drugs for high-dosage drug delivery to target cell. Aenishanslins et al. (2020) synthesized ternary cadmium sulfide-silver (CdSAg) quantum dots for bioimaging and photovoltaic applications. CdSAg quantum dots displayed high quantum yield and near IR emissions, which is a necessary condition for any molecules with fluorescent application for bioimaging applications (Aenishanslins et al. 2020). Apart from these high-end applications, metal nanoparticle has one simple applications, that is, antimicrobial agent. Metal nanoparticles such as titanium, copper, magnesium, especially silver and gold have known toxicity to bacteria, fungi, and against virus. These nanoparticles are widely used in lotions, medicines, antibacterial fabrics disinfectants, food processing, and medical devices to inhibit and kill microorganisms (Schröfel et al. 2014; Prasad et al. 2017).

Metal nanoparticle exhibits antimicrobial effect on account of their small size, shape, surface functionalization, composition of nanoparticles, and bioactive agent on surface of nanoparticles. The nanoparticle synthesized through biogenic method shows enhanced antimicrobial effect due to the presence of bioactive molecules on its surface. These bioactive molecules get deposited on ion surface during reduction of metal salts to nanoparticle in biological reduction medium. These metals nanoparticles show antimicrobial activity on account of interference in some key function of microbes necessary for human life or disrupt plasma membrane leading to leakage and ultimate death of microbes.

2.8.2 *Biomedical Applications of Algae-Synthesized Metallic Nanoparticles*

Though there are numerous applications of algal-synthesized nanoparticles, here the focus will be on biomedical and biosensor applications of nanoparticles. Biomedical applications include drug delivery, antibacterial agents, and tissue engineering. The following section outlines biomedical applications of metallic nanoparticles individually.

2.8.2.1 Silver Nanoparticle

The antimicrobial properties of silver are well known through ages. In ancient India, Kings and Queens and other high-ranking officials of royalty used to consume food and drinks in utensils made of silver. In modern application of silver, silver is not used as bulk material rather it is reduced to nanoscale in form of nanoparticles. At nanoscale, silver shows higher toxicity than at the bulk scale (Yaqoob et al. 2020).

Antibacterial Applications

Silver nanoparticles are extensively used in formulations intended for killing microbes. United States Food and Drug Administration (USFDA) has approved the use of AgNPs as an antimicrobial agent in humans, and in burns and wound-healing applications. Dhas et al. (2014) synthesized silver nanoparticle using algae and found it effective against Gram-negative bacteria *E. coli*. Similarly, silver nanoparticle synthesized by *C. pyrenoidosa* shows superior antibacterial activity against Gram-negative bacteria such as *Klebsiella pneumoniae*, *Aeromonas hydrophila*, and *Acenetobacter* sp. However, silver nanoparticles have lower antibacterial effect against Gram-positive bacteria *Staphylococcus aureus* possibly due to thicker cell wall present in Gram-positive bacteria. Thicker cell wall of Gram-positive bacteria resists penetration by silver nanoparticles. The algal-synthesized silver nanoparticles show high antibacterial activity compared to chemically synthesized nanoparticle owing to capping by protein and other biocapping agent present in algal extract. These capping agent acts in synergy and enhances antibacterial properties of silver nanoparticles. The silver nanoparticle exerts its antibacterial effects through three routes: (1) permeabilization of bacterial membrane leading to loss of intracellular content and ultimately death, (2) generation of reactive oxide species (ROS) owing to their photocatalytic activity disrupting respiratory chain and other metabolic process in bacteria, and (3) disruption of the protein structure by binding to thiol and amino groups bringing about cell disruption and ultimately death (Dhas et al. 2014; Aziz et al. 2015, 2016).

Anticancer Therapy

Several reports have suggested toward the cytotoxic and anticancer activity of silver nanoparticles. They can be used both as drug carrier for delivering anticancer drug to affected cancerous site and can also act as anticancer agent alone. The standalone silver nanoparticle exerts their anticancer activity by inducing programmed cell death through double-strand DNA breaks, oxidative stress, and chromosomal instability (Zhang et al. 2016; Aziz et al. 2019). Cortese et al. (2019) prepared hybrid clustered nanoparticle (HCNPs) loaded with colloidal suspension of silver nanoparticle and demonstrated its effectiveness against leukemia KU812 cells. The HCNPs fused with cancerous leukemic cells and released Ag^+ ions leading to production of reactive oxygen species, which facilitated killing of cancerous leukemic cells. Literature reports have suggested that biogenically synthesized nanoparticles have better efficacy in killing cancerous cells compared to chemically synthesized nanoparticles. Ahmeda et al. (2020) reported that biogenically synthesized nanoparticles were able to effectively kill T-cell leukemia (Ahmeda et al. 2020). Similarly, silver nanoparticles synthesized by aqueous extract of macroalga *Gracilaria edulis* exhibited anticancerous activity against human PC3 cell lines and breast cancer cell MCF-7 (Nallamuthu et al. 2012; Priyadarshini et al. 2014).

Theranostics Applications

Silver nanoparticles due to their unique photonic, electronic, catalytic, and therapeutic properties, and larger surface area and area:volume ratio are ideal candidates for theranostic applications. Theranostic is used to describe combinatorial treatment strategy comprising of diagnosis and treatment. Theranostics is a new technological field which develops molecular diagnostics integrated with targeted therapeutics. It comprises nanosize structures, combining targeting properties, that is, control of the spatial and temporal release of therapeutic agents, and monitoring modalities for the response to the treatment and assessing the effectiveness of these agents (Matteis et al. 2018). This strategy has been widely successful in diagnosis and treatment of cancer. Further, silver nanoparticle when conjugated with suitable contrast agents can be used for noninvasive methods of imaging, including X-ray-based computer-assisted tomography (CT), positron emission tomography (PET), single-photon emission tomography, and magnetic resonance imaging.

Biosensing Applications

Algae-synthesized silver nanoparticles can be potentially used in biosensing applications. Nanocomposite silver nanoparticles conjugated with 2D materials are reported to be effective for biosensing applications with high accuracy and sensitivity. Folic acid immobilized on the surface of glucose oxidase nanocomposites to efficiently recognize target cancer cells with folate receptors. Park et al. (2018) deposited silver and gold concurrently on cellulose fiber for nanofabrication of nanoplasmonic alloy of gold/silver nanocomposites for detection of folic acid as biomarker for cancer.

2.8.2.2 Gold Nanoparticles

Antibacterial Applications

Similar to silver nanoparticles, gold nanoparticles are also known for antibacterial properties. The antibacterial properties are dependent upon size and shape with higher activity reported for smaller and rod-shaped nanoparticles. The mode of killing of bacteria by gold nanoparticles is believed to be similar to silver nanoparticles. Gold nanoparticles employ either one or combination of the following mode to kill bacteria: (1) it interacts with SH groups of proteins and inactivates bacteria; (2) uncouples respiratory electron transport from oxidative phosphorylation, thereby inhibiting respiratory chain enzyme; (3) penetrates bacterial cell wall increasing membrane permeabilization and leaking of intracellular content of bacteria; and (4) impairment in bacterial DNA replication (Raouf et al. 2017). Similar to silver nanoparticles, gold nanoparticles too exhibited heightened antibacterial activity against gram-negative bacteria such as *Pseudomonas aeruginosa*, *Klebsiella oxytoca*, *Enterobacter faecalis*, *Klebsiella pneumoniae*, and *Vibrio cholera* (Rajathi et al. 2012).

Anticancer Therapy

Gold nanoparticles exhibit unique physicochemical properties including surface plasmon resonance and the ability to bind amine and thiol groups, allowing surface modification for functionalization of silver nanoparticles for treatment of cancer.

Gold nanoparticle is believed to exert anticancerous activity through activation of apoptotic cell death pathways. Gold nanoparticle synthesized by brown algae *Cystoseira baccata* extract, when tested on colon cell line Caco-2 and HT-29, showed anticancerous activity. In case of Caco-2, gold nanoparticle induced cell death via activation of extrinsic cell death pathways, whereas in case of HT-29, both extrinsic and intrinsic routes appear to be activated, as can be inferred by the significant activation of caspase-8 and -9. Caspase protein family is responsible for killing cancerous cell or defective cells of the body. Sun et al. (2019) also reported cytotoxic anticancer activity of biogenically synthesized gold nanoparticles against lung carcinoma.

Biosensing Applications

Gold nanoparticles (AuNPs) help in improving the sensitivity and actively target the biomarker due to high surface-to-volume ratio (Nunna et al. 2019). Algae-derived gold nanoparticles can be used as optical as well as nonoptical biosensors. In optical biosensors, AuNPs are widely used to improve the detection sensitivity of fluorescence, chemiluminescence, surface plasmon resonance, surface-enhanced Raman scattering (SERS) method, and other optical detection, whereas in nonoptical biosensors, AuNPs are mainly used in piezoelectric biosensors, electrochemical biosensors, and ICP-MS biosensors. Gold nanoparticles-oligonucleotides semiconductor conjugates can be used for diagnosis and detection of pathogenic and genetic disease through DNA hybridization (Daraee et al. 2016). Gold nanoparticle-based biosensor can be fabricated for early detection of cancer biomarker. Saeed et al. (2017) fabricated DNA biosensors based on gold nanoparticles-modified graphene oxide for the detection of breast cancer biomarkers. Gold nanoparticles were linked to graphene oxide and then grafted on to glassy carbon. The signal enhancement achieved allowed sensitive detection of breast cancer marker ERBB2 (Saeed et al. 2017).

2.8.2.3 Zinc Oxide Nanoparticles

Antibacterial Properties

Zinc oxide nanoparticle (ZnO NPs) is known to show both bactericidal as well as bacteriostatic effects. The bactericidal action is due to generation of reactive oxygen species (ROS) such as H_2O_2 in cells. These ROS damages the cell wall, enhances membrane permeability, weakens mitochondria, increases intracellular outflow, and induces gene expression of oxidative stress which caused eventual cell growth inhibition and cell death (Xie et al. 2011; Bhuyan et al. 2015; Rao and Pennathura 2017). Algal-derived ZnO nanoparticles can be used to kill Gram-positive and Gram-negative bacteria such as *Escherichia coli*, *Pseudomonas aeruginosa*, *Proteus vulgaris*, *Vibrio cholerae*, *Bacillus subtilis*, *Enterococcus faecalis*, and *Campylobacter jejuni* (Jiang et al. 2018). ZnO nanoparticle is also being explored as antibacterial coating on food packaging items to kill major food spoilage bacteria.

Anticancer Therapy

In anticancer therapy, ZnO NPs can be used both as anticancer agent as well as a delivery vehicle for delivering anticancer drugs to cancerous site. Priyadharshini et al. (2014) reported that rod-shaped ZnO nanoparticle was able to kill PC3 cancer cell line effectively. Comparison between the cytotoxicity effect of ZnO NPs and silver nanoparticles found that ZnO NPs have higher cytotoxicity activity than silver nanoparticle. It was reported that ZnO NPs induce apoptosis of cancer cell or kill cancerous cell through autophagy (Jiang et al. 2018).

2.8.2.4 Palladium Nanoparticles

Palladium nanoparticles are basically used in fabrication of sensors, biosensors, and catalytic applications. Momeni and Nabipour (2015) used algal-synthesized palladium nanoparticles to fabricate electrochemical sensor for detection of hydrogen peroxide (H_2O_2) because it acts as an important mediator in the food, pharmaceutical, medical, and chemical industries. H_2O_2 is also a major byproduct of many biological reactions catalyzed by various enzymes. Thus, rapid and reliable determination of H_2O_2 is of great importance.

2.8.2.5 Iron Oxide Nanoparticles

Algal-synthesized nanoparticles can be used in magnetic data storage, cancer treatment, biosensing, drug delivery, and bioremediation of heavy metals such as lead, copper, mercury, and silver (Dave and Chopda 2014). Iron oxide nanoparticle due to its biocompatibility and nontoxicity has attracted numerous applications in biomedical field. Iron oxide nanoparticles can be used as contrast agents for magnetic resonance imaging, drug carriers for target-specific drug delivery, gene carriers for gene therapy, therapeutic agents for hyperthermia-based cancer treatments, magnetic sensing probes for in vitro diagnostics, and nanoadjuvant for vaccine and antibody production. Kassas et al. (2016) demonstrated that the nanoparticles synthesized by algal extract reduction process can be used for bioremediation of lead from water. Iron oxide nanoparticle, when conjugated with silicon oxide, and alginate acted as adsorbent for trapping lead from water samples and facilitates its removal.

2.8.2.6 Applications of Other Algal-Synthesized Metallic Nanoparticles

Copper oxide nanoparticles (CuO NPS) are effectively used as antimicrobial agent. CuO NPS has shown antimicrobial activity against a wide number of Gram-positive and Gram-negative bacteria such as *Escherichia coli*, *Pseudomonas aeruginosa*, *Proteus mirabilis*, *Klebsiella* spp., and *Staphylococcus aureus*, drug-resistant bacteria, and eukaryotic cyanobacteria such as *Chlamydomonas reinhardtii*. These properties have led to the development of various approaches with direct applications to the biomedical field, such as tailored surfaces with antimicrobial effect, wound dressings, and modified textiles (Grigore et al. 2016). Cadmium sulfide nanoparticles can be used in photovoltaic applications such as solar cells and semiconductor devices. The biomedical applications of aluminum oxide nanoparticle include drug delivery applications, biosensing applications (sensing of bovine serum albumin and DNA in competitive bioassay), anticancer therapy, immunotherapy, antibacterial agent (against bacteria such as *Bacillus subtilis*, *Escherichia coli*, *Pseudomonas fluorescens*, and *Staphylococcus epidermidis*), and stabilization of biomolecules (Hassanpour et al. 2018).

2.9 Conclusion

The algal-mediated synthesis of various metallic nanoparticles is a cost-effective, nontoxic, and eco-friendly method. The main advantage of algal-mediated synthesis (extract or whole cell) is the rapid rate of synthesis compared to other biogenic methods such as bacteria and fungi. Both intracellular and extracellular mode can be used for synthesis, but extracellular mode should be the preferred choice due to ease of separation, purification, and cost-effectiveness of the process. Further, extracellular mode gives immense control over important process parameters such as temperature, pH, extract concentration, stirring speed, and reactant ratio, which govern the nanoparticle shape and size. Shape and size are an important aspect of nanoparticle because it determines the important characteristics and properties of nanoparticles such as optical properties, electronic properties, catalytic properties, bioactivity, antibacterial properties, nanoparticle toxicity, and biocompatibility. Biocompatibility and nontoxicity are two important attributes that any metallic nanoparticle should possess to be eligible for biomedical applications. Various biomedical applications of metallic nanoparticle such as drug delivery, gene delivery, anticancer treatment, biosensing, and bioimaging have been successfully explored and are currently being tested for newer biomedical applications. The uniqueness and ability to engineer the properties of nanoparticles at nanoscale are the main driving force in increasing application in every field imaginable. Given the future increase in demand of nanoparticle, algae-mediated synthesis can present a novel cost-effective and nontoxic method for large-scale production of metallic nanoparticles.

References

Abboud Y, Saffaj T, Chagraoui A et al (2014) Biosynthesis, characterization and antimicrobial activity of copper oxide nanoparticles (CONPs) produced using brown alga extract (*Bifurcaria bifurcata*). Appl Nanosci 4:571–576

Abdelghany TM, Rajhi AMH, Abboud MAA et al (2018) Recent advances in green synthesis of silver nanoparticles and their applications: about future directions. A review. Bio Nano Sci 8:5–16

Aboelfetoh EF, Shenody RAE, Ghobara MM (2017) Eco-friendly synthesis of silver nanoparticles using green algae (*Caulerpa serrulata*): reaction optimization, catalytic and antibacterial activities. Environ Monit Assess 189:349

Abdel-Raouf N, Al-Enazi NM, Ibraheem IBM et al (2019) Biosynthesis of silver nanoparticles by using of the marine brown alga *Padina pavonia* and their characterization. Saudi J Biol Sci 26(6):1207–1215

Aenishanslins NO, Ostuni GA, Monras JP et al (2020) Bacterial synthesis of ternary CdsAg quantum dots through cation exchange: tuning the composition and properties of biological nanoparticles for bioimaging and photovoltaic applications. Microorganisms 8(5):631

Agarwal H, Kumar SV, Rajeshkumar S (2017) A review on green synthesis of zinc oxide nanoparticles – an eco-friendly approach. Resour Eff Technol 3(4):406–413

Ahmeda A, Zangeneh A, Zangeneh MM (2020) Characterization and anti-acute T cell leukemia properties of silver nanoparticles synthesized by a green approach for bioremediation applications: introducing a new chemotherapeutic drug for clinical trial studies. Appl Organomet Chem 34(3):1–11

Anastas PT, Warner JC (1998) Green chemistry: theory and practice. Oxford University Press, New York

Amna Rao MD, Pennathura G (2017) Green Synthesis and characterization of cadmium sulphide nanoparticles from *Chlamydomonas reinhardtii* and their application as photocatalysts. Mat Res Bulle 85:64–73

Arsiya F, Sayadi MH, Sobhani S (2017) Green synthesis of palladium nanoparticles using *Chlorella vulgaris*. Mater Lett 186:113–115

Arya A, Gupta K, Chundawat TS et al (2018) Biogenic synthesis of copper and silver nanoparticles using green alga *Botryococcus braunii* and its antimicrobial activity. Bioinorg Chem Appl 7879403

Aziz N, Fatma T, Varma A, Prasad R (2014) Biogenic synthesis of silver nanoparticles using *Scenedesmus abundans* and evaluation of their antibacterial activity. Journal of Nanoparticles, Article ID 689419, 2014. https://doi.org/10.1155/2014/689419

Aziz N, Faraz M, Pandey R, Sakir M, Fatma T, Varma A, Barman I, Prasad R (2015) Facile algae-derived route to biogenic silver nanoparticles: Synthesis, antibacterial and photocatalytic properties. Langmuir 31:11605–11612. https://doi.org/10.1021/acs.langmuir.5b03081

Aziz N, Pandey R, Barman I, Prasad R (2016) Leveraging the attributes of *Mucor hiemalis*-derived silver nanoparticles for a synergistic broad-spectrum antimicrobial platform. Front Microbiol 7:1984. https://doi.org/10.3389/fmicb.2016.01984

Aziz N, Faraz M, Sherwani MA, Fatma T, Prasad R (2019) Illuminating the anticancerous efficacy of a new fungal chassis for silver nanoparticle synthesis. Front Chem 7:65. https://doi.org/10.3389/fchem.2019.00065

Azizi S, Ahmad MB, Namvar F et al (2014) Green biosynthesis and characterization of zinc oxide nanoparticles using brown marine macroalga *Sargassum muticum* aqueous extract. Mater Lett 116:275–277

Ballesteros NG, López SP, González JBR, Lastra M, Argüelles MCR (2017) Green synthesis of gold nanoparticles using brown algae *Cystoseira baccata*: its activity in colon cancer cells. Colloids Surf B Biointerfaces 153:190–198

Bhattacharya P, Swarnakar S, Ghosh S et al (2019) Disinfection of drinking water via algae mediated green synthesized copper oxide nanoparticles and its toxicity evaluation. J Environ Chem Eng 7:102867

Bhuyan T, Mishra K, Khanuja M, Prasad R, Varma A (2015) Biosynthesis of zinc oxide nanoparticles from *Azadirachta indica* for antibacterial and photocatalytic applications. Mater Sci Semicond Process 32:55–61

Brayner R, Coradin T, Vincent FF et al (2005) Algal polysaccharide capsule-templated growth of magnetic nanoparticles. New J Chem 29:681–685

Castro L, Blázquez ML, Muñoz JA et al (2013) Biological synthesis of metallic nanoparticles using algae. IET Nanobiotechnol 7:109–116

Chaudhary R, Nawaz K, Khan AK et al (2020) An overview of the algae-mediated biosynthesis of nanoparticles and their biomedical applications. Biomolecules 10(11):1–35

Cortese B, Amone S, Testini M et al (2019) Hybrid clustered nanoparticles for chemo-antibacterial combinatorial cancer therapy. Cancers (Basel) 11:1338

Dağlıoğlu Y, Öztürk BY (2019) A novel intracellular synthesis of silver nanoparticles using *Desmodesmus* sp. (Scenedesmaceae): different methods of pigment change. Rendiconti Lincei 30(3):611–621

Daraee H, Eatemadi A, Abbasi E et al (2016) Application of gold nanoparticles in biomedical and drug delivery. Artif Cells Nanomed Biotechnol 44(1):410–422

Dave PN, Chopda LV (2014) Application of iron oxide nanomaterials for the removal of heavy metals. J Nanotechnol 2014:398569

Dhas TS, Kumar VG, Karthick V, Angel KJ, Govindaraju K (2014) Facile synthesis of silver chloride nanoparticles using marine alga and its antibacterial efficacy. Spectrochim Acta Part A Mol Biomol Spectrosc 120:416–420

Edison TJI, Sethuraman M (2012) Instant green synthesis of silver nanoparticles using *Terminalia chebula* fruit extract and evaluation of their catalytic activity on reduction of methylene blue. Process Biochem 47:1351–1357

Gour A, Jain NK (2019) Advances in green synthesis of nanoparticles. Artif Cells Nanomed Biotechnol 47(1):844–851

Griffin S, Masood MI, Nasim MJ et al (2018) Natural nanoparticles: a particular matter inspired by nature. Antioxidants 7(1)

Grigore ME, Biscu ER, Holban AM, Gestal MC, Grumezescu AM (2016) Methods of synthesis, properties and biomedical applications of CuO nanoparticles. Pharmaceuticals 9:75

Hamouda RA, Hussein MH, Abo-elmagd RA, Bawazir SS (2019) Synthesis and biological characterization of silver nanoparticles derived from the cyanobacterium *Oscillatoria limnetica*. Sci Rep 9(1):1–17

Hassanpour P, Panahi Y, Ebrahimi-Kalan A et al (2018) Biomedical applications of aluminium oxide nanoparticles. Micro Nano Letters 13(9):1227–1231

Horikoshi S, Serpone N (2013) Introduction to nanoparticles. In: Microwaves in nanoparticle synthesis: fundamentals and applications, pp 1–24

Hou J, Wu Y, Li X, Wei B et al (2018) Toxic effects of different types of zinc oxide nanoparticles on algae, plants, invertebrates, vertebrates and microorganisms. Chemosphere 193:852–860

Ibrahim HM (2015) Green synthesis and characterization of silver nanoparticles using banana peel extract and their antimicrobial activity against representative micro-organisms. J Radiat Res Appl Sci 8:265–275

Jeevanandam J, Barhoum A, Chan YS, Dufresne A, Danquah MK (2018) Review on nanoparticles and nanostructured materials: history, sources, toxicity and regulations. Beilstein J Nanotech 1050–1074

Jena J, Pradhan N, Dash BP, Sukla LB, Panda PK (2013) Biosynthesis and characterization of silver nanoparticles using microalga *Chlorococcum humicola* and its antibacterial activity. Int J Nanomater Biostruct 3:1–8

Jena J, Pradhan N, Nayak RR et al (2014) Microalga *Scenedesmus* sp.: a potential low-cost green machine for silver nanoparticle synthesis. J Microbiol Biotechnol 24(4):522–533

Jiang J, Pi J, Cai J (2018) The advancing of zinc oxide nanoparticles for biomedical applications. Bioinorg Chem Appl

Kannan RRR, Arumugam R, Ramya D, Manivannan K, Anantharaman P (2013) Green synthesis of silver nanoparticles using marine macroalga *Chaetomorpha linum*. Appl Nanosci 3:229–233

Kassas H, Mohamed A, Gharib SM (2016) Green synthesis of iron oxide (Fe_3O_4) nanoparticles using two selected brown seaweeds: characterization and application for lead bioremediation. Acta Oceanol Sin

Kathiraven T, Sundaramanickam A, Shanmugam N, Balasubramanian T (2015) Green synthesis of silver nano- particles using marine algae *Caulerpa racemosa* and their antibacterial activity against some human pathogens. Appl Nanosci 5:499–504

Khan A, Rashid A, Younas R, Chong R (2016) A chemical reduction approach to the synthesis of copper nanoparticles. Int Nano Lett 6(1):21–26

Khan I, Saeed K, Khan I (2019) Nanoparticles: properties, applications and toxicities. Arab J Chem 12(7):908–931

Khanna P, Kaur A, Goyal D (2019) Algae-based metallic nanoparticles : synthesis, characterization and applications. J Microbiol Methods 163:105656

Kim DY, Saratale RG, Shinde S et al (2018) Green synthesis of silver nanoparticles using *Laminaria japonica* extract: characterization and seedling growth assessment. J Clean Prod: 2910–2918

Koopi H, Buazar F (2018) A novel one-pot biosynthesis of pure alpha aluminum oxide nanoparticles using the macroalgae *Sargassum ilicifolium*: a green marine approach. Ceram Int 44(8):8940–8945

Kumar M, Sahaya P, Ali DM et al (2017) Synthesis of nano-cuboidal gold particles for effective antimicrobial property against clinical human pathogens. Microb Pathog 113:68–73

Kumbhakar P, Ray SS, Stepanov AL (2014) Optical properties of nanoparticles and nanocomposites. J Nanomater 2014:181365

Li L, Zhang Z (2016) Biosynthesis of gold nanoparticles using green alga *Pithophora oedogonia* with their electrochemical performance for determining carbendazim in soil. Int J Electrochem Sci 11(6):4550–4559

Maddela NR, Chakraborty S, Prasad R (2021) Nanotechnology for Advances in Medical Microbiology. Springer Singapore (ISBN 978-981-15-9915-6) https://www.springer.com/gp/book/9789811599156

Matteis VD, Cascione M, Toma CC, Leporatti S (2018) Silver nanoparticles: synthetic routes, in vitro toxicity and theranostic applications for cancer disease. Nano 8(5):319

Mody VV, Siwale R, Singh A, Mody HR (2010) Introduction to metallic nanoparticles. J Pharm Bioallied Sci 2(4):282

Momeni S, Nabipour I (2015) A simple green synthesis of palladium nanoparticles with *Sargassum* alga and their electrocatalytic activities towards hydrogen peroxide. Appl Biochem Biotechnol 176(7):1937–1949

Mukherjee P, Ahmad A, Mandal D, Senapati S et al (2001) Bioreduction of $AuCl^{4-}$ Ions by the fungus, *Verticillium* sp. and surface trapping of the gold nanoparticles. Angewandte Chemie 40(19):3585–3588

Nagarajan S, Kuppusamy KA (2013) Extracellular synthesis of zinc oxide nanoparticle using seaweeds of Gulf of Mannar. Ind J Nanobiotechnol 11:39

Nallamuthu T, Venkatalakshmi RP, Chinnasamy A, Kannaiyan P (2012) Synthesis of silver nanoparticles and the antibacterial and anticancer activities of the crude extract of *sargassum polycystum* C. Agardh. Nano Biomed Eng 4(2):89–94

Nunna BB, Mandal D, Lee JU et al (2019) Detection of cancer antigens (CA-125) using gold nano particles on interdigitated electrode-based microfluidic biosensor. Nano Convergence 6(1):3

Ogi T, Saitoh N, Nomura T, Konishi Y (2010) Room-temperature synthesis of gold nanoparticles and nanoplates using *Shewanella* algae cell extract. J Nanopart Res 12(7):2531–2539

Oza G, Pandey S, Mewada A et al (2012) Facile biosynthesis of gold nanoparticles exploiting optimum PH and temperature of fresh water algae *Chlorella pyrenoidusa*. Adv Appl Sci Res 3(3):1405–1412

Parial D, Pal R (2014) Biosynthesis of monodisperse gold nanoparticles by green alga *Rhizoclonium* and associated biochemical changes. J Appl Phycol 27:975–984

Park I, Ko SH, Pan H, Grigoropoulos CP et al (2008) Nanoscale patterning and electronics on flexible substrate by direct nanoimprinting of metallic nanoparticles. Adv Mater 20(3):489–496

Park M, Hwang CSH, Jeong KH (2018) Nanoplasmonic alloy of Au/Ag nanocomposites on paper substrate for biosensing applications. ACS Appl Mater Interfaces 10(1):290–295

Patil MP, Kim GD (2018) Marine microorganisms for synthesis of metallic nanoparticles and their biomedical applications. Colloids Surf B: Biointerfaces 172:487–495

Prasad R (2014) Synthesis of silver nanoparticles in photosynthetic plants. Journal of Nanoparticles, Article ID 963961, 2014. https://doi.org/10.1155/2014/963961

Prasad R (2019a) Microbial Nanobionics: Basic Research and Applications. Springer International Publishing (ISBN 978-3-030-16534-5) https://www.springer.com/gp/book/9783030165338

Prasad R (2019b) Microbial Nanobionics: State of Art. Springer International Publishing (ISBN 978-3-030-16383-9) https://www.springer.com/gp/book/9783030163822

Prasad R (2019c) Plant Nanobionics: Advances in the Understanding of Nanomaterials Research and Applications. Springer International Publishing (ISBN 978-3-030-12495-3) https://www.springer.com/gp/book/9783030124953

Prasad R (2019d) Plant Nanobionics: Approaches in Nanoparticles Biosynthesis and Toxicity. Springer International Publishing (ISBN 978-3-030-16379-2) https://www.springer.com/gp/book/9783030163785

Prasad TNVKV, Kambala VSR, Naidu R (2013) Phyconanotechnology: synthesis of silver nanoparticles using brown marine algae *Cystophora moniliformis* and their characterisation. J Appl Phycol 25:177–182

Prasad R, Pandey R, Barman I (2016) Engineering tailored nanoparticles with microbes: quo vadis. WIREs Nanomed Nanobiotechnol 8:316–330. https://doi.org/10.1002/wnan.1363

Prasad R, Kumar V, Kumar M (2017) Nanotechnology: Food and Environmental Paradigm. Springer Nature Singapore Pte Ltd. (ISBN 978-981-10-4678-0)

Prasad R, Jha A, Prasad K (2018) Exploring the Realms of Nature for Nanosynthesis. Springer International Publishing (ISBN 978-3-319-99570-0) https://www.springer.com/978-3-319-99570-0

Prasad R, Kumar V, Kumar M, Choudhary D (2019) Nanobiotechnology in Bioformulations. Springer International Publishing (ISBN 978-3-030-17061-5) https://www.springer.com/gp/book/9783030170608

Prasad R, Siddhardha B, Dyavaiah M (2020) Nanostructures for Antimicrobial and Antibiofilm Applications. Springer International Publishing (ISBN 978-3-030-40336-2) https://www.springer.com/gp/book/9783030403362

Prathna TC, Lazar M, Chandrasekaran N, Raichur AM, Mukherjee A (2010) Biomimetic synthesis of nanoparticles: science and applicability. Biomimetics Learning from Nature, Amitava Mukherjee, IntechOpen

Priyadarshini RI, Prasannaraj G, Geetha N, Venkatachalam P (2014) Microwave-mediated extracellular synthesis of metallic silver and zinc oxide nanoparticles using macro-algae (*Gracilaria edulis*) extracts and its anticancer activity against human PC3 cell lines. Appl Biochem Biotechnol 174:2777–2790

Ramaswamy S, Narendhran S, & Sivaraj R (2016) Potentiating effect of ecofriendly synthesis of copper oxide nanoparticles using brown alga: antimicrobial and anticancer activities. Bulletin of Materials Science, 39(2): 361–364

Rafie HM, Zahran MK (2013) Green synthesis of silver nanoparticles using polysaccharides extracted from marine macro algae. Carbohydr Polym 96(2):403–410

Rahman A, Kumar S, Bafana A, Dahoumane SA, Jeffryes C (2019) Biosynthetic conversion of ag^{+} to highly stable Ag^{0} nanoparticles by wild type and cell wall deficient strains of *Chlamydomonas reinhardtii*. Molecules 24(1)

Rajathi FAAA, Parthiban C, Kumar VG, Anantharaman P (2012) Biosynthesis of antibacterial gold nanoparticles using brown alga, *Stoechospermum marginatum* (Kützing). Spectrochim Acta Part A Mol Biomol Spectrosc 99:166–173

Rajeshkumar S, Kannan C, Annadurai G (2012) Synthesis and characterization of antimicrobial silver nanoparticles using marine brown seaweed *Padina tetrastromatica*. Drug Invention Today 4(10):511–513

Rajeshkumar S, Malarkodi C, Gnanajobitha G et al (2013) Seaweed-mediated synthesis of gold nanoparticles using *Turbinaria conoides* and its characterization. J Nanostruct Chem 3(44):1–7

Ramakrishna M, Babu DR, Gengan RM, Chandra S, Rao GN (2016) Green synthesis of gold nanoparticles using marine algae and evaluation of their catalytic activity. J Nanostruct Chem 6(1):1–13

Ranjani S, Sakthivel P (2013) Biosynthesis and characterization of silver nanoparticles using *Parthenium hysterophorous* leaves. Asian J Chem 25(S):1–8

Rao MD, Pennathura G (2017) Green synthesis and characterization of cadmium sulphide nanoparticles from *Chlamydomonas reinhardtii* and their application as photocatalysts. Mater Res Bullet 85:64–73

Raouf NA, Enazi NMA, Ibraheem IBM (2017) Green biosynthesis of gold nanoparticles using *Galaxaura elongata* and characterization of their antibacterial activity. Arab J Chem 10:S3029–S3039

Renuga DJ, Jeyasundari AS, Athithan S, Jacob YBA (2020) Synthesis and characterization of copper oxide nanoparticles using *Brassica oleracea* var. *italica* extract for its antifungal application. Mater Res Express 7(4)

Roni M, Murugan K, Panneerselvam C et al (2015) Characterization and biotoxicity of *Hypnea musciformis*-synthesized silver nanoparticles as potential eco-friendly control tool against *Aedes aegypti* and *Plutella xylostella*. Ecotoxicol Environ Saf 121:31–38

Roychoudhury P, Bhattacharya A, Dasgupta A, Pal R (2016) Biogenic synthesis of gold nanoparticle using fractioned cellular components from eukaryotic algae and cyanobacteria. Phycol Res 64(3):133–140

Saeed AA, Sánchez JLA, Sullivan CK, Abbas MN (2017) DNA biosensors based on gold nanoparticles-modified graphene oxide for the detection of breast cancer biomarkers for early diagnosis. Bioelectrochemistry 118:91–99

Saglam N, Korkusuz, F, Prasad R (2021) Nanotechnology Applications in Health and Environmental Sciences. Springer International Publishing (ISBN: 978-3-030-64410-9) https://www.springer.com/gp/book/9783030644093

Sandhu RS, Aharwal RP, Kumar S (2019) Green synthesis: a novel approach for nanoparticles synthesis. Int J Pharm Sci Res 10(8):3550

Sarma H, Joshi S, Prasad R, Jampilek J (2021) Biobased Nanotechnology for Green Applications. Springer International Publishing (ISBN 978-3-030-61985-5) https://www.springer.com/gp/book/9783030619848

Sayadi MH, Salmani N, Heidari A, Rezaei MR (2018) Bio-synthesis of palladium nanoparticle using *Spirulina platensis* alga extract and its application as adsorbent. Surf Interfaces 10:136–143

Schröfel A, Kratošová G, Šafařík I et al (2014) Applications of biosynthesized metallic nanoparticles – a review. Acta Biomater 10(10):4023–4042

Selvam GG, Sivakumar K (2015) Phycosynthesis of silver nanoparticles and photocatalytic degradation of methyl orange dye using silver (Ag) nanoparticles synthesized from *Hypnea musciformis* (Wulfen) J.V. Lamouroux. Appl Nanosci 5:617–622

Shankar SS, Ahmad A, Pasricha R, Sastry M (2003) Bioreduction of chloroaurate ions by *Geranium* leaves and its endophytic fungus yields gold nanoparticles of different shapes. J Mater Chem 13(7):1822–1826

Shargh AY, Sayadi MH, Heidari A (2018) Green biosynthesis of palladium oxide nanoparticles using *Dictyota indica* seaweed and its application for adsorption. Water Environ Nanotechnol 3(4):337–347

Sharma A, Sharma S, Sharma K (2016) Algae as crucial organisms in advancing nanotechnology: a systematic review. J Appl Phycol 28:1759–1774

Siddiqui MN, Redhwi HH, Achilias DS et al (2018) Green synthesis of silver nanoparticles and study of their antimicrobial properties. J Polym Environ 26:423–433

Singaravelu G, Arockiamary JS, Ganesh Kumar V, Govindaraju K (2007) A novel extracellular synthesis of monodisperse gold nanoparticles using marine alga, *Sargassum wightii* Greville. Colloids Surf B: Biointerfaces 57(1):97–101

Singh G, Babele PK, Shahi SK, Sinha RP, Tyagi MB, Kumar A (2014) Green synthesis of silver nanoparticles using cell extracts of *Anabaena doliolum* and screening of its antibacterial and antitumor activity. J Microbiol Biotechnol 24(10):1354–1367

Singh J, Dutta T, Kim K et al (2018) Green synthesis of metals and their oxide nanoparticles: applications for environmental remediation. J Nanobiotechnol 16:84

Sinha SN, Paul D, Halder N, Dipta S, Samir KP (2015) Green synthesis of silver nanoparticles using fresh water green alga *Pithophora oedogonia* (Mont.) Wittrock and evaluation of their antibacterial activity. Appl Nanosci 5(6):703–709

Slavin YN, Asnis J, Häfeli UO, Bach H (2017) Metal nanoparticles: understanding the mechanisms behind antibacterial activity. J Nanobiotechnol 15(1):1–20

Smitha SL, Nissamudeen KM, Philip D, Gopchandran KG (2008) Studies on surface plasmon resonance and photoluminescence of silver nanoparticles. Spectrochim Acta Part A Mol Biomol Spectrosc 71(1):186–190

Srivastava S, Usmani Z, Atanasov AG, Singh VK, Singh NP, Abdel-Azeem AM, Prasad R, Gupta G, Sharma M, Bhargava A (2021) Biological nanofactories: Using living forms for metal nanoparticle synthesis. Mini-Reviews in Medicinal Chemistry 21(2):245–265

Suganya KSU, Govindaraju K, Ganesh Kumar V, Stalin Dhas T, Karthick V, Singaravelu G, Elanchezhiyan M (2015) Blue green alga mediated synthesis of gold nanoparticles and its antibacterial efficacy against gram positive organisms. Mater Sci Eng C 47:351–356

Sun B, Hu N, Han L, Pi Y, Gao Y, Chen K (2019) Anticancer activity of green synthesised gold nanoparticles from *Marsdenia tenacissima* inhibits A549 cell proliferation through the apoptotic pathway. Artif Cells Nanomed Biotechnol 47(1):4012–4019

Thakkar KN, Mhatre SS, Parikh RY (2010) Biological synthesis of metallic nanoparticles. Nanomedicine 6(2):257–262

Thangadurai D, Sangeetha J, Prasad R (2020) Functional Bionanomaterials. Springer International Publishing (ISBN 978-3-030-41464-1) https://www.springer.com/gp/book/9783030414634

Venkatesan J, Manivasagan P, Kim SK, Kirthi AV, Marimuthu S, Rahuman AA (2014) Marine algae-mediated synthesis of gold nanoparticles using a novel *Ecklonia cava*. Bioprocess Biosyst Eng 37:1591–1597

Wang Q, Zou L, Yang X et al (2019) Direct quantification of cancerous exosomes via surface plasmon resonance with dual gold nanoparticle-assisted signal amplification. Biosens Bioelectron 135:129–136

Xie J, Lee JY, Wang DIC, Ting YP (2007) Silver nanoplates: from biological to biomimetic synthesis. ACS Nano 1(5):429–439

Xie Y, He Y, Irwin PL, Jin T, Shi X (2011) Antibacterial activity and mechanism of action of zinc oxide nanoparticles against *Campylobacter jejuni*. Appl Environ Microbiol 77(7):2325–2331

Yadav L, Tripathi RM, Prasad R, Pudake RN, Mittal J (2017) Antibacterial activity of Cu nanoparticles against *E. coli, Staphylococcus aureus* and *Pseudomonas aeruginosa*. Nano Biomed Eng 9(1): 9–14. https://doi.org/10.5101/nbe.v9i1.p9-14

Yaqoob AA, Ahmad H, Parveen T et al (2020) Recent advances in metal decorated nanomaterials and their various biological applications: a review. Front Chem 8:1–23

Yousefzadi M, Rahimi Z, Ghafori V (2014) The green synthesis, characterization and antimicrobial activities of silver nanoparticles synthesized from Green Alga *Enteromorpha flexuosa* (Wulfen). J Agardh Mater Lett 137:1–4

Yuan H, Ji W, Chu S et al (2019) Mercaptopyridine-functionalized gold nanoparticles for fiber-optic surface plasmon resonance Hg^{2+} sensing. ACS Sensors 4(3):704–710

Zaragoza ZML, Guerrero-Quintanar D, Munoz-Mendoza N, Gomez-Leyava G (2020) Nanoemulsions and nanosized ingredients for food formulations. In: Jafri SM (ed) Handbook of food nanotechnology: applications and approaches. Academic Press, pp 207–256

Zhang XF, Liu ZG, Shen W, Gurunathan S (2016) Silver nanoparticles: synthesis, characterization, properties, applications, and therapeutic approaches. Int J Mol Sci 17(9)

Chapter 3
Biogenic Synthesis of Metallic Nanoparticles from Algae

Mônika Grazielle Heinemann and Daiane Dias

Contents

3.1 Introduction

Among the different nanostructured materials (nanosheets, nanorods, nanowires, nanoparticles, nanotubes, etc.), nanoparticles (NPs) stand out due to their increasing advance in synthesis methods and unique applications for their excellent physical, chemical, biological, optical, mechanical, and electronic properties when compared to macrostructured particles (Ramkumar et al. 2017). Another important factor is that the ratio between the NPs surface area and their volume increases, making this type of material more reactive. In addition, different electronic and optical behaviors can occur, due to their characteristic electronic confinement (Grassian 2008).

Several metals such as Ag, Au, Pt, and metal oxides, namely, CuO, ZnO, Fe_3O_2, TiO_2, and among others, can be used as metallic precursors in the production of

M. G. Heinemann · D. Dias (✉)
Laboratório de Eletroespectro Analítica (LEEA), Escola de Química e Alimentos, Universidade Federal do Rio Grande, Rio Grande, RS, Brazil

D. Thangadurai et al. (eds.), *Bioprospecting Algae for Nanosized Materials*, Nanotechnology in the Life Sciences,
https://doi.org/10.1007/978-3-030-81557-8_3

MNPs or MONPs for different applications (Abdel-Raouf et al. 2019; Vinosha et al. 2019; Ramkumar et al. 2017; Bhattacharya et al. 2019; Azizi et al. 2014; Mahdavi et al. 2013; Indira et al. 2015). Gold and silver nanoparticles (AuNPs and AgNPs, respectively) are the most used due to their great potential because they present antioxidant, antibacterial, antifungal, antimicrobial anticancer, and DNA cleavage activities (Firdhouse and Lalitha 2016; Abdel-Raouf et al. 2017; Minhas et al. 2018; Gulbagca et al. 2019; Vinosha et al. 2019). Additionally, these NPs can be used in the degradation of environmental pollutants, photocatalytic studies, development of electrochemical sensor, and catalysis studies (Rasheed et al. 2018; Khan et al. 2016; Saha et al. 2017). Platinum nanoparticles (PtNPs) are widely used in catalysis and biomedical applications once they present antibacterial and anticancer proprieties, integrating the role of nano-enzymes, nano-carriers, and nano-diagnostic tools (Chen and Holt-hindle 2010; Al-Radadi 2019). On the other hand, MONPs can form new antibacterial agents (Naika et al. 2015) which are effective against pathogenic microorganisms once they present biocidal properties and can be used in fertilizers (Bhattacharya et al. 2019). Additionally, they have been used in environmental remediation (Singh et al. 2018a), and they present semiconductor proprieties allowing their utilization for electronics, optics, and optoelectronic applications (Azizi et al. 2014).

Different methods are used to synthesize MNPs (or MONPs) and they are divided into three main groups: chemical, physical, and biological methods following a top-down or bottom-up approach (Salam et al. 2012). The top-down method is based on deconstructing larger structures, using manufacturing techniques to cut, grind, and shape materials until they reach the nanoscale (Salam et al. 2012). On the other hand, in the bottom-up method, chemical reactions are used to obtain nanoscale structures (Cauerhff and Castro 2013).

Chemical methods for the production of MNPs (or MONPs) generally use toxic chemicals such as sodium borohydride, hydrazine hydrate, amines, and thiols, which can cause several health and environmental problems (Hannah and Thompson 2008). Physical methods, such as laser ablation, sputter deposition, and microwave-assisted synthesis, have high energy consumption, high temperatures, radiation, and pressure, which pose a threat to organisms and the environment (Alzahrani et al. 2015). However, both methods use the chemical reduction of metallic precursors, in aqueous solutions, with or without the addition of stabilizing agents (Sharma et al. 2009).

In contrast, green nanotechnology (or nanobiotechnology) is integrated with the principles of green chemistry, which is a multidisciplinary area that is developing fast achieving synthesis with clean, safe, and environmentally friendly procedures when combined with chemical or physical methods (Kim et al. 2010). The biogenic synthesis of MNPs (or MONPs) using biological sources such as bacteria, fungi, plants, and algae (Mathivanan et al. 2019; Clarance et al. 2020; Gulbagca et al. 2019; Abdel-Raouf et al. 2019; Prasad et al. 2016, 2018; Srivastava et al. 2021) presents itself as a sustainable alternative because it uses water as solvent, the reaction occurs in one step, it is a biodegradable and biocompatible synthesis, and the use of different substances is not necessary for the reduction, capping, and stabilization of NPs (Park 2014).

The biomolecules and secondary metabolites that are present in these organisms can act as stabilizing and reducing agents, which will favor the reduction of the metal and influence the size of the synthesized biogenic metallic nanoparticles (BMNPs) (or biogenic metal oxide nanoparticles (BMONPs)) forming a stabilizing layer (capping) on their surface (Singh et al. 2013). Stabilizing agents are very important in the synthesis of NPs due to their useful properties such as controlling shape and size, avoiding aggregation or disordered growth during their production (Sinha et al. 2009).

Among the biological sources mentioned, algae have shown an exponential growth application in the green synthesis of NPs, due to their wide chemical composition of biomolecules that can act in the reduction and/or stabilization of the BMNPs (or BMONPs) obtained. Additionally, the algae abundance and availability in nature have encouraged researchers to develop biogenic and eco-friendly methods (Sharma et al. 2015; Aziz et al. 2014, 2015).

Several techniques are used to characterize BMNPs (or BMONPs), such as (i) UV–vis absorption spectroscopy (UV–vis) used to confirm the formation of NPs from the specific wavelength of each metal (or metal oxide) (Nasrollahzadeh et al. 2019), (ii) Fourier transmission infrared spectroscopy (FTIR) used to identify possible functional groups present in biomolecules (Nasrollahzadeh et al. 2019), (iii) transmission electron microscopy that provides information about the size and morphology of NPs obtained (Faria et al. 2014), (iv) energy-dispersive X-ray (EDX or EDS) for chemical or elemental composition characterization (Kumar et al. 2019), (v) X-ray diffraction (XRD) used to obtain information on the crystalline structure (Balasubramanian et al. 2020), and (vi) zeta potential for the study of stability and surface load (Kumar et al. 2019), among other techniques.

Therefore, this chapter addresses the green synthesis of BMNPs and BMONPs using different types of algae as a reducing and stabilizing agent, some explanation about the reaction mechanisms involved in this type of synthesis, and the future perspectives about this important subject.

3.2 Biogenic Synthesis from Algae

Different surfactants and polymers commercially available have been used as a stabilizing agent in the synthesis of MNPs (MONPs) where different morphologies and properties are obtained. However, there is growing concern about the removal and the complete biodegradation of these chemicals, which in many cases are toxic (Chiu et al. 2013). Therefore, several researches have been approached the methods and performing synthesis considering the principles of green chemistry (Kim et al. 2010).

Biogenic synthesis using metals or metal oxides can be performed using prokaryotic or eukaryotic organisms (plants, algae, bacteria, fungi, etc.), and the reaction can occur in the intra- or extracellular medium (Sinha et al. 2009). In the intracellular medium, the metal ions are absorbed and react with the biomolecules located inside the cell reducing and forming the BMNPs (Gericke and Pinches 2006) or they interact with the biomolecules that are released by the cell or with membrane proteins (extracellular medium) (Gericke and Pinches 2006).

As already mentioned, among the biological sources that are being used for biogenic synthesis, algae have received considerable attention due to their phytochemicals, which can reduce, cover, and stabilize BMNPs (or BMONPs) (Jena et al. 2013). Plants are widely used in the biogenic synthesis of MNPs due to their abundance, diversity, and easy availability (Iravani 2011; Prasad 2014). Bacteria and fungi are also used in the synthesis of different nanoparticles. However, their culture demands a high time-consuming process, and the nanoparticles obtained are generally polydispersed, with no controlled size distribution, shape, and crystallinity in a slow production rate (Narayanan and Sakthivel 2010).

In general, natural marine products have a wide range of applications in nanotechnology, and the utilization in the biogenic synthesis of MNPs (or MONPs) is highlighted, since it does not require the use of high temperature, high pressure, and/or the toxic reagents, normally applied in traditional chemical synthesis (Aziz et al. 2014).

Algae are important representatives of the aquatic ecosystem due to the great diversity in their chemical composition, such as proteins, phenolic compounds, polysaccharides, and lipids, among others, which can reduce and stabilize the BMNPs (or BMONPs) (Stengel et al. 2011). In addition, they provide a green, fast, inexpensive synthesis route, and they can be easily grown in the laboratory (Baker et al. 2013).

In biogenic synthesis with algae, some parameters such as temperature, pH, substrate, and metallic precursor concentration and algae morphology must be evaluated once they can affect the BMNPs (or BMONPs) proprieties (Saratale et al. 2018). Figure 3.1 shows a general outline of the synthesis of BMNPs (or BMONPs) with algae.

The algae extracts containing different biomolecules are mixed with the solutions of the metal precursors (with or without temperature), giving rise to the BMNPs (or BMONPs). Depending on the chosen metal, the suspensions formed

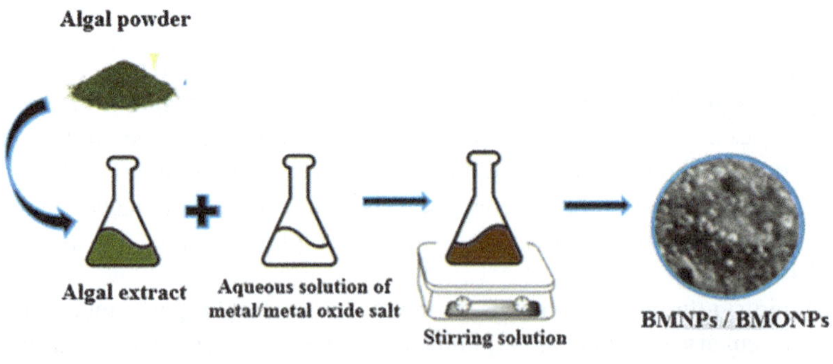

Fig. 3.1 Schematic representation of the synthesis of BMNPs and BMONPs with algae

will present specific color and absorption band characteristic to the collective excitation of the electrons from the BMNPs (BMONPs) by the phenomenon known as surface plasmon resonance (SPR) (Sharma et al. 2009). In the case of AgNPs, the SPR bands around 410–460 nm and the color change from dark to brown indicate the formation of AgNPs (Fatima et al. 2020). For AuNPs and PtNPs, the bands are around 525–555 nm and 260–280 nm, with a color change from yellow to dark red and from light yellow to dark yellowish brown, respectively (Annamalai and Nallamuthu 2015; Ramkumar et al. 2017). BMONPs, such as ZnONPs, have bands around 300–500 nm with color change from dark or brown to pale white (Nagarajan and Kuppusamy 2013), and CuONPs present bands around 265–285 nm from the formation of Cu_2ONPs and another band between 600 and 700 nm corresponding to the formation of CuONPs, with a color change from dark blue to dark red (Abboud et al. 2014).

3.3 Synthesis of Metallic and Metal Oxide Nanoparticles with Algae

Inorganic NPs comprise a very large group of different types of NPs, such as magnetic (iron), noble metals (silver, gold, platinum, palladium, etc.), and metal oxides (titanium oxide, zinc oxide, copper oxide, etc.) (Menon et al. 2017).

Magnetic NPs are synthesized from iron oxides, and the most common are magnetite (Fe_3O_4) and maghemite (Fe_2O_3). These NPs present electrical, magnetic, thermal, and mechanical properties, and they can be applied in several as ferrofluids, biosensors development, catalysts studies, separation processes, and environmental remediation (Jeyadevan et al. 2003; Miller et al. 2002; Zhang et al. 2005; Guo et al. 2012; Abd-Elsalam et al. 2019). Additionally, magnetic NPS have been applied in biomedical applications such as cell therapy, drug delivery, magnetic resonance imaging, and magnetofection (Gupta and Gupta 2005).

Noble metal nanoparticles are synthesized from their respective salts, and they present excellent electrical, optical, mechanical, and biological properties as well as exhibit excellent SPR phenomenon (Krishnaraj et al. 2014). They can be applied in several areas but have a wider range of applications in the biomedical ones (Al-Radadi 2019). Nanoparticles of metal oxides are synthesized from their metal oxides and are excellent semiconductors, and they can be applied in electrochemical cells, field emitters, high-Tc superconductors, and catalysis studies (Wu et al. 2012; Hsieh et al. 2003; Dai et al. 2000; Carnes and Klabunde 2003).

However, all these different applications of magnetic NPs, MNPs and MONPs, are dependent on their size, morphology, and surface area that present direct relation with the synthesis methods applied (Chiang et al. 2012). Therefore, based on what has been exposed, in the next sections some aspects related to the biogenic synthesis of metallic and metal oxides nanoparticles using different types of algae will be discussed as well as the mechanisms involved in these syntheses.

3.3.1 Synthesis of Gold Nanoparticles with Algae

Algae are biological sources rich in bioactive molecules with functional groups carboxyl, hydroxyl, and amine that act as reducing and capping agents to form BMNPs (Shankar et al. 2016). As shown in Table 3.1, different species of algae have been applied in the biogenic synthesis when gold is used as precursor. The algae applied in these syntheses are considered sources of bioactive compounds once they produce a wide variety of secondary metabolites and also because they present several compounds with antioxidant, antiviral, antifungal, and antimicrobial activities (Cox et al. 2010). As depicted in Table 3.1, among the algae normally used, the brown one is the most applied to the obtention of gold BMNPs.

In the gold biogenic synthesis from the red alga *Galaxaura elongata*, Abdel-Raouf et al. (2017) affirm that due to the presence of compounds such as andrographolide, alloaromadendrene oxide, glutamic acid, hexadecanoic acid, and oleic acid (and others), it is possible to obtain spherical, triangular, and hexagonal BMPs ranging from 3.85 to 77.13 nm. On the other hand, only triangular and spherical gold BMNPs (16 nm) were obtained from red alga *Halymenia dilatata* due to the presence of polysaccharides and pigments in its composition (Vinosha et al. 2019).

The green alga *Chlorella vulgaris* has been widely used in different applications, such as in the obtention of food, in the medicine area, and manufacturing industries (Faulkner 2001). These materials are rich in active compounds such as chlorophylls, carotenoids, astaxanthin, phenols, flavonoids, proteins, vitamins, and minerals (Faulkner 2001). Its phytochemicals are composed of hydroxyl, carboxyl, and amino functional groups that can be used to reduce and stabilize nanoparticles in one step (Annamalai and Nallamuthu 2015). The synthesis of AuNPs was performed by mixing an aqueous solution of chloroauric acid with the aqueous extract of *Chlorella vulgaris* obtaining stable and spherical NPs with the size from 2 to 10 nm (Annamalai and Nallamuthu 2015).

Tetraselmis kochinensis (another type of green alga) was used in the intracellular synthesis of AuNPs, causing the reduction of $AuCl_4^-$ ions and leading to the formation of AuNPs that are more concentrated in the cell wall than in the cytoplasmic membrane, making the AuNPs more accessible and interesting to catalysis, coatings, electronic applications, and drug distribution (Senapati et al. 2012).

Brown algae are important representatives of the aquatic marine ecosystem, and they have in their composition bioactive compounds such as terpenoids, steroids, and fucoidan polysaccharides (Kim et al. 2010). These materials present a wide variety of pharmacological effects such as anticancer, anti-inflammatory, and antibiotic properties (Lakshmi et al. 2006; Ammar et al. 2015); moreover, they also have been applied in the BMNPs synthesis as depicted in Table 3.1. The brown algae *Cystoseira baccata*, *Stoechospermum marginatum*, and *Padina gymnospora* present in their chemical composition compounds such as polysaccharides, polyphenols, and proteins that were able to reduce $AuCl_4^-$ ions in Au^0, forming stable and preferably spherical AuNPs (Rajathi et al. 2012; Singh et al. 2013; Ballesteros et al. 2017). AuNPs synthesized from *Sargassum plagiophyllum* alga was reduced

Table 3.1 Different species of algae for the biogenic synthesis of gold nanoparticles

Species of algae (color)	Reducing, capping, and stabilizing agents	Size (nm)/shape	Characterization techniques	Properties	References
Galaxaura elongata (red)	Andrographolide, alloaromadendrene oxide, glutamic acid, hexadecanoic acid, oleic acid, 11-eicosenoic acid, stearic acid, gallic acid, epigallocatechin catechin, and epicatechin gallate	3.85–77.13/ spherical, triangular, and hexagonal	TEM, zeta potential, FTIR, and UV–vis	Antibacterial	Abdel-Raouf et al. (2017)
Halymenia dilatata (red)	Polysaccharides and pigments	16/triangular and spherical	FTIR, UV–vis, FESEM, HRTEM, zeta potential, and XRD	Antioxidant, anticancer, and antibacterial	Vinosha et al. (2019)
Chlorella vulgaris (green)	Peptides, proteins, phenol, and flavonoid	2–10/spherical	UV–vis, SEM, TEM, XRD, and FTIR	Anti-pathogenic	Annamalai and Nallamuthu (2015)
Tetraselmis kochinensis (green)	Enzymes present in the cell wall and cytoplasmic membrane	5–35/spherical and triangular	UV–vis, XRD, and TEM	*	Senapati et al. (2012)
Cystoseira baccata (brown)	Polyphenols, polysaccharides, and proteins	8.4 ± 2.2/spherical	UV–vis, TEM, HRTEM, STEM, zeta potential, EELS, EDS, and FTIR	Activity in colon cancer cells	Ballesteros et al. (2017)
Sargassum plagiophyllum (brown)	Alcohols, phenols, carboxylic acids, and amines	65.87/ polydispersed and polymorphic in nature	UV–vis, FTIR, HRTEM, and XRD	Antimicrobial	Dhas et al. (2020)
Fucus vesiculosus (brown)	Hydroxyl groups present in the algal polysaccharides	*/spherical	XRD, SEM, EDS, FETEM, TEM, and FTIR	*	Mata et al. (2009)
Sargassum muticum (brown)	Bio-organic compounds/proteins	5.42 ± 1.18/ spherical	UV–vis, TEM, XRD, and zeta potential	*	Namvar et al. (2015)

(continued)

Table 3.1 (continued)

Species of algae (color)	Reducing, capping, and stabilizing agents	Size (nm)/shape	Characterization techniques	Properties	References
Stoechospermum marginatum (brown)	Hydroxyl groups present in the diterpenoids	18.7–93.7/spherical, hexagonal, and triangle	FTIR, SEM, TEM, and XRD	Antibacterial	Rajathi et al. (2012)
Padina gymnospora (brown)	Hydroxyl groups present in the algal polysaccharides	53–67/spherical	UV–vis, SEM, FTIR, AFM, and XRD	*	Singh et al. (2013)
Ecklonia cava (brown)	Primary amine and hydroxyl group	30 ± 0.25/spherical and triangular	FTIR, XRD, FESEM, and TEM	Antimicrobial	Venkatesan et al. (2014)

TEM Transmission electron microscopy, *FTIR* Fourier transform infrared spectroscopy, *UV–vis* Ultraviolet spectroscopy, *FESEM* Field emission scanning electron microscopy, *HRTEM* High-resolution transmission electron microscopy, *XRD* X-ray diffraction, *SEM* Scanning electron microscopy, *STEM* Scanning transmission electron microscopy, *EELS* Electron energy loss spectroscopy, *AFM* Atomic force microscopy, *EDS* Energy-dispersive spectroscopy, *information not available and stabilized through the functional group's alcohols, phenols, carboxylic acids, and amines present in this alga, with around 65.87 nm and, using the HRTEM technique, the authors observed that AuNPs were polydispersed and polymorphic in nature (Dhas et al. 2020).

In addition, these algae are well known for applications in biosorption due to their high metal uptakes when compared to fungi and other types of algae (Romera et al. 2007). These higher metal uptakes are due to its more complex cell wall, containing mucilaginous polysaccharides (alginate and sulfated fucoidans) and also molecules with functional groups containing amino, sulfhydryl, and sulfonates (Raize et al. 2004). These compounds present in brown algae are also very important in biogenic synthesis, as reported by Mata et al. (2009) where the alga *Fucus vesiculosus* was applied to produce spherical AuNPs. The hydroxyl groups present in the polysaccharides of this alga were responsible for the reduction and stabilization from Au^{+3} to Au^0.

In most of the works cited, the FTIR technique was used to assist the information regarding the functional groups present in the different types of algae. This information is important because it can help to elucidate the NPs formation mechanism through the shift of the functional group's bands, which can assist the identification of the main groups that may be responsible for the reduction of metals and capping of the MNPs (Ballesteros et al. 2017). In addition, microscopy techniques are also important for the characterization, once these techniques can provide information about the NPs shape and size that will directly influence their properties and applications (Annamalai and Nallamuthu 2015).

3.3.2 Synthesis of Silver Nanoparticles with Algae

Several species of algae, as shown in Table 3.2, can be used in the biogenic synthesis of silver nanoparticles (AgBNPs). Biological sources as brown marine algae *Padina pavonia* (Abdel-Raouf et al. 2019), *Sargassum wightii* (Selvaraj et al. 2020), and *Cystophora moniliformis* (Prasad et al. 2013) can catalyze specific reactions as a part of realistic and modern biosynthetic strategies, thus being alternatives to non-biological synthesis methods. Among the non-biologic synthesis method, the ultraviolet, irradiation, lithography, ultrasonic fields, and laser ablation are the most applied, and they are time-consuming, demand high consumption of energy, and require more than one reaction step for the synthesis (Govindaraju et al. 2008).

As already mentioned, algae present several biomolecules in their chemical composition. In the case of *Padina pavonia*, the catechin group and gallic acid act as reducing agents for silver metallic salt ($AgNO_3$), but they also influence the effective protection to provide a robust coating on AgNPs in one step (Abdel-Raouf et al. 2017, 2019). In this synthesis, the authors obtained AgNPs with different shapes (spherical, triangular, rectangle, polyhedral, and hexagonal) and the sizes (from 49.58 to 86.37 nm) through the extract of the algae with only 2 min of reaction but

Table 3.2 Biogenic silver nanoparticles synthesized with different types of algae

Species of algae (color)	Reducing and stabilizing agent	Size (nm) and shape	Characterization techniques	Properties	References
Padina pavonia (brown)	Catechin group and gallic acid	49.58–86.37/ spherical, triangular, rectangle, polyhedral, and hexagonal	UV–vis, TEM, DLS, and FTIR	*	Abdel-Raouf et al. (2019)
Sargassum wightii (brown)	Cardiac glycosides, phenolic compounds, and carbohydrates in both the algae	55–70/*	UV–vis and SEM	Antibacterial	Selvaraj et al. (2020)
Cystophora moniliformis (brown)	Phenolic compounds	50–100/ spherical	UV–vis, SEM, and XRD	*	Prasad et al. (2013)
Portieria hornemannii (red)	Carbonyl groups obtained from proteins and amino acids	70–75/ spherical	UV–vis, FTIR, XRD, SEM, TEM, EDS, and zeta potential	Antibacterial	Fatima et al. (2020)
Gelidium corneum (red)	Azo compounds and oxidation of the aldehyde group	20–50/ spherical	UV–vis, TEM, EDS, XRD, and FTIR	Antibiofilm and antimicrobial	Öztürk et al. (2020)
Gelidium amansii (red)	Carbonyl groups of amino acid	27–54/ spherical	UV–vis, FTIR, and SEM	Antimicrobial	Pugazhendhi et al. (2018)
Gelidiella acerosa (red)	Aromatic compound or alkanes or amines	22/spherical	UV–vis, XRD, TEM, SEM, and FTIR	Antifungal	Vivek et al. (2011)
Caulerpa racemosa (green)	Proteins	5–25/ spherical and triangular	UV–vis, FTIR, XRD, and TEM	Antibacterial	Kathiraven et al. (2015)
Ulva compressa (L.) Kütz. and *Cladophora glomerata* (L.) Kütz. (green)	Carbohydrates, reducing sugar, alkaloids, flavonoids, terpenoids, tannins, and amino acids	0.315 (*Cladophora glomerata*) and 66.3 (*Ulva compressa*) different and irregular shapes	UV–vis, SEM, AFM, FTIR, and XRD	Antibacterial	Minhas et al. (2018)

(continued)

Table 3.2 (continued)

Species of algae (color)	Reducing and stabilizing agent	Size (nm) and shape	Characterization techniques	Properties	References
Spirulina platensis (green)	Protein	7–16/ spherical	UV–vis, HRTEM, XRD, and FTIR	*	Govindaraju et al. (2008)

TEM Transmission electron microscopy, *FTIR* Fourier transform infrared spectroscopy, *UV–vis* Ultraviolet spectroscopy, *FESEM* Field emission scanning electron microscopy, *HRTEM* High-resolution transmission electron microscopy, *XRD* X-ray diffraction, *SEM* Scanning electron microscopy, *STEM* Scanning transmission electron microscopy, *EELS* Electron energy loss spectroscopy, *AFM* Atomic force microscopy, *EDS* Energy-dispersive spectroscopy, *DLS* Dynamic light scattering, *information not available

with the solid material (powder), the reaction was performed in 2 h (Abdel-Raouf et al. 2019).

In the synthesis of AgNPs using the brown alga *Cystophora moniliformis*, it was possible to obtain NPs after 30 min of reaction at 65 °C. Through SEM analysis, the authors verified the formation of spherical and monodispersed AgNPs, ranging from 50 to 100 nm and they suggest that the phenolic compounds contained in brown alga were responsible for the reduction and stabilization of AgNPs (Prasad et al. 2013).

The extract of the red alga *Portieria hornemannii* was used in biogenic synthesis, obtaining AgNPs that were monitored at 420 nm. FTIR analyses confirmed the great ability of the carbonyl group from the proteins and the amino acids in reducing Ag^+ ions to Ag^0, keeping them capped and stable (Fatima et al. 2020). The reduction of silver nitrate was also observed in the presence of the red alga *Gelidium corneum* from the azo compounds and oxidation of the aldehyde group that were considered effective capping and stabilizing agents, resulting in the formation of spherical AgNPs from 20 to 50 nm (Öztürk et al. 2020).

The green algae *Ulva compressa* (L.) Kütz. and *Cladophora glomerata* (L.) Kütz. are abundantly available in natural watercourses, and they have in their chemical composition carbohydrates, reducing sugar, alkaloids, flavonoids, terpenoids, tannins, and amino acids that are capable to reduce, coat, and stabilize the AgNPs (Minhas et al. 2018). It is important to mention that the nanoparticles obtained by Minhas et al. (2018) present wide size (0.315 nm for *Cladophora glomerata* and 66.3 nm for *Ulva compressa*) with different and irregular shapes.

In another study, the biomass of the green algae *Spirulina platensis* showed the ability to form spherical AgNPs (size from 7 to 16 nm) from the extracellular reduction of an $AgNO_3$ solution after 120 h of reaction at 37 °C (Govindaraju et al. 2008).

Using microscopic techniques, it is possible to verify the size and morphology of the formed AgNPs and, from the exposed works, it was observed that most of the AgNPs formed were predominantly spherical (*Padina pavonia, Ulva compressa* (L.) Kütz, and *Cladophora glomerata* (L.) Kütz.) and presented different shapes (triangular, rectangle, polyhedral, and hexagonal) (Minhas et al. 2018; Abdel-Raouf et al. 2019).

3.3.3 Synthesis of Platinum and Copper Nanoparticles with Algae

The interest in platinum nanoparticles (PtNPs) has increased a lot due to their inimitable structural, optical, and catalytic properties, which makes them excellent candidates in biomedical applications and in catalysis (Pugazhendhi et al. 2018). However, the synthesis of PtNPs with alga has not been widely studied, being the first work published in 2014, in which, for the first time, the brown alga *Padina gymnospora* was used for the biosynthesis (Shiny et al. 2014). In this work, the studies with TEM and XRD showed that the PtNPs had a size of approximately 35 nm and a spherical shape. In addition, from the FTIR analyzes, it was possible to verify that the alga extract was rich in alginate, carrageenan, laminarin, fucoidan (in the form of galactofucan), glucose, and mannose sugars with sulfate esters, which are responsible for the reduction and stabilization of platinum ions.

In another study using the same alga, the PtNPs formation was observed between 10 and 15 min of reaction from the addition of *Padina gymnospora* extract in the H_2PtCl_6 solution, in which the color changes to dark brown and the absorption bands monitored were 260 and 370 nm referring to Pt^+ and $PtCl_6^{2-}$, respectively, through the UV–vis analysis (Ramkumar et al. 2017). The size of these PtNPs varied from 5 to 50 nm with a truncated octahedral shape, and the author affirms that the functional groups such as ester, alcohol, and amine, among others, present in the material were responsible for the reduction and stabilization of the PtNPs (Ramkumar et al. 2017).

Once few studies have reported the possibility of synthesizing platinum nanoparticles with algae, this topic deserves special attention considering the fact that these materials can be used for different applications, and the reaction mechanism involved has not been explored yet.

3.3.4 Synthesis of Oxide Metallic Nanoparticles with Algae

MONPs are interesting due to their unique optical, electronic, and magnetic properties (Ramgir et al. 2013). They represent an important class of semiconductors that can be applied in magnetic storage, transformation of solar energy, electronics, sensors, and catalysis (De Montferrand et al. 2013; Jani et al. 2013; Ramgir et al. 2013). Several physical and chemical methods are described in the literature to synthesize MONPs, such as microemulsion, aerosol, sonochemical, and hydrothermal techniques, among others (Kumar et al. 2001; Zhang et al. 2006; Nassar and Husein 2007; Chiang et al. 2012). However, these methods use toxic chemicals in the synthesis procedure, which limits their applications in the clinical field (Nassar and Husein 2007). Thus, the use of algae for the development of synthesis methods clean, biocompatible, non-toxic, ecological, and economic has gradually increased, making room for further research due to the number of species of algae that have not yet been explored for biogenic synthesis (Abboud et al. 2014). Table 3.3 shows some studies that used algae to obtain BMONPs.

Table 3.3 Biogenic metal oxide nanoparticles synthesized with different types of algae

BMONPs	Species of algae (color)	Reducing and stabilizing agent	Size (nm) and shape	Characterization techniques	Properties	Reference
CuONPs	*Bifurcaria bifurcata* (brown)	Flavones and terpenoids	5–45/spherical	UV–vis, FTIR, XRD, and TEM	Antimicrobial	Abboud et al. (2014)
CuONPs	*Anabaena cylindrica* (green)	Polysaccharides	3.6/rod	UV–vis, XRD, FESEM, FTIR, EDS, and TEM	Disinfection	Bhattacharya et al. (2019)
ZnONPs	*Sargassum muticum* (brown)	Sulfate and hydroxyl moieties of polysaccharide	30–57/hexagonal wurtzite	FTIR, FESEM, and UV–vis	*	Azizi et al. (2014)
ZnONPs	*Caulerpa peltata* (green), *Hypnea valencia* (red), and *Sargassum myriocystum* (brown)	Fucoidan water-soluble pigments present in *Sargassum myriocystum*	36/spherical, triangle, radial, hexagonal, rod, and rectangle (*Sargassum myriocystum*)	UV–vis, DLS, AFM, FTIR, XRD, SEM, TEM, and EDS	Antibacterial	Nagarajan and Kuppusamy (2013)
Fe_3O_4NPs	*Sargassum muticum* (brown)	Sulfated polysaccharides	18 ± 4/cubic	FESEM, XRD, FTIR, EDXRF, VSM, and TEM	*	Mahdavi et al. (2013)

TEM Transmission electron microscopy, *FTIR* Fourier transform infrared spectroscopy, *UV–vis* Ultraviolet spectroscopy, *FESEM* Field emission scanning electron microscopy, *XRD* X-ray diffraction, *SEM* Scanning electron microscopy, *AFM* Atomic force microscopy, *EDS* Energy-dispersive spectroscopy, *DLS* Dynamic light scattering, *EDXRF* Energy-dispersive X-ray fluorescence spectrometry, *VSM* Vibrating sample magnetometer, *information not available

In the zinc oxide nanoparticles (ZnO NPs) synthesis, the dehydrated zinc acetate was mixed with the aqueous extract of the brown marine alga *Sargassum muticum* and calcined at 450 °C (Azizi et al. 2014). According to Azizi et al. (2014), the ZnO NPs were agglomerated with hexagonal structures, and particle sizes from 30 to 57 nm were obtained. Additionally, the FTIR spectra showed the involvement of the sulfate and hydroxyl groups (derived from the polysaccharides present in the *Sargassum muticum*) in the reduction and stabilization of the NPs formed. The same alga was used by Mahdavi et al. (2013) to synthesize iron oxide nanoparticles ($Fe_3O_4^-$NPs) with an average size of 18 nm and cubic shape.

ZnO NPs were also synthesized from the green alga *Caulerpa peltata*, red alga *Hypnea valencia,* and brown alga *Sargassum myriocystum* by Nagarajan and Kuppusamy (2013). From the results of the physical and chemical parameters evaluated (concentration of metal and extract, pH, temperature, and reaction time), the authors observed that only *Sargassum myriocystum* algae were able to synthesize ZnO NPs. From the results of the FTIR spectra, fucoidan water-soluble pigments present in alga leaf extract were responsible for the reduction and stabilization of the NPs (36 nm), without any visible change for 6 months showing the strong stability of the material formed.

The green microalga *Anabaena cylindrica* and the brown alga *Bifurcaria bifurcata* were used to synthesize stable copper oxide nanoparticles (CuONPs), with 3.6 nm and rod shape as well as from 5 to 45 nm with spherical shape, respectively (Abboud et al. 2014; Bhattacharya et al. 2019). Functional groups such as hydroxyl, carbonyl, amine, derived from polyols like flavones, terpenoids, and polysaccharides, were considered responsible for the reduction, capping, and stabilization of the CuONPs formed (Abboud et al. 2014; Bhattacharya et al. 2019).

The bibliography concerning the biogenic synthesis of BMONPs with algae is still scarce, and the research is slowly being shifted from conventional physical and chemical methods to methods using green synthesis. However, it is hoped that in the future, more researchers will be interested in developing MONPs, since they have numerous advantages over conventional methods that use toxic chemical reagents.

3.4 Mechanism Reaction of Metallic Biogenic Synthesis with Algae

Several sophisticated biochemical techniques and instruments are available to help the identification and the interpretation of the kind of interaction that happens between biomolecules and MNPs or MONPs in the reaction mechanism, which involves biogenic synthesis. Figure 3.2 shows a representation of a simple mechanism that explains the reduction of the metallic precursor and the stabilization of NPs from biomolecules present in algae. However, even with the modern equipment nowadays available, few studies explain the reaction mechanism of NPs with algae, and this requires further studies in this area. As depicted in Fig. 3.2, the biomolecules present in the algae composition (proteins, polyphenols, polysaccharides, etc.) will react with the metallic ions (M^+) reducing the metal (M^0), and the formation of NPs will occur from the nucleation and growth stage.

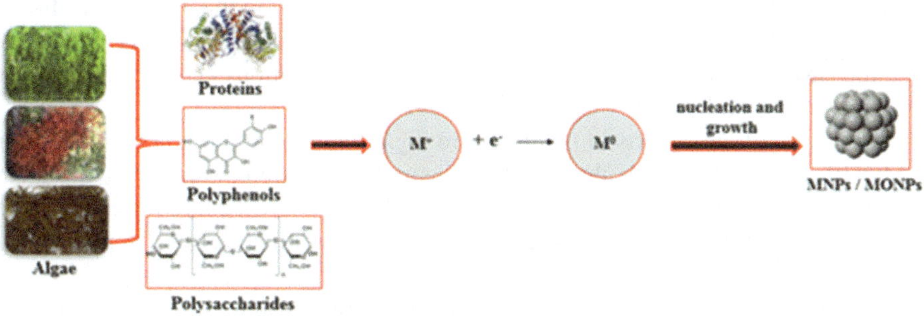

Fig. 3.2 Representation of the mechanism involving MNPs and MONPs in the presence of algae

When *Anabaena cylindrica* was used in the formation of CuONPs, the nucleation and the growth occurred within the nucleus of the extracellular polymeric fraction of the algae extract and the speed of rotation during the synthesis helped to reduce the size of the particles (Bhattacharya et al. 2019). The metal ions in hydrolysis form metallic nuclei that grow within the algae matrix forming the CuONPs.

The AgNPs synthesized by the extracellular method with the alga *Gelidium corneum* still do not have an exact mechanism (Öztürk et al. 2020). However, from the studies carried out with FTIR, it was observed that through the addition of the metal salt to the extract, the metal ions will bind quickly to functional groups such as -OH and -COOH (present in proteins and other water-soluble molecules) reducing the metal (Singh et al. 2018b). This leads to conformational changes in proteins and exposures hydrophobic residues to aqueous phases, leading to the removal of reducing agents from the extract, which facilitates the conversion of trapped metals to metal nanoparticles (Li et al. 2007).

In the work of Mata et al. (2009), the influence of pH was evaluated in the synthesis of AuNPs, and it was observed that the decrease in pH would be an indication that protons are released during an Au reduction step, according to the following reaction proposed by Kuyucak and Volesky (1989).

$$AuCl_4^- + 3R - OH \rightarrow Au^0 + 3R = O + 3H^+ + 4Cl^- \qquad (3.1)$$

According to the authors, the reduction from Au^{+3} to Au^0 occurs through the oxidation of hydroxyl (from the polysaccharides of the cell wall of the alga *Fucus vesiculosus*) to carbonyl groups. The participation of OH groups in the formation of AuNPs was confirmed by the FTIR analysis of the algae biomass before and after Au recovery (Mata et al. 2009).

3.5 Conclusion and Future Perspectives

The biogenic synthesis of MNPs and MONPs using biological compounds, such as algae, will help researchers to design safer nanomaterials and will also help to understand their safety and toxicity. Different materials, with countless applications, can be produced easily, on a reasonable scale, due to the fact that routes based on green synthesis do not require the use of toxic reagents. In this type of synthesis, only the biomolecules present in the algae are necessary to reduce, cover, and stabilize the BMNPs (or BMONPs) formed, avoiding aggregations and increasing their longevity and stability.

There are still no reports in the literature regarding the production of BMNPs or BMONPs synthesized with algae on an industrial scale, even though there are great opportunities to perform it due to their diverse applications. This limitation may be due to the high financial costs associated with their manufacture and the processing of nanoproducts.

However, green nanotechnology presents potential benefits in the development of important materials to face the various current challenges, improving the properties of the materials already used and assisting in the main global future demands of humanity, such as supplying consumption energy efficiency and reduction of ecological impacts. In addition, efforts are needed to minimize the metal's toxicity (or metal oxide) and to improve the biocompatibility and effectiveness of BMNPs (or BMONPs). Moreover, a greater understanding of the mechanisms involved in the formation of the biogenic synthesis with algae is needed.

References

Abboud Y, Saffaj T, Chagraoui A et al (2014) Biosynthesis, characterization and antimicrobial activity of copper oxide nanoparticles (CONPs) produced using brown alga extract (*Bifurcaria bifurcata*). Appl Nanosci 4:571–576. https://doi.org/10.1007/s13204-013-0233-x

Abdel-Raouf N, Al-Enazi NM, Ibraheem IBM (2017) Green biosynthesis of gold nanoparticles using *Galaxaura elongata* and characterization of their antibacterial activity. Arab J Chem 10:S3029–S3039. https://doi.org/10.1016/j.arabjc.2013.11.044

Abdel-Raouf N, Al-Enazi NM, Ibraheem IBM et al (2019) Biosynthesis of silver nanoparticles by using of the marine brown alga *Padina pavonia* and their characterization. Saudi J Biol Sci 26:1207–1215. https://doi.org/10.1016/j.sjbs.2018.01.007

Abd-Elsalam K, Mohamed AA, Prasad R (2019) Magnetic Nanostructures: Environmental and Agricultural Applications. Springer International Publishing (ISBN 978-3-030-16438-6) https://www.springer.com/gp/book/9783030164386

Al-Radadi NS (2019) Green synthesis of platinum nanoparticles using Saudi's Dates extract and their usage on the cancer cell treatment. Arab J Chem 12:330–349. https://doi.org/10.1016/j.arabjc.2018.05.008

Alzahrani E, Sharfalddin A, Alamodi M (2015) Microwave-hydrothermal synthesis of ferric oxide doped with cobalt. Adv Nanopart 04:53–60. https://doi.org/10.4236/anp.2015.42007

Ammar HH, Lajili S, Said RB et al (2015) Physico-chemical characterization and pharmacological evaluation of sulfated polysaccharides from three species of Mediterranean brown algae of the genus *Cystoseira*. DARU J Pharm Sci 23:4–11. https://doi.org/10.1186/s40199-015-0089-6

Annamalai J, Nallamuthu T (2015) Characterization of biosynthesized gold nanoparticles from aqueous extract of *Chlorella vulgaris* and their anti-pathogenic properties. Appl Nanosci 5:603–607. https://doi.org/10.1007/s13204-014-0353-y

Aziz N, Fatma T, Varma A, Prasad R (2014) Biogenic synthesis of silver nanoparticles using *Scenedesmus abundans* and evaluation of their antibacterial activity. Journal of Nanoparticles, Article ID 689419, 2014. https://doi.org/10.1155/2014/689419

Aziz N, Faraz M, Pandey R, Sakir M, Fatma T, Varma A, Barman I, Prasad R (2015) Facile algae-derived route to biogenic silver nanoparticles: Synthesis, antibacterial and photocatalytic properties. Langmuir 31:11605–11612. https://doi.org/10.1021/acs.langmuir.5b03081

Azizi S, Ahmad MB, Namvar F, Mohamad R (2014) Green biosynthesis and characterization of zinc oxide nanoparticles using brown marine macroalga *Sargassum muticum* aqueous extract. Mater Lett 116:275–277. https://doi.org/10.1016/j.matlet.2013.11.038

Baker S, Harini BP, Rakshith D, Satish S (2013) Marine microbes: invisible nanofactories. J Pharm Res 6:383–388. https://doi.org/10.1016/j.jopr.2013.03.001

Balasubramanian S, Kala SMJ, Pushparaj TL (2020) Biogenic synthesis of gold nanoparticles using *Jasminum auriculatum* leaf extract and their catalytic, antimicrobial and anticancer activities. J Drug Deliv Sci Technol 57:101620. https://doi.org/10.1016/j.jddst.2020.101620

Ballesteros N, López SP, González JBR et al (2017) Green synthesis of gold nanoparticles using brown algae *Cystoseira baccata*: its activity in colon cancer cells. Colloids Surf B Biointerfaces 153:190–198

Bhattacharya P, Swarnakar S, Ghosh S et al (2019) Disinfection of drinking water via algae mediated green synthesized copper oxide nanoparticles and its toxicity evaluation. J Environ Chem Eng 7:102867. https://doi.org/10.1016/j.jece.2018.102867

Carnes CL, Klabunde KJ (2003) The catalytic methanol synthesis over nanoparticle metal oxide catalysts. J Mol Catal A Chem 194:227–236. https://doi.org/10.1016/S1381-1169(02)00525-3

Cauerhff A, Castro GR (2013) Bionanoparticles, a green nanochemistry approach. Electron J Biotechnol 16. https://doi.org/10.2225/vol16-issue3-fulltext-3

Chen A, Holt-hindle P (2010) Platinum-based nanostructured materials: synthesis, properties and applications:3767–3804. https://doi.org/10.1021/cr9003902

Chiang CY, Aroh K, Ehrman SH (2012) Copper oxide nanoparticle made by flame spray pyrolysis for photoelectrochemical water splitting – part I. CuO nanoparticle preparation. Int J Hydrog Energy 37:4871–4879. https://doi.org/10.1016/j.ijhydene.2011.10.033

Chiu CY, Ruan L, Huang Y (2013) Biomolecular specificity controlled nanomaterial synthesis. Chem Soc Rev 42:2512–2527. https://doi.org/10.1039/c2cs35347d

Clarance P, Luvankar B, Sales J et al (2020) Green synthesis and characterization of gold nanoparticles using endophytic fungi *Fusarium solani* and its in-vitro anticancer and biomedical applications. Saudi J Biol Sci 27:706–712. https://doi.org/10.1016/j.sjbs.2019.12.026

Cox S, Abu-Ghannam N, Gupta S (2010) An assessment of the antioxidant and antimicrobial activity of six species of edible Irish seaweeds. Int Food Res J 17:205–220. https://doi.org/10.21427/D7HC92

Dai P, Mook HA, Aeppli G, Hayden SM, Dogan F (2000) Resonance as a measure of pairing correlations in the high-T_c superconductor $YBa_2Cu_3O_{6.6}$. Nature 406:965–968

De Montferrand C, Hu L, Milosevic I et al (2013) Iron oxide nanoparticles with sizes, shapes and compositions resulting in different magnetization signatures as potential labels for multiparametric detection. Acta Biomater 9:6150–6157. https://doi.org/10.1016/j.actbio.2012.11.025

Dhas TS, Sowmiya P, Kumar VG et al (2020) Antimicrobial effect of *Sargassum plagiophyllum* mediated gold nanoparticles on *Escherichia coli* and *Salmonella typhi*. Biocatal Agric Biotechnol 26:101627. https://doi.org/10.1016/j.bcab.2020.101627

Faria VW, Oliveira DGM, Kurz MHS et al (2014) Palladium nanoparticles supported in a polymeric membrane: an efficient phosphine-free "green" catalyst for Suzuki-Miyaura reactions in water. RSC Adv 4:13446–13452. https://doi.org/10.1039/c4ra01104j

Fatima R, Priya M, Indurthi L et al (2020) Biosynthesis of silver nanoparticles using red algae *Portieria hornemannii* and its antibacterial activity against fish pathogens. Microb Pathog 138:103780. https://doi.org/10.1016/j.micpath.2019.103780

Faulkner DJ (2001) Marine natural products. Nat Prod Rep 18:1–49. https://doi.org/10.1039/b006897g

Firdhouse MJ, Lalitha P (2016) Biogenic silver nanoparticles – synthesis, characterization and its potential against cancer inducing bacteria. J Mol Liq 222:1041–1050. https://doi.org/10.1016/j.molliq.2016.07.141

Gericke M, Pinches A (2006) Biological synthesis of metal nanoparticles. Hydrometallurgy 83:132–140. https://doi.org/10.1016/j.hydromet.2006.03.019

Govindaraju K, Basha SK, Kumar VG, Singaravelu G (2008) Silver, gold and bimetallic nanoparticles production using single-cell protein (*Spirulina platensis*) Geitler. J Mater Sci 43:5115–5122. https://doi.org/10.1007/s10853-008-2745-4

Grassian VH (2008) When size really matters: size-dependent properties and surface chemistry of metal and metal oxide nanoparticles in gas and liquid phase environments. J Phys Chem C 112:18303–18313. https://doi.org/10.1021/jp806073t

Gulbagca F, Ozdemir S, Gulcan M, Sen F (2019) Synthesis and characterization of *Rosa canina*-mediated biogenic silver nanoparticles for anti-oxidant, antibacterial, antifungal, and DNA cleavage activities. Heliyon 5:e02980. https://doi.org/10.1016/j.heliyon.2019.e02980

Guo J, Wang R, Tjiu WW et al (2012) Synthesis of Fe nanoparticles@graphene composites for environmental applications. J Hazard Mater 225–226:63–73. https://doi.org/10.1016/j.jhazmat.2012.04.065

Gupta AK, Gupta M (2005) Synthesis and surface engineering of iron oxide nanoparticles for biomedical applications. Biomaterials 26:3995–4021. https://doi.org/10.1016/j.biomaterials.2004.10.012

Hannah W, Thompson PB (2008) Nanotechnology, risk and the environment: a review. J Environ Monit 10:291–300. https://doi.org/10.1039/b718127m

Hsieh CT, Chen JM, Lin HH, Shih HC (2003) Field emission from various CuO nanostructures. Appl Phys Lett 83:3383–3385. https://doi.org/10.1063/1.1619229

Indira K, Mudali UK, Nishimura T, Rajendran N (2015) A review on TiO_2 nanotubes: influence of anodization parameters, formation mechanism, properties, corrosion behavior and biomedical applications. J Bio Tribo Corros 1:28. https://doi.org/10.1007/s40735-015-0024-x

Iravani S (2011) Green synthesis of metal nanoparticles using plants. Green Chem 13:2638–2650. https://doi.org/10.1039/c1gc15386b

Jani AM, Losic D, Voelcker NH (2013) Nanoporous anodic aluminium oxide: advances in surface engineering and emerging applications. Prog Mater Sci 58:636–704

Jena J, Pradhan N, Prasad DB et al (2013) Biosynthesis and characterization of silver nanoparticles using microalga *Chlorococcum humicola* and its antibacterial activity. Int J Nanomater Biostruct 3:1–8

Jeyadevan B, Chinnasamy CN, Shinoda K et al (2003) Mn-Zn ferrite with higher magnetization for temperature sensitive magnetic fluid. J Appl Phys 93:8450–8452. https://doi.org/10.1063/1.1543135

Kathiraven T, Sundaramanickam A, Shanmugam N, Balasubramanian T (2015) Green synthesis of silver nanoparticles using marine algae *Caulerpa racemosa* and their antibacterial activity against some human pathogens. Appl Nanosci 5:499–504. https://doi.org/10.1007/s13204-014-0341-2

Khan ZUH, Khan A, Shah A et al (2016) Photocatalytic, antimicrobial activities of biogenic silver nanoparticles and electrochemical degradation of water soluble dyes at glassy carbon/silver modified past electrode using buffer solution. J Photochem Photobiol B Biol 156:100–107. https://doi.org/10.1016/j.jphotobiol.2016.01.016

Kim EJ, Park SY, Lee JY, Park JHY (2010) Fucoidan present in brown algae induces apoptosis of human colon cancer cells. BMC Gastroenterol 10. https://doi.org/10.1186/1471-230X-10-96

Krishnaraj C, Muthukumaran P, Ramachandran R et al (2014) *Acalypha indica* Linn: biogenic synthesis of silver and gold nanoparticles and their cytotoxic effects against MDA-MB-231, human breast cancer cells. Biotechnol Rep 4:42–49. https://doi.org/10.1016/j.btre.2014.08.002

Kumar RV, Elgamiel R, Diamant Y et al (2001) Sonochemical preparation and characterization of nanocrystalline copper oxide embedded in poly(vinyl alcohol) and its effect on crystal growth of copper oxide. Langmuir 17:1406–1410

Kumar I, Mondal M, Meyappan V, Sakthivel N (2019) Green one-pot synthesis of gold nanoparticles using *Sansevieria roxburghiana* leaf extract for the catalytic degradation of toxic organic pollutants. Mater Res Bull 117:18–27. https://doi.org/10.1016/j.materresbull.2019.04.029

Kuyucak N, Volesky B (1989) Accumulation of gold by algal biosorbent. Biorecovery 1(3):189–204

Lakshmi V, Goel AK, Srivastava MN, Raghubir R (2006) Bioactivity of marine organisms: part X – screening of some marine fauna from the Indian coasts. Indian J Exp Biol 44:754–756. https://doi.org/10.3329/bjsir.v45i3.6537

Li S, Shen Y, Xie A et al (2007) Green synthesis of silver nanoparticles using *Capsicum annuum* L. extract. Green Chem 9:852–885. https://doi.org/10.1039/b615357g

Mahdavi M, Namvar F, Bin AM, Mohamad R (2013) Green biosynthesis and characterization of magnetic iron oxide (Fe_3O_4) nanoparticles using seaweed (*Sargassum muticum*) aqueous extract. Molecules 18:5954–5964. https://doi.org/10.3390/molecules18055954

Mata YN, Torres E, Blázquez ML et al (2009) Gold(III) biosorption and bioreduction with the brown alga *Fucus vesiculosus*. J Hazard Mater 166:612–618. https://doi.org/10.1016/j.jhazmat.2008.11.064

Mathivanan K, Selva R, Chandirika JU et al (2019) Biologically synthesized silver nanoparticles against pathogenic bacteria: synthesis, calcination and characterization. Biocatal Agric Biotechnol 22:101373. https://doi.org/10.1016/j.bcab.2019.101373

Menon S, Rajeshkumar S, Venkat KS (2017) A review on biogenic synthesis of gold nanoparticles, characterization, and its applications. Resour Effic Technol 000:1–12. https://doi.org/10.1016/j.reffit.2017.08.002

Miller MM, Prinz GA, Cheng SF, Bounnak S (2002) Detection of a micron-sized magnetic sphere using a ring-shaped anisotropic magnetoresistance-based sensor: a model for a magnetoresistance-based biosensor. Appl Phys Lett 81:2211–2213. https://doi.org/10.1063/1.1507832

Minhas FT, Arslan G, Gubbuk IH et al (2018) Evaluation of antibacterial properties on polysulfone composite membranes using synthesized biogenic silver nanoparticles with *Ulva compressa* (L.) Kütz. and *Cladophora glomerata* (L.) Kütz. extracts. Int J Biol Macromol 107:157–165. https://doi.org/10.1016/j.ijbiomac.2017.08.149

Nagarajan S, Kuppusamy A (2013) Extracellular synthesis of zinc oxide nanoparticle using seaweeds of gulf of Mannar, India. J Nanobiotechnol 11:1–11. https://doi.org/10.1186/1477-3155-11-39

Naika HR, Lingaraju K, Manjunath K et al (2015) Green synthesis of CuO nanoparticles using *Gloriosa superba* L. extract and their antibacterial activity. J Taibah Univ Sci 9:7–12. https://doi.org/10.1016/j.jtusci.2014.04.006

Namvar F, Azizi S, Ahmad MB et al (2015) Green synthesis and characterization of gold nanoparticles using the marine macroalgae *Sargassum muticum*. Res Chem Intermed 41:5723–5730. https://doi.org/10.1007/s11164-014-1696-4

Narayanan KB, Sakthivel N (2010) Biological synthesis of metal nanoparticles by microbes. Adv Colloid Interf Sci 156:1–13. https://doi.org/10.1016/j.cis.2010.02.001

Nasrollahzadeh M, Mahmoudi-Gom YS, Motahharifar N, Ghafori GM (2019) Recent developments in the plant-mediated green synthesis of Ag-based nanoparticles for environmental and catalytic applications. Chem Rec 19:2436–2479. https://doi.org/10.1002/tcr.201800202

Nassar NN, Husein MM (2007) Effect of microemulsion variables on copper oxide nanoparticle uptake by AOT microemulsions. J Colloid Interface Sci 316:442–450. https://doi.org/10.1016/j.jcis.2007.08.044

Öztürk YB, Yenice GB, Dağ İ (2020) Antibiofilm and antimicrobial activities of green synthesized silver nanoparticles using marine red algae *Gelidium corneum*. Process Biochem 89:208–219

Park Y (2014) A new paradigm shift for the green synthesis of antibacterial silver nanoparticles utilizing plant extracts. Toxicol Res 30:169–178. https://doi.org/10.5487/TR.2014.30.3.169

Prasad R (2014) Synthesis of silver nanoparticles in photosynthetic plants. Journal of Nanoparticles, Article ID 963961, 2014. https://doi.org/10.1155/2014/963961

Prasad TNVKV, Kambala VSR, Naidu R (2013) Phyconanotechnology: synthesis of silver nanoparticles using brown marine algae *Cystophora moniliformis* and their characterisation. J Appl Phycol 25:177–182. https://doi.org/10.1007/s10811-012-9851-z

Prasad R, Pandey R, Barman I (2016) Engineering tailored nanoparticles with microbes: quo vadis. WIREs Nanomed Nanobiotechnol 8:316–330. https://doi.org/10.1002/wnan.1363

Prasad R, Jha A, Prasad K (2018) Exploring the Realms of Nature for Nanosynthesis. Springer International Publishing (ISBN 978-3-319-99570-0) https://www.springer.com/978-3-319-99570-0

Pugazhendhi A, Edison TNJI, Karuppusamy I, Kathirvel B (2018) Inorganic nanoparticles: a potential cancer therapy for human welfare. Int J Pharm 539:104–111. https://doi.org/10.1016/j.ijpharm.2018.01.034

Raize O, Argaman Y, Yannai S (2004) Mechanisms of biosorption of different heavy metals by brown marine macroalgae. Biotechnol Bioeng 87:451–458. https://doi.org/10.1002/bit.20136

Rajathi FAA, Parthiban C, Kumar VG, Anantharaman P (2012) Biosynthesis of antibacterial gold nanoparticles using brown alga, *Stoechospermum marginatum* (Kützing). Spectrochim Acta – Part A Mol Biomol Spectrosc 99:166–173

Ramgir N, Datta N, Kaur M et al (2013) Metal oxide nanowires for chemiresistive gas sensors: issues, challenges and prospects. Colloids Surf A Physicochem Eng Asp 439:101–116. https://doi.org/10.1016/j.colsurfa.2013.02.029

Ramkumar VS, Pugazhendhi A, Prakash S et al (2017) Synthesis of platinum nanoparticles using seaweed *Padina gymnospora* and their catalytic activity as PVP/PtNPs nanocomposite towards biological applications. Biomed Pharmacother 92:479–490. https://doi.org/10.1016/j.biopha.2017.05.076

Rasheed T, Bilal M, Li C et al (2018) Catalytic potential of bio-synthesized silver nanoparticles using *Convolvulus arvensis* extract for the degradation of environmental pollutants. J Photochem Photobiol B Biol 181:44–52. https://doi.org/10.1016/j.jphotobiol.2018.02.024

Romera E, González F, Ballester A et al (2007) Comparative study of biosorption of heavy metals using different types of algae. Bioresour Technol 98:3344–3353. https://doi.org/10.1016/j.biortech.2006.09.026

Saha J, Begum A, Mukherjee A, Kumar S (2017) A novel green synthesis of silver nanoparticles and their catalytic action in reduction of methylene blue dye. Sustain Environ Res 27:245–250. https://doi.org/10.1016/j.serj.2017.04.003

Salam HA, Raju R, Murugesan K et al (2012) Plants: green route for nanoparticle synthesis. Int Res J Biol Sci 1:85–90

Saratale RG, Karuppusamy I, Saratale GD et al (2018) A comprehensive review on green nanomaterials using biological systems: recent perception and their future applications. Colloids Surf B Biointerfaces 170:20–35. https://doi.org/10.1016/j.colsurfb.2018.05.045

Selvaraj P, Neethu E, Rathika P et al (2020) Antibacterial potentials of methanolic extract and silver nanoparticles from marine algae. Biocatal Agric Biotechnol 28:101719. https://doi.org/10.1016/j.bcab.2020.101719

Senapati S, Syed A, Moeez S et al (2012) Intracellular synthesis of gold nanoparticles using alga *Tetraselmis kochinensis*. Mater Lett 79:116–118. https://doi.org/10.1016/j.matlet.2012.04.009

Shankar PD, Shobana S, Karuppusamy I et al (2016) A review on the biosynthesis of metallic nanoparticles (gold and silver) using bio-components of microalgae: formation mechanism and applications. Enzym Microb Technol 95:28–44. https://doi.org/10.1016/j.enzmictec.2016.10.015

Sharma VK, Yngard RA, Lin Y (2009) Silver nanoparticles: green synthesis and their antimicrobial activities. Adv Colloid Interf Sci 145:83–96. https://doi.org/10.1016/j.cis.2008.09.002

Sharma D, Kanchi S, Bisetty K (2015) Biogenic synthesis of nanoparticles: a review. Arab J Chem 12:3576–3600. https://doi.org/10.1016/j.arabjc.2015.11.002

Shiny PJ, Mukherjee A, Chandrasekaran N (2014) Haemocompatibility assessment of synthesised platinum nanoparticles and its implication in biology. Bioprocess Biosyst Eng 37:991–997. https://doi.org/10.1007/s00449-013-1069-1

Singh M, Kalaivani R, Manikandan S et al (2013) Facile green synthesis of variable metallic gold nanoparticle using *Padina gymnospora*, a brown marine macroalga. Appl Nanosci 3:145–151. https://doi.org/10.1007/s13204-012-0115-7

Singh J, Dutta T, Kim KH et al (2018a) "Green" synthesis of metals and their oxide nanoparticles: applications for environmental remediation. J Nanobiotechnol 16:1–24. https://doi.org/10.1186/s12951-018-0408-4

Singh P, Pandit S, Beshay M et al (2018b) Anti-biofilm effects of gold and silver nanoparticles synthesized by the *Rhodiola rosea* rhizome extracts. Artif Cells Nanomed Biotechnol 46:S886–S899. https://doi.org/10.1080/21691401.2018.1518909

Sinha S, Pan I, Chanda P, Sen SK (2009) Nanoparticles fabrication using ambient biological resources. J Appl Biosci 19:1113–1130

Srivastava S, Usmani Z, Atanasov AG, Singh VK, Singh NP, Abdel-Azeem AM, Prasad R, Gupta G, Sharma M, Bhargava A (2021) Biological nanofactories: Using living forms for metal nanoparticle synthesis. Mini-Reviews in Medicinal Chemistry 21(2):245–265

Stengel DB, Connan S, Popper ZA (2011) Algal chemodiversity and bioactivity: sources of natural variability and implications for commercial application. Biotechnol Adv 29:483–501. https://doi.org/10.1016/j.biotechadv.2011.05.016

Venkatesan J, Manivasagan P, Kim SK et al (2014) Marine algae-mediated synthesis of gold nanoparticles using a novel *Ecklonia cava*. Bioprocess Biosyst Eng 37:1591–1597. https://doi.org/10.1007/s00449-014-1131-7

Vinosha M, Palanisamy S, Muthukrishnan R et al (2019) Biogenic synthesis of gold nanoparticles from *Halymenia dilatata* for pharmaceutical applications: antioxidant, anti-cancer and antibacterial activities. Process Biochem 85:219–229. https://doi.org/10.1016/j.procbio.2019.07.013

Vivek M, Kumar PS, Steffi S, Sudha S (2011) Biogenic silver nanoparticles by *Gelidiella acerosa* extract and their antifungal effects. Avicenna J Med Biotechnol 3:143–148

Wu HB, Chen JS, Hng HH, Lou XW (2012) Nanostructured metal oxide-based materials as advanced anodes for lithium-ion batteries. Nanoscale 4:2526–2542. https://doi.org/10.1039/c2nr11966h

Zhang J, Wang Y, Ji H et al (2005) Magnetic nanocomposite catalysts with high activity and selectivity for selective hydrogenation of ortho-chloronitrobenzene. J Catal 229:114–118. https://doi.org/10.1016/j.jcat.2004.09.029

Zhang Y, Wang S, Li X et al (2006) CuO shuttle-like nanocrystals synthesized by oriented attachment. J Cryst Growth 291:196–201. https://doi.org/10.1016/j.jcrysgro.2006.02.044

Part III
Production of Nanomaterials and Nanoproducts from Algae

Chapter 4
Microalgal Applications in Nanotechnology: An Outstanding Tool for Nanocompounds Synthesis and Bioproducts Obtention

Juliana Botelho Moreira, Etiele Greque de Morais, Ana Claudia Araujo de Almeida, Jorge F. Pontes, Ana Grenha, Luísa Barreira, João Varela, Jorge Alberto Vieira Costa, and Michele Greque de Morais

Contents

4.1 Introduction

The scientific knowledge, the awareness of the population for a healthier life-style, and the scientific validation for the use of bioactive compounds from biological resources have been gaining momentum around the world. In this

J. B. Moreira (✉) · A. C. A. de Almeida · M. G. de Morais
Laboratory of Microbiology and Biochemistry, College of Chemistry and Food Engineering, Federal University of Rio Grande, Rio Grande, Brazil

E. G. de Morais · J. F. Pontes · A. Grenha · L. Barreira · J. Varela
Centre for Marine Sciences, University of Algarve, Faro, Portugal

J. A. V. Costa
Laboratory of Biochemical Engineering, College of Chemistry and Food Engineering, Federal University of Rio Grande, Rio Grande, Brazil

D. Thangadurai et al. (eds.), *Bioprospecting Algae for Nanosized Materials*, Nanotechnology in the Life Sciences,
https://doi.org/10.1007/978-3-030-81557-8_4

context, the search for biocompounds displaying and retaining the activity of compounds found in natural sources, such as microbial organisms, has gained a prominent status. These compounds present greater biocompatibility and decreased deleterious effects caused by their application to living beings (Centella et al. 2017).

Microalgae are photosynthetic microorganisms recognized for their ability to produce diverse compounds with biological activity. These compounds have positive characteristics for human health, present high added value, and have been used in biomedicine, cosmetics, and pharmaceuticals, as well as food and feed production. Among bioactive compounds, pigments such as carotenoids (e.g., astaxanthin, β-carotene, lutein), phycobiliproteins (phycocyanin), biopeptides, and chemicals with high market demand stand out (Molino et al. 2020).

The production of bioactive compounds by microalgae has a series of advantages. These microorganisms can double their biomass in hours, and their daily harvest has lower dependence on seasonal changes depending on the species. In addition, the biosynthesis of bioactive compounds by microalgae has been highlighted within the concept of biorefinery, enabling the production of other bioproducts such as biofuels and polymers from their biomass (Morais et al. 2015). Nonetheless, extraction of active biocompounds from microalgal biomass and their final application require an additional effort to prevent oxidation and their ensuing inactivation (Molino et al. 2020). An alternative to preserve these compounds and obtain higher efficiency in their application is the use of nanotechnology.

The combination of nanotechnology with the production of active compounds by microalgae has been used to increase the effectiveness of the application of these natural compounds (Morais et al. 2014). The nanoscale gives the materials improved mechanical, physical, and chemical characteristics by increasing the contact surface, permeability, and reactivity. Through the transformation of structures at the nanometric scale, it is possible to create protection capable of preserving target compounds through the development of polymeric nanocoats (e.g., nanospheres, nanocapsules, and nanoemulsions) and nanofibrous structures. Besides protection, these nanosystems enable the delivery of the compounds to a target (organ or cell) increasing their tissue and organ specificity. Moreover, the incorporation of microalgal compounds in nanostructures has been used in the food area to develop intelligent and active packaging, and in the tissue engineering with the development of band-aids and sensors.

In this context, this chapter addresses how nanotechnological systems, such as microalgae-derived scaffolds or nanocoats (e.g., nanocapsules, nanospheres, nanoemulsions, and nanofibers), can be used to carry and protect bioactive compounds for multiple biotechnological applications. Future perspectives and the main challenges of applying nanotechnological concepts to innovative packaging and delivery of microalgae-based bioactive products will also be discussed.

4.2 Scaffolds with Microalgae Biocompounds

Tissue engineering studies the development of biological substitutes for the recovery, regeneration, or replacement of defective tissues. Permanent implants usually cause chronic inflammation, which can lead to clinical complications. Implants developed using biomaterials can be a viable alternative to restore normal functions to damaged tissues and organs. Scaffolds are three-dimensional structures that can be used in tissue engineering to stimulate the development of tissues in situ. The success of their application depends on the growth of the tissue itself, the type of repair needed, and/or of the organ to be rebuilt (Asadi et al. 2019).

The scaffolds must have high porosity and pore interconnectivity to be effective. Thus, the specific surface area available is increased, which is relevant characteristic for cell anchorage and internal tissue growth, such as to facilitate the distribution and transport of oxygen, nutrients, and cellular wastes (Rnjak-Kovacina and Weiss 2011). Scaffolds made from polymeric nanofibers manufactured by electrospinning processes mimic the mechanical and architectural characteristics of the natural extracellular matrix, allowing the anchoring and migration of cells. In general, nanostructured scaffolds are produced using biodegradable polymers, natural or synthetic, which, at the end of the tissue regeneration process, are totally consumed/degraded and replaced by living tissues (Morais et al. 2010). Nanofibers are produced by electrospinning, and a schematic illustration of these production processes is shown in Fig. 4.1.

Drugs, antioxidants, and proteins can be incorporated into the matrix to accelerate the healing process as well as to increase biocompatibility with human tissues (Costa et al. 2020). Bioactive compounds by microalgae are an option for incorporation into scaffolds as they often have anti-inflammatory, antimicrobial, and

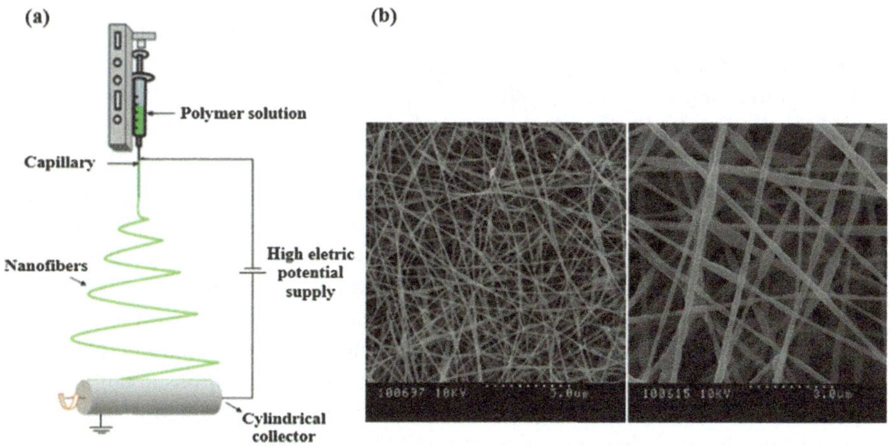

Fig. 4.1 Schematic illustration of the electrospinning (**a**) and micrographs of nanofibers produced by this nanotechnology process (**b**)

antioxidant activities. Their anti-inflammatory capacity is mainly related to polyunsaturated fatty acids, sulfur polysaccharides, and microalgal pigments (Morais et al. 2015; Costa et al. 2020). This property is especially important for application in patients with severe burns where the skin has been damaged (Steffens et al. 2013).

Despite the recognized functions of microalgal compounds, their application in scaffolds for tissue regeneration is recent and still little reported. Bioactive compounds from microalgae are generally applied as extracts rich in pigments or polyunsaturated fatty acids. However, studies in scaffolds using microalgae have also focused on the use of whole biomass to catch up synergistic effects of other microalgal compounds in tissue regeneration. It is reported that the addition of microalgal biomass improves the mechanical characteristics of the scaffold, presenting a similar biodegradability to that of natural tissue. Furthermore, scaffolds can increase the viability of cells, such as stem cells, due to the bioactive compounds constituted in the biomass (Morais et al. 2010; Steffens et al. 2013; Schmatz et al. 2016).

Morais et al. (2010) were the pioneer concerning the use of microalgal biomass for the development of scaffolds for tissue engineering. In their study, the whole biomass of *Spirulina* sp. LEB 18 was used in a blend with polyethylene oxide (PEO), a synthetic polymer, and nanofibers with diameters down to about 100 nm. This blend was used to produce nanofibers that were successfully applied to tissue engineering. Steffens et al. (2013) studied the incorporation of *Spirulina* sp. LEB 18 biomass in scaffolds of poly-(D,L-lactic acid) (PDLLA) for stem cell culture. The authors observed that, in a blend between *Spirulina* and PDLLA, the bioactive properties of the microalgal biomass improved the adherence and viability of the stem cells. Schmatz et al. (2016) developed scaffolds using poly(hydroxybutyrate-co-hydroxyvalerate) (PHB-HV) and polycaprolactone (PCL) together also with *Spirulina* sp. LEB 18, obtaining uniform fibers with diameters between 335 and 617 nm. Biodegradability tests of the fibers produced revealed that the material displayed enzymatic activity that simulated the behavior of natural tissue (Morais et al. 2014).

In addition to having bioactive compounds, microalgae can be a source of natural biodegradable polymers with potential application for the development of scaffolds. Studies comparing the fibers produced of polyhydroxybutyrate (PHB) extracted from *Spirulina* sp. LEB 18 with commercial polymers show that the former formulation performed better than the commercial product. Under the same experimental conditions of nanofibers production (20% w/v of PHB), flow rate of 150 $\mu L \cdot h^{-1}$, capillary diameter of 0.45 mm, and voltage of 24.1 kV, the commercial product formed drops, while the microalgal biomass formed fibers with a diameter of 470 nm. In another study, comparing elasticity, tensile strength, and breaking elongation of nanofibers with PHB from *Spirulina* sp. LEB 18 and with commercial PHB, it was observed that microalgal one performed better in all parameters, which are important points in the application of fibers as scaffolds in tissue engineering. The use of microalgal biomass was also better for the development of fibers with improved mechanical characteristics and with decreased diameter, when compared to fibers produced only with synthetic polymers (Schmatz et al. 2016), which may be a consequence of the presence of PHB in the microalgal biomass.

In addition to providing specific bioactive properties, the scaffolds developed by the electrospinning process protect bioactive compounds of microalgae by preventing their degradation. Kuntzler et al. (2018a) produced nanofibers through the electrospinning process using a blend of *Spirulina* sp. LEB 18 extracted phenolic compounds and PHB polymer. The antibacterial activity of these fibers became apparent, as the phenolic compounds incorporated in the nanofibers did not lose their activity after the electrospinning process. Braga et al. (2016) developed nanofibers with the addition of C-phycocyanin extracted from *Spirulina* and observed an improvement in the thermal stability of the pigment when compared to the compound in solution.

4.3 Nanocoats of Active Metabolites from Microalgae

Apart from microalgae being renewable feedstocks for scaffolding biomaterials, microalgae can also be the source of different target chemical compounds with biotechnological applicability (Khalid 2020), whose synthesis can be regulated and manipulated (Metting 1996). Possible target metabolites are pigments (Saini et al. 2020), polysaccharides (Chanda et al. 2019), lipids (Katiyar and Arora 2020), and sterols (Randhir et al. 2020), among others (Dehghani et al. 2020), which have been the focus of microalgal research over the last few decades. Indeed, the use of these compounds by the pharmaceutical (Mourelle et al. 2017; Gómez-Zorita et al. 2019; Riccio and Lauritano 2019; Dehghani et al. 2020; Rosales-Mendoza et al. 2020) and food industries (Caporgno and Mathys 2018; Bernaerts et al. 2019; Camacho et al. 2019; Dineshbabu et al. 2019) has been described. However, some of these chemicals lose their active properties upon contact with adverse conditions, such as acid pHs or extreme temperature (Maqsoudlou et al. 2020). If an application involving the delivery to humans or animals is envisaged, the use of carriers assumes importance, as they can protect the compounds of interest, apart from modulating other different features such as their release into the target cell or tissue. Several studies have been published in the literature, in which nanoemulsions (Farahin et al. 2018; Costa et al. 2019; Moreira et al. 2019a), liposomes (Machado et al. 2019), polymeric (Bezerra et al. 2019), magnetic nanoparticles (Zamani et al. 2019), and nanofibers (Kim et al. 2012) are proposed as carriers or embedding structures, resulting in either an encapsulated extract or an active matrix.

Nanoemulsions and liposomes have also been reported as the most common systems for the encapsulation of microalgal extracts. Nanoemulsions can be generated as nanoscaled structures, which can be stabilized by surfactants (Solans et al. 2005). One interesting technological aspect of this system is the possibility of using the oils extracted from microalgal biomass as a lipophilic phase, which can favor the encapsulation of more hydrophobic compounds (Costa et al. 2019). Concerning liposomes, lecithin is often used for their preparation, forming a vesicle hosting the microalgal extract to be encapsulated. Both systems are biocompatible (Solans et al. 2005; Akbarzadeh et al. 2013), further offering the possibility of using lipid components from microalgae. The processes by which nanosystems are prepared vary depending on the used materials and the final aim. Comprehensive reviews on emulsification-based methodologies, liposome production, and polymeric nanoparticle preparation processes can be found in Has and Sunthar (2019), Sheth et al. (2020), and Souto et al. (2020). Nevertheless, other techniques have been reported for the production of nanosystems, such as electrospinning (Sill and von Recum 2008) and ball milling (Morsy et al. 2019), even though the latter is not able to generate "nanocoats."

Spirulina spp. is a microalga known for its applications as functional food. It is a group of planktonic blue-green microalga (Belay et al. 1993) rich in phenolic compounds that endow the algae with antioxidant activity (Akao et al. 2009). Owing to that, *Spirulina* might play an important role as an adjuvant in cancer therapies (Abd El-Hack et al. 2019). Pectin was proposed as the matrix of nanoparticles (~84 nm) encapsulating a carotenoid-containing extract of *Spirulina* sp. LEB 18 with an efficiency of around 96%. The nanoparticles were prepared (Fig. 4.2) by solvent displacement method, and the encapsulation was six times higher than that of the control in which synthetic carotenoids were used (Bezerra et al. 2019). In the

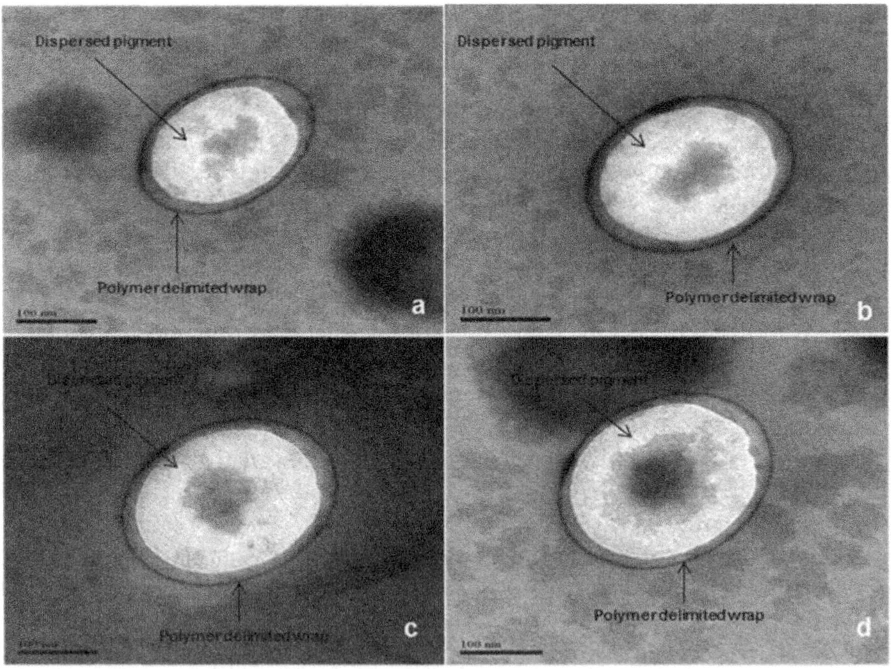

Fig. 4.2 Transmission electron microscopy (MET): (**a**) ABF-EXT; (**b**) ABF-β; (**c**) PEC-β; and (**d**) PEC-EXT (×6000) (ABF-EXT: Albedo Flour + Carotenoid extract obtained from *Spirulina* sp. LEB 18; ABF-β: Albedo Flour + β-Carotene; PEC-β: Commercial Pectin + β-Carotene; PEC-EXT: Commercial Pectin + Carotenoid extract obtained from *Spirulina* sp. LEB 18). (Copyright © Elsevier, 2019 reprinted with permission from Bezerra et al. 2019)

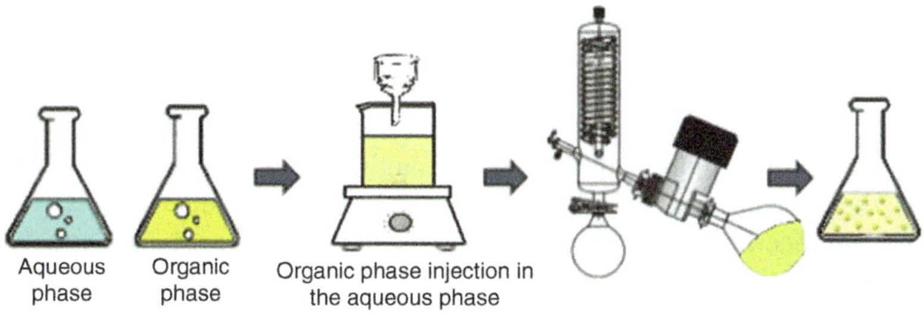

Fig. 4.3 Schematic representation of the synthesis of nanoparticles by the solvent displacement method

solvent displacement method, the organic phase consists of polymer, a component of interest, and organic solvent, while the aqueous phase is composed of water and stabilizer. The synthesis of the nanoparticles is obtained by injecting the organic phase into the aqueous phase (Fig. 4.3). Liposomes were proposed in another study as carriers of the same extract. Produced using lecithin from both rice and soy, they showed sizes around 270 nm with an encapsulation efficiency between 88% and 97%. Interestingly, it was observed higher release of the microalgal extract at higher pH values, characteristic of the intestinal area (125 μg vs. 22 μg for lower pH values) (Machado et al. 2019).

Electrospun nanofibers were also produced using *Spirulina*, along with polycaprolactone. The structure's potential to alleviate astrogliosis-derived inflammation, which ultimately leads to neuronal injuries, was evaluated on the primary culture of astrocytes. Increased cell growth, along with increased metabolism, was observed, without any signs of cytotoxicity (Kim et al. 2012). An approach in the food industry proposed the use of *Spirulina* extracts for the oxidative protection of olive oil. The presence of *Spirulina* nanoparticles (~90 nm), prepared by milling, increased the shelf life of olive oil (Morsy et al. 2019). Other applications are described elsewhere (Huang et al. 2017; Seyed Yagoubi et al. 2017; Schmatz et al. 2020), showing the multitude of processes and nanocarriers capable of protecting the antioxidant properties of *Spirulina*. Usually, the main focus is on pharmaceutical and food applications; however, other uses are yet to be researched and assessed.

Dunaliella spp. is another microalgal genus reported for encapsulation purposes. It is a biflagellate halophilic unicellular green (Chlorophyta) alga, mainly studied as a natural source of antioxidants and as functional food (Herrero et al. 2006). The potential for anticancer applications was materialized in a study reporting the encapsulation of an algal extract from *Dunaliella* in arabic gum-coated magnetic nanoparticles (Zamani et al. 2019). Magnetic nanoparticles are carriers that require a magnetic field to provide an effect (Tarhini et al. 2020). Nanoparticles of 120–340 nm and neutral zeta potential were produced, encapsulating the extract with an efficiency of around 90%. In vitro assays performed in MCF-7 and HeLa cells revealed improvement of anticancer and antioxidant properties at pH values closer to the physiological range, and in a dose-dependent effect (Zamani et al. 2019).

The use of antioxidants can also find application in cosmetics (Farahin et al. 2018). Nanoemulsions (~200 nm and negative zeta potential between −33 and −71 mV) containing the extract of another chlorophyte, the tetra-flagellate *Tetraselmis tetrathele* (Peña and Villegas 2005), revealed that they contained high content of phenolic compounds (around 1.7 mg Gallic Acid Equivalent (GAE) g^{-1} extract), thus making feasible an application to antiaging products.

Tissue engineering has also been highlighted in research involving extracts of the chlorophytes *Haematococcus pluvialis* and *Chlorella vulgaris* encapsulated in polycaprolactone-based nanofibers (~810 nm) produced by electrospinning (Cetmi et al. 2019). *H. pluvialis* is known for its high content of astaxanthin (Faraone et al. 2020), while *C. vulgaris* has long been used as a nutritional feedstock (Ambati et al. 2019). The microalgae-loaded nanofibers were evaluated regarding the impact on cardiovascular tissue growth. Cell proliferation was observed with no visible cytotoxicity, benefitting from the supportive nanofibers mesh, and the oxidative stress alleviation provided by the microalgae-derived antioxidants. These findings indicate the potential of an application to the production and maintenance of artificial tissues.

Chlorella protothecoides and *Nannochloropsis oculata*, which are reported to have anticancer properties (Caporgno et al. 2019; Ates et al. 2020), were encapsulated in nanoparticles (130–770 nm) prepared by electrospinning (Karakaş et al. 2019). Extract encapsulation efficiency of around 80% was observed. The carriers revealed the absence of cytotoxic effect on normal cells (HUVEC). Nevertheless, a cytotoxic effect was observed on two cancer cell lines, A172 and HCT116, thus unveiling a potential use in antitumoral therapies.

The cases herein mentioned evidence of the use of microalgae as a source of biomaterials and the role that these compounds can play in several different fields. Remarkably, *Spirulina* has been the most studied genus, proving its versatility in several scientific domains. However, its eukaryotic counterparts (Chlorophyta, Ochrophyta) are only starting to show their potential. Many other species still need to be explored, since marine resources have much to give to the pharmaceutical, biomedical, and food industries. All in all, the antioxidant properties of microalgae might be of interest to cosmetic formulations, especially as a complement for cancer therapies. However, it can also be a means to increment the shelf life of food products.

4.4 Microalgae Potential in Nanotechnological Food Packing

Traditional food packaging is primarily developed to provide mechanical support and protect products from external influences, such as microorganisms, moisture, oxygen, odors, and dust (Sung et al. 2013; Zhang et al. 2020). In recent years, research on food packaging has been gaining momentum due to consumer concerns about food quality and safety (Kalpana et al. 2019). In this way, industries and researchers have shown interest in developing new packaging production strategies, such as active and intelligent packaging, as well as using biodegradable materials from renewable and self-sustainable sources (Chausali et al. 2022).

Several active agents can be inserted in the matrix of packaging materials, and microalgae are promising sources of these compounds (Vaz et al. 2016). Besides, microalgae have, in their composition, macromolecules, as proteins and PHB that can be applied as wall material for encapsulation of bioactive compounds. The use of biodegradable polymers, such as protein and PHB from microalgae, also contributes to the development of environmentally friendly packaging (Moreira et al. 2018a, b, 2019b). For the development of active packaging, microalgal compounds with active biological property are used to provide antioxidants (Moreira et al. 2019b; Schmatz et al. 2020) and antimicrobial (Kuntzler et al. 2018a, b) functions to preserve products or extend the shelf life of food. In intelligent systems, compounds are incorporated into packaging materials to indicate to the consumer the physicochemical and biochemical changes that can compromise the quality of food (Moreira et al. 2018b).

Microalga *Spirulina*, for example, is widely recognized for its high protein content in its biomass (Belay 2013; Moreira et al. 2016) in addition to pigments such as phycocyanin (Morais et al. 2018). These two biocompounds have been studied for the development of nanomaterials with potential application in food packaging. In a study by Moreira et al. (2018a), polymeric nanofibers were developed with protein concentrate from *Spirulina* sp. LEB 18 (5% and 10%, w w^{-1}) and low concentration of polyethylene oxide (PEO) (0.8% w w^{-1}). Uniform and thermally stable nanofibers with an average diameter of 390 nm were produced with the highest concentration of protein concentrate. Moreover, the authors report that the use of a low concentration of PEO and the high protein content of this microalga associated with the free surface electrospinning process demonstrates the potential of protein concentrate from microalga to develop nanofibers, besides to contribute the scale-up for application in the food packaging field.

Phycocyanin is a pigment that has antioxidant properties, making it attractive for the development of active packaging based on nanotechnology. Recently, this microalgal pigment was added to polymeric solutions for the production of antioxidant nanomaterials such as nanofibers (Moreira et al. 2019b) and nanoparticles (Schmatz et al. 2020). For the production of these nanomaterials, electrospinning and electrospraying techniques can be used, respectively. These processes use the knowledge of nanotechnology, materials engineering, and physics to obtain nanomaterials (Schmatz et al. 2019). Electrospinning and electrospraying stand out for applications related to food and nutrition (Lim et al. 2019), since the processes occur under environmental conditions, therefore being suitable for encapsulating thermally unstable active compounds (Zhao et al. 2020).

In this context, Schmatz et al. (2019) demonstrated that electrospinning and electrospraying techniques can be combined to produce nanofibers with 13% polycaprolactone (PCL) or 12% poly-L-lactic acid (PLLA) (diameter 640 and 608 nm, respectively) containing nanoparticles of polyvinyl alcohol (PVA)/phycocyanin (diameter 395 nm), giving rise to a new nanomaterial with improved characteristics. Furthermore, PVA nanoparticles showed 75.1% nanoencapsulation efficiency and provided antioxidant activity by 2,2-azinobis (3-ethylbenzothiazoline-6-sulfonic acid) (ABTS), 2,2-diphenyl-1-picrylhydrazyl (DPPH), and reducing power methods.

In addition to its antioxidant property, phycocyanin is a promising compound for application in intelligent packaging systems, as it has thermal instability and sensitivity to light. Moreover, the pigment has an intense blue color, which changes depending on the pH of the medium in which it is exposed, as shown in Fig. 4.4. Therefore, based on these statements, Moreira et al. (2018c) proposed the development of pH indicators from nanofibers containing phycocyanin. In this study, it was developed poly(lactic acid) PLA/PEO nanofibers with an average diameter of 921 nm and a thickness of 68.7 μm, containing 3% (w v^{-1}) of phycocyanin for use as pH indicators with potential application in food packaging. Under these conditions, it was possible to obtain 76% of the color change values (ΔE) above 12, indicating an absolute color change. This result demonstrates that fibers containing phycocyanin are innovative materials that can be used to provide dynamic feedback to the consumer about the quality of the product.

In addition to phycocyanin, *Spirulina* has other important pigments that provide the green-blue color of this microalgal biomass. Kuntzler et al. (2020) developed a colorimetric pH indicator using PLA/PEO nanofibers combined with microalgal biomass of *Spirulina* sp. LEB 18. The authors verified that the indicators containing 1%, 2%, or 3% (w v^{-1}) of microalgal biomass provided values $\Delta E > 12$ when submitted to variation in pH, which indicates an absolute difference in color. Besides, the nanofibers had a hydrophilic character, which allows the ions to interact with the biomass, presenting a quick response in the color change.

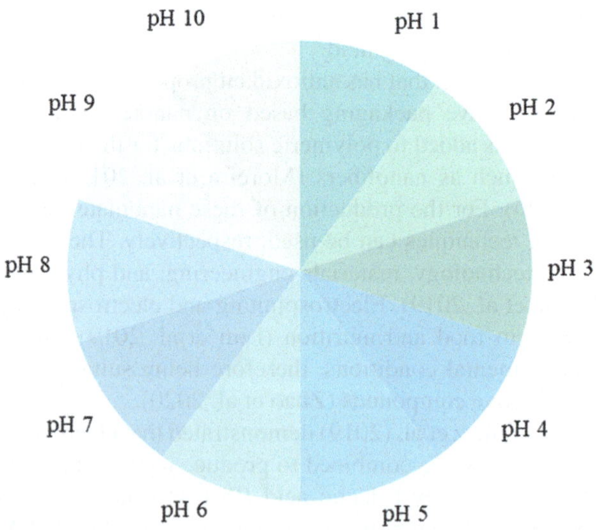

Fig. 4.4 Illustrative representation for coloration change of PLA/PEO nanofibers developed with phycocyanin according to pH variation

The antimicrobial property conferred to the microalgae *Spirulina* is related to the biocompounds that constitute its biomass, such as biopeptides and phenolic compounds. Gonçalves et al. (2017) developed PCL/biopeptide nanofibers with antioxidant activity and a diameter of 438 nm that provided greater stability to chicken meat and control of oxidative processes, ensuring the maintenance of product quality during 12 days of storage. Kuntzler et al. (2018b) produced long and continuous nanofibers containing 3% chitosan/2% PEO/1% phenolic compounds. Nanofibers with an average diameter of 214 nm showed antioxidant activity against *Staphylococcus aureus* ATCC 25923 and *Escherichia coli* ATCC 25972. Therefore, nanofibers developed in these polymeric conditions containing biopeptides or phenolic compounds of microalgal origin are promising materials for use as primary active packaging to conserve food.

Moreover, microalga *Spirulina* synthesizes bioproducts, such as polyhydroxybutyrate (PHB), which is a biodegradable polymer and has mechanical and thermal properties similar to plastics of petrochemical origin. In this sense, Kuntzler et al. (2018a) developed PHB nanofibers containing phenolic compounds, with an average diameter of 810 nm and hydrophobicity characteristic to protect food concerning the external moisture of the packaging. These nanofibers showed inhibition of the growth of *S. aureus* ATCC 25923. Furthermore, the thermal and mechanical properties of the nanofibers confirmed the potential applicability of this material as a food packaging.

4.5 Application of Nanostructures in Microalgae Culture

Microalgae are considered promising organisms in biotechnology due to their potential for sustainably obtaining compounds. The variety of procedures for the biosynthesis of nanomaterials is another advantage of using microalgae in this context (Dahoumane et al. 2016). These microorganisms help to reduce the greenhouse effect and global warming using CO_2 as a carbon source in the cultivation process and reduce the release of excess CO_2 into the environment (Tang et al. 2020).

The incorporation of nanomaterials in microalgal cultures appears as an alternative to improve the production of biomass and intracellular lipids, significantly reducing production costs. Nanostructures promote greater efficiency in collecting samples and help to improve process performance due to their ability to recover and reuse (Nguyen et al. 2020). Recent studies point to the use of nanomaterials as CO_2 adsorbents in microalgae cultivations (Vaz et al. 2019a, b, 2020).

CO_2 biofixation in microalgae crops interferes with the productivity and accumulation of metabolites of interest (Vaz et al. 2019b). However, the CO_2 (g) residence time during microalgal cultivation must be sufficient for it to be retained and the following chemical equilibrium occurs to convert CO_2 into biomass: $CO_2 (g) \leftrightarrow CO_2 (aq) \leftrightarrow H_2CO_3 \leftrightarrow HCO_3^- \leftrightarrow CO_3^{2-}$ (Rosa et al. 2015). Therefore, the addition of polymeric nanofibers as CO_2 adsorbents in microalgal cultures has been studied to increase the permanence of the gas in the culture medium, allowing a longer contact time between the microorganism and CO_2. Besides, nanofibers could be used as a source of carbon or nutrients for microalgae, according to their polymeric composition. Metallic nanoparticles can be incorporated into nanofibers to improve interaction with CO_2, due to increased porosity and specific surface area and depending on the change in functional groups (Vaz et al. 2019a, b, 2020).

Vaz et al. (2019a) developed nanofibers containing different concentrations of nanoparticles to evaluate biological CO_2 fixation in cultures of the microalga *Chlorella fusca* LEB 111. Nanofibers with iron oxide nanoparticles at 4% (w v^{-1}) were selected for application to cultivations, based on their physical characteristics and better CO_2 adsorption capacity (164.2 mg g^{-1}). Cultivations of the microalga *C. fusca* LEB 111 showed a higher rate of CO_2 biofixation with the presence of nanofibers. In a subsequent study, Vaz et al. (2019b) evaluated the CO_2 biofixation and the production of biocompounds by *C. fusca* LEB 111 cultivated with different concentrations of nanofibers in free form or retained form (Fig. 4.5), with or without nanoparticles. Cultures were grown in 15% (v v^{-1}) CO_2 with nanofibers of 0.1, 0.3, and 0.5 g L^{-1}, developed with 10% (w v^{-1}) polyacrylonitrile. Nanofibers at a concentration of 0.1 g L^{-1}, in their free form with nanoparticles, promoted the accumulation of approximately three times more carbon in the medium (46.6 mg L^{-1}), a 45% higher biofixation rate (89.2 mg L^{-1} d^{-1}), and increased the production of carbohydrates by approximately 2.3% (w w^{-1}) of that observed in cultures grown without nanofibers. Still in this line of research, Vaz et al. (2020) studied indoor and outdoor cultivations of *C. fusca* LEB 111 with different concentrations of nanofibers developed with 4% (w v^{-1}) of iron oxide nanoparticles. The outdoor experiments with 0.1 and 0.3 g L^{-1} of nanofibers showed higher CO_2 biofixation (270.6 and 310.9 mg L-1 d^{-1}, respectively). Moreover, outdoor cultivations with 0.3 g L^{-1} of nanofibers showed lipid production 10.9% higher than experiments without nanofibers. From these studies, it was stated that the use of nanofibers as CO_2 adsorbents promotes greater CO_2 biofixation in microalgal cultures. Besides, the fixation process is improved with the addition of metallic nanoparticles.

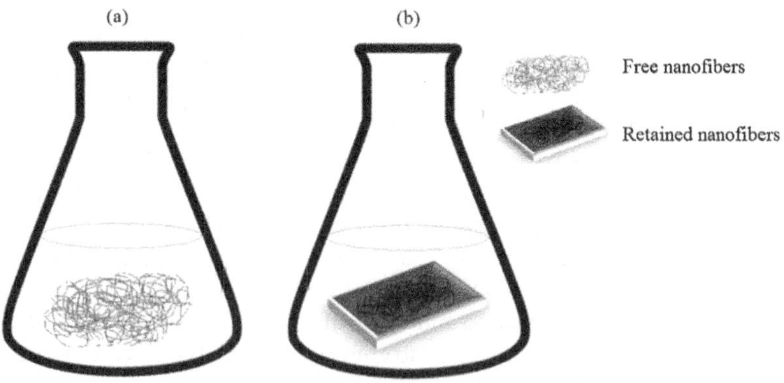

Fig. 4.5 Cultures with nanofibers arranged in their free form (**a**) and retained (**b**) in closed photobioreactor-type flasks. (Copyright © Elsevier, 2019 reprinted with permission from Vaz et al. 2019b)

Advances in biotechnology have predicted that nanoparticles synthesized from microalgae have the potential for large-scale synthesis (Aziz et al. 2014, 2015). Moreover, there are many prospects for the areas of environmental bioremediation and biomedicine fields (Jacob et al. 2020). According to Radzun et al. (2015), metallic nanoparticles present in wastewater can increase the production of biomass in microalgal cultures due to the supplementation of the culture medium with nutrients and minerals. Pattarkine and Pattarkine (2012) report that metallic nanoparticles can also be used to increase the absorption of light in microalgae cultures due to the so-called plasmonic effect on their surface. This phenomenon happens due to electromagnetic interactions that involve the electron cloud above the surface of the nanoparticles generating a dipole that modify the scattered light flow, wavelength, and intensity.

Kaliamurthi et al. (2019) analyzed the effect of different concentrations of zinc oxide nanoparticles (50, 100, and 200 mg L^{-1}) on biochemical changes, oxygen evolution, and lipid production in *Chlorella* sp. Among the studied concentrations, an increase in specific growth rate (0.72 times) and oxygen evolution capacity (0.96–0.90 nmol min^{-1}) was observed with 50 mg L^{-1} of nanoparticles. In this same condition, higher levels of total carotenoids (1.23 times), total protein (1.68 times), starch (2.26 times), neutral lipid (6.08 times), and triacylglycerol (8.21 times) were recorded on the second day of cultivation, compared to their respective controls. Therefore, treatment with 50 mg L^{-1} of zinc oxide nanoparticles improved the production of neutral lipids and triacylglycerol without complete growth inhibition.

Tayemeh et al. (2020) carried out the evaluation and comparison of the influence of silver nanoparticles (AgNPs) and silver nitrate salt ($AgNO_3$) in the composition of fatty acids, pigments, and growth rates of the microalgae *Chlorella vulgaris*. The toxicity test was performed in the culture medium using concentrations of AgNPs (0, 1, 5, 10, 50, 100, and 500 mg L^{-1}) and $AgNO_3$ (0, 1, 5, 10, 50, and 100 mg L^{-1}). Treatments with $AgNO_3$ impaired the growth of *C. vulgaris* about 2–183 times more than the respective AgNPs. Regarding pigments, on exposure to 10 mg L^{-1} of $AgNO_3$, the content of chlorophyll a, chlorophyll b, total chlorophyll, and carotenoid showed a reduction of about 5, 3, 4, and 4 times, respectively, when compared to the same respective concentration of AgNPs. The contents of saturated fatty acids, monounsaturated fatty acids, and polyunsaturated fatty acids, as well as the ratio of unsaturated to saturated acids, showed similar responses in treatments with $AgNO_3$ and AgNPs.

Pádrová et al. (2015) evaluated the biomass production, lipid accumulation, and fatty acid composition of *Desmodesmus subspicatus*, *Dunaliella salina*, *Parachlorella kessleri*, *Raphidocelis subcapitata*, *Nannochloropsis limnetica*, *Trachydiscus minutus*, and *Spirulina maxima* against the addition of zero iron nanoparticles in their cultivations. Higher concentrations of biomass and lipids were found for all microorganisms studied when they were exposed to 5.1 mg L^{-1} of zero-valent iron nanoparticles (nZVI). *P. kessleri* obtained an increase in biomass yield (29.3 g L^{-1}) and lipid content (30.9%) compared to the control assay (19.4 g L^{-1} and 27.1%). *T. minutus* showed higher biomass yield (13.7 g L^{-1}) and lipid content (35%) compared to the control (10.4 g L^{-1} and 26.1%). For the *N. limetica*, an increase in biomass production (7.4 g L^{-1}) and lipid content (48%) was obtained when compared to the untreated test (6.2 g L^{-1} and 35%). *D. subspicatus* and *D. salina* presented lipid accumulation (38% and 32%, respectively), while the control culture provided 24% of the macromolecule for both microalgae. The biomass production for these microalgae cultivated with 5.1 mg L^{-1} of nZVI reached 2.6 g L^{-1} and 5.4 g L^{-1}, respectively, while the control tests provided 1.5 g L^{-1} for *D. subspicatus* and 3.5 g L^{-1} for *D. salina. R. subcapitata* showed higher biomass production (2.3 g L^{-1}), and the total amount of lipids was increased by approximately 50% compared to the control. In the cultures of cyanobacteria *S. maxima*, only a greater accumulation of lipids was observed (21.4%) compared to the control experiment (17.6%). Regarding the fatty acid content, the most significant increase was observed in the microalgae *S. maxima* with 50.9% eicosapentaenoic acid in the culture containing 5.1 mg L^{-1} nZVI. On the other hand, the content of saturated and monounsaturated fatty acids decreased, except for the palmitoleic acid.

In this context, it is clear that studies involving the application of nanostructures in microalgal cultures show the relevance and innovation for obtaining the best-grown parameters. Moreover, it is possible to improve process efficiency, gaseous effluent fixation, and the accumulation of many different bioproducts that can be applied in several areas considering the sustainability concept.

4.6 Synthesis of Nanoparticles in Microalgae Culture

The interest in nanoparticle biosynthesis has been highlighted in the scientific community due to its unique characteristics and excellent biocompatibility (Bhunia et al. 2015). These nanomaterials can be applied in the medical field, in solar and fuel cells for efficient energy production, in water and air filters to reduce pollution. Furthermore, nanoparticles can be used as catalysts in manufacturing processes to eliminate the use of toxic materials (Agarwal et al. 2019).

Nanoparticles can be synthesized by chemical, physical, and biological methods. Chemical synthesis consists of growth in a liquid medium containing reducing and stabilizing agents, such as potassium bitartrate, sodium dodecyl benzyl sulfate, methoxypolyethylene glycol, polyvinyl pyrrolidone, or sodium borohydride. Physical methods, in turn, are based on pyrolysis and friction (Agarwal et al. 2019). However, the approach of these methods causes environmental impacts and toxicity to organisms, resulting from industrial processing. In this way, it is necessary to explore nontoxic synthesis methods, operable at low temperatures, and using renewable energy sources (Sharma et al. 2016).

Microalgae have the ability to remedy wastewater along with the production of a variety of commercially important nanomaterials. The biosynthesis of nanoparticles from these microorganisms is considered a technology with great potential since it appears as a low-cost alternative and can be performed under environmental conditions. However, for the large-scale commercial use of these biological agents, it is still necessary to develop research on the prospects for the production of nanoparticles to facilitate the increase and commercialization of technology (Jacob et al. 2020).

Several methodologies have been developed to biosynthesize metallic nanoparticles using microalgae from their corresponding aqueous saline solutions (LewisOscar et al. 2016). There are four ways to synthesize microalgae nanoparticles: direct exploration of biomolecules extracted from disrupted cells, exploration of cell-free supernatants obtained from the culture medium, biosynthesis of nanoparticles of different natures of whole cells, or using living cells (Agarwal et al. 2019).

The method of direct exploration of biomolecules extracted from disrupted cells was the pioneer in the synthesis of nanoparticles by microalgae. Usually, biomolecules are extracted by disrupting algal cells obtained from their live cultures (Dahoumane et al. 2017). Regarding the second method, the nanoparticles are synthesized from the cell-free supernatant obtained by centrifugation (Patel et al. 2015). Biosynthesis of nanoparticles using whole cells involves centrifugation/filtration or simple washing of the cells to remove the growth medium from the native culture and then be exposed to precursor saline solutions. Finally, the nanoparticle biosynthesis methodology uses live algae cultures to synthesize them without any intermediate steps. The synthesis process can take place in two different ways. Intracellular synthesis occurs inside cells, in which nanoparticles are generally released in the culture medium where they stabilize. The secretion of extracellular materials by algae in the surrounding environment can also promote the process, being called extracellular synthesis (Dahoumane et al. 2016).

Regarding the effects of operational variables on the synthesis of nanoparticles using microalgae, promising results have been obtained under high pH and temperature conditions. Higher operating temperatures result in better reactivity in the reagents, accelerating the formation of nanoparticles. As for pH, microalgal extracts have better reducing and stabilizing power under alkaline pH (Jacob et al. 2020).

Noble metal nanoparticles consist of a group of unique nano-dimensional particles that mainly include gold (AuNPs) and silver (AgNPs). These nanomaterials can be used in several sectors due to their higher stability and less cytotoxicity when compared to other metallic nanoparticles. The obtaining of AgNPs through microalgae biotechnology has been extensively studied, as it does not involve the use of reagents and solvents that may cause toxicity in the synthesis product and the residual process (Suresh Kumar et al. 2015). Besides, the biosynthesis of AgNPs simultaneously with microalgal cultures results in the production of biomass with antimicrobial properties (Aziz et al. 2014, 2015). Therefore, AgNPs have antimicrobial properties and can be used together with microalgal biomass produced to control pathogens in agriculture (Terra et al. 2019b). AuNPs biosynthesized by microalgae have stood out in the areas of biotechnology and biomedicine due to plasmon resonance and greater surface area for molecule binding. Moreover, AuNPs synthesized by microalgae have known antibacterial activity (Khan et al. 2014; Fawcett et al. 2017).

Some studies report the application of microalgae in the biosynthesis of metallic nanoparticles. Terra et al. (2019a) investigated the simultaneous biosynthesis of AgNPs and the growth of the microalgae *Spirulina* sp. LEB 18. In this study, AgNPs with a diameter of 20 nm were synthesized by the microalgae between the 5th and

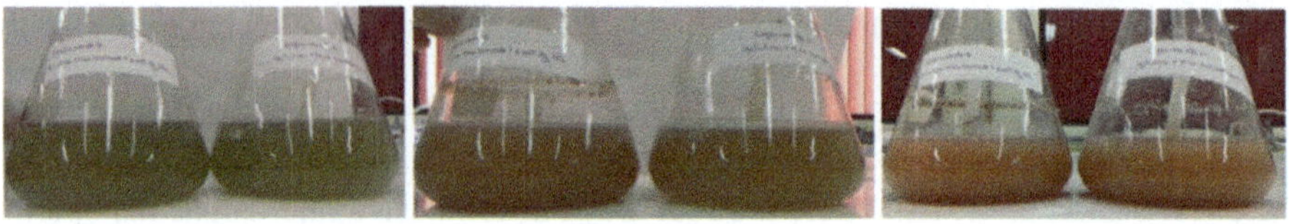

Fig. 4.6 Coloration change in microalgae cultivation during AgNPs biosynthesis

11th days of culture, with maximum cell concentration reaching 2.68 g L^{-1}. According to the authors, the biosynthesis of AgNPs simultaneously with the cultivation of microalgae is an innovative method for the bioremediation of heavy metals with the production of microalgae biomass. The biosynthesis of nanoparticles can be observed macroscopically, by changing the color of the green culture medium for an orange hue (Fig. 4.6).

Bao and Lan (2018) evaluated the influence of chlorophyll and light in obtaining AgNPs through the aqueous cellular extract of *Neochloris oleoabundans*. According to this study, the biosynthesis of AgNPs was a reaction catalyzed by light, and the addition of chlorophyll extracted with methanol can accelerate the reaction. The experiment subjected to white light illumination (control test) obtained a higher production of AgNPs (~0.007 g L^{-1}) compared to blue (450–595 nm) and purple (380–450 nm) illumination. There was no evidence of nanoparticle production under red (620–750 nm) or orange (590–620 nm) illumination.

Singh et al. (2019) proved the biosynthesis of AuNPs from aqueous extract of the microalgae *Dunaliella salina* for later application in the breast cancer cell line. In this study, AuNPs were obtained in an almost spherical format with an average size of 22.4 nm. In another study, Luangpipat et al. (2011) evaluated the biosynthesis of AuNPs by exposing live cells of *Chlorella vulgaris* to gold, palladium, ruthenium, rhodium, iridium, and platinum salts. AuNPs were obtained intracellularly with diameters from 40 to 60 nm. The exposure of *C. vulgaris* to the palladium salt resulted in the production of nanoparticles inside the cells, with diameters from 5 to 12 nm. With the ruthenium and rhodium salts, agglomerations of particles were found inside and outside the cells. Iridium was toxic to microalga, and no synthesis of nanoparticles was observed when using platinum solutions.

Considering these aspects, it is relevant to explore the potential of microalgae in synthesizing nanoparticles due to the relative ease in the harvesting of biomass and its adaptability to environmental conditions. Nanoparticles derived from microalgae have been studied in the biomedical area, due to their antibacterial and antifungal properties. Besides, nanoparticles have biocatalytic properties with a potential for degradation of dyes and organic contaminants. Thus, these nanomaterials can be applied in environmental bioremediation (Jacob et al. 2020). In this sense, it is essential to develop new research so that the biological biosynthesis of nanoparticles can be carried out on a large scale.

4.7 Future Prospects and Main Challenges in Microalgae Nanotechnology

Although technologies for obtaining products and ingredients at the nanoscale as well as studies on microalgal cultivation for the evaluation, extraction, and application of active microalgae compounds are widely studied topics, the combination of these lines of research can still be considered to be recent. The literature describes the association of some microalgae extracts to nanosystems, aiming at fostering intrinsic properties of the microalgae, mostly linked with antioxidant content. The range of systems, producing methodologies and materials used to elaborate the nanocoats, is diverse, as are the envisaged applications. The application of microalgae as nanotechnological products to cosmetics and food areas and their use in medicine as nanocoats (nanocapsules, nanospheres, and nanoemulsions) and scaffolds are still limited. Therefore, studies concerning the effect of the long-term use of these products in humans are necessary. Although some microalgal strains have been GRAS (Generally recognized as Safe) certified by the FDA (Food and Drug Administration), and by European regulatory agencies for application and development of products for human use, chemical and mechanical modifications promoted by the nano dimensions must be evaluated.

The application in food is the one where the use of nanotechnology is perhaps more unbalanced, as the required technology is expensive and possibly difficult to equilibrate their costs and benefits. In the other fields, the application in the nanoscale is more promising. However, there are many challenges ahead, and sustained research in the area is still needed. The knowledge of the stability of the extracts upon encapsulation, toxicological aspects, and finding ways to circumvent the inherent variability of the extracts, due to seasonal and geographical variations, are examples of aspects to address in nanotechnological product development. Despite this, the application of nanotechnology in the conservation of microalgal biocompounds is a promising alternative that has been showing positive results in both in vivo and in vitro studies.

References

Abd El-Hack ME, Abdelnour S, Alagawany M et al (2019) Microalgae in modern cancer therapy: current knowledge. Biomed Pharmacother 111:42–50

Agarwal P, Gupta R, Agarwal N (2019) Advances in synthesis and applications of microalgal nanoparticles for wastewater treatment. J Nanotechnol 2019:1–9

Akao Y, Ebihara T, Masuda H et al (2009) Enhancement of antitumor natural killer cell activation by orally administered *Spirulina* extract in mice. Cancer Sci 100:1494–1501

Akbarzadeh A, Rezaei-Sadabady R, Davaran S et al (2013) Liposome: classification, preparation, and applications. Nanoscale Res Lett 8:102

Ambati RR, Gogisetty D, Aswathanarayana RG et al (2019) Industrial potential of carotenoid pigments from microalgae: current trends and future prospects. Crit Rev Food Sci Nutr 59:1880–1902

Asadi N, Bakhshayesh AR, Davaran S, Akbarzadeh A (2019) Common biocompatible polymeric materials for tissue engineering and regenerative medicine. Mater Chem Phys 2019:122528

Ates M, Cimen ICC, Unal I et al (2020) Assessment of impact of α-Fe_2O_3 and γ-Fe_2O_3 nanoparticles on phytoplankton species *Selenastrum capricornutum* and *Nannochloropsis oculata*. Environ Toxicol 35:385–394

Aziz N, Fatma T, Varma A, Prasad R (2014) Biogenic synthesis of silver nanoparticles using Scenedesmus abundans and evaluation of their antibacterial activity. Journal of Nanoparticles, Article ID 689419, 2014. https://doi.org/10.1155/2014/689419

Aziz N, Faraz M, Pandey R, Sakir M, Fatma T, Varma A, Barman I, Prasad R (2015) Facile algae-derived route to biogenic silver nanoparticles: Synthesis, antibacterial and photocatalytic properties. Langmuir 31:11605–11612. https://doi.org/10.1021/acs.langmuir.5b03081

Bao Z, Lan CQ (2018) Mechanism of light-dependent biosynthesis of silver nanoparticles mediated by cell extract of *Neochloris oleoabundans*. Colloids Surf B: Biointerfaces 170:251–257

Belay A (2013) Biology and industrial production of *Arthrospira* (*Spirulina*). In: Richmond A, Hu C (eds) Handbook of microalgal culture: applied phycology and biotechnology, 2rd edn. Wiley Blackwell, Hoboken

Belay A, Ota Y, Miyakawa K, Shimamatsu H (1993) Current knowledge on potential health benefits of *Spirulina*. J Appl Phycol 5:235–241

Bernaerts TMM, Gheysen L, Foubert I, Hendrickx ME, Van Loey AM (2019) The potential of microalgae and their biopolymers as structuring ingredients in food: a review. Biotechnol Adv 37:107419

Bezerra PQM, Matos MFR, Ramos IG et al (2019) Innovative functional nanodispersion: combination of carotenoid from *Spirulina* and yellow passion fruit albedo. Food Chem 285:397–405

Bhunia AK, Samanta PK, Aich D, Saha S, Kamilya T (2015) Biocompatibility study of protein capped and uncapped silver nanoparticles on human hemoglobin. J Phys D Appl Phys 48(23):235305

Braga ARC, Figueira FS, Silveira JT, Morais MG, Costa JAV, Kalil SJ (2016) Improvement of thermal stability of C-phycocyanin by nanofiber and preservative agents. J Food Process Preserv 40:1264–1269

Camacho F, Macedo A, Malcata F (2019) Potential industrial applications and commercialization of microalgae in the functional food and feed industries: a short review. Mar Drugs 17(6):312

Caporgno MP, Mathys A (2018) Trends in microalgae incorporation into innovative food products with potential health benefits. Front Nutr 5:58

Caporgno MP, Haberkorn I, Bocker L, Mathys A (2019) Cultivation of *Chlorella protothecoides* under different growth modes and its utilisation in oil/water emulsions. Bioresour Technol 288:121476

Centella MH, Arévalo-Gallegos A, Parra-Saldivar R, Iqbal HMN (2017) Marine-derived bioactive compounds for value-added applications in bio- and non-bio sectors. J Clean Prod 168:1559–1565

Cetmi SD, Renkler NZ, Kose A, Celik C, Oncel SS (2019) Preparation of electrospun polycaprolactone nanofiber mats loaded with microalgal extracts. Eng Life Sci 19:691–699

Chanda MJ, Merghoub N, El Arroussi H (2019) Microalgae polysaccharides: the new sustainable bioactive products for the development of plant bio-stimulants? World J Microbiol Biotechnol 35:177

Chausali N, Jyoti Saxena J, Prasad R (2022) Recent trends in nanotechnology applications of bio-based packaging. Journal of Agriculture and Food Research, https://doi.org/10.1016/j.jafr.2021.100257

Costa AM, Bueno KTL, Rosa APC, Costa JAV (2019) The antioxidant activity of nanoemulsions based on lipids and peptides from *Spirulina* sp. LEB 18. LWT Food Sci Technol 99:173–178

Costa JAV, Moreira JB, Fanka LS, Kosinski RC, Morais MG (2020) Microalgal biotechnology applied in biomedicine. In: Konur O (ed) Handbook of algal science, technology and medicine. Elsevier, pp 429–439

Dahoumane SA, Mechouet M, Alvarez FJ, Agathos SN, Jeffryes C (2016) Microalgae: an outstanding tool in nanotechnology. Rev Bionatura 1(4):196–201

Dahoumane SA, Mechouet M, Wijesekera K et al (2017) Algae-mediated biosynthesis of inorganic nanomaterials as a promising route in nanobiotechnology – a review. Green Chem 19(3):552–587

Dehghani J, Adibkia K, Movafeghi A, Maleki-Kakelar H, Saeedi N, Omidi Y (2020) Towards a new avenue for producing therapeutic proteins: microalgae as a tempting green biofactory. Biotechnol Adv 40:107499

Dineshbabu G, Goswami G, Kumar R, Sinha A, Das D (2019) Microalgae–nutritious, sustainable aqua- and animal feed source. J Funct Foods 62:103545

Farahin AW, Yusoff FM, Basri M, Nagao N, Shariff M (2018) Use of microalgae: *Tetraselmis tetrathele* extract in formulation of nanoemulsions for cosmeceutical application. J Appl Phycol 31:1743–1752. https://doi.org/10.1007/s10811-018-1694-9

Faraone I, Sinisgalli C, Ostuni A et al (2020) Astaxanthin anticancer effects are mediated through multiple molecular mechanisms: a systematic review. Pharmacol Res 55:104689

Fawcett D, Verduin JJ, Shah M, Sharma SB, Poinern GEJ (2017) A review of current research into the biogenic synthesis of metal and metal oxide nanoparticles via marine algae and seagrasses. J Nanosci 4:8013850

Gómez-Zorita S, Trepiana J, González-Arceo M et al (2019) Anti-obesity effects of microalgae. Int J Mol Sci 21(1):41

Gonçalves CF, Schmatz DA, Uebel LS et al (2017) Microalgae biopeptides applied in nanofibers for the development of active packaging. Polímeros 27(4):290–297

Has C, Sunthar P (2019) A comprehensive review on recent preparation techniques of liposomes. J Liposome Res 30(4):336–365

Herrero M, Jaime L, Martín-Álvarez PJ, Cifuentes A, Ibáñez E (2006) Optimization of the extraction of antioxidants from *Dunaliella salina* microalga by pressurized liquids. J Agric Food Chem 54:5597–55603

Huang Y, He L, Song Z et al (2017) Phycocyanin-based nanocarrier as a new nanoplatform for efficient overcoming of cancer drug resistance. J Mater Chem B 5:3300–3314

Jacob JM, Ravindran R, Narayanan M, Samuel SM, Pugazhendhi A, Kumar G (2020) Microalgae: a prospective low cost green alternative for nanoparticle synthesis. Curr Opin Environ Sci Health 20:100163. https://doi.org/10.1016/j.coesh.2019.12.005

Kaliamurthi S, Selvaraj G, Cakmak ZE, Korkmaz AD, Cakmak T (2019) The relationship between *Chlorella* sp. and zinc oxide nanoparticles: changes in biochemical, oxygen evolution, and lipid production ability. Process Biochem 85:43–50

Kalpana S, Priyadarshini SR, MariaLeena M, Moses JA, Anandharamakrishnan C (2019) Intelligent packaging: trends and applications in food systems. Trends Food Sci Technol 93:145–157

Karakaş CY, Tekarslan Şahin H, İnan B, Özçimen D, Erginer YÖ (2019) In vitro cytotoxic activity of microalgal extracts loaded nano-micro particles produced via electrospraying and microemulsion methods. Biotechnol Prog 35:e2876

Katiyar R, Arora A (2020) Health promoting functional lipids from microalgae pool: a review. Algal Res 46:101800

Khalid M (2020) Nanotechnology and chemical engineering as a tool to bioprocess microalgae for its applications in therapeutics and bioresource management. Crit Rev Biotechnol 40:46–63

Khan AK, Rashid R, Murtaza G, Zahra A (2014) Gold nanoparticles: synthesis and applications in drug delivery. Trop J Pharm Res 13(7):1169–1177

Kim SH, Shin C, Min SK, Jung SM, Shin HS (2012) In vitro evaluation of the effects of eletrospun PCL nanofiber mats containing the microalgae *Spirulina* (*Arthrospira*) extract on primary astrocytes. Colloids Surf B: Biointerfaces 90:113–118

Kuntzler SG, Costa JAV, Morais MG (2018a) Development of electrospun nanofibers containing chitosan/PEO blend and phenolic compounds with antibacterial activity. Int J Biol Macromol 117:800–806

Kuntzler SG, Almeida ACA, Costa JAV, Morais MG (2018b) Polyhydroxybutyrate and phenolic compounds microalgae electrospun nanofibers: a novel nanomaterial with antibacterial activity. Int J Biol Macromol 113:1008–1014

Kuntzler SG, Costa JAV, Brizio AR, Morais MG (2020) Development of a colorimetric pH indicator using nanofibers containing *Spirulina* sp. LEB 18. Food Chem 328:126768

LewisOscar F, Vismaya S, Arunkumar M, Thajuddin N, Dhanasekaran D, Nithya C (2016) Algal nanoparticles: synthesis and biotechnological potentials. In: Thajuddin N, Dhanasekaran D (eds) Algae – organisms for imminent biotechnology. IntechOpen, Rijeka, pp 157–182

Lim LTK, Mendes AC, Chronakis IS (2019) Electrospinning and electrospraying technologies for food applications. In: Lim L-T, Rogers M (eds) Advances in food and nutrition research. Elsevier Inc., pp 167–234

Luangpipat T, Beattie IR, Chisti Y, Haverkamp RG (2011) Gold nanoparticles produced in a microalga. J Nanopart Res 13(12):6439–6445

Machado AR, Pinheiro AC, Vicente AA, Souza-Soares LA, Cerqueira MA (2019) Liposomes loaded with phenolic extracts of *Spirulina* LEB-18: physicochemical characterization and behavior under simulated gastrointestinal conditions. Food Res Int 120:656–667

Maqsoudlou A, Assadpour E, Mohebodini H, Jafari SM (2020) Improving the efficiency of natural antioxidant compounds via different nanocarriers. Adv Colloid Interf Sci 278:102122

Metting FB (1996) Biodiversity and application of microalgae. J Ind Microbiol Biotechnol 17:477–489

Molino A, Mehariya S, Di Sanzo G et al (2020) Recent developments in supercritical fluid extraction of bioactive compounds from microalgae: role of key parameters, technological achievements and challenges. J CO_2 Util 36:196–209

Morais MG, Stillings C, Dersch R et al (2010) Preparation of nanofibers containing the microalga *Spirulina* (*Arthrospira*). Bioresour Technol 101:2872–2876

Morais MG, Vaz BS, Morais EG, Costa JAV (2014) Biological effects of *Spirulina* (*Arthrospira*) biopolymers and biomass in the development of nanostructured scaffolds. Biomed Res Int 2014:1–9

Morais MG, Vaz BS, Morais EG, Costa JAV (2015) Biologically active metabolites synthesized by microalgae. Biomed Res Int 2015:1–15

Morais MG, Prates DF, Moreira JB, Duarte JH, Costa JAV (2018) Phycocyanin from microalgae: properties, extraction and purification, with some recent applications. Ind Biotechnol 14(1):30–37

Moreira JB, Terra ALM, Costa JAV, Morais MG (2016) Utilization of CO_2 in semi-continuous cultivation of *Spirulina* sp. and *Chlorella fusca* and evaluation of biomass composition. Braz J Chem Eng 33(3):691–698

Moreira JB, Lim LT, Zavareze ER, Dias ARG, Costa JAV, Morais MG (2018a) Microalgae protein heating in acid/basic solution for nanofibers production by free surface electrospinning. J Food Eng 230:49–54

Moreira JB, Morais MG, Morais EG, Vaz BS, Costa JAV (2018b) Electrospun polymeric nanofibers in food packaging. In: Grumezescu AM, Holban AM (eds) Impact of nanoscience in the food industry. Academic Press, London, pp 387–417

Moreira JB, Terra ALM, Costa JAV, Morais MG (2018c) Development of pH indicator from PLA/PEO ultrafine fibers containing pigment of microalgae origin. Int J Biol Macromol 118:1855–1862

Moreira JB, Goularte PG, Morais MG, Costa JAV (2019a) Preparation of beta-carotene nanoemulsion and evaluation of stability at a long storage period. Food Sci Technol 39(3):599–604

Moreira JB, Lim LT, Zavareze ER, Dias ARG, Costa JAV, Morais MG (2019b) Antioxidant ultrafine fibers developed with microalga compounds using a free surface electrospinning. Food Hydrocoll 93:131–136

Morsy MK, Morsy OM, Elbarbary HA, Saad MA (2019) Enhancing of oxidative stability and quality attributes of olive oil using *Spirulina* (*Arthrospira platensis*) nanoparticles. LWT Food Sci Technol 101:444–455

Mourelle ML, Gómez CP, Legido JL (2017) The potential use of marine microalgae and cyanobacteria in cosmetics and thalassotherapy. Cosmetics 4(4):46
Nguyen MK, Moon JY, Bui VKH, Oh YK, Lee YC (2020) Recent advanced applications of nanomaterials in microalgae biorefinery. Algal Res 41:101522
Pádrová K, Lukavský J, Nedbalová L et al (2015) Trace concentrations of iron nanoparticles cause overproduction of biomass and lipids during cultivation of cyanobacteria and microalgae. J Appl Phycol 27(4):1443–1451
Patel V, Berthold D, Puranik P, Gantar M (2015) Screening of cyanobacteria and microalgae for their ability to synthesize silver nanoparticles with antibacterial activity. Biotechnol Rep 5(1):112–119
Pattarkine MV, Pattarkine VM (2012) Nanotechnology for algal biofuels. In: Gordon R, Seckbach J (eds) The science of algal fuels. Cellular origin, life in extreme habitats and astrobiology, vol 25. Springer, Dordrecht, pp 147–163
Peña MR, Villegas CT (2005) Cell growth, effect of filtrate and nutritive value of the tropical prasinophyte *Tetraselmis tetrathele* (Butcher) at different phases of culture. Aquac Res 36:1500–1508
Radzun KA, Wolf J, Jakob G et al (2015) Automated nutrient screening system enables high-throughput optimisation of microalgae production conditions. Biotechnol Biofuels 8(1):1–17
Randhir A, Laird DW, Maker G, Trengove R, Moheimani NR (2020) Microalgae: a potential sustainable commercial source of sterols. Algal Res 46:101772
Riccio G, Lauritano C (2019) Microalgae with immunomodulatory activities. Mar Drugs 18(1):2
Rnjak-Kovacina J, Weiss AS (2011) Increasing the pore size of electrospun scaffolds. Tissue Eng Part B Rev 17(5):365–372
Rosa GM, Moraes L, Cardias BB, Souza MRAZ, Costa JAV (2015) Chemical absorption and CO_2 biofixation via the cultivation of *Spirulina* in semicontinuous mode with nutrient recycle. Bioresour Technol 192:321–327
Rosales-Mendoza S, Solís-Andrade KI, Márquez-Escobar VA, González-Ortega O, Bañuelos-Hernandez B (2020) Current advances in the algae-made biopharmaceuticals field. Expert Opin Biol Ther 20(7):751–766
Saini DK, Chakdar H, Pabbi S, Shukla P (2020) Enhancing production of microalgal biopigments through metabolic and genetic engineering. Crit Rev Food Sci Nutr 60:391–405
Schmatz DA, Uebel LS, Kuntzler SG, Dora CL, Costa JAV, Morais MG (2016) Scaffolds containing *Spirulina* sp. LEB 18 biomass: development, characterization and evaluation of in vitro biodegradation. J Nanosci Nanotechnol 16:1050–1059
Schmatz DA, Costa JAV, Morais MG (2019) A novel nanocomposite for food packaging developed by electrospinning and electrospraying. Food Packag Shelf Life 20:100314
Schmatz DA, Mastrantonio DJS, Costa JAV, Morais MG (2020) Encapsulation of phycocyanin by electrospraying: a promising approach for the protection of sensitive compounds. Food Bioprod Process 119:206–215
Seyed Yagoubi A, Shahidi F, Mohebbi M, Varidi M, Golmohammasdzadeh S (2017) Preparation, characterization and evaluation of physicochemical properties of phycocyanin-loaded solid lipid nanoparticles and nanostructured lipid carriers. J Food Meas Charact 12:378–385
Sharma A, Sharma S, Sharma K et al (2016) Algae as crucial organisms in advancing nanotechnology: a systematic review. J Appl Phycol 28(3):1759–1774
Sheth T, Seshadri S, Prileszky T, Helgeson ME (2020) Multiple nanoemulsions. Nat Rev Mater 5:214–228
Sill TJ, von Recum HA (2008) Electrospinning: applications in drug delivery and tissue engineering. Biomaterials 29:1989–2006
Singh AK, Tiwari R, Singh VK et al (2019) Green synthesis of gold nanoparticles from *Dunaliella salina*, its characterization and in vitro anticancer activity on breast cancer cell line. J Drug Deliv Sci Technol 51:164–176
Solans C, Izquierdo P, Nolla J, Azemar N, Garciacelma M (2005) Nano-emulsions. Curr Opin Colloid Interface Sci 10(3–4):102–110

Souto EB, Silva GF, Dias-Ferreira J et al (2020) Nanopharmaceutics: part II – production scales and clinically compliant production methods. Nanomaterials 10(3):455

Steffens D, Lersch M, Rosa A et al (2013) A new biomaterial of nanofibers with the microalga *Spirulina* as scaffolds to cultivate with stem cells for use in tissue engineering. J Biomed Nanotechnol 9(4):710–718

Sung SY, Sin LT, Tee TT et al (2013) Antimicrobial agents for food packaging applications. Trends Food Sci Technol 33:110–123

Suresh Kumar K, Dahms HU, Won EJ, Lee JS, Shin KH (2015) Microalgae – a promising tool for heavy metal remediation. Ecotoxicol Environ Saf 113:329–352

Tang DYY, Khoo KS, Chew KW, Tao Y, Ho SH, Show PL (2020) Potential utilization of bioproducts from microalgae for the quality enhancement of natural products. Bioresour Technol 304:122997

Tarhini M, Vega-Chacón J, Jafelicci M et al (2020) Structured magnetic core/silica internal shell layer and protein out layer shell (BSA@SiO_2@SME): preparation and characterization. Chem Afr 3:127–134

Tayemeh BM, Esmailbeigi M, Shirdel I et al (2020) Perturbation of fatty acid composition, pigments, and growth indices of *Chlorella vulgaris* in response to silver ions and nanoparticles: a new holistic understanding of hidden ecotoxicological aspect of pollutants. Chemosphere 238:124576

Terra ALM, Cruz ND, Henrard ASA, Costa JAV, Morais MG (2019a) Simultaneous biosynthesis of silver nanoparticles with *Spirulina* sp. LEB 18 cultivation. Ind Biotechnol 15(4):263–267

Terra ALM, Kosinski RC, Moreira JB, Costa JAV, Morais MG (2019b) Microalgae biosynthesis of silver nanoparticles for application in the control of agricultural pathogens. J Environ Sci Health B 54(8):709–716

Vaz BS, Moreira JB, Morais MG, Costa JAV (2016) Microalgae as a new source of bioactive compounds in food supplements. Curr Opin Food Sci 7:73–77

Vaz B, Costa JAV, Morais MG (2019a) Innovative nanofiber technology to improve carbon dioxide biofixation in microalgae cultivation. Bioresour Technol 273:592–598

Vaz BS, Mastrantonio DJS, Costa JAV, Morais MG (2019b) Green alga cultivation with nanofibers as physical adsorbents of carbon dioxide: evaluation of gas biofixation and macromolecule production. Bioresour Technol 287:121406

Vaz BS, Costa JAV, Morais MG (2020) Physical and biological fixation of CO_2 with polymeric nanofibers in outdoor cultivations of *Chlorella fusca* LEB 111. Int J Biol Macromol 151:1332–1339

Zamani H, Rastegari B, Varamini M (2019) Antioxidant and anti-cancer activity of *Dunaliella salina* extract and oral drug delivery potential via nano-based formulations of gum Arabic coated magnetite nanoparticles. J Drug Deliv Sci Technol 54:101278

Zhang C, Li Y, Wang P, Zhang H (2020) Electrospinning of nanofibers: potentials and perspectives for active food packaging. Compr Rev Food Sci Food Saf 19:479–502

Zhao L, Duan G, Zhang G, Yang H, He S, Jiang S (2020) Electrospun functional materials toward food packaging applications: a review. Nanomaterials 10:150

Chapter 5
Revalorization of Algae Natural Resources as Starting Materials for the Development of Nanotechnology-Based Composite Materials

Laura M. Sanchez

Contents

5.1 Introduction

Algae is a wide group of photosynthetic organisms that are considered to be responsible for supplying a significant amount of the total atmospheric oxygen (Keeling 2004; Khalid 2020). An outstanding characteristic of these organisms is their capability to grow under a variety of environmental conditions; thus, they can be found from oceans, rivers, and lakes to salt pans and even wastewaters (Sahoo and Baweja 2015; Seckbach 2015; Khan et al. 2018). As a result of this, algae could grow naturally or in artificial medium (algae culture) (Mathijsen 2020). In both cases, there is a photosynthetic conversion of atmospheric CO_2 to useful value-added resources (He 2015; Devi and Sahoo 2015; Khan et al. 2018; Khalid 2020). Sometimes, it happened that algae grow excessively due to certain environmental conditions, thus producing an algal cell accumulation usually known as algal bloom. Although in most cases algal blooms do not affect human health, they cause an alteration of their

L. M. Sanchez (✉)
Materiales Compuestos Termoplásticos (CoMP), Instituto de Investigaciones de Ciencia y Tecnología de Materiales (INTEMA) CONICET-UNMdP, Mar del Plata, Argentina
e-mail: laura.sanchez@fulbrightmail.org

D. Thangadurai et al. (eds.), *Bioprospecting Algae for Nanosized Materials*, Nanotechnology in the Life Sciences,
https://doi.org/10.1007/978-3-030-81557-8_5

immediate environment whose consequences may be diverse (He 2015; Tripathy et al. 2021).

Taking into account the immense variety of algal organisms, their appropriate classification to be further studied is recognized as a necessity. In this sense, the most commonly employed criteria are shown in Fig. 5.1. Baweja and Sahoo (2015) have recompiled algae classifications (Fig. 5.1).

While macroalgae (seaweed) are multicellular large-size algae capable to be seen by the naked eye, microalgae are microscopic single cells and may be both prokaryotic and eukaryotic. The biochemical diversity of microalgae depends on both the specific species and the growing environmental conditions. As can be expected, it opens up a broad field of possible applications. Furthermore, it is important to mention that microalgae could be considered as an eco-friendly, nontoxic, and renewable starting materials source that does not compete with traditional agriculture for arable land (Mathimani and Pugazhendhi 2019; Khalid 2020; Aziz et al. 2014, 2015). Which kind of industries could employ microalgae or microalgal derivatives as starting materials? For example, those related to the production of pharmaceutical, biomedical, agricultural, nutraceutical, cosmetics, feed, supplements, and food products (Metting 1996; Khan et al. 2018). Additionally, they could also be used for biofuel production (Aziz et al. 2017; Rawat et al. 2021) and to develop environmental remediation devices (Salama et al. 2019; Khalid 2020).

On the other hand, nanotechnology is a research and production area that has been installed a few decades ago it is here to stay. The prefix *nano* came from Greek word *Nanos* which is used to describe things of one to a hundred billionth (10^{-9} meter) in size in one dimension (Patil and Kim 2018; Saxena and Harish 2019). One of the most fascinating aspects of this area is related to its interdisciplinary, making nanoscale devices applicable in almost all lives facets. The present book chapter summarizes different ways in which algae natural resources are employed as starting materials for the development of nanotechnology-based composite materials for a broad set of applications.

5.2 Algae as Valuable Starting Materials

As it was previously mentioned, the application possibilities by using macro- and microalgae as starting materials are diverse (Fig. 5.2). Algae can be used by their selves in several sizes, or some of their components could be isolated and purified to be further employed. In the following paragraphs, algae are considered as starting materials to produce different nanotechnology-based composite materials.

Metal and metal-oxide nanoparticles are used in biomedical, pharmaceutical, and agricultural fields, and there is a variety of physical, chemical, and biological methods to prepare them. Employing algae could be considered a greener, renewable, and eco-friendly way to prepare biocompatible nanoparticles by taking advantage of its intrinsic bio-reductive properties. The intracellular reduction of metallic ions takes place by the action enzymes present on the cell wall and on the cytoplasmic membrane, whereas the extracellular chemical process involves proteins, amino

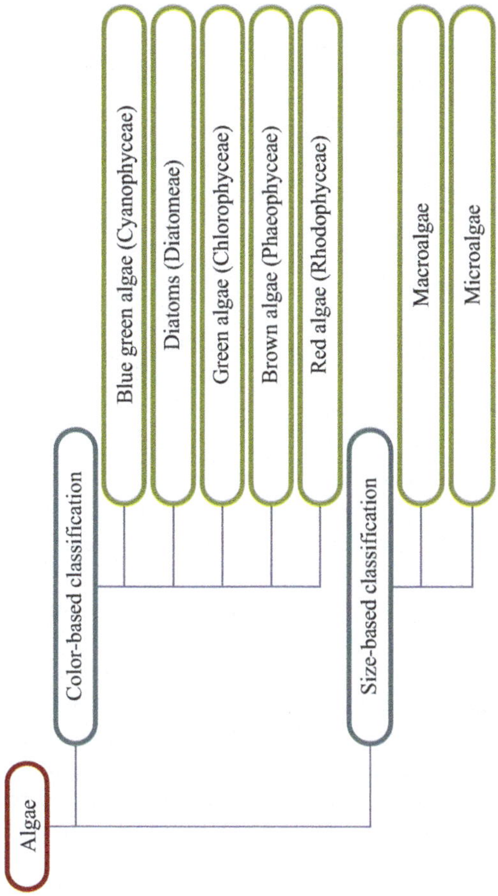

Fig. 5.1 Most commonly employed criteria to classify algae

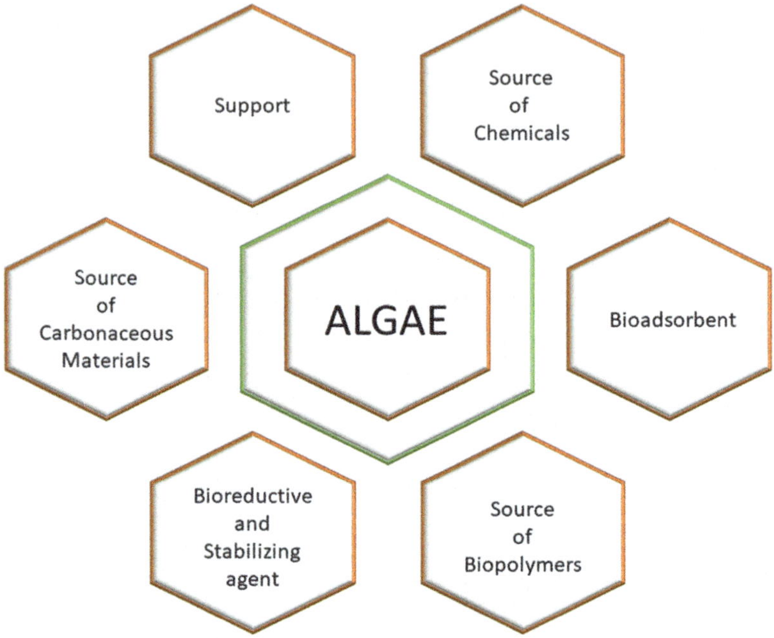

Fig. 5.2 Examples of possible algal resources exploitation

acids, and enzymes (Mohanpuria et al. 2008; Patil and Kim 2018). Cyanobacteria, diatoms, green, brown, and red algae have been employed as both bioreductor and stabilizing agents (Patil and Kim 2018). The most relevant factors affecting the final properties of the nanoparticles synthesized by the algae, as well as they collected a number of publications in this field, have been recently summarized (Saxena and Harish 2019).

It has been demonstrated that several microalgae, having large surface areas and multiple binding groups, can act by themselves as bioadsorbents to capture toxic elements such as arsenic, copper, and boron (Saavedra et al. 2018). Thus, some microalgae were conveniently treated to enhance this capability. Shen and coworkers achieve the superficial synthesis of nanosized Fe_2O_3 in microalgae to develop new composite materials useful to adsorb Cr (VI), Cd (II), Cu (II), and Pb (II) (Shen et al. 2020). Salama and collaborators have recently published a very complete review summarizing information about the employment of different algae for polluted water treatments through phytoremediation (Salama et al. 2019).

Regarding the usefulness of algae to be employed as support, the possibility to act as starting material to develop algal carbon-based materials should also be considered. It is well known that carbonaceous materials are one of the most widely employed adsorbents due to their nontoxic, high superficial area, and chemically inert characteristics (Valix et al. 2006; Hu and Hsieh 2014; Stone and Kozlov 2014;

Sanchez et al. 2018). In this sense, Lu et al. (2020) prepared algal carbon-supported composites containing flower-like sulfidated nanoscale zerovalent iron compounds potentially useful as water remediation devices through achieving the removal of bromate.

Microalgae-derived carbon was also employed by Seo et al. (2019). They developed composite structures, containing nanosized MoS_2, capable to act as anode for lithium-ion-containing batteries. In this case, the microalgae had two main functions: they were not only a source of materials but also act as templates with specific internal structures providing excellent properties to the targeted applications.

On the other hand, diatoms are single-celled microalgae that exist as a number of species who have a common structure usually named frustule. It is mainly composed of amorphous hydrated mesoporous silica that conforms a micro- and nanopatterned skeleton (Ragni et al. 2018). This structure has been used as support to develop composite materials for water remediation, drug delivery, bio-sensing, optoelectronics, and photonics, among other applications. Ragni and coworkers have presented a very complete review of this topic. Among the already used nanoreinforcements used to prepare diatom-composite materials, there are ZrO_2 (Thakkar et al. 2015), noble metal nanoparticles (Au, Pt, Ag) (Jantschke et al. 2012; Toster et al. 2013; Kong et al. 2016), TiO_2 (Losic et al. 2006; Mao et al. 2014; Gautam et al. 2016), CdS (Gutu et al. 2009), and Fe_3O_4 nanoparticles (Losic et al. 2010). Some of the above-mentioned nanoparticles were previously prepared and further included in the diatoms, while in some other cases the nanoparticles were prepared in situ due to the already demonstrated algae capability to act as both a reactant and a supporting material.

Macro- and microalgae also constitute a renewable and green source of a number of extractable chemical compounds, including vitamins, polysaccharides, proteins, lipids, polyunsaturated fatty acids, dietary fiber, and minerals (Bajpai et al. 2018; Feki et al. 2020). In order to little illustrate, among the possible chemical compounds useful in cancer treatments, there are some phytochemicals such as carotenoids and phlorotannins (Almeida et al. 2019), the pigment scytonemin (Stevenson et al. 2002), some calothrixins (Bernardo et al. 2004; Hatae et al. 2014), and Apatroxin (Luesch et al. 2006). Bajpai and collaborators (2018) not only summarized some microalgae isolated compounds but they also presented some of them that are commonly employed for effective nanoformulations development. In Fig. 5.3, there is a summarization of the most relevant polysaccharides extracted from algal resources.

Brown marine algae, such as *Laminaria digitata*, *Laminaria hyperborea*, *Ascophyllum nodosum*, *Laminaria japonica*, and also *Macrocystis pyrifera*, could be used as alginate source (Yong Lee and Mooney 2012). The alginic acid is first extracted through an alkaline treatment, and HCl, NaCl, or $CaCl_2$ is subsequently added according to the desired final alginic product form (Severino et al. 2019). The naturally occurring alginate polymer has remarkable importance due to its biodegradable and biocompatible characteristics (Aswin Kumar and Viswanathan 2020).

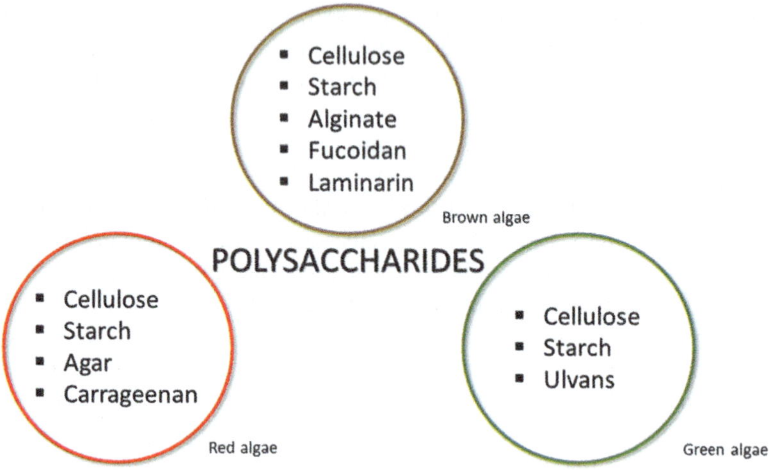

Fig. 5.3 Most relevant polysaccharides extracted from algal resources

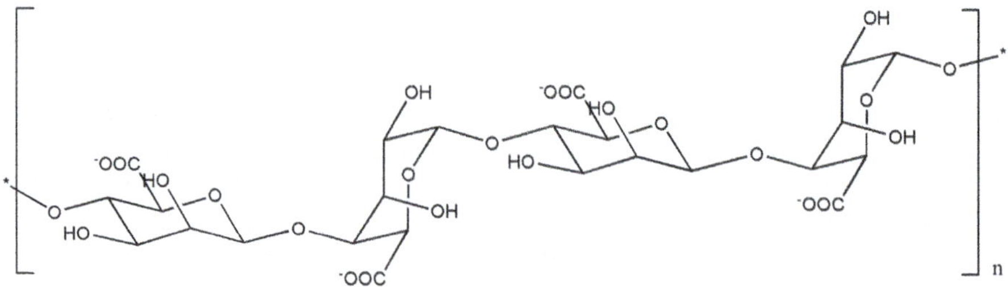

Fig. 5.4 Example of alginate chemical structure consisting of alternating M and G blocks

This biopolymer has been approved by the Food and Drug Administration (FDA) for human use (Ahmad et al. 2006). In some other cases, the alginate present in the algal materials is just partially extracted to develop useful materials, such as the beads prepared by Hamza and collaborators (2020), based on the interaction of the biomass-derived materials with polyethylenimine.

From a chemical point of view, alginate is a water-soluble linear copolymer mainly composed of mannuronic (M) and guluronic (G) acids. This anionic polysaccharide contains different blocks: consecutive M residues, consecutive G residues, and alternating M and G residues (Fig. 5.4) (Mignon et al. 2016). The properties of alginate are closely related to the G-block length, the M/G ratio, and the intrinsic macromolecule molecular weight (George and Abraham 2006). Since alginates from different sources and environmental growth conditions differ not only in the length of each of the possible blocks but also in their M and G contents, there are hundreds to thousands of different alginate structures (Tønnesen and Karlsen 2002; Severino et al. 2019).

Alginate-containing nanoparticles have been prepared in a number of scientific studies with different purposes. Severino et al. (2019) have recently published a review detailing alginate-containing nanomaterials useful as both drug delivery and targeting devices. On the other hand, alginate biopolymeric matrix was also used to develop graphene oxide nanoreinforced composite materials. Larraza et al. (2020) deeply studied this kind of systems paying special attention to the enhancement of mechanical and thermal nanocomposite properties. The employment and benefits of alginate-based nanocomposite materials for applications such as aqueous dyes removal and biomaterials development have also been summarized by several authors (Sanchez et al. 2018; Alavi and Rai 2019; Ramazani and Aghahosseini 2019; Zahran and Marei 2019; Alavi and Nokhodchi 2020; Ollier et al. 2020).

There are other polysaccharides that can be extracted from brown algae, such as fucoidans. From a chemical point of view, they contain a sulfate group attached to some of their fucose units, whose linkages are mainly 1 → 3 or alternating 1 → 3 and 1 → 4 (Olatunji 2020). These fucose backbones could be branched by other monosaccharides (such as D-xylose, D-mannose, D-galactose, and uronic acid), and it depends on the growth conditions and also on the specific species from which it is extracted. The same factors also influence both the sulfation and polymer main chain configuration, so the fucoidans can be very diverse. Their structure is exemplified in Fig. 5.5.

Although fucoidans have a variety of potential medicinal uses such as anticoagulant, antitumoral, immunomodulatory, anti-inflammatory, antiviral, antithrombotic, and antioxidant agents (Ale et al. 2011; Kimura et al. 2013; Lee et al. 2013b), their approbation to be used for clinical applications by international regulation

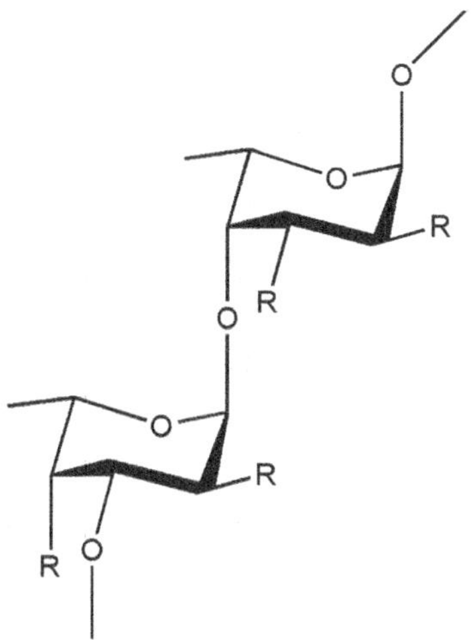

Fig. 5.5 Example of fucoidan chemical structure consisting of alternating 1 → 3 and 1 → 4 linkages, where R indicates a sulfate group or any monosaccharide possible unit found in fucoidans

organizations is still not reached (Olatunji 2020). This fact is mainly attributed to their inherent very variable chemical composition which determines their properties and possible applications.

A number of nanosystems potentially useful for clinical purposes have been developed based on the nontoxicity of fucoidans as starting materials. Lee and collaborators (2013a) have successfully loaded doxorubicin into acetylated fucoidan nanoparticles looking forward to the development of immuno- and chemotherapy agents. Manivasagan et al. (2016) prepared fucoidan-capped gold nanoparticles which were also subsequently loaded with doxorubicin for both drug delivery and photoacoustic imaging clinical applications.

Additionally, taking into account that fucoidans are negatively charged due to their sulfate groups, they can be conveniently combined with some positively charged starting materials (such as chitosan and protamine) in order to create nanocomplexes having other tuned specific properties (Tran et al. 2020). A number of fucoidan- and chitosan-based nanostructures were designed as drug delivery systems containing curcumin, insulin, amoxicillin, gentamicin, doxorubicin, paclitaxel, and ciprofloxacin, among some other active compounds (Huang and Lam 2011; Huang and Liu 2012; Huang and Li 2014; Huang et al. 2014, 2016; Huang and Kuo 2016; Huang and Yang 2016; Chen et al. 2018a; Tran et al. 2020). Furthermore, fucoidan and protamine multi-stimuli-responsive nanoparticle systems were developed by Lu et al. (2017) for doxorubicin drug delivery.

Another sulfate-containing polysaccharide present in algae resources is carrageenan. This linear polymer could be extracted from *Rhodophyceae* red seaweeds, and it is composed of galactose monosaccharide with some sulfate units attached. Both the galactose linkages and the specific location of sulfate groups determine the form of carrageenan. For example, *kappa*- and *iota*-carrageenan have alternating D-galactose and 3,6-anhydrogalactose units, and their difference has to do with their location and the number of sulfate groups, while *lambda*-carrageenan is primarily composed of D-galactose groups with alternating α-1,3 and β-1,4-glycosidic linkages and just a few of 3,6-anhydrogalactose monomers (McKim et al. 2019). Some examples of these chemical structures are illustrated in Fig. 5.6.

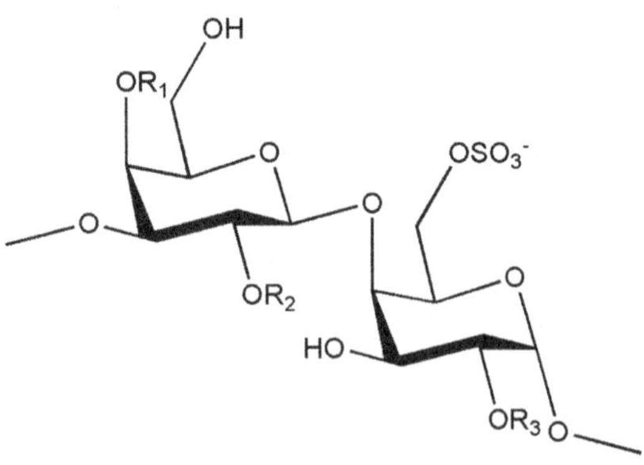

Fig. 5.6 Example of carrageenan chemical structure consisting of alternating α-1,3 and β-1,4-glycosidic linkages, where R indicates a –H or a $-SO_3^-$ substituent

Carrageenans have been employed in several food and pharmaceutical formulations since they are generally considered to be safe for human consumption (Piyakulawat et al. 2007; Torres et al. 2019). To cite, Mirza and collaborators (2020) have recently prepared a series of bone tissue-engineered materials by using nano-hydroxyapatite, Arabic Gum, and *kappa*-carrageenan that have demonstrated promising results. Otherwise, the highly charged carrageenan structure results useful as stabilizing agent. For example, it was used as capping agent to prevent gold and silver nanoparticle coagulation (Ahmed 2019).

The gel-forming ability of carrageenans under the presence of several cationic species is commonly used to develop a variety of materials (Geonzon et al. 2020). Calcium chloride was employed as crosslinking agent in the preparation of ionically crosslinked *kappa*-carrageenan/xanthan gum hydrogels containing halloysite nanotubes and carboxylated-cellulose nanocrystals and being potentially useful in soft tissue engineering (Kumar et al. 2019a). Several nanogel carriers for drug delivery applications were also developed by using carrageenans as starting materials (Alvarez-Lorenzo et al. 2013; Chiriac et al. 2019), even achieving the development of devices having not-as-usual shapes such as nanotubes (Maldonado et al. 2019). Some of them were magnetically responsive, thanks to the addition of magnetic nanoparticles (Daniel-Da-Silva et al. 2009; Sagbas et al. 2012).

Furthermore, carrageenans have also been employed as starting materials for the development of environmental remediation devices. Magnetic nanoparticle-containing gels were prepared by using *iota*-carrageenan and polyamidoamine polymer (PAMAM) through a physical crosslinking method for the removal of pollutants including Cu^{2+}, Cd^{2+}, Cr(VI), Mn^{7+}, Co^{2+}, and alphanol fast blue dye (Abdellatif et al. 2020). Ali and collaborators (2019) also prepared materials for metal cations sorption from polluted waters: they developed carrageenan beads containing nano-cellulosic reinforcements.

Some red algae species belonging to the genus *Gracilaria*, *Pyropia yezoensis*, *Gelidium,* and *Gelidiella* serve as source of other important polysaccharides mixture named agar (Olatunji 2020). The major component of agar is the polysaccharide agarose (neutral linear copolymer consisting of D-galactose and 3,6-anhydro-α-l-galactose units linked by β-1,3 and α-1,4 glycoside unions), whereas the secondary component corresponds to agaropectine (essentially it has the same backbone as agarose, but it is substituted with sulfate esters, pyruvic, and D-guluronic acids) (Jönsson et al. 2020). The general chemical structure of agar is illustrated in Fig. 5.7.

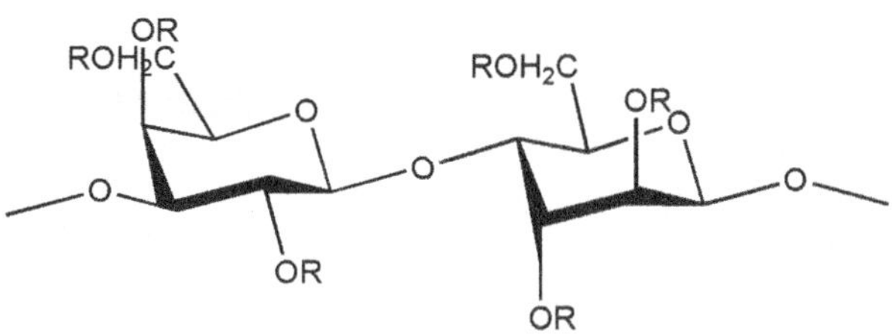

Fig. 5.7 General chemical structure of agar consisting of galactopyranose units having alternating β-1,3 and α-1,4 glycoside unions, where R could be –H, sulfate ester, methoxy ether, or pyruvic acid

Agar is commonly employed in food, cosmetics, and packaging industries since it has been approved by the FDA to be considered as a biocompatible polysaccharide (Hou et al. 2020; Olatunji 2020). One of its most outstanding and commonly exploited properties is related to its ability to form gels. For this, agarose is usually extracted and conveniently purified to be further used as starting material for hydrogel preparations. In this sense, Hou and coworkers (2020) incorporated natural humic acid and MnO_2 nanoparticles into an agarose-based hydrogel to achieve the development of devices potentially useful in tumor therapies. Bonhome-Espinosa and collaborators (2020) recently developed 3D magnetic biomaterials potentially useful as articular hyaline cartilage-like tissues by encapsulating magnetic nanoparticles and human hyaline chondrocytes within agarose-containing hydrogels. As the above-mentioned, there are thousands of nanocomposite materials developed for biomedical purposes whose hydrogel matrix contains agarose (Chiriac et al. 2019; Kazimierczak et al. 2019; Nguyen et al. 2019; Sanchez and Alvarez 2019; Sanchez et al. 2019; Olatunji 2020).

Regarding the usefulness of agar in packaging industries, it has to do with the possibility of forming eco-friendly and biodegradable films (Makwana et al. 2020). However, to overcome some of their intrinsic undesired limitations, and also to provide some useful properties, they are usually combined with different substances (Lee et al. 2019; Roy and Rhim 2019; Mostafavi and Zaeim 2020). Biofilms with antimicrobial properties were developed by Makwana et al. (2020) through the incorporation of nanosilver-modified Montmorillonite, whereas other nanoreinforcements have also been tested with the same objective (Arfat et al. 2017; Kumar et al. 2019b; Radovanovic et al. 2019).

Meanwhile, ulvans are other categories of sulfated polysaccharides (Fig. 5.8) that can be extracted from algae (Kidgell et al. 2019; Li et al. 2020; Olatunji 2020). Although they are less known than some other extractable polymers from algae natural resources, ulvans are found to be potentially useful for biomedical developments (Laza et al. 2007; Tziveleka et al. 2018). Thereby, Tziveleka and coworkers tested the antibacterial activities of a series of lysozyme/ulvan potentially useful as nanocarriers for positively charged active principles. As a result, authors reported increased activity levels against some Gram-positive organisms regarding the components by themselves (Tziveleka et al. 2018).

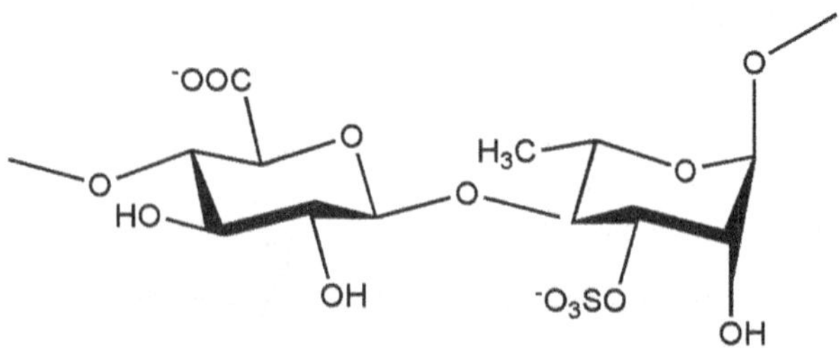

Fig. 5.8 Example of a ulvan chemical structure consisting of alternating β-1,4-D-glucuronic acid and α-1,4-L-rhamnopyranose

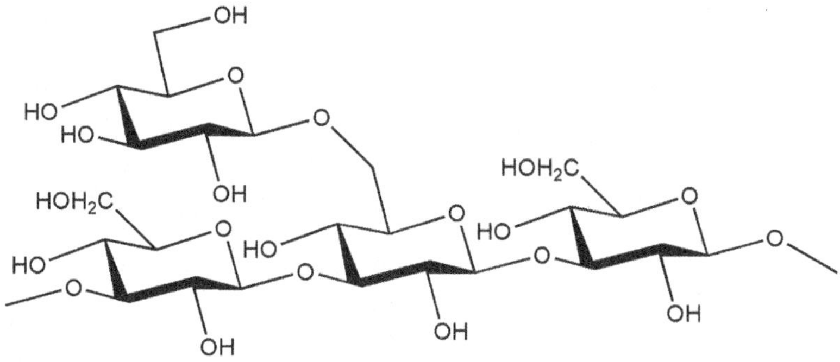

Fig. 5.9 Laminarin chemical structure

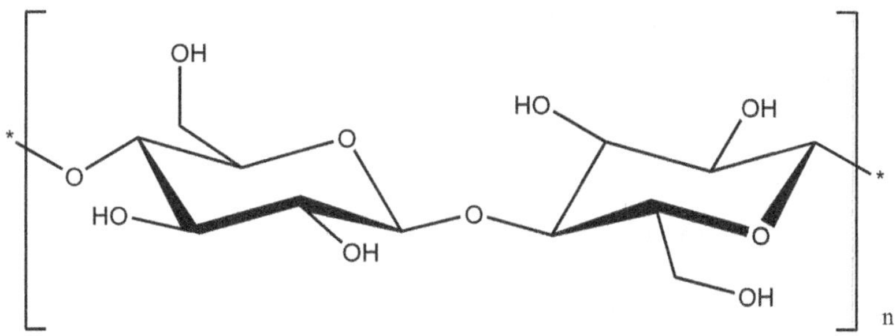

Fig. 5.10 Chemical structure of cellulose

Brown algae are also a source of laminarins (Fig. 5.9), which are short-chain polysaccharides constituted by a β-1,3-D-glucopyranose backbone with a small degree of side branching (β-1,6-D-glucopyranose units) (Jönsson et al. 2020).

As it happens with some other algae chemical extracts, laminarins have shown some important biological activities including antimicrobial, antitumoral, and anti-cancer ones (Yu et al. 2019). In this sense, Yu and collaborators (2019) developed laminarin nanoparticles potentially useful as Protoporphyrin IX delivery devices in photodynamic therapy of human breast cancer cells. However, laminarins have also been employed to develop nanomaterials capable to be used in other fields than medicine. For example, self-assembled laminarin nanostructures dried onto mica were designed by Dunstal and Goodall (2007). They pursue the immobilization of functional materials in the generated templates for novel microelectronic and sensor devices.

Cellulose and starch are polysaccharides of such importance that several books could be written about each of them. Cellulose has a remarkable abundance in nature, and it is formed by two glucose rings linked by a β-1,4-glycosidic union (Fig. 5.10) (Olatunji 2020).

Cellulose could be obtained from green, red, and brown micro- and macroalgae, and it is important to highlight that the intrinsic chemical properties of algal cellulose differ depending on the algae nature. Furthermore, they all are different from cellulose whose sources are non-aquatic ones (Olatunji 2020). Regarding the broad fields in which cellulosic materials are constantly being employed, they include textile, papermaking, medical, energy, environmental, adhesives, paint, packaging, agriculture, and food industries, among others (Sanchez et al. 2016; Prasad et al. 2017; Camara et al. 2019; Zhu 2019; Naskar and Kim 2020; Olatunji 2020; Thomas et al. 2020). Cellulose could be used as major component, in blends, and also as an additive or even as a filler in different formulations depending on the required properties for the functional material. The possibility of tuning cellulose to develop nano-cellulosic materials has open up an incredible number of possible applications (Farooq et al. 2020; Lasrado et al. 2020; Thomas et al. 2020; Tshikovhi et al. 2020; Zinge and Kandasubramanian 2020).

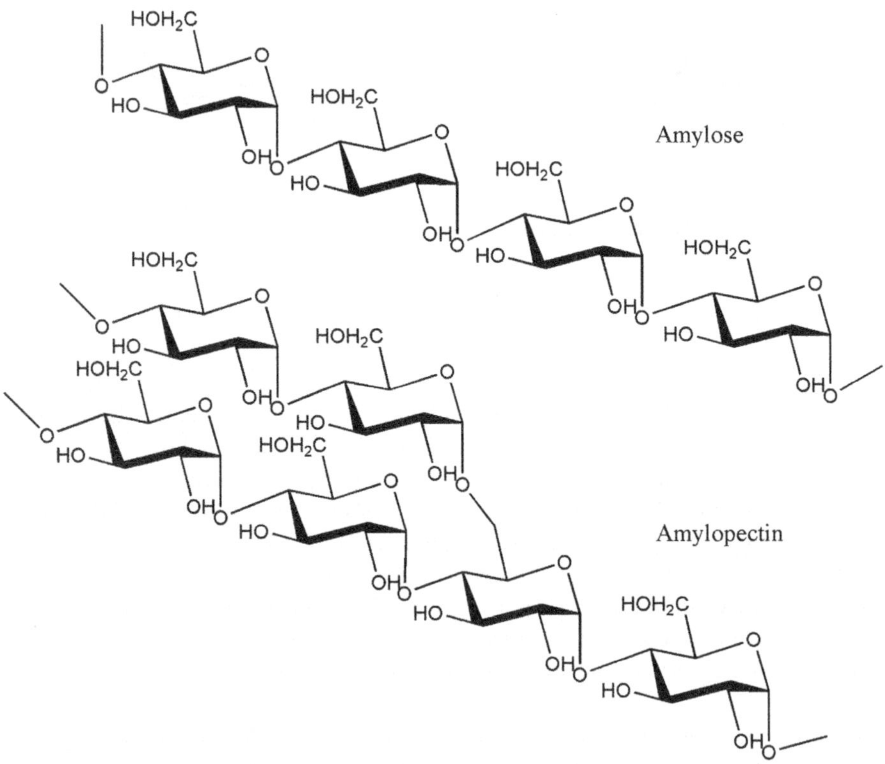

Fig. 5.11 Chemical structure of the polysaccharides found in starch

Regarding starch, it is present in green, brown, and red both micro- and macroalgae (Olatunji 2020). It is composed of two polymers whose monomeric units are D-glucose, and they are known as amylose (α-1,4 linkages) and amylopectin (α-1,4 and α-1,6 glycosidic bonds) (Fig. 5.11) (Mischnick and Momcilovic 2010; Omoregie Egharevba 2019). Although the probably most well-known use of starch is related to food industry, nowadays it is also employed by textile, medicine, paper, fine chemicals, agriculture, and packaging industries, among many other ones (Camara et al. 2019; Omoregie Egharevba 2019). For these applications, analogous to what was mentioned for cellulosic-containing materials, starch could be employed as unique/major component, combined forming blends, and also as an additive or even as a filler in different scales and formulations depending on desired material (Ghoshal and Singh 2020; Rodriguez Llanos et al. 2021). Outstandingly, nano-starch offers unique properties' combination such as biodegradability, biocompatibility, large surface area, and specific biological properties (Rostami Monjezi et al. 2019; Yang et al. 2019; Ahmad et al. 2020).

Nano-starch was employed as starting material for the development of biocatalysts, wound healing, food packaging, surface coating, drug delivery devices, and stabilizing agent for sensitive food materials (Chen et al. 2018c, 2019b; Poonguzhali et al. 2018; Suriya et al. 2018; Rostami Monjezi et al. 2019; Zarnegar et al. 2019).

5.3 Conclusions and Future Perspectives

The possibilities of revalorization of algae resources are so broad that it is hard to imagine all of their full potential. They can be used directly, or they can also act as a source of a variety of chemical entities whose applications are related to medicine, food, agriculture, packaging, textile, and several other industries. The developed materials could be found in different scales, from macro- to micro- and nano-sized ones.

Even though there are a number of well-recognized benefits related to the development and use of nanoscale composite materials, it must also be said that their potential non-desired effects as environmental pollutants are still under study. In this sense, some researchers have summarized recent reports about the employment of algae as a monitoring tool to study the effect of carbon and metal-oxide nanomaterials on a variety of species of this kind of aquatic photosynthetic organisms (Chen et al. 2018b; Saxena and Harish 2018; Saxena et al. 2020). Whereas some biological damages have been observed, there is much more work to conduct in order to find out both the nanoecotoxicological real risks and the corresponding intervening mechanisms (Kahru and Dubourguier 2010). From another point of view, since algal blooms constitute a problem very difficult to be solved in some earth planet regions, the possibility to control or mitigate them by taking advantage of these effects should also be considered (Chen et al. 2019a).

Acknowledgments LMS acknowledges the support of CONICET, ANPCyT, UNMdP, Fundación Bunge and Born, and Fulbright Argentina – Ministerio de Educación.

References

Abdellatif MM, Soliman SMA, El-Sayed NH, Abdellatif FHH (2020) Iota-carrageenan based magnetic aerogels as an efficient adsorbent for heavy metals from aqueous solutions. J Porous Mater 27:277–284. https://doi.org/10.1007/s10934-019-00812-z

Ahmad Z, Pandey R, Sharma S, Khuller GK (2006) Alginate nanoparticles as antituberculosis drug carriers: formulation development, pharmacokinetics and therapeutic potential. Indian J Chest Dis Allied Sci 48:171–176

Ahmad M, Gani A, Hassan I et al (2020) Production and characterization of starch nanoparticles by mild alkali hydrolysis and ultra-sonication process. Sci Rep 10:1–11. https://doi.org/10.1038/s41598-020-60380-0

Ahmed HB (2019) Recruitment of various biological macromolecules in fabrication of gold nanoparticles: overview for preparation and applications. Int J Biol Macromol 140:265–277. https://doi.org/10.1016/j.ijbiomac.2019.08.138

Alavi M, Nokhodchi A (2020) An overview on antimicrobial and wound healing properties of ZnO nanobiofilms, hydrogels, and bionanocomposites based on cellulose, chitosan, and alginate polymers. Carbohydr Polym 227:1–31. https://doi.org/10.1016/j.carbpol.2019.115349

Alavi M, Rai M (2019) Recent progress in nanoformulations of silver nanoparticles with cellulose, chitosan, and alginic acid biopolymers for antibacterial applications. Appl Microbiol Biotechnol 103:8669–8676. https://doi.org/10.1007/s00253-019-10126-4

Ale MT, Maruyama H, Tamauchi H et al (2011) Fucoidan from *Sargassum* sp. and *Fucus vesiculosus* reduces cell viability of lung carcinoma and melanoma cells in vitro and activates natural killer cells in mice in vivo. Int J Biol Macromol 49:331–336. https://doi.org/10.1016/j.ijbiomac.2011.05.009

Ali KA, Wahba MI, Abou-Zeid RE, Kamel S (2019) Development of carrageenan modified with nanocellulose-based materials in removing of Cu^{2+}, Pb^{2+}, Ca^{2+}, Mg^{2+}, and Fe^{2+}. Int J Environ Sci Technol 16:5569–5576. https://doi.org/10.1007/s13762-018-1936-z

Almeida TP, Ramos AA, Ferreira J et al (2019) Bioactive compounds from seaweed with anti-leukemic activity: a mini-review on carotenoids and phlorotannins. Mini Rev Med Chem 20:39–53. https://doi.org/10.2174/1389557519666190311095655

Alvarez-Lorenzo C, Blanco-Fernandez B, Puga AM, Concheiro A (2013) Crosslinked ionic polysaccharides for stimuli-sensitive drug delivery. Adv Drug Deliv Rev 65:1148–1171. https://doi.org/10.1016/j.addr.2013.04.016

Arfat YA, Ahmed J, Jacob H (2017) Preparation and characterization of agar-based nanocomposite films reinforced with bimetallic (Ag-Cu) alloy nanoparticles. Carbohydr Polym 155:382–390. https://doi.org/10.1016/j.carbpol.2016.08.097

Aswin Kumar I, Viswanathan N (2020) Fabrication of zirconium(IV) cross-linked alginate/kaolin hybrid beads for nitrate and phosphate retention. Arab J Chem 13:4111–4125. https://doi.org/10.1016/j.arabjc.2019.06.006

Aziz N, Fatma T, Varma A, Prasad R (2014) Biogenic synthesis of silver nanoparticles using Scenedesmus abundans and evaluation of their antibacterial activity. Journal of Nanoparticles, Article ID 689419, 2014. https://doi.org/10.1155/2014/689419

Aziz N, Faraz M, Pandey R, Sakir M, Fatma T, Varma A, Barman I, Prasad R (2015) Facile algae-derived route to biogenic silver nanoparticles: Synthesis, antibacterial and photocatalytic properties. Langmuir 31:11605–11612. https://doi.org/10.1021/acs.langmuir.5b03081

Aziz N, Prasad R, Ibrahim AIM, Ahmed AIS (2017) Promising applications for the production of biofuels through algae. In: Microbial Biotechnology (Patra JK, Vishnuprasad CN, Das G), Springer Nature Singapore Pte Ltd. 81–103

Bajpai VK, Shukla S, Kang SM et al (2018) Developments of cyanobacteria for nano-marine drugs: relevance of nanoformulations in cancer therapies. Mar Drugs 16:1–23. https://doi.org/10.3390/md16060179

Baweja, P. & Sahoo, D. (2015) Classification of Algae. The Algae World. 26, 31–55

Bernardo PH, Chai CLL, Heath GA et al (2004) Synthesis, electrochemistry, and bioactivity of the cyanobacterial calothrixins and related quinones. J Med Chem 47:4958–4963. https://doi.org/10.1021/jm049625o

Bonhome-Espinosa AB, Campos F, Durand-Herrera D et al (2020) In vitro characterization of a novel magnetic fibrin-agarose hydrogel for cartilage tissue engineering. J Mech Behav Biomed Mater 104:1–10. https://doi.org/10.1016/j.jmbbm.2020.103619

Camara MC, Campos EVR, Monteiro RA et al (2019) Development of stimuli-responsive nano-based pesticides: emerging opportunities for agriculture. J Nanobiotechnol 17:1–19. https://doi.org/10.1186/s12951-019-0533-8

Chen CH, Lin YS, Wu SJ, Mi FL (2018a) Mutlifunctional nanoparticles prepared from arginine-modified chitosan and thiolated fucoidan for oral delivery of hydrophobic and hydrophilic drugs. Carbohydr Polym 193:163–172. https://doi.org/10.1016/j.carbpol.2018.03.080

Chen M, Zhou S, Zhu Y et al (2018b) Toxicity of carbon nanomaterials to plants, animals and microbes: recent progress from 2015-present. Chemosphere 206:255–264. https://doi.org/10.1016/j.chemosphere.2018.05.020

Chen Q, Dong X, Zhou L et al (2018c) Nanostarch surface coating of lightweight coated paper. BioResources 13:729–739. https://doi.org/10.15376/biores.13.1.729-739

Chen F, Xiao Z, Yue L et al (2019a) Algae response to engineered nanoparticles: current understanding, mechanisms and implications. Environ Sci Nano 6:1026–1042. https://doi.org/10.1039/c8en01368c

Chen QJ, Zhou LL, Zou JQ, Gao X (2019b) The preparation and characterization of nanocomposite film reinforced by modified cellulose nanocrystals. Int J Biol Macromol 132:1155–1162. https://doi.org/10.1016/j.ijbiomac.2019.04.063

Chiriac AP, Ghilan A, Neamtu I et al (2019) Advancement in the biomedical applications of the (nano)gel structures based on particular polysaccharides. Macromol Biosci:1–20. https://doi.org/10.1002/mabi.201900187

Daniel-Da-Silva AL, Fateixa S, Guiomar AJ et al (2009) Biofunctionalized magnetic hydrogel nanospheres of magnetite and κ-carrageenan. Nanotechnology 20:1–10. https://doi.org/10.1088/0957-4484/20/35/355602

Devi SS, Sahoo D (2015) Culturing algae. In: The algae world. Springer, Dordrecht, pp 555–579

Dunstan DE, Goodall DG (2007) Terraced self assembled nano-structures from laminarin. Int J Biol Macromol 40:362–366. https://doi.org/10.1016/j.ijbiomac.2006.09.016

Farooq A, Patoary MK, Zhang M et al (2020) Cellulose from sources to nanocellulose and an overview of synthesis and properties of nanocellulose/zinc oxide nanocomposite materials. Int J Biol Macromol 154:1050–1073. https://doi.org/10.1016/j.ijbiomac.2020.03.163

Feki A, Bardaa S, Hajji S et al (2020) Falkenbergia rufolanosa polysaccharide – poly(vinyl alcohol) composite films: a promising wound healing agent against dermal laser burns in rats. Int J Biol Macromol 144:954–966. https://doi.org/10.1016/j.ijbiomac.2019.09.173

Gautam S, Kashyap M, Gupta S et al (2016) Metabolic engineering of TiO_2 nanoparticles in *Nitzschia palea* to form diatom nanotubes: an ingredient for solar cells to produce electricity and biofuel. RSC Adv 6:97276–97284. https://doi.org/10.1039/c6ra18487a

Geonzon LC, Descallar FBA, Du L et al (2020) Gelation mechanism and network structure in gels of carrageenans and their mixtures viewed at different length scales – a review. Food Hydrocoll 108:1–9. https://doi.org/10.1016/j.foodhyd.2020.106039

George M, Abraham TE (2006) Polyionic hydrocolloids for the intestinal delivery of protein drugs: alginate and chitosan – a review. J Control Release 114:1–14. https://doi.org/10.1016/j.jconrel.2006.04.017

Ghoshal G, Singh D (2020) Synthesis and characterization of starch nanocellulosic films incorporated with *Eucalyptus globulus* leaf extract. Int J Food Microbiol 332:1–13. https://doi.org/10.1016/j.ijfoodmicro.2020.108765

Gutu T, Gale DK, Jeffryes C et al (2009) Electron microscopy and optical characterization of cadmium sulphide nanocrystals deposited on the patterned surface of diatom biosilica. J Nanomater:1–7. https://doi.org/10.1155/2009/860536

Hamza MF, Lu S, Salih KAM et al (2020) As(V) sorption from aqueous solutions using quaternized algal/polyethyleneimine composite beads. Sci Total Environ 719:137396. https://doi.org/10.1016/j.scitotenv.2020.137396

Hatae N, Satoh R, Chiba H et al (2014) N-substituted calothrixin B derivatives inhibited the proliferation of HL-60 promyelocytic leukemia cells. Med Chem Res 23:4956–4961. https://doi.org/10.1007/s00044-014-1061-6

He P (2015) Harmful algal blooms. Algae World 26:339–355. https://doi.org/10.1007/978-94-017-7321-8

Hou M, Liu W, Zhang L et al (2020) Responsive agarose hydrogel incorporated with natural humic acid and MnO_2 nanoparticles for effective relief of tumor hypoxia and enhanced photo-induced tumor therapy. Biomater Sci 8:353–369. https://doi.org/10.1039/c9bm01472a

Hu S, Hsieh Y-L (2014) Preparation of activated carbon and silica particles from rice straw. ACS Sustain Chem Eng 2:726–734. https://doi.org/10.1021/sc5000539

Huang YC, Kuo TH (2016) O-carboxymethyl chitosan/fucoidan nanoparticles increase cellular curcumin uptake. Food Hydrocoll 53:261–269. https://doi.org/10.1016/j.foodhyd.2015.02.006

Huang YC, Lam UI (2011) Chitosan/fucoidan pH sensitive nanoparticles for oral delivery system. J Chin Chem Soc 58:779–785. https://doi.org/10.1002/jccs.201190121

Huang YC, Li RY (2014) Preparation and characterization of antioxidant nanoparticles composed of chitosan and fucoidan for antibiotics delivery. Mar Drugs 12:4379–4398. https://doi.org/10.3390/md12084379

Huang YC, Liu TJ (2012) Mobilization of mesenchymal stem cells by stromal cell-derived factor-1 released from chitosan/tripolyphosphate/fucoidan nanoparticles. Acta Biomater 8:1048–1056. https://doi.org/10.1016/j.actbio.2011.12.009

Huang Y-C, Yang Y-T (2016) Effect of basic fibroblast growth factor released from chitosan–fucoidan nanoparticles on neurite extension. J Tissue Eng Regen Med 10:418–427. https://doi.org/10.1002/term

Huang YC, Chen JK, Lam UI, Chen SY (2014) Preparing, characterizing, and evaluating chitosan/fucoidan nanoparticles as oral delivery carriers. J Polym Res 21:1–9. https://doi.org/10.1007/s10965-014-0415-6

Huang YC, Li RY, Chen JY, Chen JK (2016) Biphasic release of gentamicin from chitosan/fucoidan nanoparticles for pulmonary delivery. Carbohydr Polym 138:114–122. https://doi.org/10.1016/j.carbpol.2015.11.072

Jantschke A, Herrmann AK, Lesnyak V et al (2012) Decoration of diatom biosilica with noble metal and semiconductor nanoparticles (<10 nm): assembly, characterization, and applications. Chem Asian J 7:85–90. https://doi.org/10.1002/asia.201100563

Jönsson M, Allahgholi L, Sardari RRR et al (2020) Extraction and modification of macroalgal polysaccharides for current and next-generation applications. Molecules 25:1–29. https://doi.org/10.3390/molecules25040930

Kahru A, Dubourguier HC (2010) From ecotoxicology to nanoecotoxicology. Toxicology 269:105–119. https://doi.org/10.1016/j.tox.2009.08.016

Kazimierczak P, Kolmas J, Przekora A (2019) Biological response to macroporous chitosan-agarose bone scaffolds comprising Mg-and Zn-doped nano-hydroxyapatite. Int J Mol Sci 20:1–20. https://doi.org/10.3390/ijms20153835

Keeling PJ (2004) Diversity and evolutionary history of plastids and their hosts. Am J Bot 91:1481–1493. https://doi.org/10.3732/ajb.91.10.1481

Khalid M (2020) Nanotechnology and chemical engineering as a tool to bioprocess microalgae for its applications in therapeutics and bioresource management. Crit Rev Biotechnol 40:46–63. https://doi.org/10.1080/07388551.2019.1680599

Khan MI, Shin JH, Kim JD (2018) The promising future of microalgae: current status, challenges, and optimization of a sustainable and renewable industry for biofuels, feed, and other products. Microb Cell Factories 17:1–21. https://doi.org/10.1186/s12934-018-0879-x

Kidgell JT, Magnusson M, de Nys R, Glasson CRK (2019) Ulvan: a systematic review of extraction, composition and function. Algal Res 39:1–20. https://doi.org/10.1016/j.algal.2019.101422

Kimura R, Rokkaku T, Takeda S et al (2013) Cytotoxic effects of fucoidan nanoparticles against osteosarcoma. Mar Drugs 11:4267–4278. https://doi.org/10.3390/md11114267

Kong X, Squire K, Li E et al (2016) Chemical and biological sensing using diatom photonic crystal biosilica with in-situ growth plasmonic nanoparticles. IEEE Trans Nanobioscience 15:828–834. https://doi.org/10.1016/j.physbeh.2017.03.040

Kumar A, Matari IAI, Choi H et al (2019a) Development of halloysite nanotube/carboxylated-cellulose nanocrystal-reinforced and ionically-crosslinked polysaccharide hydrogels. Mater Sci Eng C 104:1–13. https://doi.org/10.1016/j.msec.2019.109983

Kumar S, Boro JC, Ray D et al (2019b) Bionanocomposite films of agar incorporated with ZnO nanoparticles as an active packaging material for shelf life extension of green grape. Heliyon 5:1–8. https://doi.org/10.1016/j.heliyon.2019.e01867

Larraza I, Ugarte L, Fayanas A et al (2020) Influence of process parameters in graphene oxide obtention on the properties of mechanically strong alginate nanocomposites. Materials (Basel) 13. https://doi.org/10.3390/ma13051081

Lasrado D, Ahankari S, Kar K (2020) Nanocellulose-based polymer composites for energy applications—a review. J Appl Polym Sci 137:1–14. https://doi.org/10.1002/app.48959

Laza AL, Jaber M, Miehé-Brendlé J et al (2007) Green nanocomposites: synthesis and characterization. J Nanosci Nanotechnol 7:3207–3213. https://doi.org/10.1166/jnn.2007.698

Lee KW, Jeong D, Na K (2013a) Doxorubicin loading fucoidan acetate nanoparticles for immune and chemotherapy in cancer treatment. Carbohydr Polym 94:850–856. https://doi.org/10.1016/j.carbpol.2013.02.018

Lee KY, Jeong MR, Choi SM et al (2013b) Synergistic effect of fucoidan with antibiotics against oral pathogenic bacteria. Arch Oral Biol 58:482–492. https://doi.org/10.1016/j.archoralbio.2012.11.002

Lee H, Rukmanikrishnan B, Lee J (2019) Rheological, morphological, mechanical, and water-barrier properties of agar/gellan gum/montmorillonite clay composite films. Int J Biol Macromol 141:538–544. https://doi.org/10.1016/j.ijbiomac.2019.09.021

Li Q, Hu F, Zhu B et al (2020) Insights into ulvan lyase: review of source, biochemical characteristics, structure and catalytic mechanism. Crit Rev Biotechnol 40:432–441. https://doi.org/10.1080/07388551.2020.1723486

Losic D, Triani G, Evans PJ et al (2006) Controlled pore structure modification of diatoms by atomic layer deposition of TiO_2. J Mater Chem 16:4029–4034. https://doi.org/10.1039/b610188g

Losic D, Yu Y, Aw MS et al (2010) Surface functionalisation of diatoms with dopamine modified iron-oxide nanoparticles: toward magnetically guided drug microcarriers with biologically derived morphologies. Chem Commun 46:6323–6325. https://doi.org/10.1039/c0cc01305f

Lu KY, Li R, Hsu CH et al (2017) Development of a new type of multifunctional fucoidan-based nanoparticles for anticancer drug delivery. Carbohydr Polym 165:410–420. https://doi.org/10.1016/j.carbpol.2017.02.065

Lu J, Zhang C, Wu J (2020) One-pot synthesis of magnetic algal carbon/sulfidated nanoscale zerovalent iron composites for removal of bromated disinfection by-product. Chemosphere 250:126257. https://doi.org/10.1016/j.chemosphere.2020.126257

Luesch H, Chanda SK, Raya RM et al (2006) A functional genomics approach to the mode of action of apratoxin A. Nat Chem Biol 2:158–167. https://doi.org/10.1038/nchembio769

Makwana D, Castaño J, Somani RS, Bajaj HC (2020) Characterization of Agar-CMC/Ag-MMT nanocomposite and evaluation of antibacterial and mechanical properties for packaging applications. Arab J Chem 13:3092–3099. https://doi.org/10.1016/j.arabjc.2018.08.017

Maldonado L, Chough S, Bonilla J et al (2019) Mechanism of fabrication and nano-mechanical properties of α-lactalbumin/chitosan and BSA/κ-carrageenan nanotubes through layer-by-layer assembly for curcumin encapsulation and determination of in vitro cytotoxicity. Food Hydrocoll 93:293–307. https://doi.org/10.1016/j.foodhyd.2019.02.040

Manivasagan P, Bharathiraja S, Bui NQ et al (2016) Doxorubicin-loaded fucoidan capped gold nanoparticles for drug delivery and photoacoustic imaging. Int J Biol Macromol 91:578–588. https://doi.org/10.1016/j.ijbiomac.2016.06.007

Mao L, Liu J, Zhu S et al (2014) Sonochemical fabrication of mesoporous TiO_2 inside diatom frustules for photocatalyst. Ultrason Sonochem 21:527–534. https://doi.org/10.1016/j.ultsonch.2013.09.001

Mathijsen D (2020) Algae to sustainably produce carbon fiber and simultaneously take CO_2 out of the atmosphere. Reinf Plast 64:50–53. https://doi.org/10.1016/j.repl.2019.04.070

Mathimani T, Pugazhendhi A (2019) Utilization of algae for biofuel, bio-products and bio-remediation. Biocatal Agric Biotechnol 17:326–330. https://doi.org/10.1016/j.bcab.2018.12.007

McKim JM, Willoughby JA, Blakemore WR, Weiner ML (2019) Clarifying the confusion between poligeenan, degraded carrageenan, and carrageenan: a review of the chemistry, nomenclature, and in vivo toxicology by the oral route. Crit Rev Food Sci Nutr 59:3054–3073. https://doi.org/10.1080/10408398.2018.1481822

Metting FB (1996) Biodiversity and application of microalgae. J Ind Microbiol Biotechnol 17:477–489. https://doi.org/10.1007/bf01574779

Mignon A, Snoeck D, Halluin KD et al (2016) Alginate biopolymers: counteracting the impact of superabsorbent polymers on mortar strength. Constr Build Mater 110:169–174. https://doi.org/10.1016/j.conbuildmat.2016.02.033

Mirza S, Jolly R, Zia I et al (2020) Bioactive gum Arabic/κ-carrageenan-incorporated nano-hydroxyapatite nanocomposites and their relative biological functionalities in bone tissue engineering. ACS Omega 5:11279–11290. https://doi.org/10.1021/acsomega.9b03761

Mischnick P, Momcilovic D (2010) Chemical structure analysis of starch and cellulose derivatives. Elsevier Inc, United States.

Mohanpuria P, Rana NK, Yadav SK (2008) Biosynthesis of nanoparticles: technological concepts and future applications. J Nanopart Res 10:507–517. https://doi.org/10.1007/s11051-007-9275-x

Mostafavi FS, Zaeim D (2020) Agar-based edible films for food packaging applications – a review. Int J Biol Macromol 159:1165–1176. https://doi.org/10.1016/j.ijbiomac.2020.05.123

Naskar A, Kim K (2020) Recent advances in nanomaterial-based wound-healing therapeutics. Pharmaceutics 12:1–20

Nguyen DHT, Nguyen NB, Nguyen LTP et al (2019) An agarose–curdlan nanogel that carries etanercept to target and neutralises TNF-α produced by dectin-1-expressing immune cells. J Electron Mater 48:6570–6582. https://doi.org/10.1007/s11664-019-07458-2

Olatunji O (2020) Aquatic Biopolymers: Understanding their Industrial Significance and Environmental Implications. Springer, Switzerland. https://doi.org/10.1007/978-3-030-34709-3

Ollier RP, Villanueva ME, Copello GJ et al (2020) Engineered nanomaterials for emerging contaminant removal from wastewater. In: Handbook of nanomaterials and nanocomposites for energy and environmental applications. Springer, Cham, pp 1–22

Omoregie Egharevba H (2019) Chemical properties of starch and its application in the food industry. In: Chemical properties of starch. IntechOpen, London, pp 1–441
Patil MP, Kim GD (2018) Marine microorganisms for synthesis of metallic nanoparticles and their biomedical applications. Colloids Surf B Biointerfaces 172:487–495. https://doi.org/10.1016/j.colsurfb.2018.09.007
Piyakulawat P, Praphairaksit N, Chantarasiri N, Muangsin N (2007) Preparation and evaluation of chitosan/carrageenan beads for controlled release of sodium diclofenac. AAPS PharmSciTech 8:E97. https://doi.org/10.1208/pt0804094
Poonguzhali R, Basha SK, Kumari VS (2018) Nanostarch reinforced with chitosan/poly (vinyl pyrrolidone) blend for in vitro wound healing application. Polym Plast Technol Eng 57:1400–1410. https://doi.org/10.1080/03602559.2017.1381255
Prasad R, Kumar V, Kumar M (2017) Nanotechnology: Food and Environmental Paradigm. Springer Nature Singapore Pte Ltd. (ISBN 978-981-10-4678-0)
Radovanovic N, Malagurski I, Levic S et al (2019) Influence of different concentrations of Zn-carbonate phase on physical-chemical properties of antimicrobial agar composite films. Mater Lett 255:126572. https://doi.org/10.1016/j.matlet.2019.126572
Ragni R, Cicco SR, Vona D, Farinola GM (2018) Multiple routes to smart nanostructured materials from diatom microalgae: a chemical perspective. Adv Mater 30:1–23. https://doi.org/10.1002/adma.201704289
Ramazani A, Aghahosseini H (2019) The biological properties of hydrogels based on natural polymers. In: Hydrogels based on natural polymers. Elsevier Inc, Amsterdam, pp 247–269
Rawat J, Gupta PK, Pandit S, Prasad R, Pande V (2021) Current perspectives on integrated approaches to enhance lipid accumulation in microalgae. 3 Biotech 11, 303. https://doi.org/10.1007/s13205-021-02851-3
Rodriguez Llanos JH, Tadini CC, Gastaldi E (2021) New strategies to fabricate starch/chitosan-based composites by extrusion. J Food Eng 290:110224. https://doi.org/10.1016/j.jfoodeng.2020.110224
Rostami Monjezi H, Zarnegar Z, Safari J (2019) Starch nanoparticles as a bio-nanocatalyst in synthesis of diheteroaryl thioethers. J Saudi Chem Soc 23:973–979. https://doi.org/10.1016/j.jscs.2019.05.003
Roy S, Rhim JW (2019) Agar-based antioxidant composite films incorporated with melanin nanoparticles. Food Hydrocoll 94:391–398. https://doi.org/10.1016/j.foodhyd.2019.03.038
Saavedra R, Muñoz R, Taboada ME et al (2018) Comparative uptake study of arsenic, boron, copper, manganese and zinc from water by different green microalgae. Bioresour Technol 263:49–57. https://doi.org/10.1016/j.biortech.2018.04.101
Sagbas S, Butun S, Sahiner N (2012) Modifiable chemically crosslinked poli(κ-carrageenan) particles. Carbohydr Polym 87:2718–2724. https://doi.org/10.1016/j.carbpol.2011.11.064
Sahoo D, Baweja P (2015) General characteristics of algae. In: The algae world. Springer, Dordrecht, pp 3–29
Salama ES, Roh HS, Dev S et al (2019) Algae as a green technology for heavy metals removal from various wastewater. World J Microbiol Biotechnol 35:1–19. https://doi.org/10.1007/s11274-019-2648-3
Sanchez LM, Alvarez VA (2019) Advances in magnetic noble metal/iron-based oxide hybrid nanoparticles as biomedical devices. Bioengineering 6:1–18. https://doi.org/10.3390/bioengineering6030075
Sanchez LM, Alvarez VA, Gonzalez JS (2016) Ferrogels: smart materials for biomedical and remediation applications. In: Handbook of composites from renewable materials, volume 8: nanocomposites: advanced applications. Wiley, New York, pp 400–430
Sanchez LM, Ollier RP, Gonzalez JS, Alvarez VA (2018) Nanocomposite materials for dyes removal. In: Handbook of nanomaterials for industrial applications. Elsevier, Amsterdam, pp 921–951

Sanchez LM, Arciniegas L, Alvarez VA, Gonzalez JS (2019) Phantom gels towards medicine improvement: uses for magnetic device tests and enhancements on magnetic-dependent clinical techniques. In: Materials for biomedical engineering: nanomaterials-based drug delivery. Elsevier Inc, Amsterdam, pp 429–450

Saxena P, Harish (2018) Nanoecotoxicological reports of engineered metal oxide nanoparticles on algae. Curr Pollut Rep 4:128–142. https://doi.org/10.1007/s40726-018-0088-6

Saxena P, Harish (2019) Phyco-nanotechnology: new horizons of gold nano-factories. Proc Natl Acad Sci India Sect B Biol Sci 89:1–11. https://doi.org/10.1007/s40011-016-0813-0

Saxena P, Sangela V, Ranjan S et al (2020) Aquatic nanotoxicology: impact of carbon nanomaterials on algal flora. Energy Ecol Environ 5:240–252. https://doi.org/10.1007/s40974-020-00151-9

Seckbach J (2015) Survey of algae in extreme environments. Algae World 26:307–315. https://doi.org/10.1007/978-94-017-7321-8

Seo JC, Umirov N, Park SB et al (2019) Microalgae-derived hollow carbon-MoS_2 composite as anode for lithium-ion batteries. J Ind Eng Chem 79:106–114. https://doi.org/10.1016/j.jiec.2019.05.046

Severino P, da Silva CF, Andrade LN et al (2019) Alginate nanoparticles for drug delivery and targeting. Curr Pharm Des 25:1312–1334. https://doi.org/10.2174/138161282566619042516342 4

Shen L, Wang J, Li Z et al (2020) A high-efficiency Fe_2O_3@microalgae composite for heavy metal removal from aqueous solution. J Water Process Eng 33:1–12. https://doi.org/10.1016/j.jwpe.2019.101026

Stevenson CS, Capper EA, Roshak AK et al (2002) Scytonemin – a marine natural product inhibitor of kinases key in hyperproliferative inflammatory diseases. Inflamm Res 51:112–114. https://doi.org/10.1007/BF02684014

Stone MT, Kozlov M (2014) Separating proteins with activated carbon. Langmuir 30:8046–8055. https://doi.org/10.1021/la501005s

Suriya M, Reddy CK, Haripriya S, Harsha N (2018) Influence of debranching and retrogradation time on behavior changes of *Amorphophallus paeoniifolius* nanostarch. Int J Biol Macromol 120:230–236. https://doi.org/10.1016/j.ijbiomac.2018.08.059

Thakkar M, Wu Z, Wei L, Mitra S (2015) Water defluoridation using a nanostructured diatom-ZrO_2 composite synthesized from algal biomass. J Colloid Interface Sci 450:239–245. https://doi.org/10.1016/j.jcis.2015.03.017

Thomas P, Duolikun T, Rumjit NP et al (2020) Comprehensive review on nanocellulose: recent developments, challenges and future prospects. J Mech Behav Biomed Mater 110:1–21. https://doi.org/10.1016/j.jmbbm.2020.103884

Tønnesen HH, Karlsen J (2002) Alginate in drug delivery systems. Drug Dev Ind Pharm 28:621–630

Torres FG, Troncoso OP, Pisani A et al (2019) Natural polysaccharide nanomaterials: an overview of their immunological properties. Int J Mol Sci 20:1–22. https://doi.org/10.3390/ijms20205092

Toster J, Zhou QL, Smith NM et al (2013) In situ coating of diatom frustules with silver nanoparticles. Green Chem 15:2060–2063. https://doi.org/10.1039/c3gc40660a

Tran PHL, Duan W, Tran TTD (2020) Fucoidan-based nanostructures: a focus on its combination with chitosan and the surface functionalization of metallic nanoparticles for drug delivery. Int J Pharm 575:118956. https://doi.org/10.1016/j.ijpharm.2019.118956

Tripathy A, More RD, Gupta S, Samuel J, Singh J, Prasad R (2021) Present and future prospect of algae: A potential candidate for sustainable pollution mitigation. The Open Biotechnology Journal 15:142–156. https://doi.org/10.2174/1874070702115010142

Tshikovhi A, Mishra SB, Mishra AK (2020) Nanocellulose-based composites for the removal of contaminants from wastewater. Int J Biol Macromol 152:616–632. https://doi.org/10.1016/j.ijbiomac.2020.02.221

Tziveleka LA, Pippa N, Georgantea P et al (2018) Marine sulfated polysaccharides as versatile polyelectrolytes for the development of drug delivery nanoplatforms: complexation of ulvan with lysozyme. Int J Biol Macromol 118:69–75. https://doi.org/10.1016/j.ijbiomac.2018.06.050

Valix M, Cheung WH, McKay G (2006) Roles of the textural and surface chemical properties of activated carbon in the adsorption of acid blue dye. Langmuir 22:4574–4582. https://doi.org/10.1021/la051711j

Yang N, Zhang W, Ye C et al (2019) Nanobiopolymers fabrication and their life cycle assessments. Biotechnol J 14:1–38. https://doi.org/10.1002/biot.201700754

Yong Lee K, Mooney DJ (2012) Alginate: properties and biomedical applications. Prog Polym Sci 37:106–126. https://doi.org/10.1016/j.progpolymsci.2011.06.003

Yu Y, Wang B, Guo C et al (2019) Protoporphyrin IX-loaded laminarin nanoparticles for anticancer treatment, their cellular behavior, ROS detection, and animal studies. Nanoscale Res Lett 14:1–12. https://doi.org/10.1186/s11671-019-3138-0

Zahran M, Marei AH (2019) Innovative natural polymer metal nanocomposites and their antimicrobial activity. Int J Biol Macromol 136:586–596. https://doi.org/10.1016/j.ijbiomac.2019.06.114

Zarnegar Z, Monjezi HR, Safari J (2019) Arginine-based surface modification of nanostarch, a catalytic carbohydrates in synthesis of heteroaryl sulfides. J Mol Struct 1193:14–23. https://doi.org/10.1016/j.molstruc.2019.05.017

Zhu F (2019) Starch based aerogels: production, properties and applications. Trends Food Sci Technol 89:1–10. https://doi.org/10.1016/j.tifs.2019.05.001

Zinge C, Kandasubramanian B (2020) Nanocellulose based biodegradable polymers. Eur Polym J 133:1–20. https://doi.org/10.1016/j.eurpolymj.2020.109758

Chapter 6
Seaweeds: A Potential Source in Progressing Nanotechnology

Nivas Desai, Umesh Pawar, Vishal Aparadh, Uttam Dethe, and Dattatraya Gaikwad

Contents

6.1 Introduction

Nanotechnology is the most vibrant field of science and technology research field making a fruitful impact and expanding day by day. Nanotechnology is an integrated field of science and technology, which refers to the processing, separation, consolidation, and deformation of material by an atom or molecules (Sharma et al. 2015a). The products of nanotechnology are nanoparticles (NPs), or nanomaterials are in the nanoscale (≤100 nm) using top-down and bottom-up strategies. The NPs are either natural NPs, incidental NPs, or engineered NPs (Buzea et al. 2007). The

N. Desai (✉) · U. Pawar · V. Aparadh · U. Dethe
PG Department of Botany, Shri Pancham Khemraj Mahavidyalay, Sawantwadi, India

D. Gaikwad
Department of Botany, Shivaji University, Kolhapur, India

D. Thangadurai et al. (eds.), *Bioprospecting Algae for Nanosized Materials*, Nanotechnology in the Life Sciences,
https://doi.org/10.1007/978-3-030-81557-8_6

recent advancement in nanotechnology is the synthesis of nanowires and nanotubes that have great applications in the medical field in the diagnosis and treatment of diseases and fabrication of medical devices (Sri Ramkumar et al. 2016). NPs are in limelight due to their smaller size, large surface-to-volume ratio, and the ability to easily interact with other particles and surfaces. Recently, NPs are being used extensively in electronics, cosmetics, biomedical, food science and technology, agriculture, and biotechnological applications (Prasad 2014; Prasad et al. 2014, 2017). The proficient crystallographic and physicochemical properties of NPs and several other features make nanotechnology an attractive tool and an excellent area of focus. The NPs can be synthesized either by physical or chemical method. The wet method is the most widely used and traditional method for NPs synthesis. The chemical synthesis includes the growth of NPs in a liquid medium containing various reactants (reducing agents). The commonly used reactants are borohydride, potassium bitartrate, methoxy polyethylene glycol, or hydrazine (Kim et al. 2007). Algae are an ecologically and economically important group among photosynthetic organisms. These are a wide range of forms from unicellular to multicellular and can live in freshwater as well as marine environment even can grow on moist rocks. Algae are either microscopic (microalgae) or macroscopic (macroalgae). They have a wide spectrum of applications in medical, pharma, agriculture, and cosmetics. A large number of various commercial products and natural dyes are prepared from algae. Recently, algal biotechnology for biofuel production is in bloom (Aziz et al. 2017; Rawat et al. 2021). From the biosynthesis of metallic NPs perspectives, seaweeds from Chlorophyceae, Phaeophyceae, Cyanophyceae, and Rhodophyceae have been largely utilized. The use of diatoms and euglenoids is also in great demand. Due to the ability to accumulate metals and to reduce metal ions, algae have become the largest contender for the biosynthesis of NPs. Additionally, algae are more convenient to handle and less toxic (Aziz et al. 2014, 2015). Algae are considered as biofactories among the biological material due to their use of both dried and live biomass for the synthesis of metallic NPs (Davis et al. 1998). Marine green algae *Ulva fasciata* was used for the synthesis of silver NPs and was found effective in growth inhibition of *Xanthomonas campestris* pv. *malvacearum* (Rajesh et al. 2012). The present chapters highlights the potential and beneficial applications of seaweed mediated synthesized NPs, their present status and future perspectives.

6.1.1 Nanoparticles: Types and Importance

NPs are of two types, that is, inorganic NPs include metal and metal oxides and organic NPs include poly-lysine, quaternary ammonium compounds, cationic quaternary polyelectrolytes, and chitosan. The inorganic NPs are potent antibacterial agents, and these are oxide NPs such as silver (Ag), iron oxide (Fe_3O_4), titanium oxide (TiO_2), copper oxide (CuO), and zinc oxide (ZnO) (Loomba and Scarabelli 2013). The organic NPs are generally less stable at high temperature and due to this, these are more preferable antimicrobial polymers (Jain et al. 2014) (Fig. 6.1).

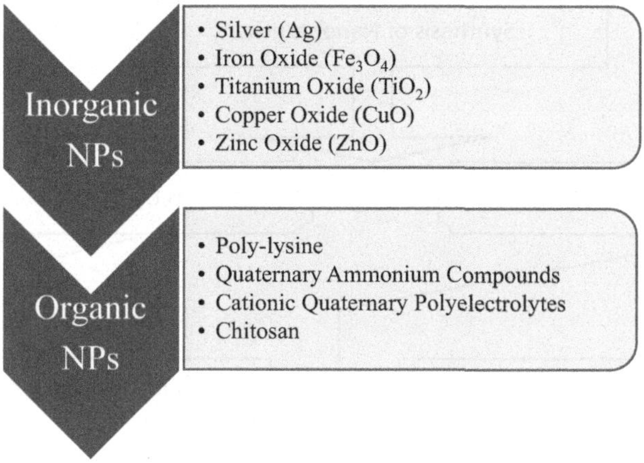

Fig. 6.1 Different types of nanoparticles

6.1.2 Methods of Synthesis of Nanoparticles

The synthesis of NPs can be done through physical, chemical, and biological methods. In the chemical methods, the hazardous substances like sodium borohydride, tetrakisphosponium chloride, poly-N-vinyl pyrrolidone, and hydroxylamine are being used, which have a toxic nature. Due to this, such type of NPs has limitations in the clinical fields. Physical methods such as laser ablation, UV irradiations, evaporation condensation, aerosol, and lithography are most widely used for the synthesis of NPs. These technique found unaffordable due to its cost and energy consumption point, as well as requirement of expensive equipment. On the other hand, biological methods are facile, economical, and more convenient to produce multiple NPs (Iravani 2011). The various methods of NPs synthesis are depicted in Fig. 6.2. The great advantage of green synthesis is that it is cost-effective and eco-friendly, and there is no need for higher pressure, energy, temperature, and even toxic chemicals. Hence, the synthesis of clean, biocompatible, non-toxic, and eco-friendly NPs produced through extracellular or intracellular substances is in huge demand. Biological resources such as plants and plant products, algae, fungi, yeasts, bacteria, actinobacteria, and viruses are now widely used to synthesize advanced nanomaterials (Sankar et al. 2013; Prasad 2014; Prasad et al. 2016, 2018; Srivastava et al. 2021) (Fig. 6.2).

6.1.3 Synthesis of NPs Using Seaweeds

Seaweeds are a rich source of bioactive compounds such as polysaccharides like alginate, laminarin, fucoidan, carrageenan, agar-agar, polyphenols, carotenoids, proteins, amino acids, vitamins, and minerals. Seaweed biomolecules are widely used in the food, medicine, cosmetics, and other manufacturing industries. The algal

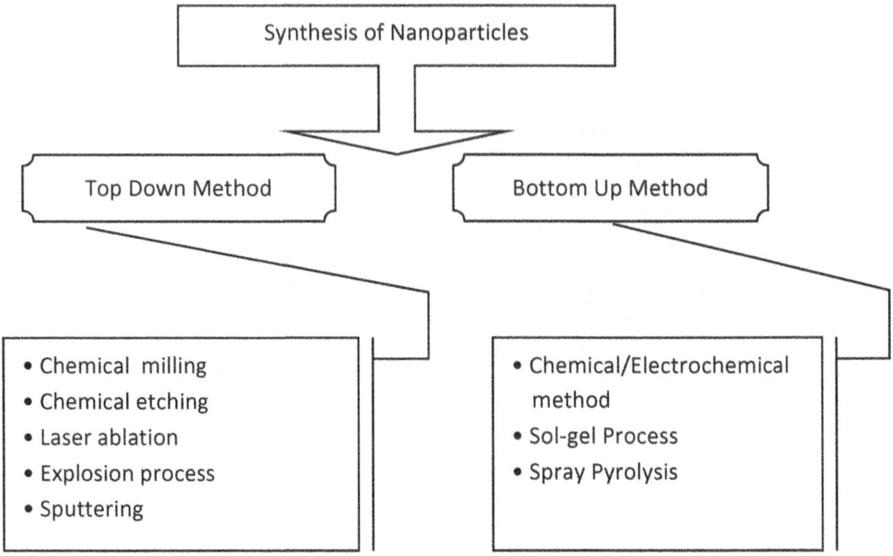

Fig. 6.2 Methods of nanoparticle synthesis

phytochemicals serve as an effective metal-reducing and capping agent to provide a robust coating on the metal NPs in one-step synthesis. Seaweeds are the group of marine plants that lives in either marine water or brackish water environment. These groups have potential bioactive properties. They are considered as a prime member of the ecological, biological, and renewable component in the marine ecosystem contributing substantial contribution to the primary production and provide a habitat for benthic communities. It was estimated that around 90% of the species of marine plants are algae, and algae derive about 50% of the global photosynthesis. Another interesting fact is that every second oxygen molecule that we breathe in is derived from algae and algae utilize every second molecule of carbon dioxide we breathe out. About 841 taxa of marine algae were recorded from both intertidal and deepwater regions of the Indian coast, and they are classified into three groups, namely, Chlorophyceae (green algae), Phaeophyceae (brown algae), and Rhodophyceae (red algae).

Seaweeds are macroscopic found attached to the bottom substratum in relatively shallow coastal water. They can grow in the intertidal, shallow, deep sea areas up to 180-meter depth. Some grow well in estuaries, rocks, dead corals, pebbles, shells, on the root and trunk of mangroves, as well as on other plants also. The lagoons, rocky shore, mudflats, and estuaries along the Indian coast provide perfect habitats for the growth of seaweeds. Seaweeds are the important and only natural source of some of the chemicals like agar, carrageenan, and alginates. Seaweeds have great advantages due to their high capacity of metal uptake and their macroscopic nature (Davis et al. 2003).

6.1.4 Synthesis Steps of Algal Nanoparticles

The abundance and easy availability of material make algae, a popular and worthwhile source for the synthesis of metallic NPs (Kannan et al. 2013). Algal nanoparticle synthesis is a stepwise process and the three important steps are as follows:

1. Preparation of algal extract in water or organic solvent by heating or boiling for a certain duration
2. Preparation of molar solutions of ionic metallic compounds
3. Incubation of algal solutions and molar solutions of ionic metallic compounds and followed by either continuous steering or without steering for a certain duration under controlled condition

The synthesis of NPs is dose-dependent and also depends on the type of algae. The reduction of metals greatly depends on the variety of biomolecules such as polysaccharides, peptides, and even pigments. The sulfated polysaccharides and proteins are used for the stabilization and capping of NPs in an aqueous solution (Singaravelu et al. 2007). The algal NPs synthesis is a time reducing than the other biosynthesizing methods (Rauwel et al. 2015). So far, several NPs are synthesized using seaweeds (Table 6.1). Marine algae are mainly explored for the synthesis of NPs (Fig. 6.3, Table 6.1).

6.2 Characterization of Nanoparticles

As nanometals show striking optical properties due to surface plasmon resonance (SPR), the biosynthesis of metal NPs in the aqueous phase is monitored using absorption spectroscopy in the UV–visible spectral range (190–1100 nm) (Focsan et al. 2011). A specific SPR absorption band (λmax) unique to nanometal crises due to the combined vibration of the free electron of metal NPs in resonance with the light wave also depends on the size and shape of NPs, the aspect ratio, and the dielectric constant of the metals (Mukherjee et al. 2001). The reducing agent that is responsible for the reduction, stabilization, and capping of the metal NPs can be revealed using Fourier transform infrared (FTIR) analysis. The various functional groups such as -C=O, -NH2, and -SH groups adhering to the surface of nanometal synthesized have been characterized through FTIR (Jena et al. 2013). The formation of NPs with various shapes and sizes can be analyzed using transmission electron microscopy (TEM) and scanning electron microscopy (SEM) combined with electron diffraction (EDX) analysis (Jena et al. 2013). The presence of metal is confirmed by energy-dispersive spectroscopy (EDS) and the crystalline nature of the nanometals can be depicted by X-ray diffraction (XRD) analysis. The dynamic light scattering (DLS) is used to characterize the surface charge with the size distribution of the NPs suspended in a colloidal solution and the zeta potential (ZP). The analysis is used to determine the stability of the particles (Krishnan et al. 2015). The

Table 6.1 Characteristics of the metal nanoparticles biosynthesized using seaweeds

Class	Species	Type and shape of NPs	Size of NPs (nm)	References
Ochrophyta	*Klebsormidium flaccidum*	Au	10–15	Sicard et al. (2010)
	Scenedesmus-24	CdS	150–175	Jena et al. (2014)
	Tetraselmisko chinensis	Au (spherical)	15	Senapati et al. (2012)
	Tetraselmis suecica	Au (spherical)	79	Shakibaie et al. (2010)
	Nannochloropsis oculata	Ag	>15	Mohseniazar et al. (2011)
Chlorophyta	*Caulerpa racemosa*	Ag (spherical and triangular)	5–25	Kathiraven et al. (2015)
	Codium capitatum	Ag	3–44	Kannan et al. (2013)
	Prasiola crispa	Au (spherical)	5–25	Sharma et al. (2015b)
	Rhizoclonium fontinale	Au	16	Parial and Pal (2015)
	Ulva compressa	Ag (spherical)	40–50	Dhanalakshmi et al. (2012)
	Ulva fasciata	Ag (spherical and polydispersal)	40.5	Rajesh et al. (2012)
	Ulva intestinalis	Au (spherical to irregular)	42.29	Parial et al. (2012)
	Ulva reticulata	Ag (spherical)	40–50	Dhanalakshmi et al. (2012)
Phaeophyta	*Bifurcaria bifurcata*	Cu	20.6	Abboud et al. (2014)
	Cystophora moniliformis	Ag	75–77	Prasad et al. (2013)
	Fucus vesiculosus	Au	Several	Mata et al. (2009)
	Saccharina (Laminaria) japonica	Au	15–20	Ghodake and Lee (2011)
	Padina gymnosperma	Ag (spherical)	25–40	Shiny et al. (2013)
	Padina gymnosperma	Au	53–67	Singh et al. (2013)
	Padina pavonica	Ag (spherical)	45–64	Sahayaraj et al. (2012)
	Padina tetrastromatica	Ag (spherical)	14	Sangeetha et al. (2012)
	Sargassum cinereum	Ag	45–76	Mohandass et al. (2013)
	Sargassum muticum	Ag	5–15	Azizi et al. (2013)
	Sargassum muticum	Fe_3O_4	18 ± 4	Mahdavi et al. (2013)
	Sargassum muticum	ZnO	30–57	Azizi et al. (2014)
	Sargassum plagiophyllum	Ag (spherical)	20–50	Dhanalakshmi et al. (2012)
	Sargassum plagiophyllum	AgCl	18–42	Dhas et al. (2014)
	Sargassum vulgare	Ag	~10	Govindaraju et al. (2015)
	Sargassum wightii	Au	8–12	Singaravelu et al. (2007)
	Turbinaria conoides	Au	60	Rajeshkumar et al. (2013)
	Turbinaria conoides	Ag	2–19	Vijayan et al. (2014)

(continued)

Table 6.1 (continued)

Class	Species	Type and shape of NPs	Size of NPs (nm)	References
Rhodophyta	*Galaxar arugosa*	Au	3.85–77.13	Abdel-Raouf et al. (2017)
	Gelidiella acerosa	Ag (spherical)	22	Marimuthu et al. (2011)
	Gracilaria corticata	Au	45–57	Naveena and Prakash (2013)
	Gracilaria dura	Ag (spherical)	6	Shukla et al. (2012)
	Kappaphycus alvarezii	Au (spherical)	12.5–40	Rajasulochana et al. (2010)
	Kappaphycus alvarezii	Ag	73	Ganesan et al. (2013)
	Lemanea fluviatilis	Au	5–15	Sharma et al. (2014)

morphology of the NPs is also characterized using field emission scanning electron microscope (FESEM), and the qualitative and quantitative elemental composition of the sample is examined using energy-dispersive X-ray spectroscopy. The high-resolution transmission electron microscope (HRTEM) is used to determine the accurate size and shape of the NPs. The crystalline structure of the synthesized NPs and their atomic coordinates in the crystalline lattice planes is obtained through selected area electron diffraction (SAED).

6.3 Applications of Seaweed Nanoparticles

The antibacterial, antifungal, anticancer, and wound-healing properties of algal NPs are making them more popular and important in medico-biological applications as summarized in Fig. 6.4.

6.3.1 Antimicrobial Activity

The synthesized NPs from algae are well known for their potent antibacterial activity. The mechanism with a graphical representation is shown in Fig. 6.5. The copper oxide NPs (5–45 nm) synthesized using brown alga *Bifurcaria bifurcata* showed better antibacterial activity against Gram-negative *Enterobacter aerogenes* and Gram-positive bacteria *S. aureus* (Senapati et al. 2012). The gold NPs synthesized

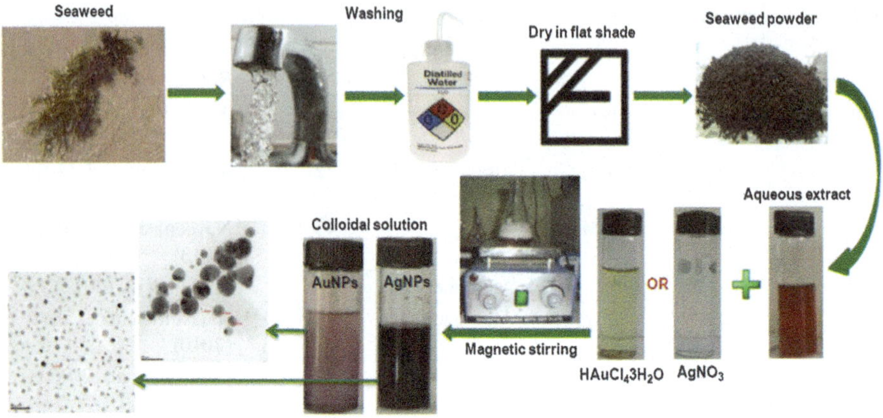

Fig. 6.3 Preparation of aqueous seaweed extract and biosynthesis of Ag and AuNPs. (Adapted from Sri Ramkumar et al. 2016)

from *Galaxaura elongata* powder were screened for their antibacterial activity against *E. coli*, *K. pneumoniae*, MRSA, *S. aureus,* and *Pseudomonas aeruginosa* (Abdel-Raouf et al. 2017). Silver chloride (AgCl) NPs synthesized using *Sargassum plagiophyllum* were analyzed using fluorescence, and electron microscopy showed bactericidal activity against *E. coli* (Dhas et al. 2014). AgNPs synthesized using *Caulerpa racemosa*, marine algae, exhibited antibacterial activity against human pathogens such as *S. aureus* and *Proteus mirabilis* (Kathiraven et al. 2015). The silver NPs obtained from *Ulva fasciata* reported having a bactericidal effect against *Xanthomonas campestris* pv. *malvacearum* (Rajesh et al. 2012). The silver NPs synthesized from *Gracilaria dura* were found effective against *B. pumilus* (HQ318731) (Shukla et al. 2012) (Fig. 6.5).

6.3.2 Antifungal Activity

The algal-synthesized NPs have potent and efficient antifungal applications. Not much work has been carried out in this aspect. The notable work includes the synthesis of AgNPs using the aqueous extract of red seaweed *Gelidiella acerosa* as the reducing agent exhibited antifungal property against *Humicola insolens* (MTCC 4520), *Fusarium dimerum* (MTCC 6583), *Mucor indicus* (MTCC 3318), and *Trichoderma reesei* (MTCC 3929) (Marimuthu et al. 2011). In another report, the effect of the algal (*Sargassum longifolium*)-mediated AgNPs against the pathogenic fungi *Aspergillus fumigatus*, *Candida albicans*, and *Fusarium* sp. was determined (Rajeshkumar et al. 2014).

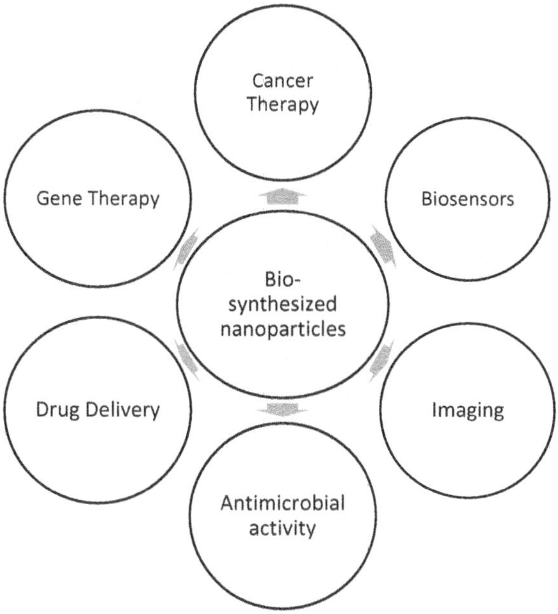

Fig. 6.4 Applications of biosynthesized algal nanoparticles

6.3.3 Anticancer Activity

The synthesized chitosan-coated silver nano-triangles (Chit-AgNPs) were used as a photothermal agent against a line of human nonsmall lung cancer cells (NCI-H460) (Boca et al. 2011). In another work, AgNPs (10 nm) were synthesized using *Sargassum vulgare*, and its ability to kill cancerous human myeloblastic leukemic cells HL60 and cervical cancer cells HeLa was tested (Govindaraju et al. 2015).

6.4 Future Applications

The emerging area of nanoscience and technology will be the use of NPs as antibiofilm agents. With the extensive use and misuse of antibiotics, many of the pathogens acquired resistance toward multiple drugs. As the bacteria are less likely to develop resistance against NPs, they can be used as a promising therapeutic agent against biofilms. The production of lipid NPs from lipid-rich seaweeds is possible. The lipid NPs can be synthesized through heating to liquefy fatty acids, incorporating active agents of pharmacological and cosmetics importance, adding a hot surfactant, and stirring or by homogenizing under high pressure by ultrasound. These lipid NPs can widely be used in foodstuffs, cosmetics, and medicines. Algal-synthesized NPs can be explored in biosensing applications. AuNPs have been proved as an important tool for hormone (HCG) detection in a pregnant woman's urine sample (Kuppusamy et al. 2014). Platinum (Pt) NPs act as a novel biosensor

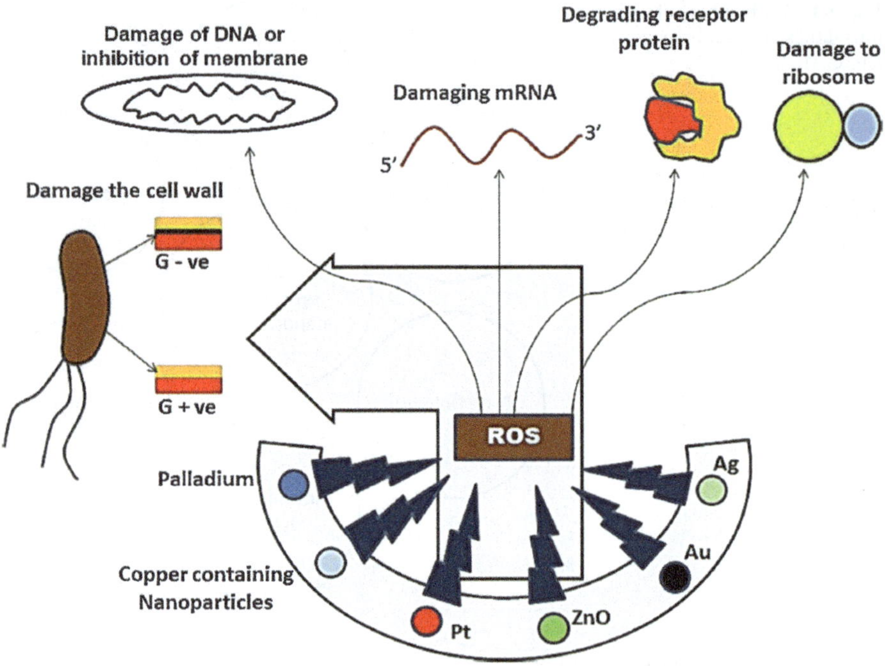

Fig. 6.5 Different nanoparticles and their mode of inhibition against bacteria. (Adapted from LewisOscar et al. 2014)

with high sensitivity for the determination of adrenaline for the treatment of allergies, heart attack, asthma, and cardiac surgery (Brondani et al. 2009).

6.5 Conclusion

Algae (micro and macro) are being extensively used in the field of nanotechnology. In the last 7–8 years, algal nanoscience is more focused on as an emerging important green nanotechnology field. Algae as a promising biological tool for the production of metallic NPs and algae-mediated synthetic metal NPs have proven to be a versatile material along with its diverse potential applications in various medico-biological fields especially in clinical diagnostics, biotechnology, agriculture, and other areas like paint, cosmetics, packaging, and coating. Despite numerous studies conducted over the last decade, there are still considerable gaps in our knowledge about the biotechnological potential of green-synthesized NPs. Furthermore, the precise basis of their antibiotic and antibiofilm activity has yet to be defined. Future research on the marine organisms-mediated synthesis of nanomaterials with their unique properties is of great importance for applications in the areas of pharmaceutical chemistry, biomedical sciences, and environmental sciences.

References

Abboud Y, Saffaj T, Chagraoui A et al (2014) Biosynthesis, characterization and antimicrobial activity of copper oxide nanoparticles (CONPs) produced using brown alga extract (*Bifurcaria bifurcata*). Appl Nanosci 4:571–576

Abdel-Raouf N, Al-Enazi NM, Ibraheem IBM (2017) Green biosynthesis of gold nanoparticles using *Galaxaura elongata* and characterization of their antibacterial activity. Arab J Chem 10(2):S3029–S3039

Aziz N, Fatma T, Varma A, Prasad R (2014) Biogenic synthesis of silver nanoparticles using Scenedesmus abundans and evaluation of their antibacterial activity. Journal of Nanoparticles, Article ID 689419, 2014. https://doi.org/10.1155/2014/689419

Aziz N, Faraz M, Pandey R, Sakir M, Fatma T, Varma A, Barman I, Prasad R (2015) Facile algae-derived route to biogenic silver nanoparticles: Synthesis, antibacterial and photocatalytic properties. Langmuir 31:11605–11612. https://doi.org/10.1021/acs.langmuir.5b03081

Aziz N, Prasad R, Ibrahim AIM, Ahmed AIS (2017) Promising applications for the production of biofuels through algae. In: Microbial Biotechnology (Patra JK, Vishnuprasad CN, Das G), Springer Nature Singapore Pte Ltd. 81–103

Azizi S, Namvar F, Mahdavi M et al (2013) Biosynthesis of silver nanoparticles using brown marine macroalga, *Sargassum muticum* aqueous extract. Materials 6:5942–5950

Azizi S, Ahmad MB, Namvar F, Mohamad R (2014) Green biosynthesis and characterization of zinc oxide nanoparticles using brown marine macroalga *Sargassum muticum* aqueous extract. Mater Lett 116:275–277

Boca SC, Potara M, Gabudean AM et al (2011) Chitosan coated triangular silver nanoparticles as a novel class of biocompatible, highly effective photothermal transducers for in vitro cancer cell therapy. Cancer Lett 311:131–140

Brondani D, Scheeren CW, Dupont J, Vieira IC (2009) Biosensor based on platinum nanoparticles dispersed in ionic liquid and laccase for determination of adrenaline. Sens Actuat B Chem 140(1):252–259

Buzea C, Pacheco II, Robbie K (2007) Nanomaterials and nanoparticles: sources and toxicity. Biointerphases 2(4):17–71

Davis SA, Patel HM, Mayes EL et al (1998) Brittle bacteria: a biomimetic approach to the formation of fibrous composite materials. Chem Mater 10:2516–2524

Davis TA, Volesky B, Mucci A (2003) A review of the biochemistry of heavy metal biosorption by brown algae. Water Res 37:4311–4330

Dhanalakshmi PK, Azeez R, Rekha R et al (2012) Synthesis of silver nanoparticles using green and brown seaweeds. Phykos 42:39–45

Dhas TS, Kumar VG, Karthick V et al (2014) Facile synthesis of silver chloride nanoparticles using marine alga and its antibacterial efficacy. Spectrochim Acta A 120:416–420

Focsan M, Ardelean II, Craciun C, Astilean S (2011) Interplay nanoparticle biosynthesis and metabolic activity of cyanobacterium *Synechocystis* sp. PCC 6803 between gold. Nanotechnology 22:1–8

Ganesan V, Aruna Devi J, Astalakshmi A et al (2013) Eco-friendly synthesis of silver nanoparticles using a sea weed, *Kappaphycus alvarezii* (Doty) Doty ex P.C.Silva. Int J Eng Adv Tech 2:559–563

Ghodake G, Lee DS (2011) Biological synthesis of gold nanoparticles using the aqueous extract of the brown algae *Laminaria japonica*. J Nanoelectron Optoelectron 6:268–271

Govindaraju K, Krishnamoorthy K, Alsagaby SA et al (2015) Green synthesis of silver nanoparticles for selective toxicity towards cancer cells. IET Nanobiotechnol 9(6):325–330

Iravani S (2011) Green synthesis of metal nanoparticles using plants. Green Chem 13:2638–2650

Jain A, Duvvuri LS, Farah S et al (2014) Antimicrobial polymers. Adv Healthc Mater 3:1969–1985

Jena J, Pradhan N, Dash BP et al (2013) Biosynthesis and characterization of silver nanoparticles using microalga *Chlorococcum humicola* and its antibacterial activity. Int J Nanomater Bios 3:1–8
Jena J, Pradhan N, Aishvarya V et al (2014) Biological sequestration and retention of cadmium as CdS nanoparticles by the microalga *Scenedesmus*-24. J Appl Phycol 27(6):2251–2260
Kannan RRR, Stirk WA, Staden JV (2013) Synthesis of silver nanoparticles using the seaweed *Codium capitatum* P.C. Silva (Chlorophyceae). S Afr J Sci Bot 86:1–4
Kathiraven T, Sundaramanickam A, Shanmugam N, Balasubramanian T (2015) Green synthesis of silver nanoparticles using marine algae *Caulerpa racemosa* and their antibacterial activity against some human pathogens. Appl Nanosci 5:499–504
Kim JS, Kuk E, Yu KN et al (2007) Antimicrobial effects of silver nanoparticles. Nanomed Nanotechnol Biomed 3:95–101
Krishnan M, Sivanandham V, Hans-Uwe D et al (2015) Antifouling assessments on biogenic nanoparticles: a field study from polluted offshore platform. Mar Pollut Bull 101:816–825
Kuppusamy P, Mashitah MY, Maniam GP, Govindan N (2014) Biosynthesized gold nanoparticle developed as a tool for detection of HCG hormone in pregnant women urine sample. Asian Pac J Trop Dis 4(30):237
Loomba L, Scarabelli T (2013) Metallic nanoparticles and their medicinal potential. Part II: aluminosilicates, nanobiomagnets, quantum dots and cochleates. Ther Deliv 4(9):1179–1196
Mahdavi M, Namvar F, Ahmad MB, Mohamad R (2013) Green biosynthesis and characterization of magnetic iron oxide (Fe_3O_4) nanoparticles using seaweed (*Sargassum muticum*) aqueous extract. Molecules 18:5954–5964
Marimuthu V, Palanisamy SK, Sesurajan S, Sellappa S (2011) Biogenic silver nanoparticles by *Gelidiella acerosa* extract and their antifungal effects. Avicenna J Med Biotechnol 3:143–148
Mata YN, Torres E, Blazquez ML et al (2009) Gold(III) biosorption and bioreduction with the brown alga *Fucus vesiculosus*. J Hazard Mater 166:612–618
Mohandass C, Vijayaraj AS, Rajasabapathy R et al (2013) Biosynthesis of silver nanoparticles from marine seaweed *Sargassum cinereum* and their antibacterial activity. Indian J Pharm Sci 75:606–610
Mohseniazar M, Barin M, Zarredar H et al (2011) Potential of microalgae and lactobacilli in biosynthesis of silver nanoparticles. Bioimpacts 1:149–152
Mukherjee P, Ahmad A, Mandal D et al (2001) Fungus mediated synthesis of silver nanoparticles and their immobilization in the mycelial matrix: a novel biological approach to nanoparticle synthesis. Nano Lett 1:515–519
Naveena BE, Prakash S (2013) Biological synthesis of gold nanoparticles using marine algae *Gracilaria corticata* and its application as a potent antimicrobial and antioxidant agent. Asian J Pharm Clin Res 6:179–182
Parial D, Pal R (2015) Biosynthesis of monodisperse gold nanoparticles by green alga *Rhizoclonium* and associated biochemical changes. J Appl Phycol 27:975–984
Parial D, Patra HK, Dasgupta AK, Pal R (2012) Screening of different algae for green synthesis of gold nanoparticles. Eur J Phycol 47:22–29
Prasad R (2014) Synthesis of silver nanoparticles in photosynthetic plants. Journal of Nanoparticles, Article ID 963961, 2014. https://doi.org/10.1155/2014/963961
Prasad R, Kumar V, Prasad KS (2014) Nanotechnology in sustainable agriculture: present concerns and future aspects. Afr J Biotechnol 13(6):705–713
Prasad R, Pandey R, Barman I (2016) Engineering tailored nanoparticles with microbes: quo vadis. WIREs Nanomed Nanobiotechnol 8:316–330. https://doi.org/10.1002/wnan.1363
Prasad R, Bhattacharyya A, Nguyen QD (2017) Nanotechnology in sustainable agriculture: Recent developments, challenges, and perspectives. Front Microbiol 8:1014. https://doi.org/10.3389/fmicb.2017.01014
Prasad R, Jha A, Prasad K (2018) Exploring the Realms of Nature for Nanosynthesis. Springer International Publishing (ISBN 978-3-319-99570-0) https://www.springer.com/978-3-319-99570-0

Rajasulochana P, Dhamotharan R, Murugakoothan P et al (2010) Biosynthesis and characterization of gold nanoparticles using the alga *Kappaphycus alvarezii*. Int J Nanosci 9:511–516

Rajesh S, Raja DP, Rathi JM, Sahayaraj K (2012) Biosynthesis of silver nanoparticles using *Ulva fasciata* (Delile) ethyl acetate extract and its activity against *Xanthomonas campestris* pv. *malvacearum*. J Biopest 5:119–128

Rajeshkumar S, Malarkodi C, Vanaja M et al (2013) Antibacterial activity of algae mediated synthesis of gold nanoparticles from *Turbinaria conoides*. Der Pharma Chemica 5:224–229

Rajeshkumar S, Malarkodi C, Paulkumar K et al (2014) Algae mediated green fabrication of silver nanoparticles and examination of its antifungal activity against clinical pathogens. Int J Met. https://doi.org/10.1155/2014/692643

Rauwel P, Kuunal S, Ferdov S, Rauwel E (2015) A review on the green synthesis of silver nanoparticles and their morphologies studied via TEM. Adv Mater Sci Eng 682749:9

Rawat J, Gupta PK, Pandit S, Prasad R, Pande V (2021) Current perspectives on integrated approaches to enhance lipid accumulation in microalgae. 3 Biotech 11, 303. https://doi.org/10.1007/s13205-021-02851-3

Sahayaraj K, Rajesh S, Rathi JM (2012) Silver nanoparticles biosynthesis using marine alga *Padina pavonica* (Linn.) and its microbicidal activity. Dig J Nanomater Biostruct 7:1557–1567

Sangeetha N, Manikandan S, Singh M, Kumaraguru AK (2012) Biosynthesis and characterization of silver nanoparticles using freshly extracted sodium alginate from the seaweed *Padina tetrastromatica* of Gulf of Mannar, India. Curr Nanosci 8:697–702

Sankar R, Karthik A, Prabu A et al (2013) *Origanum vulgare* mediated biosynthesis of silver nanoparticles for its antibacterial and anticancer activity. Colloids Surf B Biointerfaces 108:80–84

Senapati S, Syed A, Moeez S et al (2012) Intracellular synthesis of gold nanoparticles using alga *Tetraselmis kochinensis*. Mater Lett 79:116–118

Shakibaie M, Forootanfar H, Mollazadeh-Moghaddam K et al (2010) Green synthesis of gold nanoparticles by the marine microalga *Tetraselmis suecica*. Biotechnol Appl Biochem 57:71–75

Sharma B, Purkayastha DD, Hazra S et al (2014) Biosynthesis of fluorescent gold nanoparticles using an edible freshwater red alga, *Lemanea fluviatilis* (L.) C.Ag. and antioxidant activity of biomatrix loaded nanoparticles. Bioprocess Biosyst Eng 37:2559–2565

Sharma B, Purkayastha DD, Hazra S et al (2015a) Biosynthesis of gold nanoparticles using a fresh water green alga, *Prasiola crispa*. Mater Lett 116:94–97

Sharma A, Sharma S, Sharma K et al (2015b) Algae as crucial organisms in advancing nanotechnology: a systematic review. J Appl Phycol 28:1759–1774

Shiny PJ, Mukherjee A, Chandrasekaran N (2013) Marine algaemediated synthesis of the silver nanoparticles and its antibacterial efficiency. Int J Pharm Sci 5:239–241

Shukla MK, Singh RP, Reddy CRK, Jha B (2012) Synthesis and characterization of agar-based silver nanoparticles and nanocomposite film with antibacterial applications. Bioresour Technol 107:295–300

Sicard C, Brayner R, Margueritat J et al (2010) Nanogold biosynthesis by silica-encapsulated micro-algae: a "living" biohybrid material. J Mater Chem 20:9342–9347

Singaravelu G, Arockiamary JS, Kumar VG, Govindaraju K (2007) A novel extracellular synthesis of monodisperse gold nanoparticles using marine alga, *Sargassum wightii* Greville. Colloids Surf B Biointerfaces 57:97–101

Singh M, Kalaivani R, Manikandan S et al (2013) Facile green synthesis of variable metallic gold nanoparticle using *Padina gymnospora*, a brown marine macroalga. Appl Nanosci 3:145–151

Sri Ramkumar V, Prakash S, Ramasubburayan R et al (2016) Seaweeds: a resource for marine bionanotechnology. Enzym Microb Technol 95:45–57

Srivastava S, Usmani Z, Atanasov AG, Singh VK, Singh NP, Abdel-Azeem AM, Prasad R, Gupta G, Sharma M, Bhargava A (2021) Biological nanofactories: Using living forms for metal nanoparticle synthesis. Mini-Reviews in Medicinal Chemistry 21(2):245–265

Vijayan SR, Santhiyagu P, Singamuthu M et al (2014) Synthesis and characterization of silver and gold nanoparticles using aqueous extract of seaweed, *Turbinaria conoides*, and their antimicrofouling activity. Sci World J. https://doi.org/10.1155/2014/938272

LewisOscar F, Bakkiyaraj D, Nithya C, Thajuddin N (2014). Deciphering the diversity of microalgal bloom in wastewater—an attempt to construct potential consortia for bioremediation. JCPAM 3 (2):92–96

Prasad, Tollamadugu N V K V & Kambala, Venkata & Naidu, Ravi. (2013). Phyconanotechnology: synthesis of silver nanoparticles using brown marine algae *Cystophora moniliformis* and their characterisation. J App Phyc. 25:177–182. https://doi.org/10.1007/s10811-012-9851-z

Chapter 7
Pelagic *Sargassum* as Source of Quantum Dots

Héctor Uriel Castañeda-Serna, Georgina Calderón-Domínguez, Ma. De la Paz Salgado-Cruz, Arturo García-Bórquez, and Reynold Ramón Farrera-Rebollo

Contents

7.1 Introduction

Since 2014, unusual pelagic *Sargassum* beaching events have occurred in the coasts of different countries of the Caribbean Sea and the Atlantic Ocean, becoming an important environmental issue (Rodríguez-Martínez et al. 2020; Wang et al. 2019b); this has caught the attention of many scientists around the world, with different

H. U. Castañeda-Serna · G. Calderón-Domínguez (✉) · R. R. Farrera-Rebollo
Instituto Politécnico Nacional, Escuela Nacional de Ciencias Biológicas, Departamento de Ingeniería Bioquímica, Ciudad de México, Mexico
e-mail: gcalderon@ipn.mx

M. De la Paz Salgado-Cruz
Instituto Politécnico Nacional, Escuela Nacional de Ciencias Biológicas, Departamento de Ingeniería Bioquímica, Ciudad de México, Mexico

Consejo Nacional de Ciencia y Tecnología (CONACYT), Ciudad de México, Mexico

A. García-Bórquez
Instituto Politécnico Nacional, Escuela Superior de Físico Matemáticas, Ciudad de México, Mexico

D. Thangadurai et al. (eds.), *Bioprospecting Algae for Nanosized Materials*, Nanotechnology in the Life Sciences,
https://doi.org/10.1007/978-3-030-81557-8_7

approaches being proposed for its solution. Some of these proposals take into account the chemical composition of this type of algae, while others focus on its soluble and insoluble components: proteins, carbohydrates, and pigments, as a source of novel high-value raw materials such as carbon nanoparticles known as carbon quantum dots (CQDs) (Milledge et al. 2020; Oyesiku and Egunyomi 2014; Ríos 2012).

These types of nanoparticles have been obtained from diverse organic sources, including plants (Arumugham et al. 2020; Vandarkuzhali et al. 2017; Wang et al. 2019a), organic residues (Atchudan et al. 2021; Rodríguez-Padrón et al. 2018), and different seaweed species, such as *Ulva lactuca* (Singh et al. 2018), *Enteromorpha prolifera* (Xu et al. 2017), *Chlorella pyrenoidosa, Scenedesmus obliquus* (Zhang et al. 2018), and pelagic *Sargassum* (Godavarthi et al. 2017).

Nanoparticles are produced by different methods, with the most common being hydrothermal or microwave irradiation processes (Guan et al. 2014; Mehta et al. 2014; Singh et al. 2018; Xu et al. 2017; Zhang et al. 2018; Zhu et al. 2012). These are considered ecofriendly methods, due to their low energy costs and not requiring the application of corrosive chemicals (Namdari et al. 2017).

The usage of these nanoparticles as simple molecules or as complex biological systems has been proposed in medicine to act as a sensor for relevant molecules or as an agent for the delivery of drugs or genes to target cells (Dong et al. 2013; Lai et al. 2016; Liu et al. 2008; Sarkar et al. 2017). In the environmental sciences field, the usage of nanoparticles has been related to the application of CQDs to detect metallic ions during the treatment of wastewater (Fan et al. 2020; Gonçalves et al. 2010; Wang et al. 2019a), while in biological sciences they have been used to stain cells or DNA (Atchudan et al. 2021; Godavarthi et al. 2017; Lai et al. 2016). Due to the non-cytotoxic effects of CQDs, their use has become more common every day.

This chapter presents basic and recent information about the different approaches used to convert pelagic *Sargassum* into useful materials at the macro, micro, and nanolevels for technological applications, paying special attention to ecofriendly extraction processes and focusing on proteins, carbohydrates, and pigments (Gao et al. 2018; Geetha et al. 2015; Godavarthi et al. 2017; Huang and Lam 2011).

7.2 Pelagic *Sargassum* Classification

Marine algae (seaweed) are a highly diverse group (approximately 20,000 species) of photosynthetic autotrophic organisms, with little to no cell differentiation, complex tissues, and a simple structure. They are thallophytes, including micro and macroalgae (Levine 2016; Norton et al. 1996). The latter are taxonomically classified into three groups: Chlorophyta, Phaeophyta, and Rhodophyta (Coppejans et al. 2001), corresponding to green, brown, and red algae, respectively, due to the predominant pigments that each one of them present (Miyashita et al. 2013).

The genus *Sargassum*, which includes 150 species, is a type of brown algae, tropical and sub-tropical, characteristic of the Sargasso Sea, a section of the North

Atlantic Ocean, ranging between 20° and 40° N and 30° and 75° W (Gower and King 2011; Huffard et al. 2014; Olabarria et al. 2009) approximately in subtidal and intertidal areas. It is commonly found in the surface waters of the Sargasso Sea and the northwestern Gulf of Mexico (GLFMEX); these are areas where ocean eddies pull, retain and consolidate surface drifters (Franks et al. 2016). Because of these characteristics, they are referred to as pelagic *Sargassum* and it is estimated that they represent approximately 7–10 × 10^6 tons of biomass (Lapointe et al. 2014).

The pelagic species of *Sargassum*, which have predominated in the algal collections of the Sargasso Sea, the Caribbean Sea, and the Gulf of Mexico (Lapointe et al. 2014) are: *S. natans* and *S. fluitans* (Fine 1970; Huffard et al. 2014; Niermann 1986; Parr 1939), and have been studied since the 1830s (Gower and King 2011). Both species are characterized by having numerous blades, a highly branched thallus, and air bladders. While their morphological differences are spines in the air bladder and the long-stemmed bladder of *S. natans*, *S. fluitans* does not present them. However, the latter has a short stem with a surrounding wing tissue (Oyesiku and Egunyomi 2014). These species coexist, and together, they create a floating habitat on the ocean surface; these mats often form linear "weed lines" that accumulate at convergence zones in parallel rows as a result, due to the movement of the wind, known as "Langmuir circulation" (Lapointe et al. 2014). Figure 7.1 shows a dried sample of pelagic *Sargassum*.

Due to wind and surface currents, pelagic *Sargassum* can be transported across the western North Atlantic Ocean and link the population of the Sargasso Sea with the Caribbean Sea and the Gulf of Mexico (Lapointe et al. 2014). In recent years, there has been an unusual bloom of *Sargassum* in the Caribbean, which could be considered as having a negative impact. In the summer of 2011, the flow of pelagic *Sargassum* increased in the Caribbean and Africa (Rodríguez-Martínez et al. 2017), in subsequent years, it spread over the Atlantic Ocean, triggering environmental problems, interfering in the seaward journeys of juvenile turtles, affecting the sea turtle nests, and altering the trophic structure of the sea urchin, *Diadema antillarum*, in coastal marine systems (Rodríguez-Martínez et al. 2020).

The population of these brown pelagic algae has grown in the latest years as a mix of *S. natans* and *S. fluitans*, having a negative impact on fishing activity and tourism in different countries close to the Caribean Sea (Milledge et al. 2020; Rodríguez-Martínez et al. 2017; Rodríguez-Martínez et al. 2020). Moreover, the

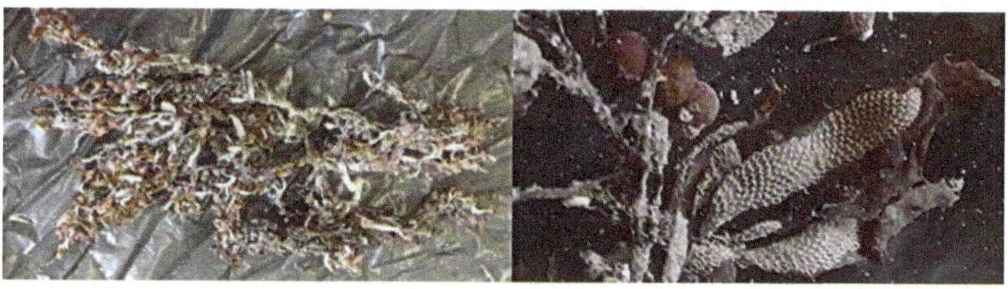

Fig. 7.1 Sample of pelagic *Sargassum* recollected in Mahahual beaches

problem has grown to the point that scientists describe it as a new phenomenon of colossal dimensions, acquiring its own identity as "The New Sargasso Sea" (Anderson et al. 2019) or the "Great Atlantic Sargassum Belt" (Wang et al. 2019b). Therefore, it is necessary to review the research that currently prevails in the possible uses and exploitation of algae biomass.

7.3 Pelagic *Sargassum* Chemical Composition, Compounds of Interest and Applications

Different types of seaweed have been used for multiple purposes, mainly as a direct source of food for human consumption, medicine, forage, as fertilizers (compost or soil improvement), and for the extraction of phycocolloids, generally alginates, agar-agar, and carrageenans. They have also been used in methane gas production, paper industry, cosmetics, and medical research (Cruz-Suárez et al. 2010; Maia et al. 2016; Peña-Rodríguez et al. 2011). However, these applications depend on the chemical composition of the algae. Tropical macroalgae contain different types of natural compounds, for example polysaccharides, which can be present in a range between 4% and76%, and are located in the cell wall, a structure that is composed of cellulose and hemicellulose in a 2–10% proportion and where the most abundant components are non-starch carbohydrates, such as carrageenan and alginate (Pal et al. 2014; Stiger-Pouvreau and Zubia 2020). Other molecules of importance are proteins (e.g., mycosporines such as amino acids-porphyra-334 and taurine or peptides-Kahalalide F), lipids (e.g., turbinaric acid, sargahydroquinoic acid- meroterpene, and palmitic acid), phenolic compounds (e.g., diphlorethol, 2,4-bromophenol, quercetin, coumarin, and triphlorethol A), pigments (e.g., phycoerythrobilin, fucoxanthin, and caulerpenyn), and others (Stiger-Pouvreau and Zubia 2020).

The composition of *S. fluitans*, *S. natans*, and mixed *Sargassum* species, reported by different authors, is shown in Table 7.1. Regarding proteins, some authors mention that the content is variable (3.25–15.4%) and describe that it changes according to the time of year, with the protein content being lower from July to October and

Table 7.1 Nutrients content on a dry basis in some pelagic *Sargassum* species

Substance	*S. fluitans*[a]	*S. fluitans*[b]	*S. natans VIII*[b]	*S. natans I*[b]	*Sargassum* mix[b]	*S. natans* and *S. fluitans*[c]
Carbohydrate	67.92	27.40	21.76	18.97	11.68	57.3
Protein	7.61	3.25	2.99	3.81	4.19	15.4
Lipids	0.26	4.56	3.58	4.51	3.88	2.5
Ash	12.6	33.63	34.26	35.71	33.31	8.65
Fiber	12.15	31.15	37.41	37.00	46.94	7.15

[a]Ríos (2012)
[b]Milledge et al. 2020
[c]Oyesiku and Egunyomi 2014

higher from November to May (Gorham and Lewey 1984; Oyesiku and Egunyomi 2014; Peng et al. 2015). However, these studies followed different methodologies to analyze the samples: Ríos (2012) and Oyesiku and Egunyomi (2014) employed Kjeldahl analysis, while Milledge et al. (2020) calculated the protein amount as the sum of amino acids quantified by ion-exchange chromatography. Milledge et al. (2018) cited that the recommended nitrogen-protein conversion factor for brown algae is 4.11 instead of 6.25.

Regarding lipids, the content is relatively low (0.26–4.56%, Table 7.1) (Maeda et al. 2008), since they are affected by the salinity of seawater, which increases the content of salts and decreases when they are in low salinity conditions (Peng et al. 2015).

In the case of fiber, Ríos (2012) and Oyesiku and Egunyomi (2014) reported crude fiber, while Milledge et al. (2020) obtained total dietary fiber. The difference between these methods is that crude fiber considers only cellulose, hemicellulose, lignin, and other types of non-soluble polysaccharides, while total dietary fiber takes into account the soluble and non-soluble carbohydrates as alginate (Dai and Chau 2017). Hence, these variations in fiber values will be noticed in the carbohydrate content results, which are calculated by difference.

Specifically, Rhein-Knudsen et al. (2017) reported that *S. natans* contain: alginate, cellulose, and fucoidan, by the presence of 16% of guluronic acid, 14% of mannuronic acid, 7% of glucose, 6% of fucose, and 6% of other monosaccharides as well as rhamnose, arabinose, galactose, xylose, mannose, and glucuronic acid (Brown 2003; Li et al. 2008; Pal et al. 2014).

About ashes, the ratio of nitrogen, phosphorus, and potassium is relatively small (1:10:3), and for this reason, seaweed has been proposed to be used as fertilizers for lawns and seedling establishment (Oyesiku and Egunyomi 2014). However, due to the high concentration of alkaline metals (Na, Ca, Mg, and K), these could salinize the soil (Fernández et al. 2017). Conversely, *Sargassum* seaweed does not have considerable high values of As, Cd, Cu, Mo, Ni, Pb, Se, and Zn, neither toxic elements such as Be, Cr, Co, and Hg (Fernández et al. 2017).

Other natural components from the Rhodophyceae, Phaeophyceae, and Chlorophyceae macroalgae classes have been isolated (Deepak et al. 2018; Milledge and Harvey 2016; Paula et al. 2011; Zubia et al. 2008). Among these components, phenolic compounds, essential amino acids, polyunsaturated fatty acids (Milledge et al. 2020; Oyesiku and Egunyomi 2014), mannitol and phlorotannins (Thompson et al. 2020; Vivekanand et al. 2011), sulfated polysaccharide, phenolics, plastoquinone, phlorotannins, fucoxanthin, fucoidan, sargaquinoic acid, sargachromenol, steroids, terpenoids, and flavonoids (Milledge and Harvey 2016; Rhein-Knudsen et al. 2017) have been reported.

Therefore, although there are important advances in the extraction and analysis of the compounds present in brown algae, interest has currently arisen in nanoparticles, since they are considered important because they exhibit higher surface to volume ratio.

7.4 Extraction of Nanoparticles from Pelagic *Sargassum*

Nanoparticles can be synthesized from various types of materials, such as carbohydrate polymers, which is why most *Sargassum* compounds, like alginate, cellulose, and fucoidan, could be used to obtain these nanoparticles (Elsabahy and Wooley 2012).

Other compounds present in brown algae that have been used for the synthesis of nanoparticles are antioxidants such as polyphenols, especially phlorotannins that are the largest group (Miyashita et al. 2013). Additionally, pelagic *Sargassum* has relatively high content of flavonoids (750 mg/100 g) as well saponins (525.01 mg/100 g) and tannins (122.5 mg/100 g) (Oyesiku and Egunyomi 2014), chlorophyll a, c, and fucoxanthin (Pal et al. 2014).

These pigments and antioxidants could be a source to prepare carbon quantum dots. The advantage of using these molecules is that they are water soluble and easily extracted from algae (Namdari et al. 2017), which could be related to an eco-friendly extraction process. In addition, the protein content in pelagic *Sargassum* could provide the components required to develop nitrogen-doped carbon quantum dots (NCQDs) (Godavarthi et al. 2017).

In this sense, the green synthesis of carbon-based materials from biomass sources, specifically from marine algae, is a recent and interesting research area due to its renewable nature, abundance, and chemical and morphological diversity.

7.4.1 Green Synthesis of Carbon Quantum Dots (CQDs)

Quantum dots are semiconductor nanoparticles between 1 and 12 nm of diameter, have size-selective photoluminescence due to their quantum confinement, and are constructed not only out of metallic compounds such as CdS, CdSe, PbSe, or Ag_2S (Gonçalves et al. 2010; Li et al. 2011a; Liang et al. 2013), but also with carbon and nitrogen.

Quantum dots have applications in biomedical fields, biolabeling (Godavarthi et al. 2017; Lai et al. 2016), optical sensors (Li et al. 2011b; Tang et al. 2014), drug or gene delivery (Das et al. 2015; Sarkar et al. 2017), and bioimaging (Dong et al. 2013; Ray et al. 2009), due to their small size as well as biocompatibility (Namdari et al. 2017). In this regard, quantum dots (QD) are commonly used to visualize biological events in a non-invasive way (Liu et al. 2014), due to their unique and tunable photoluminescence properties (Cui et al. 2014; Pan et al. 2010; Wang et al. 2015; Wang et al. 2019a), easy functionalization, high photostability (Bera et al. 2010; Smith and Nie 2010), the capacity of being passivated with organic compounds (Ma et al. 2007; Wang et al. 2006), low environmental impact, and lower cytotoxic effects (Hardman 2006; Wang et al. 2019a), when compared with fluorophores like azo, cyanine, and imidazole, since these latter substances have limitations such as low solubility rates in aqueous media, low photostability, and high toxicity (Godavarthi et al. 2017).

The QDs employed in biological imaging are four types: semiconductor quantum dots (semiconductor QDs), carbon nanodots (CDs), silicon quantum dots (SiQDs), and polymer dots (Pdots) (Gao et al. 2018). Additionally, Namdari et al. (2017) mention graphene quantum dots (GQD) (N) and nitrogen-doped carbon quantum dots (NCQDs), which present a similar structure to graphene, but with the substitution of nitrogen atoms (Tang et al. 2014).

Carbon quantum dots (CQDs) have many applications in the fields of chemical sensing, biosensing, bioimaging, nanomedicine, photocatalysis, and electrocatalysis (Lim et al. 2014). The source of carbon to produce CQDs, their areas of application, and examples are shown in Table 7.2. This type of quantum dots was discovered in 2004 (Xu et al. 2004). Two years later, Sun's group defined them as Carbon Quantum Dots (CQD), carbon nanoparticles, or carbogenic quantum dots (Zhang and Yu 2016).

There are different carbon sources for the synthesis of CQDs, from nanocarbon materials, flake graphite, and soot (Gonçalves et al. 2010; Li et al. 2011b; Shi et al.

Table 7.2 Application examples of CQDs

Application	Carbon source	Example	References
Biolabeling	Ammonium citrate	Staining of *Escherichia coli* and tumor cells	Lai et al. (2016)
	S. fluitans	Staining of DNA	Godavarthi et al. (2017)
Bioimaging	Citric acid and L-cysteine	Labeling of HeLa cells	Dong et al. (2013)
	Diethylene glycol, sucrose, and H_2SO_4	Labeling of glioma cells	Liu et al. (2014)
	Banana peel waste	Labeling of nematodes cells	Atchudan et al. (2021)
Sensor	Carbon targets	Monitoring metallic ions (Hg, Cu, Cd, Ni, Zn, or Ca)	Gonçalves et al. (2010)
	Nanocarbon material	Monitoring nucleic acid	Li et al. (2011b)
	Candle soot	Monitoring glucose	Shi et al. (2011)
	Biomass residues	Fe^{3+} detection	Wang et al. (2019a)
	Sodium alginate and ethylendiamine	Cu^{2+} detection	Fan et al. (2020)
Drug delivery	Graphitic flake	Deliver the drug SN38 for colon cancer treatment	Liu et al. (2008)
	Aloe vera plant	Deliver the drug vancomycin to the gastrointestinal tract	Sarkar et al. (2017)
Gene delivery	Glycerol and polyethyleneimine	Deliver pDNA to HeLa and PC-3 cells	Kim et al. (2013)
	Polyethylene glycol and polyethyleneimine	Deliver dsRNA to mosquito genes (SNF7 and SRC)	Das et al. (2015)

2011) to the carbon union with different compounds as molecules (e.g., amino acids, organic acids, or carbohydrates), waste, or biological specimens (Atchudan et al. 2021; Fan et al. 2020; Liu et al. 2014; Wang et al. 2019a). In the latest case, *S. fluitans* is as an example of a species of pelagic *Sargassum* (Godavarthi et al. 2017) that shows the possibility of being used to elaborate carbon nanoparticles.

Carbon quantum dots (CQD) are a relatively new class of nanomaterial, with "Zero dimension" within the carbon family, they possess an excellent nonblinking luminescence, wide excitation spectrum, high dispersion in water, low toxicity for the biological specimens, and a fluorescence peak which is independent of excitation wavelength (Gonçalves et al. 2010; Liu et al. 2014; Zhang et al. 2016). They also present good conductivity, high chemical stability, low environmental impact, and low production cost, being considered better than metal quantum dots (Namdari et al. 2017). However, these particles exhibit relatively low quantum yield (QY), usually less than 50% (Dong et al. 2013).

CQDs are nanoparticles of semiconductor materials, quasi-spherical, with sizes below 10 nm and interplanar spacing between 0.18 nm and 0.24 nm (Zheng et al. 2015). They present an amorphous graphite structure, with a carbon core and a carbonaceous surface, that also contains oxygen (Deng et al. 2014), whereas the core bonds contain oxygen in the carboxyl, hydroxyl, and aldehyde groups (Zhang and Yu 2016). The CQD can be doped with nitrogen, nitrogen and sulfur, or phosphorus (NCQD, NSCQD, and PCQDs), through a process named surface passivation. This change of electronic states offers more active sites and increases their optical properties, such as the quantum yield and luminescence lifetime (Dong et al. 2013; Tang et al. 2014; Yang et al. 2014; Zhou et al. 2014). For example, using citric acid as a source of carbon and L-cysteine for nitrogen and sulfur, N-S CQDs with a QY of 73% can be obtained (Anilkumar et al. 2011; Dong et al. 2013).

CQD synthesis can be carried out through various methods, such as chemical ablation, laser ablation, electrochemical oxidation, thermic decomposition or oxidation (hydrothermal), and microwave pyrolysis (Liu et al. 2014). Nonetheless, the reported information to prepare CQDs from seaweed biomass mainly applies hydrothermal, microwave, and thermal processes (Table 7.3).

It is important to mention that microwave-assisted and solvothermal/hydrothermal CQD methods are considered ecofriendly routes to obtain CQDs (Godavarthi et al. 2017; Liang et al. 2013; Liu et al. 2014; Namdari et al. 2017; Vandarkuzhali et al. 2017; Xu et al. 2017) because the required reagents are not toxic and obtained from renewable sources. Also, the subproduct wastes are few to none, the experiments are carried out with miniaturized methods (usually 10–60 mL for reaction), and the treatment is not dangerous for the operator or the environment (Gałuszka et al. 2013; Gilbertson et al. 2015; Tang et al. 2005).

Table 7.3 Carbon sources, methods, and conditions used to obtain carbon quantum dots (CQDs)

Carbon source	Type of synthesis	Conditions	Type of quantum dot	References
Soy milk	Hydrothermal	180 °C, 3 h	NCQDs	Zhu et al. (2012)
Solanum tuberosum	Hydrothermal	170 °C, 12 h	CQDs	Mehta et al. (2014)
Folic acid and diethylene glycol	Microwave	750 W, 40 s	NCQDs	Guan et al. (2014)
Bovine serum albumin	Hydrothermal	195 °C, 6 h	NCQDs	Abu-Ghosh et al. (2017)
Microalgae	Hydrothermal	180 °C, 10 h	CQDs	Guo et al. (2017)
S. fluitans extract	Hydrothermal	180 °C, 5 h	NCQDs	Godavarthi et al. (2017)
D-glucose and ethane-1,2-diamine	Microwave	700 W, 4 min	NCQDs	Singh et al. (2017)
Pseudo stem of banana plant	Hydrothermal	180 °C, 2 h	CQDs	Vandarkuzhali et al. (2017)
Enteromorpha prolifera	Hydrothermal	180 °C, 3 h	NSCQDs	Xu et al. (2017)
Lignocellulose residue	Microwave	100 °C, 1.5 h	NCQDs	Rodríguez-Padrón et al. (2018)
Ulva lactuca	Thermal treatment with hot air	90 °C, 12 h	CQDs*	Singh et al. (2018)
Eichhornia crassipes	Hydrothermal	110, 140, 180 and 220 °C, 6 h	NCQDs	Yang et al. (2018)
C. pyrenoidosa and *S. obliquus*	Biosynthesis	12 days incubation with Na_2SeO_3 and $Cd(NO_3)_2$	Protein functionalizated CdSe QDs	Zhang et al. (2018)
Orange peel, *Ginkgo biloba* leaves, *Paulownia* leaves and *Magnolia* flower	Hydrothermal	200 °C, 8 h	NSCQDs	Wang et al. (2019a)
Catharanthus roseus leaves	Hydrothermal	200 °C, 4 h	CQDs	Arumugham et al. (2020)
Banana juice and ethanol	Hydrothermal	150 °C, 3 h	NSCQDs	Chaudhary et al. (2020)
Banana peel waste	Hydrothermal	200 °C, 24 h	CQDs	Atchudan et al. (2021)

*Although the synthesized particles were referred to as CQDs, the size is out of range than reported.

7.4.1.1 Microwave Irradiation Method

Microwave irradiation methodology is cheap and rapid. Besides, for laboratory applications, it is possible to use simple microwave ovens and biomolecules in a solution as raw material, biomolecules such as sucrose, guanidine hydrochloride, diamine tetraacetate, glucose with ammonia hydroxide, lysine, and others (Liu et al. 2014; Prekodravac et al. 2019; Tang et al. 2014). The particles sizes generated lie in an interval between 5 nm and 10 nm, although under power conditions of 100–200 W, the diameter is over 10 nm, the most commonly used potency is 750 W, with a time process of 4–6 min, and a resulting QY in a range of 2–35%. Such results often depend on the material being synthesized and the surface functionalization.

7.4.1.2 Solvothermal/Hydrothermal Method

This methodology is cheap, ecofriendly, and nontoxic. CQDs are fabricated using organic precursors in high-boiling solvents (e.g., gelatin, extract, algae biomass, or banana pseudostem); this mixture is placed in a sealed vessel which, while turning, is put in an oven heated up to 120–200 °C. The most common/most reported condition for this process is 180 °C (Godavarthi et al. 2017; Liang et al. 2013; Namdari et al. 2017; Vandarkuzhali et al. 2017; Xu et al. 2017).

From both CQD synthesis methods (Table 7.3), the hydrothermal method has been used with more complex carbon sources (e.g., marine algae, biomass, wastes, leaves, food, or extracts) requiring higher temperatures and longer reaction times than microwaving, where simpler raw materials and softer process conditions are used. Furthermore, the type of CQDs to be synthesized are related to the reactants. As an example, if nitrogen or sulfur atoms are used, they will link in the nanoparticle surface obtaining NCQDs or NSCQDs.

Regarding the use of complex carbon sources to produce CQDs, it has been reported (Godavarthi et al. 2017; Guo et al. 2017; Singh et al. 2018; Xu et al. 2017; Zhang et al. 2018) that some extracts from seaweed species such as *U. lactuca*, *E. prolifera*, *C. pyrenoidosa*, *S. obliquus*, or *S. fluitans*, can be used to synthesize these particles. However, pelagic *Sargassum* has not been used for this purpose, opening thus an opportunity for technological development due to the chemical composition of the algae and the beaching events occurring on the coasts of different countries making large quantities of algae available for harvesting.

Besides carbon nanodots, pelagic *Sargassum* can be used to produce other types of particles like cellulose, alginate, and fucoidan, which can be extracted from leaves or the secondary byproducts from CQDs synthesis. Considering this approach, the production of these compounds can provide the algae's integral employment, thus reducing their environmental impact.

7.5 Conclusions and Future Perspectives

The massive beaching events of pelagic *Sargassum* on Caribbean coasts, can represent an opportunity to develop new technological approaches, related to nanoscience, such as the production of CQDs with simple and low-cost technology for large-scale synthesis. Furthermore, these types of products could be applied in several novel studies for the further applications as low cost and non-toxic therapies or sensors. However, a profound studies of the pelagic *Sargassum* species is necessary to corroborate that these possess the chemical characteristics that allow the obtaining of nanoparticles with the best physicochemical characteristics or conjugating bioactive compounds to apply in imaging processes. Hereby, the use of pelagic *Sargassum* as a source of CQDs is a potential line of study.

Acknowledgment Héctor Uriel Castañeda Serna thanks to CONACyT and BEIFI-IPN for the scholarships provided and the financial support for this work. This research was funded through the projects: 20195500, 20210024, 20211381 and 20210918 from the Instituto Politécnico Nacional (IPN, Mexico) and 1668 from CONACyT.

References

Abu-Ghosh S, Kumar VB, Fixler D, Dubinsky Z, Gedanken A, Iluz D (2017) Nitrogen-doped carbon dots prepared from bovine serum albumin to enhance algal astaxanthin production. Algal Res 23:161–165. https://doi.org/10.1016/j.algal.2017.01.011

Anderson CR, Berdalet E, Kudela RM, Cusack CK, Silke J, O'Rourke E, Dugan D, McCammon M, Newton JA, Moore SK, Paige K, Ruberg S, Morrison JR, Kirkpatrick B, Hubbard K, Morell J (2019) Scaling up from regional case studies to a global harmful algal bloom observing system. Front Mar Sci 6:250. https://doi.org/10.3389/fmars.2019.00250

Anilkumar P, Wang X, Cao L, Sahu S, Liu JH, Wang P, Korch K, Tackett IIKN, Parenzan A, Sun YP (2011) Toward quantitatively fluorescent carbon-based "quantum" dots. Nanoscale 3(5):2023–2027. https://doi.org/10.1039/c0nr00962h

Arumugham T, Alagumuthu M, Amimodu RG, Munusamy S, Iyer SK (2020) A sustainable synthesis of green carbon quantum dot (CQD) from *Catharanthus roseus* (white flowering plant) leaves and investigation of its dual fluorescence responsive behavior in multi-ion detection and biological applications. Sustain Mater Technol 23:e00138. https://doi.org/10.1016/j.susmat.2019.e00138

Atchudan R, Jebakumar Immanuel Edison TN, Shanmugam M, Perumal S, Somanathan T, Lee YR (2021) Sustainable synthesis of carbon quantum dots from banana peel waste using hydrothermal process for in vivo bioimaging. Physica E 126:114417. https://doi.org/10.1016/j.physe.2020.114417

Bera D, Qian L, Tseng TK, Holloway PH (2010) Quantum dots and their multimodal applications: a review. Materials 3(4):2260–2345. https://doi.org/10.3390/ma3042260

Brown RM (2003) Cellulose structure and biosynthesis: what is in store for the 21st century? J Polym Sci A Polym Chem 42:487–495. https://doi.org/10.1002/pola.10877

Chaudhary N, Gupta PK, Eremin S, Solanki PR (2020) One-step green approach to synthesize highly fluorescent carbon quantum dots from banana juice for selective detection of copper ions. J Environ Chem Eng 8(3):103720. https://doi.org/10.1016/j.jece.2020.103720

Coppejans E, De Clerck O, Leliaert F, Dargent O (2001) Progress of the taxonomic research on the macroalgae (Chlorophyta, Phaeophyta and Rhodophyta) along the East African coast. In: Richmond MD, Francis J (eds) Marine science development in Tanzania and Eastern Africa. Proceedings of the 20th Anniversary Conference on Advances in Marine Science in Tanzania. 28 June – 1 July, 1999, 1st edn. Western Indian Ocean Marine Science Association (WIOMSA), Tanzania, pp 401–418

Cruz-Suárez LE, León A, Peña-Rodríguez A, Rodríguez-Peña G, Moll B, Ricque-Marie D (2010) Shrimp/*Ulva* co-culture: a sustainable alternative to diminish the need for artificial feed and improve shrimp quality. Aquaculture 301(1–4):64–68. https://doi.org/10.1016/j.aquaculture.2010.01.021

Cui WX, Hu S, Guo Q, Wang HF, Zhang S (2014) Spin-based scheme for implementing an N-qubit tunable controlled phase gate in quantum dots by interference of polarized photons. Laser Phys 24(4):045204. https://doi.org/10.1088/1054-660x/24/4/045204

Dai FJ, Chau CF (2017) Classification and regulatory perspectives of dietary fiber. J Food Drug Anal 25(1):37–42. https://doi.org/10.1016/j.jfda.2016.09.006

Das S, Debnath N, Cui Y, Unrine J, Palli SR (2015) Chitosan, carbon quantum dot, and silica nanoparticle mediated dsRNA delivery for gene silencing in *Aedes aegypti*: a comparative analysis. ACS Appl Mater Interfaces 7(35):19530–19535. https://doi.org/10.1021/acsami.5b05232

Deepak P, Amutha V, Birundha R, Sowmiya R, Kamaraj C, Balasubramanian V, Balasubramani G, Aiswarya D, Arul D, Perumal P (2018) Facile green synthesis of nanoparticles from brown seaweed *Sargassum wightii* and its biological application potential. Adv Nat Sci Nanosci Nanotechnol 9(3):035019. https://doi.org/10.1088/2043-6254/aadc4a

Deng J, Lu Q, Mi N, Li H, Liu M, Xu M, Tan L, Xie Q, Zhang Y, Yao S (2014) Electrochemical synthesis of carbon nanodots directly from alcohols. Chem Eur J 20:4993–4999. https://doi.org/10.1002/chem.201304869

Dong Y, Pang H, Yang HB, Guo C, Shao J, Chi Y, Li CM, Yu T (2013) Carbon-based dots co-doped with nitrogen and sulfur for high quantum yield and excitation-independent emission. Angew Chem 125:7954–7958. https://doi.org/10.1002/anie.201301114

Elsabahy M, Wooley KL (2012) Design of polymeric nanoparticles for biomedical delivery applications. Chem Soc Rev 41:2545–2561. https://doi.org/10.1039/c2cs15327k

Fan Q, Li J, Zhu Y, Yang Z, Shen T, Guo Y, Wang L, Mei T, Wang J, Wang X (2020) Functional carbon quantum dots towards highly sensitive graphene transistors for Cu^{2+} ion detection. ACS Appl Mater Interfaces 12(4):4797–4803. https://doi.org/10.1021/acsami.9b20785

Fernández F, Boluda CJ, Olivera J, Guillermo LA, Gómez B, Echavarría E, Mendis Gómez A (2017) Análisis elemental prospectivo de la biomasa algal acumulada en las costas de la república dominicana durante 2015. Centro Azúcar 44(1):11–22

Fine ML (1970) Faunal variation on pelagic *Sargassum*. Mar Biol 7(2):112–122. https://doi.org/10.1007/bf00354914

Franks JS, Johnson DR, Ko DS (2016) Pelagic *Sargassum* in the tropical north Atlantic. Gulf Caribb Res 27:SC6-11. https://doi.org/10.18785/gcr.2701.08

Gałuszka A, Migaszewski Z, Namieśnik J (2013) The 12 principles of green analytical chemistry and the SIGNIFICANCE mnemonic of green analytical practices. TrAC Trends Anal Chem 50:78–84. https://doi.org/10.1016/j.trac.2013.04.010

Gao H, Duan B, Lu A, Deng H, Du Y, Shi X, Zhang L (2018) Fabrication of cellulose nanofibers from waste brown algae and their potential application as milk thickeners. Food Hydrocoll 79:473–481. https://doi.org/10.1016/j.foodhyd.2018.01.023

Geetha P, Latha MS, Pillai SS, Koshy M (2015) Nanoalginate based biosorbent for the removal of lead ions from aqueous solutions: equilibrium and kinetic studies. Ecotoxicol Environ Saf 122:17–23. https://doi.org/10.1016/j.ecoenv.2015.06.032

Gilbertson LM, Zimmerman JB, Plata DL, Hutchison JE, Anastas PT (2015) Designing nanomaterials to maximize performance and minimize undesirable implications guided by the Principles of Green Chemistry. Chem Soc Rev 44(16):5758–5777. https://doi.org/10.1039/c4cs00445k

Godavarthi S, Mohan-Kumar K, Vázquez-Vélez EV, Hernandez-Eligio A, Mahendhiran M, Hernandez-Como N, Aleman M, Gomez LM (2017) Nitrogen doped carbon dots derived from *Sargassum fluitans* as fluorophore for DNA detection. J Photochem Photobiol B Biol 172:36–41. https://doi.org/10.1016/j.jphotobiol.2017.05.014

Gonçalves H, Jorge PAS, Fernandes JRA, Esteves da Silva JCG (2010) Hg(II) sensing based on functionalized carbon dots obtained by direct laser ablation. Sensors Actuators B 145:702–707. https://doi.org/10.1016/j.snb.2010.01.031

Gorham J, Lewey SA (1984) Seasonal changes in the chemical composition of *Sargassum muticum*. Mar Biol 80:103–107. https://doi.org/10.1007/bf00393133

Gower JFR, King SA (2011) Distribution of floating *Sargassum* in the Gulf of Mexico and the Atlantic Ocean mapped using MERIS. Int J Remote Sens 32(7):1917–1929. https://doi.org/10.1080/01431161003639660

Guan W, Gu W, Ye L, Guo C, Su S, Xu P, Xue M (2014) Microwave-assisted polyol synthesis of carbon nitride dots from folic acid for cell imaging. Int J Nanomedicine 9(1):5071–5078. https://doi.org/10.2147/ijn.s68575

Guo LP, Zhang Y, Li WC (2017) Sustainable microalgae for the simultaneous synthesis of carbon quantum dots for cellular imaging and porous carbon for CO_2 capture. J Colloid Interface Sci 493:257–264. https://doi.org/10.1016/j.jcis.2017.01.003

Hardman R (2006) A toxicologic review of quantum dots: toxicity depends on physicochemical and environmental factors. Environ Health Perspect 114:165–172. https://doi.org/10.1289/ehp.8284

Huang YC, Lam UI (2011) Chitosan/fucoidan pH sensitive nanoparticles for oral delivery system. J Chin Chem Soc 58(6):779–785. https://doi.org/10.1002/jccs.201190121

Huffard CL, von Thun S, Sherman AD, Sealey K, Smith KL (2014) Pelagic *Sargassum* community change over a 40-year period: temporal and spatial variability. Mar Biol 161:2735–2751. https://doi.org/10.1007/s00227-014-2539-y

Kim J, Park J, Kim H, Singha K, Kim WJ (2013) Transfection and intracellular trafficking properties of carbon dot-gold nanoparticle molecular assembly conjugated with PEI-pDNA. Biomaterials 34(29):7168–7180. https://doi.org/10.1016/j.biomaterials.2013.05.072

Lai IPJ, Harroun SG, Chen SY, Unnikrishnan B, Li YJ, Huang CC (2016) Solid-state synthesis of self-functional carbon quantum dots for detection of bacteria and tumor cells. Sensors Actuators B Chem 228:465–470. https://doi.org/10.1016/j.snb.2016.01.062

Lapointe BE, West LE, Sutton TT, Hu C (2014) Ryther revisited: nutrient excretions by fishes enhance productivity of pelagic *Sargassum* in the western North Atlantic Ocean. J Exp Mar Biol Ecol 458:46–56. https://doi.org/10.1016/j.jembe.2014.05.002

Levine I (2016) Algae. Seaweed in health and disease prevention. In: Fleurence J, Levine I (eds) Seaweed in health and disease prevention, 1st edn. Academic Press, San Diego, pp 1–5

Li B, Lu F, Wei X, Zhao R (2008) Fucoidan: structure and bioactivity. Molecules 13:1671–1695. https://doi.org/10.3390/molecules13081671

Li H, He X, Liu Y, Huang H, Lian S, Lee ST, Kang Z (2011a) One-step ultrasonic synthesis of water-soluble carbon nanoparticles with excellent photoluminescent properties. Carbon 49(2):605–609. https://doi.org/10.1016/j.carbon.2010.10.004

Li H, Zhang Y, Wang L, Tian J, Sun X (2011b) Nucleic acid detection using carbon nanoparticles as a fluorescent sensing platform. Chem Commun 47(3):961–963. https://doi.org/10.1039/c0cc04326e

Liang Q, Ma W, Shi Y, Li Z, Yang X (2013) Easy synthesis of highly fluorescent carbon quantum dots from gelatin and their luminescent properties and applications. Carbon 60:421–428. https://doi.org/10.1016/j.carbon.2013.04.055

Lim JS, López R, Sánchez D (2014) Orbital caloritronic transport in strongly interacting quantum dots. New J Phys 16(1):015003. https://doi.org/10.1088/1367-2630/16/1/015003

Liu Z, Robinson J, Sun X, Dai H (2008) PEGylated nanographene oxide for delivery of water-insoluble cancer drugs. J Am Chem Soc 130(33):10876–10877. https://doi.org/10.1021/ja803688x

Liu Y, Xiao N, Gong N, Wang H, Shi X, Gu W, Ye L (2014) One-step microwave-assisted polyol synthesis of green luminescent carbon dots as optical nanoprobes. Carbon 68:258–264. https://doi.org/10.1016/j.carbon.2013.10.086

Ma J, Chen JY, Zhang Y, Wang PN, Guo J, Yang WL, Wang CC (2007) Photochemical instability of thiol-capped CdTe quantum dots in aqueous solution and living cells: process and mechanism. J Phys Chem B 111:12012–12016. https://doi.org/10.1021/jp073351

Maeda H, Tsukui T, Sashima T, Hosokawa M, Miyashita K (2008) Seaweed carotenoid, fucoxanthin, as multi-functional nutrient. Asia Pac J Clin Nutr 17:196–199

Maia M, Fonseca A, Oliveira HM, Mendonça C, Cabrita ARJ (2016) The potential role of seaweeds in the natural manipulation of rumen fermentation and methane production. Sci Rep 6:32321. https://doi.org/10.1038/srep32321

Mehta VN, Jha S, Singhal RK, Kailasa SK (2014) Preparation of multicolor emitting carbon dots for HeLa cell imaging. New J Chem 38(12):6152–6160. https://doi.org/10.1039/c4nj00840e

Milledge JJ, Harvey PJ (2016) Ensilage and anaerobic digestion of *Sargassum muticum*. J Appl Phycol 28(5):3021–3030. https://doi.org/10.1007/s10811-016-0804-9

Milledge JJ, Nielsen BV, Sadek MS, Harvey PJ (2018) Effect of freshwater washing pretreatment on Sargassum *muticum* as a feedstock for biogas production. Energies 11(7):1771. https://doi.org/10.3390/en11071771

Milledge JJ, Maneein S, Arribas López E, Bartlett D (2020) *Sargassum* inundations in turks and caicos: methane potential and proximate, ultimate, lipid, amino acid, metal and metalloid analyses. Energies 13(6):1523. https://doi.org/10.3390/en13061523

Miyashita K, Mikamia N, Hosokawa M (2013) Chemical and nutritional characteristics of brown seaweed lipids: a review. J Funct Foods 5(4):1507–1517. https://doi.org/10.1016/j.jff.2013.09.019

Namdari P, Negahdari B, Eatemadi A (2017) Synthesis, properties and biomedical applications of carbon-based quantum dots: an updated review. Biomed Pharmacother 87:209–222. https://doi.org/10.1016/j.biopha.2016.12.108

Niermann U (1986) Distribution of *Sargassum natans* and some of its epibionts in the Sargasso Sea. Helgoländer Meeresuntersuchungen 40(4):343–353. https://doi.org/10.1007/BF01983817

Norton TA, Melkonian M, Andersen RA (1996) Algal biodiversity. Phycologia 35(4):308–326. https://doi.org/10.2216/i0031-8884-35-4-308.1

Olabarria C, Rodil IF, Incera M, Troncoso JS (2009) Limited impact of *Sargassum muticum* on native algal assemblages from rocky intertidal shores. Mar Environ Res 67(3):153–158. https://doi.org/10.1016/j.marenvres.2008.12.007

Oyesiku OO, Egunyomi A (2014) Identification and chemical studies of pelagic masses of *Sargassum natans* (Linnaeus) Gaillon and *S. fluitans* (Borgessen) Borgesen (brown algae), found offshore in Ondo State Nigeria. Afr J Biotechnol 13(10):1188–1193. https://doi.org/10.5897/AJB2013.12335

Pal A, Kamthania MC, Kumar A (2014) Bioactive compounds and properties of seaweeds a review. Open Access Library J 1:e752. https://doi.org/10.4236/oalib.1100752

Pan D, Zhang J, Li Z, Wu M (2010) Hydrothermal route for cutting graphene sheets into blue-luminescent graphene quantum dots. Adv Mater 22(6):734–738. https://doi.org/10.1002/adma.200902825

Parr A (1939) Quantitative observations on the pelagic *Sargassum* vegetation of the western North Atlantic. Bull Bingham Oceanogr Coll 6:1–94

Paula JC, Vallim MA, Teixeira VL (2011) What are and where are the bioactive terpenoids metabolites from *Dictyotaceae* (Phaeophyceae). Rev Bras 21(2):216–228. https://doi.org/10.1590/s0102-695x2011005000079

Peña-Rodríguez A, Mawhinney TP, Ricque-Marie D, Cruz-Suárez LD (2011) Chemical composition of cultivated seaweed *Ulva clathrata* (Roth) C. Agardh. Food Chem 129(2):491–498. https://doi.org/10.1016/j.foodchem.2011.04.104

Peng Y, Hu J, Yang B, Lin XP, Zhou XF, Yang XW, Liu Y (2015) Chemical composition of seaweeds. In: Tiwari BK, Troy DJ (eds) Seaweed sustainability, 1st edn. Academic Press, San Diego, pp 79–124

Prekodravac J, Vasiljević B, Marković Z, Jovanović D, Kleut D, Špitalský Z, Mičušik M, Danko M, Bajuk-Bogdanović D, Todorović-Marković B (2019) Green and facile microwave assisted synthesis of (metal-free) N-doped carbon quantum dots for catalytic applications. Ceram Int 45(14):17006–17013. https://doi.org/10.1016/j.ceramint.2019.05.250

Ray SC, Saha A, Jana NR, Sarkar R (2009) Fluorescent carbon nanoparticles: synthesis, characterization, and bioimaging application. J Phys Chem C 113(43):18546–18551. https://doi.org/10.1021/jp905912n

Rhein-Knudsen N, Ale MT, Ajalloueian F, Meyer AS (2017) Characterization of alginates from Ghanaian brown seaweeds: *Sargassum* spp. and *Padina* spp. Food Hydrocoll 71:236–244. https://doi.org/10.1016/j.foodhyd.2017.05.016

Ríos DZ (2012) Macrofitobentos de interés nutricional en cayo Coco al norte de Ciego de Ávila, Cuba. Serie Oceanológica 10:54–61

Rodríguez-Martínez RE, Van Tussenbroek B, Jordán-Dahlgren E (2017) Afluencia masiva de sargazo pelágico a la costa del Caribe Mexicano (2014-2015). In: García-Mendoza E, Quijano-Scheggia SI, Olivos-Ortiz A, Núñez-Vázquez EJ (eds) Florecimientos algales nocivos en México, 1st edn. Centro de Investigación Científica y de Educación Superior de Ensenada Baja California, Ensenada, pp 352–365

Rodríguez-Martínez RE, Roy PD, Torrescano-Valle N, Cabanillas-Terán N, Carrillo-Domínguez S, Collado-Vides L, García-Sánchez M, van Tussenbroek BI (2020) Element concentrations in pelagic *Sargassum* along the Mexican Caribbean coast in 2018–2019. PeerJ 8:e8667. https://doi.org/10.7717/peerj.8667

Rodríguez-Padrón D, Algarra M, Tarelho LAC, Frade JR, Franco A, de Miguel G, Jimenez J, Rodriguez-Castellon E, Luque R (2018) Catalyzed microwave-assisted preparation of carbon quantum dots from lignocellulosic residues. ACS Sustainable Chem Eng 6(6):7200–7205. https://doi.org/10.1021/acssuschemeng.7b03848

Sarkar N, Sahoo G, Das R, Prusty G, Swain SK (2017) Carbon quantum dot tailored calcium alginate hydrogel for pH responsive controlled delivery of vancomycin. Eur J Pharm Sci 109:359–371. https://doi.org/10.1016/j.ejps.2017.08.015

Shi W, Wang Q, Long Y, Cheng Z, Chen S, Zheng H, Huang Y (2011) Carbon nanodots as peroxidase mimetics and their applications to glucose detection. Chem Commun 47(23):6695–6697. https://doi.org/10.1039/c1cc11943e

Singh V, Kumar V, Yadav U, Srivastava RK, Singh VN, Banerjee A, Chakraborty S, Shukla AK, Misra DK, Ahuja R, Srivastava A, Saxena PS (2017) Sensitive and selective detection of copper ions using low cost nitrogen doped carbon quantum dots as a fluorescent sensing plateform. J Micro Smart Syst 6(2):109–117. https://doi.org/10.1007/s41683-017-0011-1

Singh S, Nigam P, Pednekar A, Mukherjee S, Mishra A (2018) Carbon quantum dots functionalized agarose gel matrix for in solution detection of nonylphenol. Environ Technol 38:1–7. https://doi.org/10.1080/09593330.2018.1498133

Smith AM, Nie S (2010) Bright and compact alloyed quantum dots with broadly tunable near-infrared absorption and fluorescence spectra through mercury cation exchange. J Am Chem Soc 133(1):24–26. https://doi.org/10.1021/ja108482a

Stiger-Pouvreau V, Zubia M (2020) Macroalgal diversity for sustainable biotechnological development in French tropical overseas territories. Bot Mar 63(1):17–41. https://doi.org/10.1515/bot-2019-0032

Tang B, Niu J, Yu C, Zhuo L, Ge J (2005) Highly luminescent water-soluble CdTe nanowires as fluorescent probe to detect copper (II). Chem Commun 33:4184–4186. https://doi.org/10.1039/b502978c

Tang Y, Su Y, Yang N, Zhang L, Lv Y (2014) Carbon nitride quantum dots: a novel chemiluminescence system for selective detection of free chlorine in water. Anal Chem 86:4528–4535. https://doi.org/10.1021/ac5005162

Thompson DA, Lehmler HJ, Kolpin D, Hladik ML, Vargo JD, Schilling KE, LeFevre GH, Peeples TL, Poch MC, LaDuca LE, Cwierty DM, Field RW (2020) A critical review on the potential impacts of neonicotinoid insecticide use: current knowledge of environmental fate, toxicity and implications for human health. Environ Sci Processes Impacts 22:1315–1346. https://doi.org/10.1039/c9em00586b

Vandarkuzhali SAA, Jeyalakshmi V, Sivaraman G, Singaravadivel S, Krishnamurthy KR, Viswanathan B (2017) Highly fluorescent carbon dots from Pseudo-stem of banana plant: applications as nanosensor and bio-imaging agents. Sensors Actuators B Chem 252:894–900. https://doi.org/10.1016/j.snb.2017.06.088

Vivekanand V, Dwivedi P, Pareek N, Singh RP (2011) Banana peel: a potential substrate for laccase production by *Aspergillus fumigatus* Vkj 2.4.5 in solid-state fermentation. Appl Biochem Biotecnol 165(1):204–220. https://doi.org/10.1007/s12010-011-9244-9

Wang Q, Kuo Y, Wang Y, Shin G, Ruengruglikit C, Huang Q (2006) Luminescent properties of water-soluble denatured bovine serum albumin-coated CdTe quantum dots. J Phys Chem B 110(34):16860–16866. https://doi.org/10.1021/jp062279x

Wang D, Zhu L, Chen JF, Dai L (2015) Can graphene quantum dots cause DNA damage in cells? Nanoscale 7(21):9894–9901. https://doi.org/10.1039/c5nr01734c

Wang C, Shi H, Yang M, Yan Y, Liu E, Ji Z, Fan J (2019a) Facile synthesis of novel carbon quantum dots from biomass waste for highly sensitive detection of iron ions. Mater Res Bull 124:110730. https://doi.org/10.1016/j.materresbull.2019.110730

Wang M, Hu C, Barnes BB, Mitchum G, Lapointe B, Montoya JP (2019b) The great Atlantic *Sargassum* belt. Science 365(6448):83–87. https://doi.org/10.1126/science.aaw7912

Xu X, Ray R, Gu Y, Ploehn HJ, Gearheart L, Raker K, Scrivens WA (2004) Electrophoretic analysis and purification of fluorescent single-walled carbon nanotube fragments. J Am Chem Soc 126(40):12736–12737. https://doi.org/10.1021/ja040082h

Xu Y, Li D, Liu M, Niu F, Liu J, Wang E (2017) Enhanced-quantum yield sulfur/nitrogen co-doped fluorescent carbon nanodots produced from biomass *Enteromorpha prolifera*: synthesis, posttreatment, applications and mechanism study. Sci Rep 7(1):4499. https://doi.org/10.1038/s41598-017-04754-x

Yang M, Jiang B, Xie J, Xiang Y, Yuan R, Chai Y (2014) Electrochemiluminescence recovery-based aptasensor for sensitive Ochratoxin A detection via exonuclease-catalyzed target recycling amplification. Talanta 125(1):45–50. https://doi.org/10.1016/j.talanta.2014.02.061

Yang Q, Duan J, Yang W, Li X, Mo J, Yang P, Tang Q (2018) Nitrogen-doped carbon quantum dots from biomass via simple one-pot method and exploration of their application. Appl Surf Sci 434:1079–1085. https://doi.org/10.1016/j.apsusc.2017.11.040

Zhang J, Yu SH (2016) Carbon dots: large-scale synthesis, sensing and bioimaging. Mater Today 19(7):382–393. https://doi.org/10.1016/j.mattod.2015.11.008

Zhang F, Liu F, Wang C, Li X, Mo J, Yang P, Tang Q (2016) Effect of lateral size of graphene quantum dots on their properties and application. ACS Appl Mater Interfaces 8:2104–2110. https://doi.org/10.1016/j.apsusc.2017.11.040

Zhang Z, Chen J, Yang Q, Lan K, Yan Z, Chen J (2018) Eco-friendly intracellular microalgae synthesis of fluorescent CdSe QDs as a sensitive nanoprobe for determination of imatinib. Sensors Actuators B Chem 263:625–633. https://doi.org/10.1016/j.snb.2018.02.169

Zheng XT, Ananthanarayanan A, Luo KQ, Chen P (2015) Glowing graphene quantum dots and carbon dots: properties, syntheses, and biological applications. Small 11(14):1620–1636. https://doi.org/10.1002/smll.201402648

Zhou Y, Qu Z, Zeng Y, Zhou T, Shi G (2014) A novel composite of graphene quantum dots and molecularly imprinted polymer for fluorescent detection of paranitrophenol. Biosens Bioelectron 52:317–323. https://doi.org/10.1016/j.bios.2013.09.022

Zhu C, Zhai J, Dong S (2012) Bifunctional fluorescent carbon nanodots: green synthesis via soy milk and application as metal-free electrocatalysts for oxygen reduction. Chem Commun 48(75):9367–9369. https://doi.org/10.1039/c2cc33844k

Zubia M, Payri C, Deslandes E (2008) Alginate, mannitol, phenolic compounds and biological activities of two range-extending brown algae, *Sargassum mangarevense* and *Turbinaria ornata* (Phaeophyta: Fucales), from Tahiti (French Polynesia). J Appl Phycol 20(6):1033–1043. https://doi.org/10.1007/s10811-007-9303-3

Part IV
Biomedical Applications of Algal Nanomaterials

Chapter 8
Algal-Mediated Biosynthesis of Nanoparticles and Their Potential Therapeutic Applications

Khadija Nawaz, Rimsha Chaudhary, Amna Komal Khan, Christophe Hano, Bilal Haider Abbasi, and Sumaira Anjum

Contents

Khadija Nawaz, Rimsha Chaudhary and Amna Komal Khan contributed equally with all other contributors.

K. Nawaz · R. Chaudhary · A. K. Khan · S. Anjum (✉)
Department of Biotechnology, Kinnaird College for Women, Lahore, Pakistan
e-mail: sumaira.anjum@kinnaird.edu.pk

C. Hano
Laboratoire de Biologie des Ligneux et des Grandes Cultures (LBLGC), INRAE USC1328, Université d'Orléans, Chartres, France

B. H. Abbasi
Department of Biotechnology, Quaid-i-Azam University, Islamabad, Pakistan

D. Thangadurai et al. (eds.), *Bioprospecting Algae for Nanosized Materials*, Nanotechnology in the Life Sciences,
https://doi.org/10.1007/978-3-030-81557-8_8

8.1 Introduction

The importance of nanotechnology can be witnessed from accelerated development made in the field of nanotechnology in the last few decades due to the incomparable physicochemical and optoelectronic properties of nanoparticles (NPs). In order to generate advanced substitutes for solving various research-oriented bottlenecks, nanotechnology finds its applications in every branch of science (Tran et al. 2012). NPs can be defined as materials that have internal or external surface structures in one or more dimensions falling between 1 and 100 nm (Guo and Wang 2011). Metallic or metal oxide NPs are the most popular class of nanomaterials because they have high electrical properties, high surface area, high magnetic, high optical, high mechanical properties as well as high thermal stability (Yaqoob et al. 2020). These NPs have attracted huge attention in the field of biological sciences for the treatment, diagnosis and drug delivery as they are biocompatible and considered non-toxic by the scientific community (Yamada et al. 2015; Prasad et al. 2018, 2019, 2020; Maddela et al. 2021; Saglam et al. 2021).

Recently, metallic NPs and their derivatives like gold, silver, titanium, copper and zinc oxide NPs have received great attention for their antimicrobial properties. The surface charge, unique shape and nanosize of these NPs have been associated with the enhanced bactericidal activity of NPs (Dizaj et al. 2014). The NPs are also used to encapsulate drugs for an efficient delivery of drugs, thereby lowering therapeutic dosage while enhancing therapeutic efficacy especially against cancer (Prasad et al. 2017; Vissers et al. 2019). Moreover, NPs possess great absorption of near-infrared radiation, which has made them exponentially helpful in in vivo imaging therapies. Hence, these NPs are considered as active contrasting agents (Conde et al. 2012).

Despite their widespread applications, the fabrication of metallic or metal oxide NPs is often associated with hazardous chemical or physical synthesis methods. The chemical procedure of NPs generation involves the utilization of harmful and harsh chemicals, whereas in physical methods, parameters such as high temperature, energy and special setup are involved (Sharma et al. 2016). Therefore, production methods that are non-toxic, environmentally friendly, simple and cost-effective are required (Azizi et al. 2014). For this reason, green nanotechnology has emerged which includes the application of biological organisms like bacteria, fungi, plants and algae for the production of NPs (Kalaiarasi et al. 2010; Sinha et al. 2015; Prasad 2014; Prasad et al. 2016; Srivastava et al. 2021). These organisms possess metabolites such as few functional groups like hydroxyl, carbonyl, amine and thiol as well as proteins and peptides which behave as reducing agents as well as capping during the mediation of NPs (Pillai and Kamat 2004; Aziz et al. 2014, 2015).

Algae are a great shift towards the greener production of nanoparticles because of the presence of secondary metabolites, proteins, peptides and other pigments in a great amount which serves as nano-biofactories (Patel et al. 2015; Aziz et al. 2014, 2015). Algae are an extremely diverse category of autotrophic microorganisms inhabiting almost every ecosystem (Ven Den Hoeck et al. 1995). They have long been used for commercial and industrial applications such as food, feed, additives, pharmaceuticals, cosmetics, fertilizers and biofuels (Borowitzka 2013; Johansen 2012; Sing et al. 2013; Aziz et al. 2017; Rawat et al. 2021). In addition to these conventional uses, now there is a trend towards merging nanotechnology with phycology leading to a field called phyconanotechnology. The NPs are applied for the algal cultivation and harvesting for algal biofuel production (Mandotra et al. 2018; Aziz et al. 2017). On the contrary, the algae are employed for the biosynthesis of nanoparticles. Various algal species, including blue-green algae (*Cyanophyceae*), red algae (*Rhodophyceae*), green algae (*Chlorophyceae*) and brown algae (*Phaeophyceae*), have been reported to biosynthesize different kinds of metallic nanoparticles involving silver NPs (AgNPs), gold NPs (AuNPs) and bimetallic NPs (Sharma et al. 2016). The faster rate of growth, cost-effective scale-up and simple cultivation make algae an attractive participant for the biological generation of NPs (Chandran et al. 2011). Moreover, algae naturally have the potential to assemble heavy metals greatly and transmute them into nanoparticles under the right condition (Dhillon et al. 2012).

The significance of using algae for the biological production of NPs is that it does not require any external reducing, capping or stabilizing agents but in contrast, it is a low-cost and simpler reaction with low energy input. However, this area of research is still in its infancy and further developments are required for the success of their industrial uses. Henceforth, this chapter reviews the capability of algae for NPs biosynthesis, mechanism of synthesis and the therapeutic applications of NPs synthesized from algae in a comprehensive manner.

8.2 Mechanism Involved in Algae-Mediated Biosynthesis of Nanoparticles

The early moves included in the algae-moderated biosynthesis of NPs involve the algal extract formation in a suitable solvent, which is incubated with molar solutions of ionic metallic precursors under controlled conditions (Al-Naamani et al. 2017). The production of nanoparticles takes place in a dose-dependent manner, either extracellularly or intracellularly, in different types of algae (Dahoumane et al. 2014a). The external synthesis of NPs takes place when the metallic ions bind on the surfaces of the algal cells or because of the existence of reducing agents in the aqueous algal extract. The metabolites like antioxidants, pigments, chlorophyll, phycobilins, carbohydrates, oils, proteins and other compounds aid in decreasing the charge of metallic ions into zero-valent position and precipitate them into NPs (Selvam and Sivakumar 2015). Whereas, the intracellular production of NPs depends on the algal metabolism, in which nicotinamide adenine dinucleotide phosphate (NADPH)-dependent reductase acts as a reducing agent during respiration or photosynthetic processes (Kumar et al. 2012a; Vadlapudi and Amanchy 2017). As for the reduction of gold ions into gold NPs in an intracellular way, nitrogenase is an enzyme that has also been recorded (Oza et al. 2012). It has also been observed that NPs synthesis takes place in an intracellular fashion through the interiorization of metal ions in a dose-dependent way, followed by the reduction of metal ions and release of stable nanoparticles in culture media (Dahoumane et al. 2014a).

The biosynthesis of nanoparticles is a three-stage process. The early stage is an activating period, in which the metallic ions are reduced and nucleated due to the bioreducing agents released through algal cells resulting in the colour change of the solution. The second step or phase is the growth phase that involves the amalgamation of nucleated metal elements to form NPs of a particular size and shape that are thermodynamically stable. Last is the ending phase that mediates the final shape and form of NPs (Al-Naamani et al. 2017). In most cases, colour variation of the reaction mixture is observed visually, which indicates the synthesis of NPs. The colour variation takes place because of the phenomenon of surface plasmon resonance (SPR), which is the oscillation of electronic fields of a light wave, transmitting to a close colloidal NP that interacts with the free electrons giving rise to a concerted oscillation of e-charge that resonates with the frequency of visible light (Jana et al. 2016). For example, ruby red colour change marks the formation of AuNPs and brown colour indicates AgNPs synthesis (Dahoumane et al. 2014b; Jena et al. 2013).

Some of the initial physical parameters, which include the starting amount of salt precursor, metal type, pH, temperature, exposure period, type plus the concentration of reducing agents, influence the morphology (size and shape) of nanoparticles as well as prevent the aggregation of NPs (Oza et al. 2012). A higher level of pH enhances the decreasing capacity of the functional group which prevents the agglomeration of NPs in the reaction (Mosulishvili et al. 2007; Yoon et al. 2006). The basic level of pH promotes the stability and capping of nanoparticles by interrelating the amine groups to their residual amino acids present on the surface-attached proteins (Khan et al. 2019).

Once the NPs are biosynthesized, they are characterized for various features like shape, size and purity, detection of various functional groups accountable for the depletion plus capping of NPs by employing different techniques. Electron diffraction X-ray (EDX) analysis together with scanning electron microscopy (SEM) reveals the size, shape and elemental composition of NPs formed (Ahmed et al. 2015). On the other hand, Fourier transform infrared spectroscopy (FTIR) analysis is helpful in elucidating functional groups acting as capping, decreasing and stabilizing agents during the biosynthesis of NPs from the characteristic FTIR peaks (Garibo et al. 2020). Once characterized, the NPs can be used for various biological applications.

8.3 Algal-Mediated Biosynthesis of AgNPs

Algal-mediated biosynthesis of metallic nanoparticles seems to be the latest trend in nanotechnology these days. Among all metallic NPs, AgNPs are most widely used and synthesized evidently from more than half of literature reports presented on the production of AgNPs (Azizi et al. 2014; Ghodake and Lee 2011; Liu et al. 2005; Rajeshkumar et al. 2012; Rajeshkumar et al. 2013). They are the most commercialized among all other metallic nanoparticles, around 500 tons of AgNPs are being synthesized per year from different biosources with an optimal temperature, pressure, pH and energy conditions (Thuesombat et al. 2014). AgNPs possess unique physiochemical characteristics which make them exponentially beneficial in various industrial sites like therapeutics, drug delivery, dental alloys, jewellery, wound healing, textile, paint, water purification and many others (Maneerung et al. 2008; Mohanpuria et al. 2008; Vaidyanathan et al. 2010; Prasad 2014). Their high sensitivity makes them useful for biomedical applications and is used as a biosensor in biomolecular detectors, medicines and catalysis. Biosynthesized AgNPs also display effective antifungal, antibacterial, anti-inflammatory, antidiabetic, antioxidant, anti-angiogenesis plus many other activities (Larue et al. 2014). The bactericidal properties of silver nanoparticles in opposition to many Gram-positive and Gram-negative bacterial species vary in accordance with their size and shape (Kathiraven et al. 2015). Herein, the literature on the biosynthesis of AgNPs from the different divisions of algae has been critically explained in detail and summarized in Table 8.1.

Algal-mediated synthesis of AgNPs is more attractive, as it is simple, easy and cost-effective. As per the literature survey, it is stated that on average the total biomass growth rate of algae is 10 times faster as compared to other higher plants. Algae are majorly classified into brown algae (*Phaeophytes*), green algae (*Chlorophytes*) and red algae (*Rhodophytes*) and on the basis of the type of pigmentation they release (Prasad et al. 2013; Sahayaraj et al. 2012). Currently, many AgNPs have been generated from various algal types, as described in Table 8.1.

Brown alga is an important division of algae due to the presence of important elements of sterols, for example, fucosterols and cholesterols as well as other functional groups that behave as both stabilizing and reducing agents in AgNPs generation. It belongs to the order Fucales and the family Sargassaceae (Kumar et al. 2011). A variety of species, such as *Turbinaria conoides*, *Padina pavonica*, *Cystophora moniliformis*, *Gelidiella acerosa* and *Sargassum polycystum*, have been used in the biosynthesis of AgNPs (Azizi et al. 2014; Ghodake and Lee 2011; Liu et al. 2005; Rajeshkumar et al. 2012; Rajeshkumar et al. 2013). One of the studies recorded the extracellular production of silver nanoparticles (96 nm) from *T. conoides* with antibacterial performance in opposition to the *Escherichia coli*, *Staphylococcus epidermis*, *Candida albicans*, *Staphylococcus aureus*, *Pseudomonas aeruginosa* and *Aspergillus niger* (Rajeshkumar et al. 2012). Another study recorded the presence of organic elements and functional groups in *T. ornata* and *T. conoides* that behave as both capping and reducing agents (Khalil et al. 2014; Kumar-Krishnan et al. 2015). Kumar and his coworker recorded the production of extracellular spherical AgNPs from *Gelidium amansii* by incubating the reaction mixture at room temperature for 48 h and centrifuging at 14,000 rpm (revolutions per minute) (Kumar et al. 2013a).

Red algae belong to the family *Rhodophyta*. They are rich in various important vitamins and proteins that behave as the best contenders in reduction and stabilization for silver nanoparticle generation (Ramakritinan et al. 2010; Singaravelu et al. 2007; Yoon et al. 2006). Numerous red algae, such as *Porphyra vietnamensis*, *Gelidiella acerosa*, *Gracilaria corticata*, *Kappaphycus* sp. and *Kappaphycus*

Table 8.1 Algae-mediated biosynthesis of AgNPs

Algal species	Location	Morphology	Optimum conditions	Characterization	Reducing agent	References
Turbinaria conoides	Extracellular	Spherical, 96 nm	1 h incubation in dark	FTIR, XRD and TEM, SEM	Polyamines, carbonyl groups	Rajeshkumar et al. (2012)
Gracilaria edulis	Extracellular	Spherical, 12.5–100 nm	Incubation of 150 min and temperature of 40 °C on an orbital shaker	FTIR, SEM, TEM and UV-Vis	Terpenoids, proteins	Pugazhendhi et al. (2018)
Gelidium amansii	Extracellular	Spherical	48 h of incubation at room temperature with centrifugation of 13,000 rpm	UV-Vis	–	Kumar et al. (2013a)
Sargassum polycystum	Extracellular	Spherical, 7 nm	–	FTIR, SEM, XRD, TEM and UV-Vis	Extracellular terpenoids	Khanna et al. (2019)
Turbinaria ornata	Extracellular	Spherical, polydispersed, 22 nm	24 h incubation	EDS, FTIR, UV-Vis, XRD and FESEM	Organic moieties	Khalil et al. (2014)
Sargassum muticum	Extracellular	Spherical, crystalline, 43–79 nm	2 h incubation at room temperature	SEM, EDX, UV-Vis, XRD and FTIR	Hydroxyl groups, sulphates	Madhiyazhagan et al. (2015)
Gelidiella acerosa	Extracellular	FCC and spherical, 22 nm	48 h of incubation at 120 rpm continuous stirring	SEM, TEM, UV-Vis, FTIR and XRD	Aromatic compounds	Vivek et al. (2011)
Acanthophora spicifera	Extracellular	Spherical, 48 nm	Incubation at 60 °C for 20 min	TEM, FTIR and UV-Vis	Carboxylic acid, alcoholic and phenolic compounds	Kumar et al. (2012b)
Gracilaria dura	Extracellular	Spherical, 6 nm	Incubation period (1, 4 and 48 h), temperature (25, 60 and 100 °C) and pH 6	XRD, TEM, TGA, SAED, XRD and EDX	Polymers	Shukla et al. (2012)

Algal species	Location	Morphology	Optimum conditions	Characterization	Reducing agent	References
Gelidiella sp.	Extracellular	Spherical shape, 50 nm in size	Incubation at a temperature of 121 °C for 10 min, centrifugation at 10,000 rpm	SEM, UV-Vis, FTIR, EDS and XRD	Proteins	Devi et al. (2012)
Gelidiella acerosa	Extracellular	Spherical shape, 18–46 nm in size	–	HR-TEM, SEM, UV-Vis and XRD	Metabolites	Dahoumane et al. (2017)
Kappaphycus sp.	Extracellular	52–104 nm	–	UV-Vis, AFM and FTIR		Baskar (2013)
Gracilaria corticata	Extracellular	46 nm	20 min of incubation at a temperature of 60 °C, heating mantle	Zeta-potential, TEM, DLS, UV-Vis and FTIR	Amides and phenolic	Kumar et al. (2013b)
Kappaphycus alverazii	Extracellular	FCC, 73 nm	Incubation at 27 °C for 96 h with 250 rpm orbital shaking	SEM, FTIR, UV-Vis and EDX	Functional groups and polysaccharides	Ganesan et al. (2013)
Pterocladia capillacae	Extracellular	Spherical, 7 nm	Magnetic stirring at a temperature of 70 °C for 20 min	GLC, FTIR, UV-Vis and TEM	Sulphate, carboxylic groups	El-Rafie et al. (2013)
Gracilaria birdiae	Extracellular	Spherical, 20.3 nm	Incubation at 0 °C and pH 10 for a duration of 30 min	Zeta-potential, UV-Vis, TEM, FTIR and DLS	Carbonyl and hydroxyl groups	de Aragao et al. (2019)
Jania rubins	Extracellular	Spherical, 12 nm	Magnetic stirring at a temperature of 70 °C for 20 min	GLC, TEM, FTIR and UV-Vis	Carbonyl groups	El-Rafie et al. (2013)
Acanthophora spicifera	Extracellular	Cubic, 81 nm	Incubation at room temperature	XRD and FTIR	Uronic acids and polysaccharides	Ibraheem et al. (2016)
Amphiroa fragilissima	Extracellular	Crystalline shape	20 min of incubation with centrifugation at an rpm of 12,000	FTIR, XRD and UV-Vis	Peptides	Sajidha Parveen and Lakshmi (2016)

(continued)

Table 8.1 (continued)

Algal species	Location	Morphology	Optimum conditions	Characterization	Reducing agent	References
Porphyra vietnamensis	Extracellular	13 + 3 nm	Incubation period of 15 min at temp of 70 °C and pH 11	FTIR	Polysaccharides	Venkatpurwar and Pokharkar (2011)
Oscillato riawillei	Extracellular	Spherical shape, 10–25 nm in size	Incubation period of 1 day that is, 24 h	EDS, SEM, UV-Vis and FTIR	Tryptophan	Ali et al. (2011)
Spirulina platensis	Extracellular	Spherical shape, 2–8 nm in size	Incubation period of 2 days that is, 48 h	UV-Vis and SEM	Phytochemicals	Doshi et al. (2007)
Sargassum wightii Greville	Extracellular	Spherical, 8–27 nm	120 h of incubation at 37 °C, 5.6 pH	FTIR, UV-Vis, XRD and HR-TEM	Oxidation of the alcoholic group	Govindaraju et al. (2008)
Sargassum ilicifolium	Extracellular	Spherical, 33–40 nm	–	SEM, TEM and UV-Vis	Bioactive compounds	Li (2011)
Sargassum polycystum	Extracellular	Spherical and FCC, 5–7 nm in size	24 h of incubation at room temperature	GC-MS, UV-Vis, FTIR and HR-TEM	Octadecanol, hexadecane	Thangaraju et al. (2012)
Sargassum plagiophyllum	Intracellular	Spherical shape, 20–50 nm in size	–	SEM, FTIR and UV-Vis	Benzene rings, carboxylic	Dhanalakshmi et al. (2012)
Padina pavonica	Extracellular	Polydispersed and spherical, 10–72 nm	24 h of incubation at room temperature	XRD, TEM, SEM and UV-Vis	Terpenoids	Sahayaraj et al. (2012)
Padina tetrastromatica	Extracellular	Spherical, 4 nm	Magnetic stirring for 48 h at 60 °C	UV-Vis, TEM and FTIR	Alkanes	Bhuyar et al. (2020)
Padina gymnospora	Extracellular	Spherical shape, 25–40 nm in size	–	UV-Vis and TEM	Aqueous extract	Shiny et al. (2013)
Colpomenia sinuosa	Extracellular	Spherical, 20 nm	Stirring continuously for a duration of 20 min at 70 °C with pH 10	TEM, FTIR, GLC-C-O and UV-Vis	Polysaccharides	El-Rafie et al. (2013)

Algal species	Location	Morphology	Optimum conditions	Characterization	Reducing agent	References
Cystophora moniliformis	Extracellular	FCC, 75 nm	Incubation period of half an hour at a temperature of 65 °C	Zeta-potential, EDX, DLS and SEM	Phenolic compounds, metabolites	Prasad et al. (2013)
Sargassum cinereum	Extracellular	Spherical and triangular shape,45–76 nm in size	3 h of incubation with centrifugation at 13,000 rpm	SEM and UV-Vis	–	Mohandass et al. (2013)
Sargassum longifolium	Extracellular	Cubical shape, 30 nm in size	–	EDS, SEM, FTIR and UV-Vis	Carboxylic groups, ketones	Devi et al. (2013)
Scaberia agardhii	Extracellular	Polydispersed, 40–50 nm	24 h of incubation	EDAX, FTIR, UV-Vis and SEM	Proteins	Prasad and Elumalai (2013)
Sargassum polycystum C. Agardh	Extracellular	Spherical	–	FTIR, XRD and UV-Vis	Proteins and metabolites	Asha et al. (2015)
Sargassum vulgare	Extracellular	Spherical shape, 10 nm in size	–	FACS, EDX, TEM, HR-TEM and XRD	OH groups (Secondary)	Govindaraju et al. (2015)
Turbinaria ornata	Extracellular	Polydispersed and spherical shape, 22 nm in size	Incubation at a temperature of 70 °C in a hot water bath, centrifugation at an rpm of 10,000	FESEM, UV-Vis, XRD and SEM	Organic moieties	Krishnan et al. (2015)
Plectonema boryanum	Intracellular	Octahedral, 200 nm	Incubation at 25–100 °C for 28 days	EDS, XPS and TEM	–	Lengke et al. (2007)
Oscillitoria sp. and *Aphanothece* sp.	Extracellular	Spherical shape, 44–79 nm in size	–	SEM, EDX and UV-Vis	Biologically active compounds	Sudha et al. (2013)
Microchaete	Extracellular	Polydispersed, spherical shape, 80 nm in size	Incubation time of 60 min at a temperature of 60 °C and pH 5.6	TEM	Metabolites found in cells	Husain et al. (2019)

(continued)

Table 8.1 (continued)

Algal species	Location	Morphology	Optimum conditions	Characterization	Reducing agent	References
Cylindrospermum stagnale	Extracellular	Pentagon shaped, 38–88 nm in size	Incubation period of 45 h at a temperature of 40 °C	SEM and UV-Vis	Extracellular proteins	Husain et al. (2015)
Scenedesmus sp.	Extracellular	15–20 nm of size	Incubating at a temperature of 37 °C on a shaker at an rpm of 2000	UV-Vis, SEM, FTIR and XRD	–	Vigneshwaran et al. (2007)
Pithophora oedogonia	Extracellular	Hexagonal and cubical in shape, 24–55 nm	Room temperature incubation, centrifugation at an rpm of 15,000 for a duration of 15 min	DLS, EDS, SEM and UV-Vis	Proteins, functional groups	Sinha et al. (2015)
Plectonema boryanum	Intracellular	Smaller than 10 nm	Incubation at a temperature of 25 °C	TEM-ED, XPS and TEM	Nitrate ions	Khanna et al. (2019)
Chlorococcum humicola	Intracellular	Spherical shape, size of 16 nm	Incubation time of 72 h at a temperature of 28 °C, centrifugation at an rpm of 12,000	FTIR, TEM, EDX, SEM and UV-Vis	Proteins found intracellularly	Jena et al. (2014)
Enteromorpha flexuosa	Extracellular	Spherical shape, 2–32 nm in size	Incubation period of 1 h	FTIR, TEM, EDS, EDX, XRD and UV-Vis	Functional groups	Yousefzadi et al. (2014)
Chlorella vulgaris	Extracellular	Triangular shaped, 28 nm in size	12 h of incubation time at room temperature	FESEM, HOUR-TEM and UV-Vis	Carbonyl groups	Xie et al. (2007b)
Chlorella vulgaris and *Nannochloropsis oculata*	Extracellular	FCC shaped, 15 nm in size	Incubation period of 1–2 days	TEM and XRD	Proteins	Mohseniazar et al. (2011)
Chlamydomonas reinhardtii	Extracellular	Rounded and rectangular with 1–15 nm	Incubation period of 5–10 h	UV-Vis, SEM, EDAX and MALDI-TOF-MS	Histone-like cellular proteins	Barwal et al. (2011)
Chlorococcum humicola	Intracellular	Spherical, 16 nm	48 h of incubation at 24 °C	XRD, UV-Vis, EDX, FTIR and SEM	Intracellular proteins	Jena et al. (2013)

Algal species	Location	Morphology	Optimum conditions	Characterization	Reducing agent	References
Euglena gracilis	Extracellular	Polydispersed, spherical shaped, 15–60 nm size	Incubation time of 1–4 h, pH 7.5	TEM,UV-Vis, ICP-AES, FTIR and EDAX	Proteins (primary)	Li et al. (2015)
Chlorella pyrenoidosa	Extracellular	FCC shaped, 5–20 nm in size	Incubation period of 1 day at a temperature of 28 °C, centrifugation at an rpm of 10,000 for 15 min	FTIR, TEM, UV-Vis, EDX, SEM and XRD	Peptides (cyclic)	Aziz et al. (2015)
Euglena intermedia	Extracellular	Polydispersed, spherical shaped, 24 nm in size	Incubation time of 1–4 h, pH 7.5	ICP-AES, FTIR, TEM, SEM, EDAX and UV-Vis	Proteins	Li et al. (2015)
Enteromorpha flexuosa	Extracellular	Circular in shape, 15 nm in size	Incubation period of 1 h	FTIR, XRD, EDS, TEM and UV-Vis	Amines, peptides	Yousefzadi et al. (2014)
Ulva lactuca	Extracellular	Spherical, 20–50 nm	–	UV-Vis, TGA, XRD EDAX, TEM and SEM	Protein molecules	Bhimba and Kumari (2014)
Caulerpa racemosa	Extracellular	FCC, 5–25 nm	24 h of incubation at room temperature	TEM, SEM, XRD and FTIR	Cyclic peptides	Kathiraven et al. (2015)
Pithophora oedogonia	Extracellular	Hexagonal and cubical, 25–44 nm	Incubation at room temperature, centrifugation at 15,000 rpm for 15 min	EDS, FTIR, UV-Vis and DLS	Proteins and phytochemicals	Sinha et al. (2015)
Caulerpa serrulata	Extracellular	Spherical and FCC, 10 nm	Different pH and time ranges	HR-TEM, UV-Vis, XRD and FTIR	–	Aboelfetoh et al. (2017)
Chaetomorpha linum	Extracellular	Coalescence, 3–44 nm	Incubation for 0.5 h	UV-Vis, FTIR, TEM, SEM and EDX	–	Kannan et al. (2013b)
Codium capitatum	Extracellular	Coalescence, 3–44 nm	Incubation for 48 h at room temperature	UV-Vis, TEM, DLS and FTIR	Carboxylic group	Kannan et al. (2013a)

(continued)

Table 8.1 (continued)

Algal species	Location	Morphology	Optimum conditions	Characterization	Reducing agent	References
Acutodesmus dimorphus	Extracellular	Spherical, 2–20 nm	24 h of incubation centrifugation (rpm 16,000, time 20 min, temp 4 °C)	UV-Vis, TEM and FTIR	Proteins	Chokshi et al. (2016)
Botryococcus braunii	Extracellular	Spherical, 16 nm	3 h of continuous stirring at room temperature, dry AgNPs at temp 55 °C for a time of 5 h	–	Amines and peptides	Arya et al. (2019)
Ulva reticulata	Extracellular	Spherical, 40–50 nm	Incubation at room temperature	UV-Vis, XRD, FTIR and SEM	Benzene rings, fluoroalkane and carboxylic acids	Kannan et al. (2013a)
Enteromorpha compressa	Extracellular	Spherical, 0–50 nm	–	UV-Vis, FTIR, SEM and XRD	Hydrogen-bonded alcohols and benzene rings	Kannan et al. (2013a)
Codium capitatum	Extracellular	3–44 nm	Incubation at room temperature in dark	FTIR, UV-Vis and EDX	Sulphate, amine and peptides	Priyadharshini et al. (2014)
Urospora sp.	Extracellular	Spherical and FCC, 20–30 nm	Incubate at 70 °C in dark on magnetic stirrer	FTIR, UV-Vis, HR-TEM and XRD	Carbonyl and hydroxyl groups	Suriya et al. (2012)
Chaetomorpha linum	Extracellular	3–44 nm	Incubate at 37 °C at (static)	SEM, FTIR and UV-Vis	Peptides, terpenoids, flavonoids	Kannan et al. (2013b)
Ulva fasciata	Extracellular	Spherical, 7–20 nm	Continuous stirring for 20 min at temp 70 °C	TEM, GLC, UV-Vis and FTIR	Polyphenolic groups	El-Rafie et al. (2013)
Ulva reticulata	Extracellular	Spherical, 40–50 nm	–	HR-TEM, UV-Vis, FTIR and XRD	–	Dhanalakshmi et al. (2012)
Spirogyra varians	Extracellular	Spherical	20 min of incubation in dark	UV-Vis	Proteins	Salari et al. (2016)

Algal species	Location	Morphology	Optimum conditions	Characterization	Reducing agent	References
Ulva flexousa	Extracellular	Circular and FCC, 2–32 nm	Incubate for 24 h under illuminated conditions, centrifuge at 5000 rpm	XRD, UV-Vis, TEM and FTIR	Peptides	Rahimi et al. (2014)
Spirogyra varians	Extracellular	FCC, 17.6 nm	Incubate for 20 min in dark	UV-Vis, FTIR and SEM	Amino, hydroxyl and carboxylic groups	Salari et al. (2016)
Spirogyra insignis	Extracellular	Spherical, 28–41 nm	pH 2, 4 and 10	UV-Vis, TEM and FTIR	Proteins	Castro et al. (2013)
Ulva fasciata	Extracellular	–	–	UV-Vis	Proteins	El-Rafie et al. (2013)

alvarezii, have been reviewed in AgNPs biosynthesis. *Porphyra vietnamensis* is the most evident specie of red algae because of its robust reducing agent like sulphated polysaccharides that contain anionic disaccharide elements (Kumar et al. 2012a; Rao et al. 2007; Venkatpurwar and Pokharkar 2011). The literature reviewed that most of the silver nanoparticles which are spherical in shape are produced from red algae and range from 21 to 60 nm in size (Pugazhendhi et al. 2018) and these extracellularly synthesized AgNPs have anti-microfouling activity in medical procedures (Pugazhendhi et al. 2018; Vadlapudi and Amanchy 2017). Intracellular synthesis of AgNPs requires more stages in order to evacuate microfouling, as compared to the extracellular method (Nabikhan et al. 2010). One more vital category of red algae named *Gelidium amansii* was recorded in both the biosynthesis of silver nanoparticles and mitigation of microfouling with the help of the 96-well method (Kumar et al. 2013a; Selvam and Sivakumar 2015). In one report, *Acanthophora spicifera* synthesized AgNPs in 20 min of incubation at 60 °C by an extracellular process (Kumar et al. 2012b).

Another type of algae is blue-green algae, which are classified in the order *Chroococcales*, which is further divided into two classes, namely *Entophysalidaceae* and *Chroococcaceae* (Mosulishvili et al. 2007). Members of these algae are photoautotrophic in nature that contain chlorophyll *a* and carotene like phototrophic pigments that aid in photosynthesis (Khan et al. 2019). Many blue-green algae like *Microchaete diplosiphon*, *Nostoc ellipsosporum*, *Spirulina platensis*, *Phormidium tenue*, *Anabaena* sp., *Oscillatoria willei* and many others are reported in the biosynthesis of AgNPs. *S. platensis* is one of the most prominent blue-green algae majorly reported in AgNPs synthesis (Mukherjee et al. 2002). It is a free-floating filamentous type of cyanobacteria that contain 60–70% of vegetable proteins, essential amino acids and vitamins for capping and reduction of AgNPs (Doshi et al. 2007). Synthesis of spherically shaped 2–8 nm sized AgNPs by *S. platensis* in the literature is being effectively used in food, pharmaceutical and health sectors. In addition to *S. platensis,* other species of blue-green algae, including *Cylindrospermum stagnale* (pentagonal, size of 38–88 nm), *Plectonema boryanum* (octahedral, 200 nm), *Oscillato riawillei* (spherical, 10–25 nm), *Microchaete* (polydispersed, 80 nm) and *Plectonema boryanum* (octahedral, 200 nm), also contributed to AgNPs synthesis (Husain et al. 2015). *Microchaete* synthesized spherical and polydispersed AgNPs of 80 nm within 1 h of incubation at 60 °C with pH 5.6, while the presence of cellular metabolites inside *Microchaete* behaved as capping and reducing agents (Husain et al. 2019).

Green algae or chlorophytes are divided into two major groups of micro- and macro-algae on the basis of their habitats (Kim et al. 2011). Microgreen algae come under the order *Cladophorales* that contain high quantities of various essential components such as carbohydrates, flavonoids, phenols and alkaloids. They are used as stabilizing and reducing agents during the in vitro biosynthesis of AgNPs (Yousefzadi et al. 2014; Aziz et al. 2014, 2015). Almost 20 types of microgreen algae have been used in the biosynthesis of AgNPs of different sizes, shapes and physicochemical properties. The synthesized NPs are then characterized by X-ray powder diffraction (XRD), EDX, FTIR, SEM and transmission electron microscope (TEM) (Jena et al.

2014; Khanna et al. 2019; Lengke et al. 2006b; Sinha et al. 2015; Vigneshwaran et al. 2007). Species of microgreen algae like *Chlorella vulgaris* (triangular, 28 nm), *Enteromorpha flexuosa* (circular, 15 nm), *Scenedesmus* sp. (15–20 nm), *Pithophora oedogonia* (hexagonal and cubical, 24–55 nm), *Plectonema boryanum* (intracellular, 10 nm) and *Chlorococcum humicola* (spherical, 16 nm) have generated AgNPs of various sizes (Barwal et al. 2011; Jena et al. 2014; Xie et al. 2007b; Yousefzadi et al. 2014). Green macro-algae have synthesized AgNPs with antiviral, anticancer and antioxidant characteristics (Priyadharshini et al. 2014; Salari et al. 2016; Singh et al. 2013). One such dominant green macro-alga is *Ulva fasciata* which synthesizes nanosized silver colloids that are helpful in the evaluation of antimicrobial efficacy on cotton fabrics in the absence of citric acid (El-Rafie et al. 2013). Another study on algae-synthesized AgNPs reported the generation of spherically shaped 55–99 nm-sized AgNPs from *Gracilaria edulis* that were centrifuged at 10,000 rpm (Madhiyazhagan et al. 2017; Priyadharshini et al. 2014). The seaweed algae *Chaetomorpha linum* was also found effective in the synthesis of extracellularly 3–44 nm sized AgNPs by incubating the reaction mixture at 37 °C under static conditions and in the presence of flavonoids, terpenoids and peptides as reducing agents (Kannan et al. 2013b; Priyadharshini et al. 2014).

8.4 Algal-Mediated Biosynthesis of AuNPs

Among all metallic nanoparticles, silver and gold NPs are considered as the most attractive ones due to their unique morphology and imperative bioactivities like antibacterial, antifouling, antioxidant and anticoagulant properties (Ghodake and Lee 2011; Khanna et al. 2019; Kushnerova et al. 2010). Biologically synthesized AuNPs have red wine coloured solution, with a size range of 1 nm–8 μm opting different shapes of spherical, octahedral, triangular, multiple twined and many others (Singh et al. 2018). They show great binding for organic moieties with low toxicity levels and so are used as therapeutic agents or vaccine carriers for enhanced drug efficacy (Khan et al. 2014). Due to their exceptional optoelectronic and physicochemical properties, they are widely used in the indication of colourful probes and in the sensing of various analytes (Chithourani et al. 2010). With the vast range in size and shape of AuNPs, they are excelling in various biomedical fields of photodynamic therapy (PDT), photothermal therapy (PTT), tumour imaging, DNA labelling, cancer therapy (Khan et al. 2014) and immune-chromatographic detection of various pathogens (Deb et al. 2011). The currently reported different algal-mediated AuNPs are discussed in detail in Table 8.2.

Like AgNPs, brown algae are also extensively reported in the synthesis of AuNPs by various strains. *T. conoides* are considered to be the most reported brown algae for the biosynthesis of AuNPs. They generate a variety of shapes such as rectangular, triangular and spherical forms of AuNPs from chloroauric acid salt precursor by the extracellular method (Khodashenas and Ghorbani 2019). Along with *T. conoides*, the synthesis of AuNPs is greatly reported by *Laminaria japonica*. They

Table 8.2 Algal-mediated biosynthesis of AuNPs

Algal species	Location	Morphology	Conditions for synthesis	Characterization	Reducing agent	References
Rhizoclonium fontinale	Intracellular	Spherical (5–20 nm), nanotriangles (15–88 nm)	Incubation 72 h pH 9	EDAX, HR-TEM, DLS and UV-Vis	Intracellular synthesis	Dhanalakshmi et al. (2012)
Rhizoclonium fontinale	Intracellular	Spherical (5–20 nm) and hexagonal (34 nm)	–	UV-Vis, HR-TEM, SEM, EDAX and DLS	Intracellular proteins	Parial and Pal (2015)
Chlorella vulgaris	Extracellular	9–20 nm	Incubation 3 h Temp 45–90 °C	UV-Vis, FTIR, XRD and SEM	Biomolecules (amino, carboxylic, phosphate, thiol)	Wei and Qian (2008)
Phormidium tenue	Extracellular	Irregular and spherical shaped 14.84 nm	–	XRD, UV-Vis and TEM	Cellular metabolites	Lengke et al. (2006b)
Tetraselmis kochinensis	Intracellular	Triangular, FCC and spherical, 5–35 nm	Incubation 48 h RPM 200 (shaker) Temp 28–29 °C	TEM and UV-Vis	Cell wall and cytoplasmic enzymes	Senapati et al. (2012)
Coelastrella sp.	Intracellular	Spherical, 30 nm	Incubation 1 h in dark	EDAX, FTIR, UV-Vis, TEM and SEM	Cytoplasm's microproteins	MubarakAli et al. (2013)
Cosmarium impressulum	Extracellular	Spherical, 5.7 nm	Incubation (16 h light 8 h dark) Temp 20 °C	TEM and UV-Vis	Enzymes	Dahoumane et al. (2012)
Spirogyra submaxima	Intracellular	Triangular and spherical	–	UV-Vis, zeta-potential, TEM and XRD	Cytoplasm and chloroplast metabolites	Roychoudhury and Pal (2014)
Prasiola crispa	Extracellular	Spherical, 5–25 nm	–	UV-Vis, XRD, HR-TEM, DLS and FTIR	Proteins	Parial and Pal (2014)
Rhizoclonium fontinale	Intracellular	Spherical, 5–20 nm and hexagonal, 34 nm	Incubation 72 h pH 9	UV-Vis, SEM, HR-TEM, DLS and EDAX	Intracellular proteins	Parial and Pal (2015)

Algal species	Location	Morphology	Conditions for synthesis	Characterization	Reducing agent	References
Chlorella vulgaris	Extracellular	Triangular and truncated	Incubation 48 h	FTIR, AFM, UV-Vis, FESEM and SAED	Proteins	Xie et al. (2007a)
Klebsormidium flaccidum	Intracellular	10–20 nm	Total incubation 100 h	TEM, UV-Vis, PAM, SERS and XPS	In thylakoids	Sicard et al. (2010)
Chlorella pyrenoidusa	Extracellular	Icosahedral and spherical, 25–30 nm	Incubation Temp 100 °C at 100 rpm pH 8	UV-Vis, XRD and HR-TEM	Enzymes	Oza et al. (2012)
Tetraselmis suecica	Extracellular	Spherical, 120 nm	Incubation 5 min at temp 90 °C	XRD, UV-Vis and FTIR	Hydroxyl, carbonyl and nitrate function groups	Shakibaie et al. (2010)
Prasiola crispa	Extracellular	FCC, facile and spherical, 25 nm	Incubation 12 h Centrifugation at 10,000 rpm	UV-Vis, DLS, XRD and FTIR	Proteins	Sharma et al. (2014a)
Nostoc ellipsosporum	Extracellular	Decahedral and icosahedral, 20–40 nm	Incubation 3 h pH 5	FTIR, UV-Vis and SEM	Carboxylate groups and proteins	Parial et al. (2016)
Spirulina platensis	Extracellular	Monodispersed and spherical, 2–8 nm	–	EDAX, UV-Vis, FTIR and HR-TEM	Proteins	Thota and Crans (2018)
Spirulina platensis	Extracellular	Spherical, 20–30 nm	Incubation 40 h	TEM, UV-Vis, FTIR, XRD, SEM-EDAX, NAA and AAS	Carboxylic, phosphate, thiol and biomolecules (amino)	Kalabegishvili et al. (2012)
Spirulina platensis	Extracellular	Octahedral and cubic	Incubation 48 h, Centrifugation at 10,000 rpm	UV-Vis and SEM	Peptides and proteins	Suganya et al. (2015)
Plectonema boryanum	Extracellular	20–25 nm	Incubation 24 h Temp 200 °C	SEM, TEM, XPS and TOF-SIMS	–	Lengke et al. (2006a)

(continued)

Table 8.2 (continued)

Algal species	Location	Morphology	Conditions for synthesis	Characterization	Reducing agent	References
Synechocystis sp.	Extracellular	Spherical, 3–13 nm	Incubation 16 h Temp 20 °C in light (50 μmol photons m^{-2} s^{-1}) and 4 °C in dark	SERS, zeta-potential and TEM	Carboxylic groups and polysaccharides	Focsan et al. (2011)
Lyngbya majuscula and *Spirulina subsalsa*	Intracellular	Spherical, 20 nm	Incubation 72 h pH 6, 7 and 8	TEM	–	Chakraborty et al. (2009)
Phormidium tenue	Extracellular	Spherical and irregular, 14.84 nm	72 h of exposure, pH 7 and 1	TEM, XRD and UV-Vis	Metabolites and enzymes	Parial et al. (2012)
Phormidium sp.	Intracellular	Triangular, 25 nm	Incubation at room temperature	UV-Vis, HOUR-SEM, EDX, TEM and FTIR	Cytoplasmic proteins	MubarakAli et al. (2012)
Phormidium sp.	Intracellular	Monodispersed, triangular	–	SEM, UV-Vis and TEM	Proteins	Mosulishvili et al. (2002)
Phormidium valderianum	Extracellular	Spherical, hexagonal and FCC, 24 nm	Incubation 72 h Temp 20 °C	TEM, XRD and UV-Vis	Metabolites	Parial et al. (2012)
Lyngbya majuscula	Extracellular	Hexagonal and spherical, 2–25 nm	Incubation at room temperature and centrifugation 10,000 rpm	FTIR, TEM, UV-Vis and MLDLS	Proteins	Parial and Pal (2014)
Anabaena sp.	Extracellular	Spherical, 10 nm	Incubation 4–40 h	TEM and XRD	Protein moieties	Rösken et al. (2016)
Spirulina subsalsa	Extracellular	Spherical, 30 nm	–	XRD, MLDLS, UV-Vis and FTIR	Extracellular proteins	Parial and Pal (2014)
Kappaphycus alvarezii	Extracellular	Polydispersed, 10–40 nm	–	TEM, UV-Vis, XRD, FTIR and FAAS	Polyphenolic compounds	Rajasulochana et al. (2012)

Algal species	Location	Morphology	Conditions for synthesis	Characterization	Reducing agent	References
Galaxaura elongata	Extracellular	Rod, truncated and triangular shaped, 3.85–77.13 nm	Incubation (10 min–12 h) Continuous stirring 120 rpm	GC-MS TEM, zeta-potential and HPLC	Palmitic acid	Abdel-Raouf et al. (2017)
Chondrus crispus	Extracellular	Spherical and polyhedral, 30–50 nm	Room temperature Stirring, pH 2, 4 and 10	FTIR, EDS, SEM, UV-Vis, TEM and FAAS	Proteins	Castro et al. (2013)
Lemanea fluviatilis	Extracellular	Polydispersed, FCC and spherical, 5–15 nm	Incubation 12 h Room temperature Stirring	UV-Vis, XRD, TEM, FTIR and DLS	Organic molecules and proteins	Sharma et al. (2014b)
Gracilaria corticata	Extracellular	Spherical, 45–57 nm	Room temperature Stirring	SEM and UV-Vis		Naveena and Prakash (2013)
Corallina officinalis	Extracellular		–		Functional groups	El-Kassas and El-Sheekh (2014)
Sargassum spp.	Extracellular	Longest edges, hexagonal and triangular, 300–400 nm	pH neutral Incubation 5 h Room temperature	AFM, FTIR, UV-Vis and XRD	Functional groups like hydroxyl and amine	Liu et al. (2005)
Fucus vesiculosus	Extracellular	Spherical, 20–50 nm	Incubation time (1–8 h) pH range (2–9)	XRD, FTIR, TEM, SEM and EDS	Hydroxyl groups present in polysaccharides	Mata et al. (2009)
Laminaria japonica	Extracellular	Spherical, 15–20 nm	–	UV-Vis, XRD, TEM and FTIR	Polysaccharides	Ghodake and Lee (2011)
Turbinaria conoides	Extracellular	Polydispersed, rectangular, triangular, 6–10 nm	Incubation 24 h	TEM, UV-Vis and FTIR	Polyphenolic groups, carboxylic and fucoidan	Rajeshkumar et al. (2013)

(continued)

Table 8.2 (continued)

Algal species	Location	Morphology	Conditions for synthesis	Characterization	Reducing agent	References
Ecklonia cava	Extracellular	Spherical, 20–50 nm	Incubation 1 min Temp 80 °C magnetic stirrer	FTIR, XRD, UV-Vis, FESEM and TEM	Phenolic and hydroxyl groups	Venkatesan et al. (2014)
Sargassum wightii	Extracellular	Thin planner, 8–12 nm	12–15 h continuous stirring	HR-TEM, UV-Vis and XRD	–	Singaravelu et al. (2007)
Sargassum myriocystum	Extracellular	Polydispersed, spherical and triangular, 15 nm	Incubation 15 min Temp 72 °C water bath	GC-MS, FTIR, XRD, EDAX and UV-Vis	1-cyclopentyl group	Dhas et al. (2012)
Turbinaria conoides	Extracellular	Rectangular, square and triangular shaped, 15 nm	1 h of continuous stirring Incubation at room temperature	SEM and EDS	Biochemical groups	Rajeshkumar et al. (2012)
Stoechospermum marginatum	Extracellular	Hexagonal, triangular and spherical, 18.7–93.7 nm	Incubation 2 h	WDXRF, FTIR, XRD, TEM and SEM	Hydroxyl groups	Rajathi et al. (2012)
Padina gymnospora	Extracellular	Spherical, 8–21 nm	Temperature 24–95 °C, Incubation of 0–20 h, pH 4–12	AFM, XRD, UV-Vis, FTIR and TEM	Fucoxanthin and flavonoids	Singh et al. (2013)
Dictyota bartayresiana	Extracellular	Polydispersed and spherical nanoparticles	–	FTIR, UV-Vis and SEM	Phenolics, amine and carboxylic groups	Varun et al. (2014)
Ecklonia cava	Extracellular	Spherical, FCC and triangular, 20–50 nm	Incubation 1 min Temp 80 °C	FTIR, UV-Vis, FESEM and EDX	Hydroxyl and amine and groups	Venkatesan et al. (2014)
Padina gymnospora	Extracellular	Spherical, 8–21 nm	Temperature range of 24–95 °C, incubation of 0–20 h pH 4–12	UV-Vis, HR-TEM, AFM and XRD	Flavonoids and fucoxanthin	Singh et al. (2013)

Algal species	Location	Morphology	Conditions for synthesis	Characterization	Reducing agent	References
Sargassum muticum	Extracellular	FCC and spherical, 5.42 nm	Incubation 1 h stirring at temp 45 °C	Zeta-potential, TEM and UV-Vis	Bio-organic compounds and proteins	Namvar et al. (2015)
Sargassum tenerrimum	Extracellular	Anisotropic and polydispersed, 4–45 nm	Incubation 1.5 h	HR-TEM, zeta-potential, UV-Vis DLS and FTIR	Secondary metabolites	Ramakrishna et al. (2016)
Cystoseira baccata	Extracellular	Polycrystalline and spherical, 8.4 ± 2.2 nm	Incubation 24 h stirring at room temperature	UV-Vis, zeta-potential, TEM, FTIR, EDS and EELS	Polyphenols and polysaccharides	González-Ballesteros et al. (2017)

contain a rich amount of reducing and stabilizing agents such as peptides, phenols and vitamins (Kushnerova et al. 2010). One study reported the generation of spherically shaped AuNPs (15–20 nm) by the reduction of phytochemicals and polysaccharides (Ghodake and Lee 2011). Spherically shaped AuNPs (20–50 nm) were synthesized from *Ecklonia cava* with incubation for 1 min in a water bath of 80 °C on a continuous magnetic stirrer (Venkatesan et al. 2014). Many other brown algae like *Sargassum myriocystum*, *Cystoseira baccata*, *Padina gymnospora*, and *Sargassum wightii*, mentioned in Table 8.2, are reported for AuNPs biosynthesis.

Very little literature is reported on the biosynthesis of AuNPs from red algae. *Lemanea fluviatilis*, marine red algae, was used to generate 5.9 nm FCC (face-centred cubic), crystalline and polydispersed AuNPs from chloroauric acid metal precursor with 12 h of incubation at room temperature (Murugesan et al. 2017; Singh et al. 2015). Another study showed the biosynthesis of spherically and FCC-shaped AuNPs (30–50 nm) from *Chondrus crispus* within 24 h of incubation on continuous stirring at pH 2, 4 and 10 and was further analysed with FTIR, TEM, UV-vis, XRD and SEM (Castro et al. 2013). Along with the above-mentioned species, many other red algal species like *Galaxaura elongata*, *Kappaphycus alvarezii*, *Gracilaria corticata* and *Corallina officinalis* are involved in the biosynthesis of AuNPs. Apart from monometallic AuNPs, red algal species *Gracilaria edulis* also showed its involvement in the biogeneration of Ag-Au bimetallic NPs with the solutions of different molarities (i.e., 1:1, 1:3 and 3:1) of chloroauric acid ($HAuCl_4$) and silver nitrate ($AgNO_3$) (metal salt precursor) and reported their anticancerous activity against human breast cancer lines (Al-Naamani et al. 2017).

Spirulina platensis is a type of blue-green alga that has also shown its contribution to the biogeneration of AuNPs by the reduction and stabilization with peptides and protein. The biosynthesis of spherical, cubic and octahedral-shaped AuNPs by *S. platensis* is reported. The solution was centrifuged at 10000 rpm after 40–48 h incubation at room temperature (Iravani et al. 2018; Kalabegishvili et al. 2012). The blue-green alga, *Phormidium valderianum*, which is also known to be alkalotolerant was experimented to synthesize spherical and hexagonal-shaped AuNPs (24 nm) with 72 h incubation by bioreduction with extracellular metabolites (Parial et al. 2012). *Anabaena* sp. generated spherical AuNPs (10 nm) with incubation of 4–40 h at room temperature (Rösken et al. 2016). In another study, by using UV-Vis spectroscopy monodispersive AuNPs were analysed at an absorbance of 1.897 and a wavelength of 530 nm (Mosulishvili et al. 2002). Along with monometallic AuNPs, *S. platensis* has also shown its role in the bimetallic core-shell Ag-AuNPs biosynthesis with the aid of extracellular proteins (Govindaraju et al. 2008; Mahdieh et al. 2012).

Like other types of algae, both green micro- and macro-algae have shown their contributions to the AuNPs synthesis by overcoming the challenge of their non-reproducibility in perfect size and shape (Dhanalakshmi et al. 2012; Parial and Pal 2014; Priyadharshini et al. 2014). *Prasiola crispa* is one of the green macro-algae, inhabiting higher altitudes that generate spherical AuNPs by the extracellular pathway (Parial and Pal 2014). The intracellular and extracellular syntheses of spherical

AuNPs (25 nm) with 12 h of incubation at room temperature with the involvement of proteins, peptides and functional groups present inside *Prasiola crispa* are reported (Beganskienė et al. 2004; Castro et al. 2013; Kannan et al. 2013a; Xie et al. 2007a). Besides *P. crispa*, many other green macro-algae such as *Rhizoclonium fontinale*, *Cosmarium impressulum*, *Coelastrella* sp. and *Tetraselmis kochinensis* are reported in AuNPs biosynthesis, as mentioned in Table 8.2.

8.5 Algal-Mediated Biosynthesis of Other NPs

Along with Au and AgNPs, algal-mediated biosynthesis of other monometallic, bimetallic and metal oxide NPs is also being exponentially practiced these days. Bimetallic nanoparticles are the combination of two metallic nanoparticles with enhanced physicochemical properties (Singh et al. 2013). Brown algae are also well documented in the biosynthesis of titanium oxide nanoparticles (TiO_2NPs) and zinc oxide nanoparticles (ZnONPs) like metallic oxide nanoparticles (Sirelkhatim et al. 2015). One study reported the biosynthesis of ZnONPs from *S. muticum* by mixing the dried algal powder in distilled water under continuous heating at 45 °C till complete homogenization. Then the salt solution of added zinc acetate was added and it was heated again for 4 h till the generation of hexagonal ZnONPs (935–957 nm) (Azizi et al. 2014; Azizi et al. 2017). Along with brown, green micro- and macroalgae species are also reported for the green synthesis of many bimetallic NPs. *Chlamydomonas reinhardtii*, a freshwater green microalgal species was reported for the biosynthesis of cadmium sulphide nanoparticles (CdSNPs). It is a semiconductor that comes under the groups II–VI and has unique optoelectronic characteristics (Rao and Pennathur 2017). Green microalgae are also useful in the synthesis of silicon nanoparticles (SiNPs) that act as bioindicator in many industrial and nonindustrial wastes for the identification of noxious compounds. *C. vulgaris* is one such type of green microalgae reported for the generation of SiNPs by the polycondensation and hydrolysis of silicon alkaloids in the presence of proteins and peptides, present in it (Wei and Qian 2008). *Gracilaria edulis* is a species of green macroalgae reported in AgNPs as well as ZnONPs synthesis (Madhiyazhagan et al. 2017; Priyadharshini et al. 2014). The algal-mediated biosynthesis of various other NPs is summarized in Table 8.3.

8.6 Therapeutic Applications of Algae-Mediated NPs

The application of NPs in humans should be free of any toxicity, so NPs synthesized from algae are preferable, as they are free of impurities and chemical agents. Therefore, green-synthesized NPs find their application predominantly in therapeutics discussed herein.

Table 8.3 Algal-mediated biosynthesis of various other metallic, bimetallic and metal oxide NPs

Algal species	NPs	Location	Morphology	Conditions for synthesis	Characterization	Reducing agent	References
Sargassum muticum	ZnO	Extracellular	Hexagonal, 30–57 nm	Heating at 450 °C for 4 h	XRD, UV-Vis, FESEM, SEM and FTIR	Sulphate and hydroxyl groups	Azizi et al. (2014)
Sargassum muticum	Fe_3O_4	Extracellular	Crystalline and spherical, 43–79 nm	1.5 h of incubation	FTIR, XRD, UV-Vis, SEM and EDX	Hydroxyl and sulphate groups	Mahdavi et al. (2013)
Chitosan from algae	ZnO	Extracellular	Crystalline	Overnight incubation, pH 6, temp 60 °C	TEM, UV-Vis, SEM and FTIR	Functional groups	Al-Naamani et al. (2017)
Spirulina platensis	Si	Extracellular	Crystalline	Incubation 24 h pH 7, temp 25 °C	UV-Vis	–	Mahdieh et al. (2012)
Spirulina platensis	Au-Ag	Extracellular	Core-shell	Incubation 120 h pH 5.6, temp 37 °C	UV-Vis and SEM	Bio-active compounds	Govindaraju et al. (2008)
Chlamydomonas reinhardtii	CdSNPs	Extracellular	–	Incubation at 65 °C, centrifugation 8720 g	SEM and UV-Vis	Extracellular proteins	Rao and Pennathur (2017)
Cholera vulgaris	Si	Extracellular	Spherical	Slow incubation for 13 days	UV-Vis, TEM and SEM	Proteins and peptides	Beganskienė et al. (2004)
Gracilaria edulis	Ag and ZnO	Extracellular	55–99 nm and 66–95 nm	Centrifugation at 10 k RPM	UV-Vis, EDX, FESEM, XRD and FTIR	Alcohol, amides and nitro groups	Priyadharshini et al. (2014)

8.6.1 Antibacterial Activity

In order to control and treat bacterial infections, widespread antibiotics usage has caused bacterial species to become multidrug resistant. So, now to develop a safe, economic and proficient treatment against multidrug-resistant strains of bacteria has become a health concern worldwide. Globally, a new shift has emerged with nanoparticles' usage as an alternative source of antibacterial agent, having the potential of superior and efficient bactericidal activities. Since NPs work on the principle of killing bacteria by penetrating cell walls thus disrupting their cell membrane and by interacting with molecular proteins and DNA, it generates reactive oxygen species (ROS) that enhance inhibitory actions against Gram-positive and Gram-negative bacterial species (Wang et al. 2017; Swamy and Prasad 2012; Prasad and Swamy 2013; Aziz et al. 2016; Joshi et al. 2018).

NPs synthesized by algal strains are known to have antibiotic properties against many bacterial species. Various studies have shown that algal AgNPs have great antibacterial properties against Gram-negative as well as Gram-positive bacteria because of their large surface areas. However, comparatively Gram-positive bacteria show more resistance towards AgNPs because of the presence of a thick peptidoglycan layer, making it difficult to penetrate the cell wall. In contrast, Gram-negative bacteria having a thin peptidoglycan layer allow an easy attachment of AgNPs on the surface of the cell and thus fail to prevent them from penetrating the cell wall to reach the cell membrane (González-Ballesteros et al. 2020). AgNPs biosynthesized via green alga *Neochloris oleoabundans* cell extract were reported to having a higher bacterial resistance against *Escherichia coli* in a concentration ranging from 0 to 100 μg/ml. The total number of colonies was counted to check the cell survival rate during 6 h culture, the highest inhibitory action of AgNPs was recorded at 100 μg/ml concentration after 5 h which decreases with a decrease in NP concentration. However, the survival rate of NPs at 100 μg/ml was 1.4%, as compared to that of control (1.28×10^9/ml) (Bao et al. 2019). In another report, AgNPs synthesized by dried biomass of a green alga *Chlorella ellipsoidea* have shown antibacterial activities against *Staphylococcus aureus* (Gram-positive bacteria), *Klebsiella pneumoniae*, *Pseudomonas aeruginosa* and *E. coli* (Gram-negative bacteria) in the range of 5, 15, 10 and 20 μg/ml, respectively. The zone of inhibition after 24 h incubation was maximum of 16 mm against *P. aeruginosa*, 15 mm against *S. aureus*, 9 mm against *E. coli* and minimum of 8 mm against *K. pneumoniae* at all concentrations (Borah et al. 2020). Another study reported the stable colloidal AgNPs synthesized from the aqueous extract of *Caulerpa serrulata* (green marine algae), which showed remarkable bacterial resistance ability, even at low concentrations against *Pseudomonas aeruginosa*, *Staphylococcus aureus*, *Shigella* sp., *Salmonella typhi* and *Escherichia coli*. Seventy-five microlitres of AgNPs formed the highest zone of inhibition (21 mm) against *E. coli*, while at 50 μl, AgNPs formed the smallest zone of inhibition (10 mm) against *S. typhi* (Aboelfetoh et al. 2017). Similarly, AgNPs synthesized from marine macro-alga *Padina* sp. have shown exceptional bacterial resistance against two Gram-positive bacterial strains, namely *Bacillus subtilis* and *Staphylococcus aureus*, and against three strains of Gram-negative bacteria, namely *E. coli*, *Salmonella typhi* and *Pseudomonas aeruginosa*.

The biggest inhibition zone was found to be of *S. aureus*, which was 15.17 ± 0.58 mm, then it was of *B. subtilis* (12.67 ± 0.76 nm). However, in the case of Gram-negative strains, the highest inhibition zone was found in *P. aeruginosa* as 13.33 mm ± 0.76, followed by the inhibitory zone of *E. coli* that was 12.67 ± 0.58 mm (Bhuyar et al. 2020). AgNPs mediated by brown macroalga *Sargassum muticum* have exhibited antibacterial potential against *Staphylococcus aureus*, *Staphylococcus saprophyticus* and *Enterococcus faecalis* and have shown minimum inhibitory concentrations of 13.50 μg/ml against *E. faecalis*, 6.75 μg/ml against *S. saprophyticus* and 3.38 μg/ml against *S. aureus* (González-Ballesteros et al. 2020).

Algal-mediated AuNPs have also exhibited potential antibacterial activities against various bacterial strains. *Sargassum plagiophyllum* mediated AuNPs have been used against *E. coli* and *S. typhi* in a concentration of 10 μg/ml and live/dead cells were stained. The stained cells were examined via epifluorescence microscopy, which clearly showed the antibacterial activity of AuNPs showing complete membrane lysis of bacterial strains (Dhas et al. 2020). In another report, the antibacterial activity of AuNPs biogenerated by marine alga *Gelidiella acerosa* was examined against *Klebsiella pneumoniae*, *Serratia marcescens*, *Escherichia coli* and *Bacillus subtilis* using the agar diffusion method. The inhibitory zone for each bacterial strain was measured and the maximum antibacterial activity was recorded against *B. subtilis*, following that were the zones of *E. coli*, *S. marcescens* and *K. pneumoniae* (Senthilkumar et al. 2019). Similarly, the agar diffusion method was adopted to estimate the antibacterial potential of red algae *Acanthophora spicifera* mediated AuNPs against *V. harveyi* and *S. aureus*. AuNPs were treated in a concentration of 25, 50, 75 and 100 μg/ml. The antibacterial activity showed the highest zone of inhibition of 22 ± 0.3 mm for *V. harveyi* and 18.7 ± 0.5 mm for *S. aureus* at 100 μg/ml and decreased with decrease in AuNPs concentration (Babu et al. 2020). These results show that algal-based NPs have promising applications in treating antibacterial infections.

8.6.2 Antifungal Activity

The unavailability and emerging resistance against antifungal drugs and towards have increased global health concern regarding fungal infections. There is a great incentive towards the development of innovative, strong and effective antifungal drugs. In order to treat fungal infections, nanoparticles can serve as novel treatment options due to their exemplary fungicidal activities (Mallmann et al. 2015). AgNPs synthesized via the green approach have shown the most efficient antifungal potential so far.

A study reported that the biosynthesis of *Neodesmus pupukensis* mediated synthesized AgNPs has been examined for their antifungal abilities against 48-h-old antifungal cultures. The synthesized AgNPs were incubated with fungal strains at 28 °C for 72 h. Their antifungal potential was confirmed by measuring mycelial growth inhibition of 57.1%, 65.4%, 69.8%, 79.4% and 80.6% against *Aspergillus fumigatus*, *Fusarium solani, Candida albicans*, *A. niger* and *A. flavus* (Omomowo et al. 2020). Similarly, another investigation has been performed, which made use of the disc diffusion method in order to evaluate the antifungal potential of *Spatoglossum asperum* mediated AgNPs against pure cultures of fungal strains. After completing the incubation period, the synthesized AgNPs showed the highest antifungal activity against *Candida albicans* which formed around a 20.67 mm zone of inhibition at 100 μg/ml concentration. Whereas the 17.67 mm inhibition zone was formed against *Candida tropicalis*, 17.33 mm against *Trichophyton mentagrophytes* and 12.67 mm against *Aspergillus flavus* (Subbiah et al. 2019). Literature reviewed on biosynthesized AgNPs from *Ulva lactuca* (green algae) and *Hypnea muciformis* (red algae) has shown effective activity on the growth retardation of *A. niger*, *Candida parapsilosis* and *C. albicans* (Dhavale et al. 2019). In another report, *Desmodesmus* sp. mediated AgNPs have shown efficient antifungal potential against *Candida parapsilosis* which have shown a minimum inhibitory concentration of 0.78125 μl at 100 μg/ml concentration of AgNPs (Öztürk 2019). These studies have clearly shown the fungicidal behaviour of Ag-NPs against antifungal strains.

Only a few literature regarding the antifungal activity of algal-mediated AuNPs have been investigated. One such experimental research was performed on AuNPs which were synthesized from an aqueous extract of the brown seaweed *Dictyota bartayresiana.* The NPs made out of brown seaweed showed antifungal activity against the soft rot fungi *Fusarium dimerum* and *Humicola insolens* (El-Sheekh and El-Kassas 2016). Similarly, *Neodesmus pupukensis* mediated synthesized AuNPs have been incubated with 48-h-old fungal strains at 28 °C for about 72 h which resulted in mycelial growth inhibition of 44.3%, 54.9%, 66.4%, 75.4% and 79.4% against *Aspergillus fumigatus*, *Fusarium solani*, *Candida albicans*, *A. flavus* and *A. niger* (Omomowo et al. 2020). All these algal-mediated AgNPs and AuNPs have proved to be efficient antifungal agents against pathogenic fungi.

8.6.3 Antioxidant Activity

Algal-mediated nanoparticles have exhibited tremendous antioxidant activities. This is attributed to the large surface area which has a vital role to play in intensifying the absorption of polyphenolic and bioactive molecules (Babu et al. 2020). Antioxidants are significant in the human body, as they boost up the immune system of the body against free radical oxidative damage by playing a free radical scavenger role. During the synthesis mechanism, these antioxidants came to the surface of NPs which eventually increased their biogenic potential for various therapeutic applications (Zayadi and Bakar 2020; Aziz et al. 2019). NPs' antioxidant potential has been examined via DPPH (2,2-diphenyl-2-picrylhydrazyl hydrate) which is a stable organic free radical and when it accepts the electron from NPs, a colour change from purple to yellow occurs. Both AuNPs and AgNPs synthesized via algae species have shown antioxidant activities reported below.

In a report, AuNPs biologically synthesized by *Chlorella*, a microalga and *Spirulina*, blue-green algae, have exhibited significant antioxidant properties which were studied using the reagent DPPH (2,2-diphenyl-2-picrylhydrazyl hydrate) assay. The algal AuNPs were mixed with DPPH and incubated in a dark room for a period of 30 min. This was followed by recording an absorbance band at 517 nm using UV-vis spectrophotometer. In the case of *Spirulina*-mediated AuNPs, the percentage of radical scavenging activity was reported to be higher at 60.8% ± 2×10^{-3}, as compared to 42.6% ± 7×10^{-3} of *Chlorella*-mediated AuNPs. The higher percentage of *Spirulina*-mediated AuNPs suggested the presence of more antioxidants initially in samples (Zayadi and Bakar 2020). Similarly, the DPPH assay has been used to examine antioxidant activities of *Acanthophora spicifera* mediated AuNPs where higher inhibitory abilities of radical scavenging were reported as 62.8% for algal AuNPs. This was also because of the large surface area of NPs (Babu et al. 2020). In a similar study, microalga *Neodesmus pupukensis* synthesized AuNPs and AgNPs antioxidant abilities were evaluated via DPPH assay following incubation and measuring the absorbance peak at the wavelength of 517 nm. Next, the ability of algal NPs to scavenge free radicals against DPPH was calculated in percentage (%) inhibition which was higher at 68.96% for AuNPs and 41.21% for AgNPs (Omomowo et al. 2020). Likewise, AuNPs and AgNPs synthesized via algal extract of *Sargassum muticum* have been observed for their antioxidant activities using *Sargassum muticum* extract by the DPPH assay. The results showed the DPPH scavenging ability of *Sargassum muticum* extract as EC50 of 69.50 ± 0.02 mg/mL, which is the concentration required to significantly reduce the initial DPPH concentration by half (González-Ballesteros et al. 2020). *Ecklonia cava* mediated AgNPs have been examined to check their antioxidant potential using the DPPH assay. Both AgNPs and *Ecklonia cava* extract have shown a remarkable value of 50% of DPPH scavenging activities at similar concentrations of 100, 250 and 500 μg/ml. As for the algal extract, the great antioxidant potential is possibly due to the initially present polyphenolic compounds and for AgNPs, the high antioxidant activity is due to the *Ecklonia cava* extract present on the surface of the particles (Venkatesan et al. 2016). These reports have veritably confirmed the antioxidant potential algal-mediated NPs which overall enhance their biogenic properties and can be a great candidate as nutraceuticals and pharmaceutical products.

8.6.4 Anticancerous Activity

NPs for cancer treatment and targeted delivery of anticancerous drugs materialize to be one of the most dynamic research areas in the field of nanotechnology (Cheung et al. 2015). Several research studies exhibited the significant anticancerous potential of nanoparticles including *Sargassum vulgare* mediated synthesized AgNPs (10 nm), which have shown remarkable anticancerous activity by treating with human myeloblastic leukaemic cells HL60 and HeLa cells (LewisOscar et al. 2016). Moreover, in vitro cytotoxic effects of *Sargassum muticum* synthesized AgNPs have been observed against the MCF7 breast cancer cell line. The AgNPs were treated for 48 h in a concentration ranging from 3 to 50 μg/ml and the highest viability rate was shown at 12.5 μg/ml, which was 100.36%. These algal-mediated nanoparticles induced intracellular reactive oxygen species, which eventually resulted in great apoptotic rate and cell death (Supraja et al. 2018; Aziz et al. 2019). However, *Spirulina platensis* mediated AgNPs have shown cytotoxic activity against the human HepG2 cell line in a dose-dependent manner observed via MTT (3-(4,5-dimethylthiazol-2-yl)-2,5-diphenyltetrazolium bromide) assay. These algal-synthesized nanoparticles were treated in a concentration ranging from 7.8 to 1000 μg/ml and the viability rate decreased with an increase in nanoparticles' concentration. MTT assay showed the highest viability of 59.2% at 7.8 μg/ml of AgNPs (Namasivayam et al. 2015). Similarly, *Gracilaria corticata* mediated AgNPs have shown anticancerous potential against HepG2 human cell line in a dose-dependent manner. AgNPs were exposed in a concentration ranging from 0.25 μg/ml to 100 μg/ml and viability percentage was the highest (99.2%) at 0.25 μg/ml (Supraja et al. 2016). AgNPs synthesized by *Gracilaria corticata* have also shown cytotoxic effects against the human cervical cancer cell line and have shown results in a dose-dependent manner. The highest viability rate was recorded as 97% at 12.5 μg/ml concentration, instead of 100 μg/ml. In another report, algal-derived chitosan polymer (Chit-AgNPs) coated silver nanotriangles have the potential to serve as photothermal agents against non-small lung cancer cell line of humans (NCI-H460) (Parveen 2017). Despite AgNPs, AuNPs also have the potential to serve as cytotoxic agents against cancer cell lines, however less research has been done. According to a report, *Acanthophora spicifera* mediated AuNPs have been examined for their anticancerous activity against the colorectal adenocarcinoma HT-29 cell line. These synthesized AuNPs were incubated for 24 h with HT-29 cells in concentrations of 1.88, 3.75, 7.5, 15 and 30 μg/ml and the cytotoxic effects were observed via MTT assay. The Au-NPs have shown a maximum inhibitory concentration of 21.86 μg/ml, which veritably confirmed its cytotoxic nature, thus resulting in apoptosis, shrinkage of cells and lack of morphological structure (Babu et al. 2020). Polyethylene glycol coated AuNPs have the potential to destroy maximum tumour cells in comparison to an anticancer cytokine tumour necrosis factor-alpha (Cai et al. 2008; van Horssen et al. 2006). These studies have depicted the cytotoxic effects of algal-derived nanoparticles and their potential to act as an anticancerous drug.

8.6.5 Antidiabetic Activity

Diabetes is characterized as chronic hyperglycaemic metabolic disorder due to insulin resistance or insulin deficiency. Two major digestive enzymes, α-amylase and α-glucosidase, have a significant role to play in carbohydrate digestion. The major antidiabetic approach goes around the inhibition of these digestive enzymes (Manam and Murugesan 2014). Therefore, in order to cure diabetic and other inflammatory disorders, many researchers are in need of developing a new drug for which great focus is on natural marine flora which contains rich nutritional supplements for controlling diabetes. By incorporating seaweeds in a balanced diet helps in controlling 10–15% of diabetes in many Western countries. Hence, marine flora-mediated nanoparticles have gained much attention to be used as nanomedicines in the recent era (Venkatraman et al. 2018).

AgNPs synthesized by marine red alga *Halymenia poryphyroides* exhibited remarkable antidiabetic properties in a dose-dependent manner significantly ($P < 0.005$). The α-amylase inhibition activity increases with increase in NPs concentration, as a value of 26.20 ± 0.02% was observed at 0.2 ml, however a value of 91.30 ± 0.02% was reported at 1.0 ml. Similarly for α-glucosidase, the dose dependence significantly increased with percentage inhibition activity, which was recorded as 33.20 ± 0.01% at 0.2 ml and 89.10 ± 0.01% at 1.0 ml (Manam and Murugesan 2013). In another report, AgNPs synthesized via *Gracilaria edulis* and *Syringodium isoetifolium* have shown antidiabetic activities at 500 μg/ml concentration. The increase in α-amylase inhibition occurred with an increase in concentration, the highest inhibition recorded was 98.75% in the case of *Gracilaria edulis* and 77.25% in the case of *Syringodium isoetifolium* (Abideen and Sankar 2015). However, in the case of marine alga *Colpomenia sinuosa*, significantly increased percentage inhibition activity was observed for both α-amylase and α-glucosidase enzymes. Algal AgNPs were used in varied concentrations from 0.2 mg to 1.0 mg and the highest inhibition activity was recorded at 1.0 mg, which was 94.30 ± 0.10% for α-amylase and 90.50 ± 0.10% for α-glucosidase (Manam and Murugesan 2014). Despite antidiabetic algal AgNPs, other algal-mediated nanoparticles have also shown tremendous antidiabetic potential to be used as great candidates for nanodrug. *Turbinaria conoides* mediated AuNPs have exhibited antidiabetic activities at various concentrations from 50 to 200 μg/ml. The results veritably confirmed the highest α-amylase inhibition activity of 57.3% at 200 μg/ml, as compared to lower concentrations (Venkatraman et al. 2018). In another study, Au-NPs synthesized via marine alga *Gelidiella acerosa* have been evaluated for their antidiabetic potential with acarbose as control. The increase in α-amylase inhibition activity was recorded as IC50 value, which was highest 2.1 ± 0.01 for Au-NPs, as compared to 1.7 ± 0.02 in the case of acarbose control (Senthilkumar et al. 2019). All these studies clearly exhibited antidiabetic activities of algal-mediated nanoparticles to inhibit α-amylase activity, which can eventually control carbohydrate metabolism and decrease glucose absorption.

8.7 Conclusions

Biosynthesis of NPs from the algae continues to be a promising area of nanotechnology, as it ensures eco-friendly, non-toxic, cost-effective and reliable synthesis of NPs. Large amounts of literature have been reported in this regard and intensive research activities are going on. However, there are still several issues associated with green synthesis of NPs from algae such as low yield of NPs, non-uniform morphology of NPs and successful scale-up of algal bioreactors for the synthesis of NPs. In order to counter these issues, strategies such as optimization of synthesis conditions and choosing a high yielding algal strain could be a future direction for research. Moreover, a proficient piece of experimental data analysis and research work is mandatory for the identification and establishment of the roles for specific biomolecules that cause not only reduction but also capping of NPs during the algal-mediated biosynthesis process. Apart from these, AuNPs and AgNPs, other NPs such as palladium, zinc oxide and carbon-based NPs could also be synthesized from algae. Studies involving the use of green algae, cyanobacteria and brown algae in order to carry out the biosynthesis of NPs are widely reported; however, reports on red algae and algae inhabiting extreme habitats are scanty or non-existent, which needs further exploration. Once these factors are addressed, the algal-synthesized NPs could be commercially used for therapeutic applications in the near future.

References

Abdel-Raouf N, Al-Enazi NM, Ibraheem IB (2017) Green biosynthesis of gold nanoparticles using *Galaxaura elongata* and characterization of their antibacterial activity. Arab J Chem 10:S3029–S3039

Abideen S, Sankar M (2015) In-vitro screening of antidiabetic and antimicrobial activity against green synthesized $AgNO_3$ using seaweeds. J Nanomed Nanotechnol 10:2157–7439

Aboelfetoh EF, El-Shenody RA, Ghobara MM (2017) Eco-friendly synthesis of silver nanoparticles using green algae (*Caulerpa serrulata*): reaction optimization, catalytic and antibacterial activities. Environ Monit Assess 189(7):349

Ahmed MJ, Murtaza G, Mehmood A, Bhatti TM (2015) Green synthesis of silver nanoparticles using leaves extract of *Skimmia laureola*: characterization and antibacterial activity. Mater Lett 153:10–13

Ali A, Ali MA, Ali MU, Mohammad S (2011) Hospital outcomes of obstetrical-related acute renal failure in a tertiary care teaching hospital. Ren Fail 33(3):285–290

Al-Naamani L, Dobretsov S, Dutta J, Burgess JG (2017) Chitosan-zinc oxide nanocomposite coatings for the prevention of marine biofouling. Chemosphere 168:408–417

Arya A, Mishoura V, Chundawat TS (2019) Green synthesis of silver nanoparticles from green algae (*Botryococcus braunii*) and its catalytic behavior for the synthesis of benzimidazoles. Chem Data Collect 20:100190

Asha K, Johnson M, Chandra P, Shibila T, Revathy I (2015) Extracellular synthesis of silver nanoparticles from a marine alga, *Sargassum polycystum* C. Agardh and their biopotentials. WJPPS 4:1388–1400

Aziz N, Fatma T, Varma A, Prasad R (2014) Biogenic synthesis of silver nanoparticles using Scenedesmus abundans and evaluation of their antibacterial activity. Journal of Nanoparticles, Article ID 689419, 2014. https://doi.org/10.1155/2014/689419

Aziz N, Faraz M, Pandey R, Sakir M, Fatma T, Varma A, Barman I, Prasad R (2015) Facile algae-derived route to biogenic silver nanoparticles: Synthesis, antibacterial and photocatalytic properties. Langmuir 31:11605–11612. https://doi.org/10.1021/acs.langmuir.5b03081

Aziz N, Pandey R, Barman I, Prasad R (2016) Leveraging the attributes of *Mucor hiemalis*-derived silver nanoparticles for a synergistic broad-spectrum antimicrobial platform. Front Microbiol 7:1984. https://doi.org/10.3389/fmicb.2016.01984

Aziz N, Prasad R, Ibrahim AIM, Ahmed AIS (2017) Promising applications for the production of biofuels through algae. In: Microbial Biotechnology (Patra JK, Vishnuprasad CN, Das G), Springer Nature Singapore Pte Ltd. 81–103

Aziz N, Faraz M, Sherwani MA, Fatma T, Prasad R (2019) Illuminating the anticancerous efficacy of a new fungal chassis for silver nanoparticle synthesis. Front Chem 7:65. https://doi.org/10.3389/fchem.2019.00065

Azizi S, Ahmad MB, Namvar F, Mohamad R (2014) Green biosynthesis and characterization of zinc oxide nanoparticles using brown marine macroalga *Sargassum muticum* aqueous extract. Mater Lett 116:275–277

Azizi S, Mahdavi Shahouri M, Mohamad R (2017) Green synthesis of zinc oxide nanoparticles for enhanced adsorption of lead ions from aqueous solutions: equilibrium, kinetic and thermodynamic studies. Molecules 22(6):831

Babu B, Palanisamy S, Vinosha M, Anjali R, Kumar P, Pandi B, Tabarsa M, You S, Prabhu NM (2020) Bioengineered gold nanoparticles from marine seaweed *Acanthophora spicifera* for pharmaceutical uses: antioxidant, antibacterial, and anticancer activities. Bioprocess Biosyst Eng 43(12):2231–2242

Bao Z, Cao J, Kang G, Lan CQ (2019) Effects of reaction conditions on light-dependent silver nanoparticle biosynthesis mediated by cell extract of green alga *Neochloris oleoabundans*. Environ Sci Pollut Res 26(3):2873–2881

Barwal I, Ranjan P, Kateriya S, Yadav SC (2011) Cellular oxido-reductive proteins of *Chlamydomonas reinhardtii* control the biosynthesis of silver nanoparticles. J Nanobiotechnol 9(1):56

Baskar BB (2013) Biosynthesis of silver nanoparticles using *Kappa phycus* species. Int J Res Pharm Sci 3(3):55–63

Beganskienė A, Sirutkaitis V, Kurtinaitienė M, Juškėnas R, Kareiva A (2004) FTIR, TEM and NMR investigations of Stöber silica nanoparticles. Mater Sci (Medžiagotyra) 10:287–290

Bhimba B, Kumari P (2014) Phytosynthesis of silver nanoparticles from the extracts of seaweed *Ulva lactuca* and its antimicrobial activity. Int J Pharm BioSci 5:666–677

Bhuyar P, Rahim MHA, Sundararaju S, Ramaraj R, Maniam GP, Govindan N (2020) Synthesis of silver nanoparticles using marine macroalgae *Padina* sp. and its antibacterial activity towards pathogenic bacteria. Beni-Suef Univ J Basic Appl Sci 9(1):1–15

Borah D, Das N, Das N, Bhattacharjee A, Sarmah P, Ghosh K, Chandel M, Rout J, Pandey P, Ghosh NN (2020) Alga-mediated facile green synthesis of silver nanoparticles: Photophysical, catalytic and antibacterial activity. Appl Organomet Chem 34(5):e5597

Borowitzka MA (2013) High-value products from microalgae—their development and commercialisation. J Appl Phycol 25(3):743–756

Cai W, Gao T, Hong H, Sun J (2008) Applications of gold nanoparticles in cancer nanotechnology. Nanotechnol Sci Appl 1:17

Castro L, Blázquez ML, Muñoz JA, González F, Ballester A (2013) Biological synthesis of metallic nanoparticles using algae. IET Nanobiotechnol 7(3):109–116

Chakraborty N, Banerjee A, Lahiri S, Panda A, Ghosh AN, Pal R (2009) Biorecovery of gold using cyanobacteria and an eukaryotic alga with special reference to nanogold formation – a novel phenomenon. J Appl Phycol 21(1):145

Chandran PR, Naseer M, Udupa N, Sandhyarani N (2011) Size controlled synthesis of biocompatible gold nanoparticles and their activity in the oxidation of NADH. Nanotechnology 23(1):015602

Cheung RCF, Ng TB, Wong JH, Chan WY (2015) Chitosan: an update on potential biomedical and pharmaceutical applications. Mar Drugs 13(8):5156–5186

Chithourani DB, Dunne M, Stewart J, Allen C, Jaffray DA (2010) Cellular uptake and transport of gold nanoparticles incorporated in a liposomal carrier. Nanomedicine 6(1):161–169

Chokshi K, Pancha I, Ghosh T, Paliwal C, Maurya R, Ghosh A, Mishoura S (2016) Green synthesis, characterization and antioxidant potential of silver nanoparticles biosynthesized from de-oiled biomass of thermotolerant oleaginous microalgae *Acutodesmus dimorphus*. RSC Adv 6(76):72269–72274

Conde J, Doria G, Baptista P (2012) Noble metal nanoparticles applications in cancer. J Drug Delivery 2012:1–12

Dahoumane SA, Djediat C, Yéprémian C, Couté A, Fiévet F, Coradin T, Brayner R (2012) Species selection for the design of gold nanobioreactor by photosynthetic organisms. J Nanopart Res 14(6):883

Dahoumane SA, Wijesekera K, Filipe CD, Brennan JD (2014a) Stoichiometrically controlled production of bimetallic gold-silver alloy colloids using micro-alga cultures. J Colloid Interface Sci 416:67–72

Dahoumane SA, Yéprémian C, Djédiat C, Couté A, Fiévet F, Coradin T, Brayner R (2014b) A global approach of the mechanism involved in the biosynthesis of gold colloids using micro-algae. J Nanopart Res 16(10):2607

Dahoumane SA, Mechouet M, Wijesekera K, Filipe CD, Sicard C, Bazylinski DA, Jeffryes C (2017) Algae-mediated biosynthesis of inorganic nanomaterials as a promising route in nanobiotechnology – a review. Green Chem 19(3):552–587

de Aragao AP, de Oliveira TM, Quelemes PV, Perfeito MLG, Araujo MC, Santiago JAS, Cardoso VS, Quaresma P, de Almeida JR, da Silva DA (2019) Green synthesis of silver nanoparticles using the seaweed *Gracilaria birdiae* and their antibacterial activity. Arab J Chem 12(8):4182–4188

Deb S, Patra HK, Lahiri P, Dasgupta AK, Chakrabarti K, Chaudhuri U (2011) Multistability in platelets and their response to gold nanoparticles. Nanomedicine 7(4):376–384

Devi JS, Bhimba BV, Ratnam K (2012) In vitro anticancer activity of silver nanoparticles synthesized using the extract of *Gelidiella* sp. Int J Pharm Pharm Sci 4(4):710–715

Devi JS, Bhimba BV, Peter DM (2013) Production of biogenic silver nanoparticles using *Sargassum longifolium* and its applications. Indian J Geo-Marine Sci 42(1):125–130

Dhanalakshmi P, Azeez R, Rekha R, Poonkodi S, Nallamuthu T (2012) Synthesis of silver nanoparticles using green and brown seaweeds. Phykos 42(2):39–45

Dhas TS, Kumar VG, Abraham LS, Karthick V, Govindaraju K (2012) *Sargassum myriocystum* mediated biosynthesis of gold nanoparticles. Spectrochim Acta A Mol Biomol Spectrosc 99:97–101

Dhas TS, Sowmiya P, Kumar VG, Ravi M, Suthindhiran K, Borgio JF, Narendrakumar V, Kumar VR, Karthick V, Kumar CV (2020) Antimicrobial effect of *Sargassum plagiophyllum* mediated gold nanoparticles on *Escherichia coli* and *Salmonella typhi*. Biocatal Agric Biotechnol 26:101627

Dhavale R, Jadhav S, Sibi G (2019) Microalgae mediated silver nanoparticles (Ag-NPs) synthesis and their biological activities. J Crit Rev 7(2):2020

Dhillon GS, Brar SK, Kaur S, Verma M (2012) Green approach for nanoparticle biosynthesis by fungi: current trends and applications. Crit Rev Biotechnol 32(1):49–73

Dizaj SM, Lotfipour F, Barzegar-Jalali M, Zarrintan MH, Adibkia K (2014) Antimicrobial activity of the metals and metal oxide nanoparticles. Mater Sci Eng C 44:278–284

Doshi H, Ray A, Kothari I (2007) Bioremediation potential of live and dead Spirulina: spectroscopic, kinetics and SEM studies. Biotechnol Bioeng 96(6):1051–1063

El-Kassas HY, El-Sheekh MM (2014) Cytotoxic activity of biosynthesized gold nanoparticles with an extract of the red seaweed *Corallina officinalis* on the MCF-7 human breast cancer cell line. Asian Pac J Cancer Prev 15(10):4311–4317

El-Rafie H, El-Rafie M, Zahouran M (2013) Green synthesis of silver nanoparticles using polysaccharides extracted from marine macro algae. Carbohydr Polym 96(2):403–410

El-Sheekh MM, El-Kassas HY (2016) Algal production of nano-silver and gold: their antimicrobial and cytotoxic activities: a review. J Gen Eng Biotechnol 14(2):299–310

Focsan M, Ardelean II, Craciun C, Astilean S (2011) Interplay between gold nanoparticle biosynthesis and metabolic activity of cyanobacterium *Synechocystis* sp. PCC 6803. Nanotechnology 22(48):485101

Ganesan V, Aruna Devi J, Astalakshmi A, Nima P, Thangaraja A (2013) Eco-friendly synthesis of silver nanoparticles using a sea weed, *Kappaphycus alvarezii* (Doty) Doty ex PC Silva. Int J Eng Adv Technol 2:559–563

Garibo D, Borbón-Nuñez HA, de León JND, Mendoza EG, Estrada I, Toledano-Magaña Y, Tiznado H, Ovalle-Marroquin M, Soto-Ramos AG, Blanco A (2020) Green synthesis of silver nanoparticles using *Lysiloma acapulcensis* exhibit high-antimicrobial activity. Sci Rep 10(1):1–11

Ghodake G, Lee DS (2011) Biological synthesis of gold nanoparticles using the aqueous extract of the brown algae *Laminaria japonica*. J Nanoelectron Optoelectron 6(3):268–271

González-Ballesteros N, Prado-López S, Rodríguez-González J, Lastra M, Rodríguez-Argüelles M (2017) Green synthesis of gold nanoparticles using brown algae *Cystoseira baccata*: its activity in colon cancer cells. Colloids Surf B: Biointerfaces 153:190–198

González-Ballesteros N, Rodríguez-Argüelles M, Lastra-Valdor M, González-Mediero G, Rey-Cao S, Grimaldi M, Cavazza A, Bigi F (2020) Synthesis of silver and gold nanoparticles by *Sargassum muticum* biomolecules and evaluation of their antioxidant activity and antibacterial properties. J Nanostruct Chem 10(4):317–330

Govindaraju K, Basha SK, Kumar VG, Singaravelu G (2008) Silver, gold and bimetallic nanoparticles production using single-cell protein (*Spirulina platensis*) Geitler. J Mater Sci 43(15):5115–5122

Govindaraju K, Krishnamoorthy K, Alsagaby SA, Singaravelu G, Premanathan M (2015) Green synthesis of silver nanoparticles for selective toxicity towards cancer cells. IET Nanobiotechnol 9(6):325–330

Guo S, Wang E (2011) Noble metal nanomaterials: controllable synthesis and application in fuel cells and analytical sensors. Nano Today 6(3):240–264

Husain S, Sardar M, Fatma T (2015) Screening of cyanobacterial extracts for synthesis of silver nanoparticles. World J Microbiol Biotechnol 31(8):1279–1283

Husain S, Afreen S, Yasin D, Afzal B, Fatma T (2019) Cyanobacteria as a bioreactor for synthesis of silver nanoparticles - an effect of different reaction conditions on the size of nanoparticles and their dye decolorization ability. J Microbiol Methods 162:77–82

Ibraheem I, Abd-Elaziz B, Saad W, Fathy W (2016) Green biosynthesis of silver nanoparticles using marine Red Algae *Acanthophora specifera* and its antimicrobial activity. J Nanomed Nanotechnol 7(409):1–4

Iravani S, Thota S, Crans D (2018) Methods for preparation of metal nanoparticles. Metal Nanopart 63:15–32

Jana J, Ganguly M, Pal T (2016) Enlightening surface plasmon resonance effect of metal nanoparticles for practical spectroscopic application. RSC Adv 6(89):86174–86211

Jena J, Pradhan N, Dash BP, Sukla LB, Panda PK (2013) Biosynthesis and characterization of silver nanoparticles using microalga *Chlorococcum humicola* and its antibacterial activity. Int J Nanomater Biostruct 3(1):1–8

Jena J, Pradhan N, Nayak RR, Dash BP, Sukla LB, Panda PK, Mishoura BK (2014) Microalga *Scenedesmus* sp.: a potential low-cost green machine for silver nanoparticle synthesis. J Microbiol Biotechnol 24(4):522–533

Johansen MN (2012) Microalgae: biotechnology, microbiology, and energy. Nova Science Publisher

Joshi N, Jain N, Pathak A, Singh J, Prasad R, Upadhyaya CP (2018) Biosynthesis of silver nanoparticles using Carissa carandas berries and its potential antibacterial activities. J Sol-Gel Sci Techn 86(3):682–689. https://doi.org/10.1007/s10971-018-4666-2

Kalabegishvili TL, Kirkesali EI, Rcheulishvili AN, Ginturi EN, Murusidze IG, Pataraya DT, Gurielidze MA, Tsertsvadze GI, Gabunia VN, Lomidze LG (2012) Synthesis of gold nanoparticles by some strains of *Arthrobacter* genera. J Mater Sci Eng A 2(2):164–173

Kalaiarasi R, Jayallakshmi N, Venkatachalam P (2010) Phytosynthesis of nanoparticles and its applications. Plant Cell Biotechnol Mol Biol 11(1/4):1–16

Kannan R, Stirk W, Van Staden J (2013a) Synthesis of silver nanoparticles using the seaweed *Codium capitatum* PC Silva (Chlorophyceae). S Afr J Bot 86:1–4

Kannan RRR, Arumugam R, Ramya D, Manivannan K, Anantharaman P (2013b) Green synthesis of silver nanoparticles using marine macroalga *Chaetomorpha linum*. Appl Nanosci 3(3):229–233

Kathiraven T, Sundaramanickam A, Shanmugam N, Balasubramanian T (2015) Green synthesis of silver nanoparticles using marine algae *Caulerpa racemosa* and their antibacterial activity against some human pathogens. Appl Nanosci 5(4):499–504

Khalil MM, Ismail EH, El-Baghdady KZ, Mohamed D (2014) Green synthesis of silver nanoparticles using olive leaf extract and its antibacterial activity. Arab J Chem 7(6):1131–1139

Khan A, Rashid R, Murtaza G, Zahoura A (2014) Gold nanoparticles: synthesis and applications in drug delivery. Trop J Pharm Res 13(7):1169–1177

Khan AU, Khan M, Malik N, Cho MH, Khan MM (2019) Recent progress of algae and blue–green algae-assisted synthesis of gold nanoparticles for various applications. Bioprocess Biosyst Eng 42(1):1–15

Khanna P, Kaur A, Goyal D (2019) Algae-based metallic nanoparticles: synthesis, characterization and applications. J Microbiol Methods 163:105656

Khodashenas B, Ghorbani H (2019) Synthesis of silver nanoparticles with different shapes. Arab J Chem 12(8):1823–1838

Kim S-K, Thomas NV, Li X (2011) Anticancer compounds from marine macroalgae and their application as medicinal foods. In: Advances in food and nutrition research, vol 64. Elsevier, pp 213–224

Krishnan M, Sivanandham V, Hans-Uwe D, Murugaiah SG, Seeni P, Gopalan S, Rathinam AJ (2015) Antifouling assessments on biogenic nanoparticles: a field study from polluted offshore platform. Mar Pollut Bull 101(2):816–825

Kumar SS, Kumar Y, Khan M, Anbu J, De Clercq E (2011) Antihistaminic and antiviral activities of steroids of *Turbinaria conoides*. Nat Prod Res 25(7):723–729

Kumar P, Senthamil Selvi S, Lakshmi Prabha A, Prem Kumar K, Ganeshkumar R, Govindaraju M (2012a) Synthesis of silver nanoparticles from *Sargassum tenerrimum* and screening phytochemicals for its antibacterial activity. Nano Biomed Eng 4(1):12–16

Kumar P, Senthamilselvi S, Lakshmipraba A, Premkumar K, Muthukumaran R, Visvanathan P, Ganeshkumar R, Govindaraju M (2012b) Efficacy of bio-synthesized silver nanoparticles using *Acanthophora spicifera* to encumber biofilm formation. Dig J Nanomater Biostruct 7(2):511–522

Kumar P, Govindaraju M, Senthamilselvi S, Premkumar K (2013a) Photocatalytic degradation of methyl orange dye using silver (Ag) nanoparticles synthesized from *Ulva lactuca*. Colloids Surf B: Biointerfaces 103:658–661

Kumar P, Selvi SS, Govindaraju M (2013b) Seaweed-mediated biosynthesis of silver nanoparticles using *Gracilaria corticata* for its antifungal activity against *Candida* spp. Appl Nanosci 3(6):495–500

Kumar-Krishnan S, Prokhorov E, Hernández-Iturriaga M, Mota-Morales JD, Vázquez-Lepe M, Kovalenko Y, Sanchez IC, Luna-Bárcenas G (2015) Chitosan/silver nanocomposites: synergistic antibacterial action of silver nanoparticles and silver ions. Eur Polym J 67:242–251

Kushnerova N, Fomenko S, Sprygin V, Kushnerova T, Khotimchenko YS, Kondrateva E, Drugova L (2010) An extract from the brown alga *Laminaria japonica*: a promising stress-protective preparation. Russ J Mar Biol 36(3):209–214

Larue C, Castillo-Michel H, Sobanska S, Cécillon L, Bureau S, Barthès V, Ouerdane L, Carrière M, Sarret G (2014) Foliar exposure of the crop *Lactuca sativa* to silver nanoparticles: evidence for internalization and changes in Ag speciation. J Hazard Mater 264:98–106

Lengke MF, Fleet ME, Southam G (2006a) Morphology of gold nanoparticles synthesized by filamentous cyanobacteria from gold (I)– thiosulfate and gold (III)– chloride complexes. Langmuir 22(6):2780–2787

Lengke MF, Ravel B, Fleet ME, Wanger G, Gordon RA, Southam G (2006b) Mechanisms of gold bioaccumulation by filamentous cyanobacteria from gold (III)– chloride complex. Environ Sci Technol 40(20):6304–6309

Lengke MF, Fleet ME, Southam G (2007) Biosynthesis of silver nanoparticles by filamentous cyanobacteria from a silver (I) nitrate complex. Langmuir 23(5):2694–2699

LewisOscar F, Vismaya S, Arunkumar M, Thajuddin N, Dhanasekaran D, Nithya C (2016) Algal nanoparticles: synthesis and biotechnological potentials. Algae–Organisms Imminent Biotechnol 7:157–182

Li X (2011) Green energy for sustainability and energy security. In: Green Energy. Springer, pp 1–16

Li X, Schirmer K, Bernard L, Sigg L, Pillai S, Behoura R (2015) Silver nanoparticle toxicity and association with the alga *Euglena gracilis*. Environ Sci Nano 2(6):594–602

Liu B, Xie J, Lee J, Ting Y, Chen JP (2005) Optimization of high-yield biological synthesis of single-crystalline gold nanoplates. J Phys Chem B 109(32):15256–15263

Maddela NR, Chakraborty S, Prasad R (2021) Nanotechnology for Advances in Medical Microbiology. Springer Singapore (ISBN 978-981-15-9915-6) https://www.springer.com/gp/book/9789811599156

Madhiyazhagan P, Murugan K, Kumar AN, Nataraj T, Dinesh D, Panneerselvam C, Subramaniam J, Kumar PM, Suresh U, Roni M (2015) *Sargassum muticum*-synthesized silver nanoparticles: an effective control tool against mosquito vectors and bacterial pathogens. Parasitol Res 114(11):4305–4317

Madhiyazhagan P, Murugan K, Kumar AN, Nataraj T, Subramaniam J, Chandramohan B, Panneerselvam C, Dinesh D, Suresh U, Nicoletti M (2017) One pot synthesis of silver nanocrystals using the seaweed *Gracilaria edulis*: biophysical characterization and potential against the filariasis vector *Culex quinquefasciatus* and the midge *Chironomus circumdatus*. J Appl Phycol 29(1):649–659

Mahdavi M, Namvar F, Ahmad MB, Mohamad R (2013) Green biosynthesis and characterization of magnetic iron oxide (Fe_3O_4) nanoparticles using seaweed (*Sargassum muticum*) aqueous extract. Molecules 18(5):5954–5964

Mahdieh M, Zolanvari A, Azimee A (2012) Green biosynthesis of silver nanoparticles by *Spirulina platensis*. Sci Iran 19(3):926–929

Mallmann EJJ, Cunha FA, Castro BN, Maciel AM, Menezes EA, Fechine PBA (2015) Antifungal activity of silver nanoparticles obtained by green synthesis. Rev Inst Med Trop Sao Paulo 57(2):165–167

Manam DVK, Murugesan S (2013) Biogenic silver nanoparticles by *Halymenia poryphyroides* and its in vitro anti-diabetic efficacy. J Chem Pharm Res 5(12):1001–1008

Manam DVK, Murugesan S (2014) Biological synthesis of silver nanoparticles from marine alga *Colpomenia sinuosa* and its in vitro anti-diabetic activity. Am J Bio-pharmacol Biochem Life Sci 3(01):01–07

Mandotra SK, Kumar R, Upadhyay SK, Ramteke PW (2018) Nanotechnology: a new tool for biofuel production. In: Green nanotechnology for biofuel production. Springer, pp 17–28

Maneerung T, Tokura S, Rujiravanit R (2008) Impregnation of silver nanoparticles into bacterial cellulose for antimicrobial wound dressing. Carbohydr Polym 72(1):43–51

Mata Y, Torres E, Blazquez M, Ballester A, González F, Munoz J (2009) Gold (III) biosorption and bioreduction with the brown alga *Fucus vesiculosus*. J Hazard Mater 166(2–3):612–618

Mohandass C, Vijayaraj A, Rajasabapathy R, Satheeshbabu S, Rao S, Shiva C, De-Mello I (2013) Biosynthesis of silver nanoparticles from marine seaweed *Sargassum cinereum* and their antibacterial activity. Indian J Pharm Sci 75(5):606

Mohanpuria P, Rana NK, Yadav SK (2008) Biosynthesis of nanoparticles: technological concepts and future applications. J Nanopart Res 10(3):507–517

Mohseniazar M, Barin M, Zarredar H, Alizadeh S, Shanehbandi D (2011) Potential of microalgae and lactobacilli in biosynthesis of silver nanoparticles. Bioimpacts 1(3):149

Mosulishvili L, Kirkesali E, Belokobylsky A, Khizanishvili A, Frontasyeva M, Pavlov S, Gundorina S (2002) Experimental substantiation of the possibility of developing selenium- and iodine-containing pharmaceuticals based on blue–green algae *Spirulina platensis*. J Pharm Biomed Anal 30(1):87–97

Mosulishvili L, Belokobylsky A, Kirkesali E, Frontasyeva M, Pavlov S, Aksenova N (2007) Neutron activation analysis for studying Cr uptake in the blue-green microalga *Spirulina platensis*. J Neutron Res 15(1):49–54

MubarakAli D, Gopinath V, Rameshbabu N, Thajuddin N (2012) Synthesis and characterization of CdS nanoparticles using C-phycoerythrin from the marine cyanobacteria. Mater Lett 74:8–11

MubarakAli D, Arunkumar J, Nag KH, SheikSyedIshack K, Baldev E, Pandiaraj D, Thajuddin N (2013) Gold nanoparticles from Pro and eukaryotic photosynthetic microorganisms—Comparative studies on synthesis and its application on biolabelling. Colloids Surf B: Biointerfaces 103:166–173

Mukherjee P, Senapati S, Mandal D, Ahmad A, Khan MI, Kumar R, Sastry M (2002) Extracellular synthesis of gold nanoparticles by the fungus *Fusarium oxysporum*. Chembiochem 3(5):461–463

Murugesan S, Bhuvaneswari S, Sivamurugan V (2017) Green synthesis, characterization of silver nanoparticles of a marine red alga *Spyridia fusiformis* and their antibacterial activity. Int J Pharm Pharm Sci 9(5):192–197

Nabikhan A, Kandasamy K, Raj A, Alikunhi NM (2010) Synthesis of antimicrobial silver nanoparticles by callus and leaf extracts from saltmarsh plant, *Sesuvium portulacastrum* L. Colloids Surf B: Biointerfaces 79(2):488–493

Namasivayam SKR, Jayakumar D, Kumar R, Bharani RSA (2015) Antibacterial and anticancerous biocompatible silver nanoparticles synthesised from the cold-tolerant strain of *Spirulina platensis*. J Coastal Life Med 3(4):265–272

Namvar F, Azizi S, Ahmad MB, Shameli K, Mohamad R, Mahdavi M, Tahir PM (2015) Green synthesis and characterization of gold nanoparticles using the marine macroalgae *Sargassum muticum*. Res Chem Intermed 41(8):5723–5730

Naveena BE, Prakash S (2013) Biological synthesis of gold nanoparticles using marine algae *Gracilaria corticata* and its application as a potent antimicrobial and antioxidant agent. Asian J Pharm Clin Res 6(2):179–182

Omomowo I, Adenigba V, Ogunsona S, Adeyinka G, Oluyide O, Adedayo A, Fatukasi B (2020) Antimicrobial and antioxidant activities of algal-mediated silver and gold nanoparticles. In: IOP conference series: materials science and engineering, vol 805. IOP Publishing, p 012010

Oza G, Pandey S, Mewada A, Kalita G, Sharon M, Phata J, Ambernath W, Sharon M (2012) Facile biosynthesis of gold nanoparticles exploiting optimum pH and temperature of fresh water algae *Chlorella pyrenoidusa*. Adv Appl Sci Res 3(3):1405–1412

Öztürk BY (2019) Intracellular and extracellular green synthesis of silver nanoparticles using *Desmodesmus* sp.: their antibacterial and antifungal effects. Caryol Int J Cytol Cytosystem Cytogenet 72(1):29–43

Parial D, Pal R (2014) Green synthesis of gold nanoparticles using cyanobacteria and their characterization. Indian J Appl Res 4:69–72

Parial D, Pal R (2015) Biosynthesis of monodisperse gold nanoparticles by green alga *Rhizoclonium* and associated biochemical changes. J Appl Phycol 27(2):975–984

Parial D, Patra HK, Dasgupta AK, Pal R (2012) Screening of different algae for green synthesis of gold nanoparticles. Eur J Phycol 47(1):22–29

Parial D, Gopal PK, Paul S, Pal R (2016) Gold (III) bioreduction by cyanobacteria with special reference to in vitro biosafety assay of gold nanoparticles. J Appl Phycol 28(6):3395–3406

Parveen KJ (2017) Multi-functional silver nanoparticles for drug delivery: a review. Int J Curr Pharm Rev Res 9:1–5

Patel V, Berthold D, Puranik P, Gantar M (2015) Screening of cyanobacteria and microalgae for their ability to synthesize silver nanoparticles with antibacterial activity. Biotechnol Rep 5:112–119

Pillai ZS, Kamat PV (2004) What factors control the size and shape of silver nanoparticles in the citrate ion reduction method? J Phys Chem B 108(3):945–951

Prasad R (2014) Synthesis of silver nanoparticles in photosynthetic plants. Journal of Nanoparticles, Article ID 963961, 2014. https://doi.org/10.1155/2014/963961

Prasad R, Swamy VS (2013) Antibacterial activity of silver nanoparticles synthesized by bark extract of Syzygium cumini. Journal of Nanoparticles 2013. https://doi.org/10.1155/2013/431218

Prasad R, Pandey R, Barman I (2016) Engineering tailored nanoparticles with microbes: quo vadis. WIREs Nanomed Nanobiotechnol 8:316–330. https://doi.org/10.1002/wnan.1363

Prasad R, Pandey R, Varma A, Barman I (2017) Polymer based nanoparticles for drug delivery systems and cancer therapeutics. In: Natural Polymers for Drug Delivery (eds. Kharkwal H and Janaswamy S), CAB International, UK 53–70

Prasad R, Jha A, Prasad K (2018) Exploring the Realms of Nature for Nanosynthesis. Springer International Publishing (ISBN 978-3-319-99570-0) https://www.springer.com/978-3-319-99570-0

Prasad R, Kumar V, Kumar M, Choudhary D (2019) Nanobiotechnology in Bioformulations. Springer International Publishing (ISBN 978-3-030-17061-5) https://www.springer.com/gp/book/9783030170608

Prasad R, Siddhardha B, Dyavaiah M (2020) Nanostructures for Antimicrobial and Antibiofilm Applications. Springer International Publishing (ISBN 978-3-030-40336-2) https://www.springer.com/gp/book/9783030403362

Prasad T, Elumalai E (2013) Marine algae mediated synthesis of silver nanopaticles using *Scaberia agardhii* Greville. J Biol Sci 13(6):566

Prasad TN, Kambala VSR, Naidu R (2013) Phyconanotechnology: synthesis of silver nanoparticles using brown marine algae *Cystophora moniliformis* and their characterisation. J Appl Phycol 25(1):177–182

Priyadharshini RI, Prasannaraj G, Geetha N, Venkatachalam P (2014) Microwave-mediated extracellular synthesis of metallic silver and zinc oxide nanoparticles using macro-algae (*Gracilaria edulis*) extracts and its anticancer activity against human PC3 cell lines. Appl Biochem Biotechnol 174(8):2777–2790

Pugazhendhi A, Prabakar D, Jacob JM, Karuppusamy I, Saratale RG (2018) Synthesis and characterization of silver nanoparticles using *Gelidium amansii* and its antimicrobial property against various pathogenic bacteria. Microb Pathog 114:41–45

Rahimi Z, Yousefzadi M, Noori A, Akbarzadeh A (2014) Green synthesis of silver nanoparticles using *Ulva flexousa* from the Persian Gulf, Iran. J Persian Gulf 5(15):9–16

Rajasulochana P, Krishnamoorthy P, Dhamotharan R (2012) Potential application of *Kappaphycus alvarezii* in agricultural and pharmaceutical industry. J Chem Pharm Res 4(1):33–37

Rajathi FAA, Parthiban C, Kumar VG, Anantharaman P (2012) Biosynthesis of antibacterial gold nanoparticles using brown alga, *Stoechospermum marginatum* (Kützing). Spectrochim Acta A Mol Biomol Spectrosc 99:166–173

Rajeshkumar S, Kannan C, Annadurai G (2012) Green synthesis of silver nanoparticles using marine brown algae *Turbinaria conoides* and its antibacterial activity. Int J Pharm Bio Sci 3(4):502–510

Rajeshkumar S, Malarkodi C, Gnanajobitha G, Paulkumar K, Vanaja M, Kannan C, Annadurai G (2013) Seaweed-mediated synthesis of gold nanoparticles using *Turbinaria conoides* and its characterization. J Nanostruct Chem 3(1):44

Ramakrishna M, Babu DR, Gengan RM, Chandra S, Rao GN (2016) Green synthesis of gold nanoparticles using marine algae and evaluation of their catalytic activity. J Nanostruct Chem 6(1):1–13

Ramakritinan C, Shankar S, Anand M, Kumaraguru A (2010) Biosynthesis of silver, gold and bimetallic alloy (Ag: Au) nanoparticles from green alga, *Lyngpya* sp. In: Proc. 3rd Nat. Conf. nanaomaterials and nanotechnology, vol 21. Amity University, Lucknow, pp 174–187

Rao MD, Pennathur G (2017) Green synthesis and characterization of cadmium sulphide nanoparticles from *Chlamydomonas reinhardtii* and their application as photocatalysts. Mater Res Bull 85:64–73

Rao PS, Mantri VA, Ganesan K (2007) Mineral composition of edible seaweed *Porphyra vietnamensis*. Food Chem 102(1):215–218

Rawat J, Gupta PK, Pandit S, Prasad R, Pande V (2021) Current perspectives on integrated approaches to enhance lipid accumulation in microalgae. 3 Biotech 11, 303. https://doi.org/10.1007/s13205-021-02851-3

Rösken LM, Cappel F, Körsten S, Fischer CB, Schönleber A, van Smaalen S, Geimer S, Beresko C, Ankerhold G, Wehner S (2016) Time-dependent growth of crystalline Au^0-nanoparticles in cyanobacteria as self-reproducing bioreactors: 2. *Anabaena cylindrica*. Beilstein J Nanotechnol 7(1):312–327

Roychoudhury P, Pal R (2014) *Spirogyra submaxima* – a green alga for nanogold production. J Algal Biomass Utln 5(1):15–19

Saglam N, Korkusuz, F, Prasad R (2021) Nanotechnology Applications in Health and Environmental Sciences. Springer International Publishing (ISBN: 978-3-030-64410-9) https://www.springer.com/gp/book/9783030644093

Sahayaraj K, Rajesh S, Rathi J (2012) Silver nanoparticles biosynthesis using marine alga *Padina pavonica* (Linn.) and its microbicidal activity. Dig J Nanomater Biostruct 7(4):1557–1567

Sajidha Parveen K, Lakshmi D (2016) Biosynthesis of silver nanoparticles using red algae, *Amphiroa fragilissima* and its antibacterial potential against gram positive and gram negative bacteria. Int J Curr Sci 19(3):93–100

Salari Z, Danafar F, Dabaghi S, Ataei SA (2016) Sustainable synthesis of silver nanoparticles using macroalgae *Spirogyra varians* and analysis of their antibacterial activity. J Saudi Chem Soc 20(4):459–464

Selvam GG, Sivakumar K (2015) Phycosynthesis of silver nanoparticles and photocatalytic degradation of methyl orange dye using silver (Ag) nanoparticles synthesized from *Hypnea musciformis* (Wulfen) JV Lamouroux. Appl Nanosci 5(5):617–622

Senapati S, Syed A, Moeez S, Kumar A, Ahmad A (2012) Intracellular synthesis of gold nanoparticles using alga *Tetraselmis kochinensis*. Mater Lett 79:116–118

Senthilkumar P, Surendran L, Sudhagar B, Kumar DRS (2019) Facile green synthesis of gold nanoparticles from marine algae *Gelidiella acerosa* and evaluation of its biological potential. SN Appl Sci 1(4):284

Shakibaie M, Forootanfar H, Mollazadeh-Moghaddam K, Bagherzadeh Z, Nafissi-Varcheh N, Shahverdi AR, Faramarzi MA (2010) Green synthesis of gold nanoparticles by the marine microalga *Tetraselmis suecica*. Biotechnol Appl Biochem 57(2):71–75

Sharma B, Purkayastha DD, Hazra S, Gogoi L, Bhattacharjee CR, Ghosh NN, Rout J (2014a) Biosynthesis of gold nanoparticles using a freshwater green alga, *Prasiola crispa*. Mater Lett 116:94–97

Sharma B, Purkayastha DD, Hazra S, Thajamanbi M, Bhattacharjee CR, Ghosh NN, Rout J (2014b) Biosynthesis of fluorescent gold nanoparticles using an edible freshwater red alga, *Lemanea fluviatilis* (L.) C. Ag. and antioxidant activity of biomatrix loaded nanoparticles. Bioprocess Biosyst Eng 37(12):2559–2565

Sharma A, Sharma S, Sharma K, Chetri SP, Vashishtha A, Singh P, Kumar R, Rathi B, Agrawal V (2016) Algae as crucial organisms in advancing nanotechnology: a systematic review. J Appl Phycol 28(3):1759–1774

Shiny P, Mukherjee A, Chandrasekaran N (2013) Marine algae mediated synthesis of the silver nanoparticles and its antibacterial efficiency. Int J Pharm Pharm Sci 5:239–241

Shukla MK, Singh RP, Reddy C, Jha B (2012) Synthesis and characterization of agar-based silver nanoparticles and nanocomposite film with antibacterial applications. Bioresour Technol 107:295–300

Sicard C, Brayner R, Margueritat J, Hémadi M, Couté A, Yéprémian C, Djediat C, Aubard J, Fiévet F, Livage J (2010) Nano-gold biosynthesis by silica-encapsulated micro-algae: a "living" bio-hybrid material. J Mater Chem 20(42):9342–9347

Sing SF, Isdepsky A, Borowitzka MA, Moheimani NR (2013) Production of biofuels from microalgae. Mitig Adapt Strateg Glob Chang 18(1):47–72

Singaravelu G, Arockiamary J, Kumar VG, Govindaraju K (2007) A novel extracellular synthesis of monodisperse gold nanoparticles using marine alga, *Sargassum wightii* Greville. Colloids Surf B: Biointerfaces 57(1):97–101

Singh M, Kalaivani R, Manikandan S, Sangeetha N, Kumaraguru A (2013) Facile green synthesis of variable metallic gold nanoparticle using *Padina gymnospora*, a brown marine macroalga. Appl Nanosci 3(2):145–151

Singh CR, Kathiresan K, Anandhan S (2015) A review on marine based nanoparticles and their potential applications. Afr J Biotechnol 14(18):1525–1532

Singh P, Pandit S, Mokkapati V, Garg A, Ravikumar V, Mijakovic I (2018) Gold nanoparticles in diagnostics and therapeutics for human cancer. Int J Mol Sci 19(7):1979

Sinha SN, Paul D, Halder N, Sengupta D, Patra SK (2015) Green synthesis of silver nanoparticles using fresh water green alga *Pithophora oedogonia* (Mont.) Wittrock and evaluation of their antibacterial activity. Appl Nanosci 5(6):703–709

Sirelkhatim A, Mahmud S, Seeni A, Kaus NHM, Ann LC, Bakhori SKM, Hasan H, Mohamad D (2015) Review on zinc oxide nanoparticles: antibacterial activity and toxicity mechanism. Nano-Micro Lett 7(3):219–242

Srivastava S, Usmani Z, Atanasov AG, Singh VK, Singh NP, Abdel-Azeem AM, Prasad R, Gupta G, Sharma M, Bhargava A (2021) Biological nanofactories: Using living forms for metal nanoparticle synthesis. Mini-Reviews in Medicinal Chemistry 21(2):245–265

Subbiah M, Pandithurai M, Vajiravelu S (2019) *Spatoglossum asperum* J. Agardh mediated synthesis of silver nanoparticles, characterization and evaluation antifungal activities. J Pharmacogn Phytochem 8(1):1991–1995

Sudha S, Rajamanickam K, Rengaramanujam J (2013) Microalgae mediated synthesis of silver nanoparticles and their antibacterial activity against pathogenic bacteria. Indian J Exp Biol 51(5):393–399

Suganya KU, Govindaraju K, Kumar VG, Dhas TS, Karthick V, Singaravelu G, Elanchezhiyan M (2015) Blue green alga mediated synthesis of gold nanoparticles and its antibacterial efficacy against Gram positive organisms. Mater Sci Eng C 47:351–356

Supraja N, Prasad T, Soundariya M, Babujanarthanam R (2016) Synthesis, characterization and dose dependent antimicrobial and anti-cancerous activity of phycogenic silver nanoparticles against human hepatic carcinoma (HepG2) cell line. AIMS Bioeng 3(4):425–440

Supraja N, Dhivya J, Prasad T, David E (2018) Synthesis, characterization and dose dependent antimicrobial and anticancerous efficacy of phycogenic (*Sargassum muticum*) silver nanoparticles against Breast Cancer Cells (MCF 7) cell line. Adv Nano Res 6(2):183

Suriya J, Raja SB, Sekar V, Rajasekaran R (2012) Biosynthesis of silver nanoparticles and its antibacterial activity using seaweed *Urospora* sp. Afr J Biotechnol 11(58):12192–12198

Swamy VS, Prasad R (2012) Green synthesis of silver nanoparticles from the leaf extract of Santalum album and its antimicrobial activity. J Optoelectronic and Biomedical Materials 4(3):53–59

Thangaraju N, Venkatalakshmi R, Chinnasamy A, Kannaiyan P (2012) Synthesis of silver nanoparticles and the antibacterial and anticancer activities of the crude extract of *Sargassum polycystum* C. Agardh. Nano Biomed Eng 4(2):89–94

Thota S, Crans DC (2018) Metal nanoparticles: synthesis and applications in pharmaceutical sciences. John Wiley & Sons

Thuesombat P, Hannongbua S, Akasit S, Chadchawan S (2014) Effect of silver nanoparticles on rice (*Oryza sativa* L. cv. KDML 105) seed germination and seedling growth. Ecotoxicol Environ Saf 104:302–309

Tran PH-L, Tran TT-D, Van Vo T, Lee B-J (2012) Promising iron oxide-based magnetic nanoparticles in biomedical engineering. Arch Pharm Res 35(12):2045–2061
Vadlapudi V, Amanchy R (2017) Synthesis, characterization and antibacterial activity of silver nanoparticles from red algae, *Hypnea musciformis*. Adv Biol Res 11(5):242–249
Vaidyanathan R, Kalishwaralal K, Gopalram S, Gurunathan S (2010) Nanosilver – the burgeoning therapeutic molecule and its green synthesis. Biotechnol Adv 28(6):940–940
van Horssen R, Ten Hagen TL, Eggermont AM (2006) TNF-α in cancer treatment: molecular insights, antitumor effects, and clinical utility. Oncologist 11(4):397–408
Varun S, Sudha S, Senthil Kumar P (2014) Biosynthesis of gold nanoparticles from aqueous extract of *Dictyota bartayresiana* and their anti-fungal activity. Indian J Adv Chem Sci 2:190–193
Ven Den Hoeck C, Mann D, Jahns H (1995) Algae: an introduction to phycology. Cambridge University Press, Cambridge
Venkatesan J, Manivasagan P, Kim S-K, Kirthi AV, Marimuthu S, Rahuman AA (2014) Marine algae-mediated synthesis of gold nanoparticles using a novel *Ecklonia cava*. Bioprocess Biosyst Eng 37(8):1591–1597
Venkatesan J, Kim S-K, Shim MS (2016) Antimicrobial, antioxidant, and anticancer activities of biosynthesized silver nanoparticles using marine algae *Ecklonia cava*. Nano 6(12):235
Venkatpurwar V, Pokharkar V (2011) Green synthesis of silver nanoparticles using marine polysaccharide: study of in-vitro antibacterial activity. Mater Lett 65(6):999–1002
Venkatraman A, Yahoob S, Nagarajan Y, Harikrishnan S, Vasudevan S, Murugasamy T (2018) Pharmacological activity of biosynthesized gold nanoparticles from brown algae - seaweed *Turbinaria conoides*. Nanoworld J 4:17–22
Vigneshwaran N, Ashtaputre N, Varadarajan P, Nachane R, Paralikar K, Balasubramanya R (2007) Biological synthesis of silver nanoparticles using the fungus *Aspergillus flavus*. Mater Lett 61(6):1413–1418
Vissers C, G-l M, Song H (2019) Nanoparticle technology and stem cell therapy team up against neurodegenerative disorders. Adv Drug Deliv Rev 148:239–251
Vivek M, Kumar PS, Steffi S, Sudha S (2011) Biogenic silver nanoparticles by *Gelidiella acerosa* extract and their antifungal effects. Avicenna J Med Biotechnol 3(3):143
Wang L, Hu C, Shao L (2017) The antimicrobial activity of nanoparticles: present situation and prospects for the future. Int J Nanomedicine 12:1227
Wei D, Qian W (2008) Facile synthesis of Ag and Au nanoparticles utilizing chitosan as a mediator agent. Colloids Surf B: Biointerfaces 62(1):136–142
Xie J, Lee JY, Wang DI, Ting YP (2007a) Identification of active biomolecules in the high-yield synthesis of single-crystalline gold nanoplates in algal solutions. Small 3(4):672–682
Xie J, Lee JY, Wang DI, Ting YP (2007b) Silver nanoplates: from biological to biomimetic synthesis. ACS Nano 1(5):429–439
Yamada M, Foote M, Prow TW (2015) Therapeutic gold, silver, and platinum nanoparticles. Wiley Interdiscip Rev Nanomed Nanobiotechnol 7(3):428–445
Yaqoob AA, Parveen T, Umar K, Mohamad Ibrahim MN (2020) Role of nanomaterials in the treatment of wastewater: a review. Water 12(2):495
Yoon HS, Müller KM, Sheath RG, Ott FD, Bhattacharya D (2006) Defining the major lineages of red algae (Rhodophyta) 1. J Phycol 42(2):482–492
Yousefzadi M, Rahimi Z, Ghafori V (2014) The green synthesis, characterization and antimicrobial activities of silver nanoparticles synthesized from green alga *Enteromorpha flexuosa* (Wulfen). J Agardh. Mater Lett 137:1–4
Zayadi RA, Bakar FA (2020) Comparative study on stability, antioxidant and catalytic activities of bio-stabilized colloidal gold nanoparticles using microalgae and cyanobacteria. J Environ Chem Eng 8(4):103843

Chapter 9
Strategies for Nanoencapsulation of Algal Proteins, Protein Hydrolysates and Bioactive Peptides: The Effect of Encapsulation Techniques on Bioactive Properties

Aysun Yucetepe

Contents

A. Yucetepe (✉)
Department of Food Engineering, Faculty of Engineering, Aksaray University, Aksaray, Turkey

D. Thangadurai et al. (eds.), *Bioprospecting Algae for Nanosized Materials*, Nanotechnology in the Life Sciences,
https://doi.org/10.1007/978-3-030-81557-8_9

9.1 Introduction

Food and Agriculture Organization of the United Nations (FAO) expects that the world population will reach 9 billion by 2050. Due to the expected growth in the world's population, a 70% increase in food demand and even the danger of famine, which is worried about its emergence in the future, is predicted (Apostolidis and McLeay 2016). Therefore, intense efforts in the field of food science have been made to obtain alternative proteins from non-conventional, sustainable and cheap protein sources such as yeast, fungi, bacteria, insect and algae in recent years (Spolaore et al. 2006). In particular, algae could be utilized as an alternative protein source because of their high protein contents (Becker 2004; Spolaore et al. 2006). Furthermore, algal proteins and protein-derived peptides possess significant nutraceutical and medical potentials due to their bioactive properties such as antimicrobial, antiviral, antitumor, antioxidative, antihypertensive, anticoagulant and antidiabetic activities (Zhang et al. 2012; Fan et al. 2014; Jo et al. 2017; Ovando et al. 2018; Aziz et al. 2017; Rawat et al. 2021).

Algae, which are multicellular or unicellular organisms, are named as either macroalgae or microalgae based on their size (Sathasivam et al. 2019). Due to their beneficial effects on human health, they have been consumed as food by indigenous populations of Asia, Americas and Africa for centuries. Microalgae as huge biological sources are a group of microscopic organisms with approximately 200,000 different species, which live in oceans and fresh water (Samarakoon and Jeon 2012; Bleakley and Hayes 2017). Some of them are generally regarded as safe (GRAS). Therefore, they can be exploited as functional foods (Jainab et al. 2019; Shchelik et al. 2020). Among microalgae species, in particular *Chlorella*, *Spirulina*, *Dunalliela*, *Botryococcus*, *Phaeodactylum*, *Porphyridium*, *Chaetoceros*, *Crypthecodinium*, *Isochrysis*, *Nannochloris*, *Nitzschia*, *Schizochytrium*, *Tetraselmis* and *Skeletonema* have been grown commercially and utilized as sustainable and cheap protein sources with their remarkable protein contents (up to 50–60% of dry weight, dw) (Zepka et al. 2008; Sathasivam et al. 2019).

On the other hand, macroalgae (or seaweeds) are described as photosynthetic, unicellular or multicellular organisms with a great diversity of forms and complex metabolic pathways to survive in highly competitive marine environments (Harnedy and FitzGerald 2011). Depending on the pigment type, macroalgae can be classified into three taxonomic groups as red (Rhodophyta), brown (Phaeophyta) or green algae (Chlorophyta) (Gressler et al. 2010). Brown algae are approximately 1500–2000 species. In particular *Undaria pinnatifida*, *Hizikia fusiformis* and *Laminaria japonica* brown algae have been consumed as food by human since ancient times. Approximately 8000 species of green algae are estimated to be in existence. Especially, *Ulva* spp. including *Ulva lactuca*, *Ulva intestinalis* and *Ulva compressa* are the most commonly consumed species of green algae. There are approximately 6000 species of red algae. Among these species, *Porphyra tenera* and *Palmaria palmata* with high protein content and delicious taste are the most consumed species of red algae in Asia, as well as Western countries (Bleakley and Hayes 2017).

Although macro- and microalgae have been consumed as food by various communities in the world for a long time due to their high protein contents, there are some limitations about their usages as food or food supplement by food industry. The major obstacle which limits incorporation of algal proteins into conventional foods is their slightly fishy smell (Becker 2007). In this context, encapsulation of algal proteins or bioactive peptides derived from macro- and microalgae could be a proper strategy to mask unpleasant taste and smell. Encapsulation also offers some benefits such as reduction of hygroscopicity and reactivity and increase of shelf-life of bioactive peptides (Hernández-Ledesma et al. 2011). Although bioactive properties of algal proteins and peptides have been well described in previous studies, only several studies have been reported on encapsulation of algal bioactive peptides. To the best of my knowledge, there is no a review study about encapsulation of proteins and bioactive peptides derived from algae. Therefore, this chapter contains information on chemical composition of macro- and microalgae (i), bioactive properties of peptides derived from algal proteins (ii) and the effect of encapsulation techniques applied on their bioactivities (iii).

9.2 Chemical Composition of Macro- and Microalgae

Macro- and microalgae possess a variety of substances including fatty acids, proteins, carbohydrates, vitamins and pigments with variable concentration and composition depending on strain, season and the surrounding growth conditions such as pH value, saltiness, temperature of water (Bilgin and Ertan 2002; Renaud et al. 2002; Becker 2004; Spolaore et al. 2006; Fan and Chen 2007; Zepka et al. 2008; Sheih et al. 2009a; Aziz et al. 2017; Rawat et al. 2021). For instance, the chemical composition of microalgae differs in their proportions of protein (6–52%), carbohydrate (5–23%) and lipid (7–23%) by depending on their growing conditions (Brown et al. 1997). Much of microalgae contain high-value compounds such as phycobiliproteins, polyunsaturated fatty acids (PUFA), carotenoids, polysaccharides, etc. They have been commonly exploited to produce high-protein supplements for human nutrition and nutraceutical purposes (Sathasivam et al. 2019). More in particular, *Chlorella*, *Spirulina*, *Aphanizomenon* have been widely used as food and feed. Hence, microalgae with the high protein contents can consider as novel protein sources (Niccolai et al. 2019).

Macroalgae are an attractive source of protein (1–50% of dw), dietary fibre (up to 76% of dw) and minerals (11–55% of dw). Apart from the low fat content (0.3–5% of dw), they are also rich in long-chain polyunsaturated fatty acids (LC-PUFA) such as eicosapentaenoic acid (EPA, 20:5 n-3) and fat-soluble vitamins (Pimentel et al. 2019). Macroalgae are rich in several types of polysaccharides such as cellulose, galactans, xylans, fucoidan, laminarin, alginates, carrageenans and floridean starch, as well (Bleakley and Hayes 2017). Brown algae have fucoxanthin, which is a pigment with brown colour. Green algae are characterized by the pigments chlorophyll a and b, β-carotene and xanthophylls. Red algae contain phycobilins with red colour (Bleakley and Hayes 2017). Typically, protein content of red, brown and green algae is up to 47%, 15% and 26%, respectively (Fleurence et al. 2018). For this reason, similar to microalgae, a number of macroalgae can be considered as an alternative source of protein instead of conventional proteins. In order to give a general perspective about the chemical composition of selected macro- and microalgae species, their major constituents are compiled in Table 9.1.

9.3 Proteins from Macro- and Microalgae

Proteins possess a great significance for food industry because of their functional and nutritional properties (Pimentel et al. 2019). As mentioned above, macro- and microalgae are rich sources of proteins located in variety forms either as a component of cell wall or connected with pigments or carbohydrates. In addition, certain macro- and microalgae species have protein amounts similar to those of conventional protein sources, such as meat, egg, soybean and milk, as shown in Table 9.1 (Bleakley and Hayes 2017). Red macroalgae generally contain protein in higher amounts (5–44% of dw) than green macroalgae (7–29% of dw) and brown macroalgae (6–29% of dw) (Pimentel et al. 2019). Specially, *Porphyra tenera* and *Palmaria palmata* red algae have protein content of 47.5% and 30%, respectively (Galland-Irmouli et al. 1999). *Palmaria palmata* was also described as one of the best candidates for usage in human food and animal feed since it is rich in essential amino acids, and its proteins fulfill the demands of a complete protein defined by FAO and World Health Organization (WHO) (Mæhre et al. 2016; Bjarnadóttir et al. 2018). Protein contents of *Gracilaria* spp. and *Laurencia* spp. from red macroalgae are 5.6–24.0% and 2.7–24.5%, respectively. In the study of Valente et al. (2006), protein contents of *Gracilaria bursa-pastoris*, *Ulva rigida* and *Gracilaria cornea* have been determined as 30.2%, 29.5% and 11% of dw, respectively. As else a red macroalgae, protein content of *Acanthophora delilei* was founded as 16.87% in the study conducted by Mageswaran and Sivasubramaniam (1984). On the other hand, *Laminaria digitata*, *Undaria pinnatifida* and *Chondrus crispus* brown macroalgae have protein contents of 8–15%, 11–24% and 21.5% of dw, respectively (Garcia-Vaquero and Hayes 2016; Fleurence et al. 2018). Protein contents of *Ulva rigida*, *Ulva lactuca* and *Ulva pertusa* from green macroalgae are 17.8%, 8.7–32.7% and 17.5–26.0% of dw, respectively (Taboada et al. 2010; Fleurence et al. 2018). On the other hand, microalgae possess higher protein content than macroalgae. In particular, among microalgae, *Chlorella vulgaris* with protein content of 51–58% and *Spirulina platensis* with protein content of 60–70% of dw are the most commonly exploited by food manufacturers. Therefore, in recent years, many studies on proteins or peptides derived from microalgae have been conducted to produce functional food or food supplement (Devi and Venkataraman 1984; Benelhadj et al. 2016; Bleakley and Hayes 2017; Lupatini et al. 2017).

Table 9.1 Chemical composition (% of dry matter) of some macro- and microalgae and foods

Source	Protein	Carbohydrate	Lipid	References
Meat	43	1	34	Becker (2004)
Milk	26	38	28	Becker (2004)
Egg	47	4	41	Becker (2004)
Soybean	37	30	20	Becker (2004)
Microalgae				
Chlorella vulgaris	42–58	12–55	5–40	Safi et al. (2014)
Dunaliella salina	57	32	6	Becker (2007)
Spirulina platensis	60–75	13.5–15	5–6	Yücetepe and Özçelik (2016)
Spirulina maxima	60–71	13–16	6–7	Becker (2007)
Scenedesmus obliquus	32	37	16	El-Chaghaby et al. (2019)
Pseudochoricystis ellipsoidea	25	7	67	Biller and Ross (2014)
Chlorogloeopsis fritschii	56	49	8	Biller and Ross (2014)
Nannochloropsis occulata	63	9	35	Biller and Ross (2014)
Chlorella vulgaris	42–58	12–55	5–40	Safi et al. (2014)
Macroalgae				
Gracillaria corticata	6	43	5	Narasimman and Murugaiyan (2012)
Sargassum longifolium	19	17	8	Narasimman and Murugaiyan (2012)
Turbinaria conoides	16	15	3	Narasimman and Murugaiyan (2012)
Sargassum muticum	17	49	2	Rodrigues et al. (2015)
Grateloupia turuturu	23–28	43–60	2–3	Denis et al. (2010); Rodrigues et al. (2015)
Gracilaria gracilis	20	47	1	Rodrigues et al. (2015)
Fucus vesiculosus	13	51	4	Lorenzo et al. (2017)
Saccorhiza polyschides	14	46	1	Rodrigues et al. (2015)
Osmundea pinnatifida	20–24	32–34	1	Patarra et al. (2011); Rodrigues et al. (2015)
Ascophyllum nodosum	9	45	4	Lorenzo et al. (2017)
Codium tomentosum	19	33	4	Rodrigues et al. (2015)
Bifurcaria bifurcate	9	44	7	Lorenzo et al. (2017)
Stypopodium schimperii	15	56	12	Polat and Ozogul (2008)
Spyridia filamentosa	14	58	1	Polat and Ozogul (2008)
Halymenia floresii	2	59	2	Polat and Ozogul (2008)
Gelidium microdon	15–23	40–57	2	Paiva et al. (2017); Patarra et al. (2011)
Acanthophora nayadiformis	13	60	2	Polat and Ozogul (2008)
Gracilaria bursa-pastoris	30		1	Valente et al. (2006)
Pterocladiella capillacea	20	41	4	Paiva et al. (2017)
Caulerpa lentillifera	10	39	1	Taboada et al. (2010)

(continued)

Table 9.1 (continued)

Source	Protein	Carbohydrate	Lipid	References
Ulva rigida	16–30	55	1	Paiva et al. (2017); Valente et al. (2006)
Ulva compressa	16–27	34–41	2	Paiva et al. (2017); Patarra et al. (2011)

Values are given by rounding

9.4 Bioactive Properties of Peptides Derived from Algal Proteins

Proteins compose of bioactive peptides sequences, which consist of 2–20 amino acid residues per peptide molecule. Peptides are released from the proteins either by enzymatic hydrolysis during intestinal digestion or by exogenous proteases during stages of food processing such as fermentation and ripening. Several beneficial effects of bioactive peptides have been proved in prevention or treatment of some diseases by many studies carried out in recent years (Zhang et al. 2012; Fan et al. 2014; Jo et al. 2017; Ovando et al. 2018). The physiological effects of bioactive peptides can reveal through mimicking hormone or drug activities (Daskaya-Dikmen et al. 2017). The bioactive peptides from natural sources are superior to synthetic pharmaceutical drugs because their accumulation is less likely in body tissues. This effect leads to low toxicity with a limited risk of side effects (Goodwin et al. 2012; Li-Chan 2015). Hence, bioactive peptides derived from macro- and microalgae can be considered as a good alternative to synthetic drugs due to their wide bioactive properties such as antioxidant, antihypertensive, anticancer and anti-inflammatory activities.

9.4.1 Antihypertensive Effect

Hypertension is described as an important risk factor associated with cardiovascular diseases-related high mortality. Angiotensin I-converting enzyme (ACE; EC 3.4.15.1) has a key role in regulating blood pressure. The ACE-I catalyses the conversion of angiotensin-II from angiotensin-I. Angiotensin-II acts as an efficient vasoconstrictive molecule (Sheih et al. 2009a; Samarakoon et al. 2013; Daskaya-Dikmen et al. 2017). Therefore, the inhibition of ACE is a crucial approach for regulating blood pressure and consequently the treatment of hypertension. Some bioactive peptides named as antihypertensive peptides have the effect of angiotensin converting enzyme inhibition. This group of bioactive peptides with potential beneficial effects related to hypertension can be isolated from macro- and microalgae (Sheih et al. 2009a). Sheih et al. (2009a) investigated the potential of ACE inhibition of a hendeca-peptide derived from protein waste of *Chlorella vulgaris* by using pepsin enzyme. Amino acid sequence of the isolated peptide was

Val-Glu-Cys-Tyr-Gly-Pro-Asn-Arg-Pro-Gln-Phe and it exhibited ACE inhibitory activity with the half maximal inhibitory concentration (IC_{50}) value of 29.6 μM. Additionally, the stability of the ACE inhibitory peptide retained at wide range of pH (from 2 to 10) and temperature (from 40 to 100 °C) against gastrointestinal protease (Sheih et al. 2009a). In another study conducted by Samarakoon et al. (2013), ACE inhibitory peptides from cultured biomass of microalgae Nannochloropsis oculata have been obtained using proteases including pepsin, trypsin, α-chymotrypsin, papain, alcalase and neutrase. Among proteases, the hydrolysates obtained by pepsin showed the highest ACE inhibitory activity. The two purified peptides exhibited the IC_{50} values of 123 and 173 μM. The amino acid sequences of the peptides were Gly-Met-Asn-Asn-Leu-Thr-Pro and Leu-Glu-Gln. According to Samarakoon et al. (2013), the purified peptides from *N. oculata* proteins can be used commercially to control high blood pressure as novel peptides in functional foods market. Sato et al. (2002) investigated ACE inhibitory activity of purified and identified peptides isolated from hydrolysates of macroalgae *Undaria pinnatifida* by protease in spontaneously hypertensive rats. Three peptides with amino acid sequences of Val-Tyr, Ile-Tyr, Phe-Tyr and Ile-Trp decreased the high blood pressure in a dose of 1 mg/kg of body weight in rats (Sato et al. 2002). In the study of Yücetepe et al. (2018a), the changes in ACE inhibitory activity of protein hydrolysates derived from *Spirulina platensis* protein (SPPHs) using trypsin were investigated during in vitro gastrointestinal digestion. In this work, the highest ACE inhibitory activity value was reported as 21.79 ± 1.52% for initial SPPHs and those of the digested SPPHs by gastric and intestinal proteases did not remain generally stable after in vitro digestion (Yücetepe et al. 2018a).

9.4.2 *Antioxidant Effect*

Normally, there is a homeostatic balance, originating from enzymatic and non-enzymatic mechanisms, between oxidants and antioxidants in body. Cells are protected against degenerative effects, which are caused by free radicals and reactive oxygen species (ROSs), by cellular antioxidant capacity (Sheih et al. 2009b; Harnedy and FitzGerald 2011). When the balance disrupts, imbalance phenomenon takes place in homeostasis and oxidation products accumulate in the cells. As a result of the accumulation of free radicals and ROSs in the cells, cellular compounds such as proteins, lipids and nucleic acids damage due to oxidation (Sheih et al. 2009a, b; Aprioku 2013). Oxidation products can lead to many degenerative diseases such as cancer, heart disease, stroke, arteriosclerosis and diabetes (Sheih et al. 2009b). Moreover, the loss in food quality, arising from oxidation of lipids, is an important problem for food manufacturers due to causing a decrease in shelf-life of foods (Harnedy and FitzGerald 2011). In this point, the addition of antioxidant compounds to foods during their production plays an important role in reducing oxidative reactions in foods (Sheih et al. 2009b; Harnedy and FitzGerald 2011; Samarakoon and Jeon 2012; Yücetepe et al., 2021). In recent years, protein hydrolysates and peptides derived from many algal sources have demonstrated to have antioxidant activity in vitro (Harnedy and FitzGerald 2011). For instance, the peptide named as carnosine (beta-alanyl-L- histidine), isolated from *Acanthophora delilei*, generally present in animal muscle, was described as an antioxidant and possessing protective functions against free radical effects (Fleurence et al. 2018). In a study conducted by Yücetepe et al. (2018b), antioxidant activity of protein extracts from *Spirulina platensis* was determined as 11.32 mg Trolox equivalent/g dry weight under extraction conditions (temperature 45 °C, pH 7.46 and time 120 min). In another work, *Palmaria palmata* protein hydrolysates generated with corolase exhibited higher antioxidant activity than the hydrolysates generated with alcalase enzyme (Harnedy and FitzGerald 2013).

9.4.3 Anticancer Effect

Nowadays, cancer is one of important health problems and it is the largest single cause of death for human in the whole world (Sheih et al. 2010; Siegel et al. 2020). Chemotherapy and radiotherapy commonly apply as promising approach in cancer treatment because of reducing mortality of cancer by inhibiting the growth and proliferation of rapidly dividing cancer cells (Sheih et al. 2010; Shahin et al. 2013; Roy et al. 2021). However, chemicals or drugs used in chemotherapy and radiotherapy treatments can lead to side-effects on normal cells due to their broad-spectrum activity (Shahin et al. 2013). Protein hydrolysates and peptides generated from macro- and microalgae as natural anticancer agents have been reported to demonstrate anticancer/antitumor activity on cancer cells (Fan et al. 2014; El-Chaghaby et al. 2019). Mao et al. (2017) investigated anticancer activity of red macroalgae *Pyropia haitanensis* proteins and protein hydrolysates generated by using pepsin and papain enzymes on human breast cancer cells (MCF-7), human liver cancer cells (HepG-2) and human lung cancer cells (A549). A new peptide obtained by papain hydrolysis, possessing amino acid sequence of QTDDNHSNVLWAGFSR, exhibited maximum inhibition on HepG-2 at IC_{50} value of 500 μg/mL, but low cytotoxicity on normal liver cells (LO-2) (Mao et al. 2017). In another work conducted by Fan et al. (2017), two new peptides (VPGTPKNLDSPR and MPAPSCALPRSVVPPR) from tryptic hydrolysates of *Pyropia haitanensis* showed an effective antiproliferation activity with the IC_{50} values of 200.97 μg/mL and 276.85 μg/mL on MCF-7 and HepG-2, respectively. Moreover, they exhibited low inhibition (7.1%) on LO-2 at 500 μg/mL (Fan et al. 2017). Zhang and Zhang (2013) identified a new polypeptide Y2 from tryptic hydrolysate of *Spirulina platensis* proteins. This antitumor polypeptide exhibited cytotoxic action on MCF-7 and HepG-2 (Zhang and Zhang 2013). Antitumor activity of a polypeptide from *Chlorella pyrenoidosa* microalgae was shown on HepG2 cells (Wang and Zhang 2013).

9.4.4 Anti-inflammatory Effect

Inflammation is an innate response of immune system of body against tissue damage. It is conducted by cytokines, which are small polypeptides and released at the site of inflammation, and other mediators. However, uncontrolled and excessive production of cytokines can damage to tissues that lead to some chronic inflammation-derived diseases like cardiovascular diseases, rheumatoid arthritis, bronchitis and cancers. In this regard, the suppression of inflammation mediators is an important strategy in the treatment of the diseases associated with excessive and uncontrolled inflammation (Petersen and Pedersen 2005; Heo et al. 2010; Pimentel et al. 2019). In vitro anti-inflammatory activity of the peptide PPY1 (KAQAD) generated from red macroalgae *Pyropia yezoensis* was determined in the mouse macrophage cell line RAW 264.7 (Lee et al. 2015). The authors have reported that the peptide PPY1 inhibited production of nitric oxide (NO) by 67.7%. Moreover, it demonstrated a potent inhibition effect on the expression of the inflammatory mediators, inducible nitric oxide synthase (iNOS) and cyclooxygenase-2 (COX-2) in the macrophages without a negative effect on cell viability (Lee et al. 2015). Vo et al. (2013) investigated anti-inflammatory effect of two peptides (LDAVNR and MMLDF) from microalgae *Spirulina maxima*. In this study, the peptides inhibited release and production of histamine from antigen-stimulated RBL-2H3 mast cells (Vo et al. 2013). Similarly, anti-inflammatory activity of a glycoprotein from brown macroalgae *Undaria pinnatifida* was assessed in the macrophages RAW 264.7 by Rafiquzzaman et al. (2015) by measuring NO concentration in the cell line and by determining the inhibition of cyclooxygenase (COX). The glycoprotein isolated from *U. pinnatifida* exhibited the inhibitory effect by 94.2% on the production of NO with cell viability of 94.6% at a concentration of 100 μg/mL. It also inhibited COX-1 and COX-2 at IC_{50} values of 53.03 ± 1.03 μg/mL and 193.35 ± 3.08 μg/mL, respectively (Rafiquzzaman et al. 2015).

9.5 Encapsulation of Macro- and Microalgal Proteins, Protein Hydrolysates and Peptides

Proteins and peptides could be incorporated into foods to improve nutritional and functional features because of their huge potential regarding with bioactive properties mentioned above in detail. However, bioactive peptides are sensitive against low pH values of stomach during gastro-intestinal digestion and they could easily degrade in the ways of chemical or biochemical. For this reason, several encapsulation methods can be applied to protect them from degradation against environment conditions such as low pH values and high temperature. Encapsulation as terminologically is a process of coating solid, liquid or gaseous bioactive compounds in capsules consisting of different wall materials. Encapsulation may also mask the bitter taste that caused by hydrophobic amino acid residues generated from protein hydrolysis (Mohan et al. 2015; Zhang et al. 2016). However, the number of studies in the field of encapsulation of proteins and peptides from macroalgae is quite limited. Studies on the encapsulation of algae proteins and peptides are mostly on proteins and peptides derived from microalgae in particular *Spirulina platensis*. Therefore, in this section of the chapter, the effect of encapsulation techniques was discussed on bioactive properties of macro- and microalgal proteins and peptides by reviewing a few present studies.

9.5.1 Film Hydration

Liposomal encapsulation is mainly based on film hydration technique. It is a technique of generating submicroscopic foams called liposomes, which can be used as carriers with particle sizes ranging from 30 nm to several micrometres in encapsulation studies. Liposomes can be made from cholesterol and natural and/or synthetic phospholipids. Liposome formation occurs through hydrophilic-hydrophobic interaction between phospholipids and water molecules (Fig. 9.1). For this reason, they could trap bioactive compounds with hydrophobic and hydrophilic character and release the entrapped at designated targets. Liposomal encapsulation has been widely applied to develop delivery systems for unstable bioactive compounds such as peptides, protein hydrolysates and essential oils in the field of food science. Stability of bioactive peptides and protein hydrolysates encapsulated by liposomal encapsulation can be protected under acidic conditions such as stomach media during digestion. Hence, bioavailability of encapsulated peptides can enhance in gastrointestinal tract (da Silva et al. 2010; Fang and Bhandari 2010; Akbarzadeh et al. 2013; Zhang 2017). Several studies have been conducted on

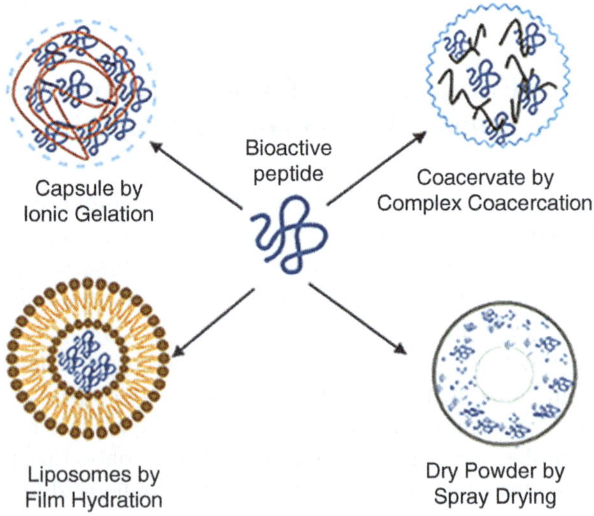

Fig. 9.1 Schematic illustration of encapsulants formed with encapsulation methods applied to bioactive peptides derived from algal proteins

liposomal encapsulation with the goal of increasing the stability of macro-and microalgae proteins, protein hydrolysates and peptides. In one of these studies, the influence of liposomal encapsulation of C-phycocyanin, which is a biliprotein derived from cyanobacteria and eukaryotic algae, was investigated on its anti-inflammatory activity by Manconia et al. (2009). According to results of the study, liposomal encapsulation improved its anti-inflammatory activity. Therefore, C-phycocyanin has the usage potential as an anti-inflammatory drug (Manconia et al. 2009). In another work, Machado et al. (2014) investigated the effect of sonication and homogenization treatments in liposomal encapsulation of proteins from *Spirulina platensis* microalgae by using thin-film technology. As a result of the study, there was not any difference between sonication and homogenization processes applied in terms of particle size. However, the liposomes subjected to homogenization were more uniform when compared to the liposomes subjected to sonication (Machado et al. 2014).

9.5.2 Coacervation

Complex coacervation is an encapsulation technique including two liquid phases in an aqueous colloidal media that caused by the interaction of two oppositely charged biopolymers. The coacervate is named as the more concentrated phase in the mixture, while the other phase is defined equilibrium solution (Fig. 9.1). The behaviour of two oppositely charged biopolymers in a solution mainly depends on the pH of the solution. It was reported that three critical pH values (pH_c, $pH_{\phi 1}$ and $pH_{\phi 2}$) determine the formation of soluble and insoluble complexes. The pH_c corresponds to the formation of the initial soluble complexes. These soluble complexes aggregated into the coacervate phase at $pH_{\phi 1}$ and $pH_{\phi 2}$ cause the dissociation of complexes into a soluble state due to charge loss of biopolymer. The values of pH_c and pH_ϕ depend on isoelectric point of the protein, the biopolymer ratio, the ionic strength and the molar masses of the biopolymers. Biopolymer complexes or coacervates can improve dispersion stability, texture, interfacial properties and sensorial characteristics of bioactive substances. Furthermore, it is possible to modify the functional properties of the proteins in the protein-polysaccharide mixtures with the formation of complexes or coacervates. Therefore, coacervation technique has a great usage potential in food industry (Yücetepe et al. 2020). The number of studies is a few on encapsulation of algal proteins or peptides by complex coaservation. For instance, Wang and Zhang (2013) examined micro- and nanoencapsulation of antitumor polypeptide from *Chlorella pyrenoidosa* by using complex coacervation and ionotropic gelation methods. The in vitro release tests revealed that the encapsulated polypeptide was well preserved from gastric digestion and the slowly controlled release in the gastrointestinal tract could be potentially achieved (Wang and Zhang 2013). In another study, Yücetepe et al. (2020) optimized pH and the ratio between *Spirulina platensis* protein concentrate (SPPC) and chitosan for the formation of complex coacervation. In this work, the strongest interaction between SPPC and chitosan occurred at pH of 5.5 and SPPC-chitosan ratio of 7.5:1 (Yücetepe et al. 2020).

9.5.3 Spray Drying

Spray drying is an encapsulation technique commonly used because it has numerous advantages such as simplicity, versatility and low production cost. Therefore, approximately 90% of encapsulates are prepared by spray drying in food industry. Liquids are transformed into powders to improve their stability against factors such as pH and temperature by spray drying (Đorđević et al. 2015; Nair et al. 2019). Modified starch, maltodextrin, gum or other substances are used as wall materials in this technique. In spray drying process, first of all, the mixture containing wall and core materials is atomized with a nozzle or spinning wheel by a hot gas. Then, water evaporation occurs by the hot air contacting the atomized material. Finally, the capsules with a mean size range of 10–100 μm in powder forms are obtained (Fig. 9.1) (Fang and Bhandari 2010; Gonçalves et al. 2017). One of the studies about encapsulation with spray drying of algal proteins or peptides was conducted by Pruchyathawornkul and Chantanawarangoon (2008). They examined the effects of encapsulation on storage stability of C-phycocyanin from *Spirulina platensis* microalgae at different storage temperatures. For this purpose, C-phycocyanin was encapsulated using maltodextrin as wall material by spray drying. According to results of the study, the stability of C-phycocyanin increased through spray drying (Pruchyathawornkul and Chantanawarangoon 2008). In another study conducted by Aquino et al. (2020), a series of peptides isolated from the hydrolysed highly water-soluble *Spirulina* extract were encapsulated with chitosan and mannitol as wall materials using spray drying. In this work, the produced microparticles exhibited good flow properties and stability up to 6 months storage (Aquino et al. 2020).

9.5.4 Ionic Gelation

Ionic gelation process includes atomization, extrusion and coextrusion or electrostatic deposition steps. In this encapsulation method, a polymeric or hydrocolloid solution is dripped or atomized into an ionic solution, during constant agitation. Then, the core material is dissolved in the polymeric solution. The drops in the ionic solution immediately form spherical gel structures that contain the active dispersed in the polysaccharide matrix (Fig. 9.1) (Kurozawa and Hubinger 2017). Rajmohan and Bellmer (2019) used two methods (internal and external gelations) to encapsulate *Spirulina* proteins in alginate by ionic gelation technique. They also examined the effect of different concentrations of sodium alginate and calcium chloride on particle size, texture, morphology and crude protein content of the beads. According to results of the texture analysis, the hardness of the external gelation beads was higher that of the internal gelation beads. Furthermore, an increase in the alginate percentage in both gelation methods caused an increase in the firmness of the beads. The crude protein content of beads obtained with external gelation was significantly higher than that of the internal gelation (Rajmohan and Bellmer 2019).

9.6 Conclusion and Future Directions

In recent years, there has been an increase in the demand for new protein sources as parallel increase in the world's population. With the purpose of meeting the demand on non-conventional protein sources, there has recently been a growing interest about finding sustainable and cheap protein sources in the field of food science. In this context, this chapter focus on revealing potential of algal proteins and bioactive peptides derived from them as a new protein source for food industry. From this point of view, this chapter aims to give information about chemical composition of macro- and microalgae, bioactive properties of algal proteins-derived peptides and encapsulation applications on algal proteins, protein hydrolysates or peptides. There have been several studies on the effect encapsulation on bioactive properties of algal proteins, protein hydrolysates or peptides, and most of the studies have been conducted on proteins or peptides from microalgae such as *Spirulina platensis* and *Chlorella vulgaris*. Hence, encapsulation of proteins and peptides from macroalgae as cheap and sustainable protein sources and their incorporation into functional foods should be investigated in future studies.

References

Akbarzadeh A, Rezaei-Sadabady R, Davaran S, Joo SW, Zarghami N, Hanifehpour Y, Samiei M, Kouhi M, Nejati-Koshki K (2013) Liposome: classification, preparation, and applications. Nanoscale Res Lett 8(1):102

Apostolidis C, McLeay F (2016) Should we stop meating like this? Reducing meat consumption through substitution. Food Policy 65:74–89

Aprioku JS (2013) Pharmacology of free radicals and the impact of reactive oxygen species on the testis. J Reprod Infert 14(4):158

Aquino RP, Auriemma G, Conte GM, Esposito T, Sommella E, Campiglia P, Sansone F (2020) Development of chitosan/mannitol microparticles as delivery system for the oral administration of a spirulina bioactive peptide extract. Molecules 25(9):2086

Aziz N, Prasad R, Ibrahim AIM, Ahmed AIS (2017) Promising applications for the production of biofuels through algae. In: Microbial Biotechnology (Patra JK, Vishnuprasad CN, Das G), Springer Nature Singapore Pte Ltd. 81–103

Becker E (2007) Micro-algae as a source of protein. Biotechnol Adv 25(2):207–210

Becker W (2004) 18 microalgae in human and animal nutrition. In: Richmond A (ed) Handbook of microalgal culture, 1st edn. Blackwell Science, Iowa, pp 312–351

Benelhadj S, Gharsallaoui A, Degraeve P, Attia H, Ghorbel D (2016) Effect of pH on the functional properties of *Arthrospira* (*Spirulina*) *platensis* protein isolate. Food Chem 194:1056–1063

Bilgin Ş, Ertan ÖO (2002) Selected chemical constituents and their seasonal variations in *Flabellia petiolata* (Turra) Nizam. and *Halimeda tuna* (Ellis and Sol.) JV Lamour. in the gulf of Antalya (North-eastern Mediterranean). Turk J Bot 26(2):87–90

Bjarnadóttir M, Aðalbjörnsson BV, Nilsson A, Slizyte R, Roleda MY, Hreggviðsson GÓ, Friðjónsson ÓH, Jónsdóttir R (2018) *Palmaria palmata* as an alternative protein source: enzymatic protein extraction, amino acid composition, and nitrogen-to-protein conversion factor. J Appl Phycol 30(3):2061–2070

Biller P, Ross AB (2014) Pyrolysis GC–MS as a novel analysis technique to determine the biochemical composition of microalgae. Algal Res 6:91–97.

Bleakley S, Hayes M (2017) Algal proteins: extraction, application, and challenges concerning production. Foods 6(5):33

Brown M, Jeffrey S, Volkman J, Dunstan G (1997) Nutritional properties of microalgae for mariculture. Aquaculture 151(1–4):315–331

da Silva MP, Daroit DJ, Brandelli A (2010) Food applications of liposome-encapsulated antimicrobial peptides? Trends Food Sci Technol 21(6):284–292

Daskaya-Dikmen C, Yucetepe A, Karbancioglu-Guler F, Daskaya H, Ozcelik B (2017) Angiotensin-I-converting enzyme (ACE)-inhibitory peptides from plants. Nutrients 9(4):316

Denis C, Morançais M, Li M, Deniaud E, Gaudin P, Wielgosz-Collin G, Barnathan G, Jaouen P, Fleurence J (2010) Study of the chemical composition of edible red macroalgae *Grateloupia turuturu* from Brittany (France). Food Chem 119(3):913–917

Devi MA, Venkataraman L (1984) Functional properties of protein products of mass cultivated blue-green alga *Spirulina platensis*. J Food Sci 49(1):24–27

Đorđević V, Balanč B, Belščak-Cvitanović A, Lević S, Trifković K, Kalušević A, Kostić I, Komes D, Bugarski B, Nedović V (2015) Trends in encapsulation technologies for delivery of food bioactive compounds. Food Eng Rev 7(4):452–490

El-Chaghaby GA, Rashad S, Abdel-Kader SF, Rawash EA, Moneem MA (2019) Assessment of phytochemical components, proximate composition and antioxidant properties of *Scenedesmus obliquus*, *Chlorella vulgaris* and *Spirulina platensis* algae extracts. Egypt J Aquat Biol Fish 23(4):521–526

Fan KW, Chen F (2007) Production of high-value products by marine microalgae thraustochytrids. In: Yang ST (ed) Bioprocessing for value-added products from renewable resources, 1st edn. Elsevier Science and Technology Books, Amsterdam, pp 293–323

Fan X, Bai L, Mao X, Zhang X (2017) Novel peptides with anti-proliferation activity from the *Porphyra haitanesis* hydrolysate. Process Biochem 60:98–107

Fan X, Bai L, Zhu L, Yang L, Zhang X (2014) Marine algae-derived bioactive peptides for human nutrition and health. J Agric Food Chem 62(38):9211–9222

Fang Z, Bhandari B (2010) Encapsulation of polyphenols–a review. Trends Food Sci Technol 21(10):510–523

Fleurence J, Morançais M, Dumay J (2018) Seaweed proteins. In: Yada RY (ed) Proteins in food processing, 2nd edn. Woodhead Publishing, Duxford, pp 245–262

Galland-Irmouli AV, Fleurence J, Lamghari R, Luçon M, Rouxel C, Barbaroux O, Bronowicki JP, Villaume C, Guéant JL (1999) Nutritional value of proteins from edible seaweed *Palmaria palmata* (Dulse). J Nutr Biochem 10(6):353–359

Garcia-Vaquero M, Hayes M (2016) Red and green macroalgae for fish and animal feed and human functional food development. Food Rev Intl 32(1):15–45

Gonçalves A, Estevinho BN, Rocha F (2017) Design and characterization of controlled-release vitamin a microparticles prepared by a spray-drying process. Powder Technol 305:411–417

Goodwin D, Simerska P, Toth I (2012) Peptides as therapeutics with enhanced bioactivity. Curr Med Chem 19(26):4451–4461

Gressler V, Yokoya NS, Fujii MT, Colepicolo P, Mancini Filho J, Torres RP, Pinto E (2010) Lipid, fatty acid, protein, amino acid and ash contents in four Brazilian red algae species. Food Chem 120(2):585–590

Harnedy PA, FitzGerald RJ (2011) Bioactive proteins, peptides, and amino acids from macroalgae. J Phycol 47(2):218–232
Harnedy PA, FitzGerald RJ (2013) In vitro assessment of the cardioprotective, anti-diabetic and antioxidant potential of *Palmaria palmata* protein hydrolysates. J Appl Phycol 25(6):1793–1803
Heo SJ, Yoon WJ, Kim KN, Ahn GN, Kang SM, Kang DH, Oh C, Jung WK, Jeon YJ (2010) Evaluation of anti-inflammatory effect of fucoxanthin isolated from brown algae in lipopolysaccharide-stimulated RAW 264.7 macrophages. Food Chem Toxicol 48(8–9):2045–2051
Hernández-Ledesma B, del Mar CM, Recio I (2011) Antihypertensive peptides: production, bioavailability and incorporation into foods. Adv Colloid Interf Sci 165(1):23–35
Jo C, Khan FF, Khan MI, Iqbal J (2017) Marine bioactive peptides: types, structures, and physiological functions. Food Rev Intl 33(1):44–61
Kurozawa LE, Hubinger MD (2017) Hydrophilic food compounds encapsulation by ionic gelation. Curr Opin Food Sci 15:50–55
Lee HA, Kim IH, Nam TJ (2015) Bioactive peptide from *Pyropia yezoensis* and its anti-inflammatory activities. Int J Mol Med 36(6):1701–1706
Li-Chan EC (2015) Bioactive peptides and protein hydrolysates: research trends and challenges for application as nutraceuticals and functional food ingredients. Curr Opin Food Sci 1:28–37
Lorenzo JM, Agregán R, Munekata PE, Franco D, Carballo J, Şahin S, Lacomba R, Barba FJ (2017) Proximate composition and nutritional value of three macroalgae: *Ascophyllum nodosum*, *Fucus vesiculosus* and *Bifurcaria bifurcata*. Mar Drugs 15(11):360
Lupatini AL, Colla LM, Canan C, Colla E (2017) Potential application of microalga *Spirulina platensis* as a protein source. J Sci Food Agric 97(3):724–732
Machado AR, Assis LM, Costa JAV, Badiale-Furlong E, Motta AS, Micheletto YMS, Souza-Soares LA (2014) Application of sonication and mixing for nanoencapsulation of the cyanobacterium *Spirulina platensis* in liposomes. Int Food Res J 21:2201–2206
Mæhre HK, Jensen IJ, Eilertsen KE (2016) Enzymatic pre-treatment increases the protein bioaccessibility and extractability in Dulse (*Palmaria palmata*). Mar Drugs 14(11):196
Mageswaran R, Sivasubramaniam S (1984) Mineral and protein contents of some marine algae from the coastal areas of northern Sri Lanka. J Natl Sci Found 12(2):179–189
Manconia M, Pendás J, Ledón N, Moreira T, Sinico C, Saso L, Fadda AM (2009) Phycocyanin liposomes for topical anti-inflammatory activity: in vitro in vivo studies. J Pharm Pharmacol 61(4):423–430
Mao X, Bai L, Fan X, Zhang X (2017) Anti-proliferation peptides from protein hydrolysates of *Pyropia haitanensis*. J Appl Phycol 29(3):1623–1633
Mohan A, Rajendran SR, He QS, Bazinet L, Udenigwe CC (2015) Encapsulation of food protein hydrolysates and peptides: a review. RSC Adv 5(97):79270–79278
Nair A, Khunt D, Misra M (2019) Application of quality by design for optimization of spray drying process used in drying of risperidone nanosuspension. Powder Technol 342:156–165
Narasimman S, Murugaiyan K (2012) Proximate composition of certain selected marine macroalgae form Mandapam coastal region (Gulf of Mannar), southeast coast of Tamil Nadu. Int J Pharm Biol Arch 3(4):918–921
Niccolai A, Zittelli GC, Rodolfi L, Biondi N, Tredici MR (2019) Microalgae of interest as food source: biochemical composition and digestibility. Algal Res 42:101617
Ovando CA, Carvalho JC, Vinícius de Melo Pereira G, Jacques P, Soccol VT, Soccol CR (2018) Functional properties and health benefits of bioactive peptides derived from *Spirulina*: a review. Food Rev Intl 34(1):34–51
Paiva L, Lima E, Neto AI, Marcone M, Baptista J (2017) Nutritional and functional bioactivity value of selected Azorean macroalgae: *Ulva compressa*, *Ulva rigida*, *Gelidium microdon*, and *Pterocladiella capillacea*. J Food Sci 82(7):1757–1764
Patarra RF, Paiva L, Neto AI, Lima E, Baptista J (2011) Nutritional value of selected macroalgae. J Appl Phycol 23(2):205–208
Petersen AMW, Pedersen BK (2005) The anti-inflammatory effect of exercise. J Appl Physiol 98(4):1154–1162

Pimentel FB, Alves RC, Harnedy PA, FitzGerald RJ, Oliveira MBP (2019) Macroalgal-derived protein hydrolysates and bioactive peptides: enzymatic release and potential health enhancing properties. Trends Food Sci Technol 93:106–124

Polat S, Ozogul Y (2008) Biochemical composition of some red and brown macro algae from the Northeastern Mediterranean Sea. Int J Food Sci Nutr 59(7–8):566–572

Pruchyathawornkul T, Chantanawarangoon S (2008) Effect of spray-dried temperature on C-phycocyanin and antioxidant capacity of spirulina (*Spirulina platensis*). In: Abstracts of the 46th Annual Conference, Kasetsart University, Bangkok (Thailand), 29 Jan 1 Feb 2008

Rafiquzzaman S, Kim EY, Lee JM, Mohibbullah M, Alam MB, Moon IS, Kim JM, Kong IS (2015) Anti-Alzheimers and anti-inflammatory activities of a glycoprotein purified from the edible brown alga *Undaria pinnatifida*. Food Res Int 77:118–124

Rajmohan D, Bellmer D (2019, 2019) Characterization of *Spirulina*-alginate beads formed using ionic gelation. Int J Food Sci:1–7. https://doi.org/10.1155/2019/7101279

Rawat J, Gupta PK, Pandit S, Prasad R, Pande V (2021) Current perspectives on integrated approaches to enhance lipid accumulation in microalgae. 3 Biotech 11, 303. https://doi.org/10.1007/s13205-021-02851-3

Renaud SM, Thinh LV, Lambrinidis G, Parry DL (2002) Effect of temperature on growth, chemical composition and fatty acid composition of tropical Australian microalgae grown in batch cultures. Aquaculture 211(1–4):195–214

Rodrigues D, Freitas AC, Pereira L, Rocha-Santos TA, Vasconcelos MW, Roriz M, Rodríguez-Alcalá LM, Gomes AM, Duarte AC (2015) Chemical composition of red, brown and green macroalgae from Buarcos bay in Central West Coast of Portugal. Food Chem 183:197–207

Roy A, Datta S, Bhatia KS, Bhumika, Jha P, Prasad R (2021) Role of plant derived bioactive compounds against cancer. South African Journal of Botany https://doi.org/10.1016/j.sajb.2021.10.015

Safi C, Zebib B, Merah O, Pontalier PY, Vaca-Garcia C (2014) Morphology, composition, production, processing and applications of *Chlorella vulgaris*: A review. Renewable and Sustainable Energy Reviews 35:265–278.

Samarakoon K, Jeon YJ (2012) Bio-functionalities of proteins derived from marine algae – a review. Food Res Int 48(2):948–960

Samarakoon KW, Kwon ON, Ko JY, Lee JH, Kang MC, Kim D, Lee JB, Lee JS, Jeon YJ (2013) Purification and identification of novel angiotensin-I converting enzyme (ACE) inhibitory peptides from cultured marine microalgae (*Nannochloropsis oculata*) protein hydrolysate. J Appl Phycol 25(5):1595–1606

Sathasivam R, Radhakrishnan R, Hashem A, Elsayed FA (2019) Microalgae metabolites: a rich source for food and medicine. Saudi J Biol Sci 26(4):709–722

Sato M, Hosokawa T, Yamaguchi T, Nakano T, Muramoto K, Kahara T, Funayama K, Kobayashi A, Nakano T (2002) Angiotensin I-converting enzyme inhibitory peptides derived from wakame (*Undaria pinnatifida*) and their antihypertensive effect in spontaneously hypertensive rats. J Agric Food Chem 50(21):6245–6252

Shahin M, Soudy R, Aliabadi HM, Kneteman N, Kaur K, Lavasanifar A (2013) Engineered breast tumor targeting peptide ligand modified liposomal doxorubicin and the effect of peptide density on anticancer activity. Biomaterials 34(16):4089–4097

Sheih IC, Fang TJ, Wu TK (2009a) Isolation and characterisation of a novel angiotensin I-converting enzyme (ACE) inhibitory peptide from the algae protein waste. Food Chem 115(1):279–284

Sheih IC, Wu TK, Fang TJ (2009b) Antioxidant properties of a new antioxidative peptide from algae protein waste hydrolysate in different oxidation systems. Bioresour Technol 100(13):3419–3425

Sheih IC, Fang TJ, Wu TK, Lin PH (2010) Anticancer and antioxidant activities of the peptide fraction from algae protein waste. J Agric Food Chem 58(2):1202–1207

Siegel RL, Miller KD, Jemal A (2020) Cancer statistics, 2020. CA Cancer J Clin 70(1):7–30

Shchelik I, Sieber S, Gademann K (2020) Green algae as a drug delivery system for the controlled release of antibiotics. Chem Eur J 26(70):16644–16648

Jainab SB, Azeez A, Fathima A, Kumar RR (2019) GC-MS analysis of the marine algae *Halymenia dilatata* Zanardini a potential source of fish feed in future. Indian Hydrobiol 18(1–2):164–169

Spolaore P, Joannis-Cassan C, Duran E, Isambert A (2006) Commercial applications of microalgae. J Biosci Bioeng 101(2):87–96

Taboada C, Millán R, Míguez I (2010) Composition, nutritional aspects and effect on serum parameters of marine algae *Ulva rigida*. J Sci Food Agric 90(3):445–449

Valente L, Gouveia A, Rema P, Matos J, Gomes E, Pinto I (2006) Evaluation of three seaweeds *Gracilaria bursa-pastoris*, *Ulva rigida* and *Gracilaria cornea* as dietary ingredients in European sea bass (*Dicentrarchus labrax*) juveniles. Aquaculture 252(1):85–91

Vo TS, Ryu B, Kim SK (2013) Purification of novel anti-inflammatory peptides from enzymatic hydrolysate of the edible microalgal *Spirulina maxima*. J Funct Foods 5(3):1336–1346

Wang X, Zhang X (2013) Separation, antitumor activities, and encapsulation of polypeptide from *Chlorella pyrenoidosa*. Biotechnol Prog 29(3):681–687

Yücetepe A, Kasapoğlu KN, Özçelik B (2018a) Angiotensin-i-converting enzyme inhibitory and antioxidant activity of tryptic *Spirulina platensis* protein hydrolysates: effect of hydrolysis and in vitro gastrointestinal digestion. Ecol Life Sci 13(3):151–162

Yücetepe A, Saroğlu O, Bildik F, Ozcelik B, Daskaya-Dikmen C (2018b) Optimisation of ultrasound-assisted extraction of protein from Spirulina platensis using RSM. Czech J Food Sci 36(1):98–108

Yücetepe A, Yavuz-Düzgün M, Şensu E, Bildik F, Demircan E, Özçelik B (2020) The impact of pH and biopolymer ratio on the complex coacervation of *Spirulina platensis* protein concentrate with chitosan. J Food Sci Technol 58:1274–1285

Yücetepe A, Altin G, Özçelik B (2021) A novel antioxidant source: evaluation of in vitro bioaccessibility, antioxidant activity and polyphenol profile of phenolic extract from black radish peel wastes (*Raphanus sativus* L. var. *niger*) during simulated gastrointestinal digestion. Inter J Food Sci Technol 56(3):1376–1384

Yücetepe A, Özçelik B (2016) Bioactive peptides isolated from microalgae *Spirulina platensis* and their biofunctional activities. Akademik Gıda 14(4):412–417

Zepka LQ, Jacob-Lopes E, Goldbeck R, Queiroz MI (2008) Production and biochemical profile of the microalgae *Aphanothece microscopica* Nägeli submitted to different drying conditions. Chem Eng Process Process Intensif 47(8):1305–1310

Zhang B, Zhang X (2013) Separation and nanoencapsulation of antitumor polypeptide from *Spirulina platensis*. Biotechnol Prog 29(5):1230–1238

Zhang C, Li X, Kim SK (2012) Application of marine biomaterials for nutraceuticals and functional foods. Food Sci Biotechnol 21(3):625–631

Zhang Z, Zhang R, Zou L, DJ MC (2016) Protein encapsulation in alginate hydrogel beads: effect of pH on microgel stability, protein retention and protein release. Food Hydrocoll 58:308–315

Zhang H (2017) Thin-film hydration followed by extrusion method for liposome preparation. In: D'Souza GGM (ed) Liposomes, vol 2nd. Humana Press, New York, pp 17–22

Chapter 10
Nanoformulations Loaded with Microalgal Bioactive Compounds for Disease Therapy

Heena Tabassum and Iffat Zareen Ahmad

Contents

H. Tabassum · I. Z. Ahmad (✉)
Department of Bioengineering, Integral University, Dasauli, Lucknow, Uttar Pradesh, India
e-mail: iffat@iul.ac.in

D. Thangadurai et al. (eds.), *Bioprospecting Algae for Nanosized Materials*,
Nanotechnology in the Life Sciences,
https://doi.org/10.1007/978-3-030-81557-8_10

10.1 Introduction

Nanoparticles may be defined as particles which have a size range between 1 and 100 nm, and there has been an enormous increase of activities in the field of nanometer-scale science and technology in the past few years. In fact, many nanocarrier-containing products were available commercially (Ostiguy et al. 2010; Sebastian et al. 2014). In areas such as development of drug, water treatment information technology and the synthesis of new materials, nanocarriers show high potential (Salata 2004). In chemical reactions, a higher surface-to-volume ratio increases the probability of the substrate binding with the nanocarrier (Pareek et al. 2017). In biology or medicine, nanocarriers are used as fluorescent biological markers, drug and gene delivery, pathogens biodetection, protein identification, study of DNA structure, tissue engineering, tumour treatment by heating (hyperthermia), separation and purification of biomolecules and phagokinetic investigations (Salata 2004; Prasad et al. 2019, 2020; Saglam et al. 2021; Maddela et al. 2021).

Cyanobacteria, relatively small and can be single or multicellular, are aquatic. They create colonies that are sufficiently large to imagine. They are the largest set of Gram-negative photosynthetic prokaryotes that are closely related to higher plants in terms of protein assembly and thylakoid membrane lipid composition (Los and Murata 1994). This microorganism has a few bacterial characteristics and a few higher plants that can photosynthesize and fix nitrogen because of them. Such species provide terrestrial habitats with a large amount of nitrogen and may also play a crucial role in successive processes (Zahran 1999). In a habitat with an elevated degree of solar irradiance, cyanobacteria are also primary; they resist erosion and preserve soil water (Booth 1988). However, in recent decades, a major depletion of the stratospheric ozone layer due to anthropogenically discharged pollutants has led to intense study on the effect of increased solar ultraviolet (UV) radiation on the earth's surface (Bais et al. 2018). Besides, cyanobacteria are rich storehouse of numerous bioactive compounds which have been reported with pharmacological activities. Some of the nanoformulations encapsulating these compounds have been reported. Uses of nanoparticles for the development of an optimized nanoformulation for clinical drug applications are shown in Fig. 10.1.

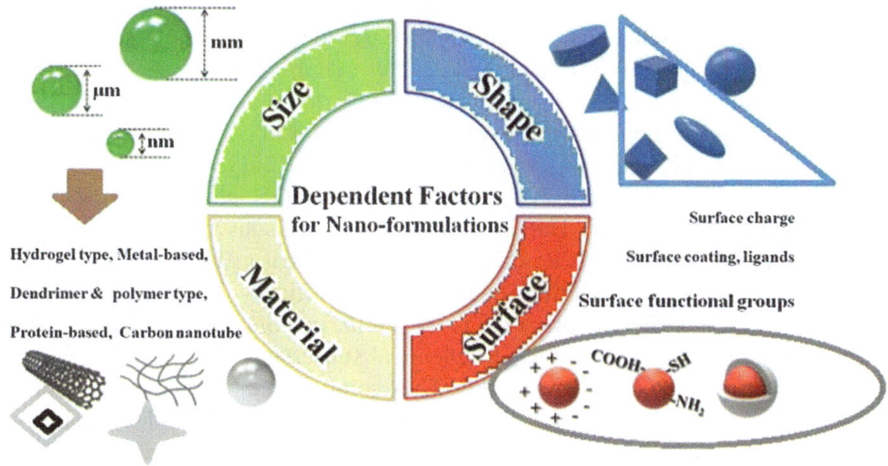

Fig. 10.1 Overview of regulative parameters of nanoparticles use for devising an optimized nanoformulation for clinical drug application. (Bajpai et al. 2018)

UV-B radiation is an extremely active portion of solar radiation that can affect the normal metabolic processes of all organisms, including cyanobacteria (Sinha et al. 2008), either indirectly through the development of reactive oxygen species (ROS) or directly through protein, DNA and lipid molecules (Garcia-Pichel et al. 1993). To withstand the deleterious effects of solar UVR, these prokaryotes have established various defence mechanisms including antioxidant production, protein re-synthesis, DNA repair, avoidance, programmed cell death and the formation of UV-absorbing compounds.

Scytonemin has been present in numerous terrestrial cyanobacteria (Bultel-Poncé et al. 2004). This compound is present in the EPS sheath of many cyanobacteria, where it acts as a passive sunscreen to protect against UVR. Three new SCY derivatives from *Scytonema* sp. have been developed. The structure and UV-protective efficacy of scytonemin is now well-known (Proteau et al. 1993). The complexity of SCY ring structures generates a particular pattern of UV absorbance (Sinha et al. 2008). In general, the maximum in vivo absorption of SCY is 370 nm, while the maximum absorption of purified SCY is 386 nm, although it is also important at 252, 278 and 300 nm. SCY protects cyanobacteria in the atmosphere from deadly UV-A/B radiation caused by strong sunlight (Sinha et al. 2008). SCY is a stable compound, and it acts without additional energy expenses, after continued physiological inactivity (Fleming and Castenholz 2007). Therefore, it is a suitable molecule to protect cyanobacteria from UVR when other protective mechanisms are insufficient (Karsten 2008). SCY acts as a potent UV filter and also as an antioxidant molecule, which can be used in cosmetics and pharmaceutical industries.

UVR has two types of adverse effects on normal skin: acute reactions (sunburn, tanning, etc.) and delayed reactions (photocarcinogenesis and photoalteration) (Edlich et al. 2004). Mainly, one of the activities to minimize exposure to the UVR has been defined by The American Academy of Dermatology (Edlich et al. 2004). Because of this, the application of sunscreens on skin in combination with other drugs is the most common way to prevent skin damage caused by UVR (Bernerd et al. 2003; Serpone et al. 2007). In sunscreen for moulds, physical and chemical agents are normally present. The physical and chemical sunscreen agents work by either dispersing the solar UVR and absorbing or attenuating it (Edlich et al. 2004; Serpone et al. 2007). In addition, in assessing the efficacy of sunscreen, the sun protection factor (SPF) also plays the most significant role (Geoffrey et al. 2019). There can be different side effects of chemical sunscreens, such as oedema, erythema and discomfort (Kale et al. 2011). The evolution in the field of nanotechnology and preparations for nanosized titanium dioxide and zinc oxide has allowed formulators to manufacture more powerful sunscreen cream that is cosmetically appropriate with minimal local side effects besides free radicals that may be carcinogenic (Smijs and Pavel 2011). In addition to physical sunscreen, the addition of a natural sunscreen agent is also a healthier choice to improve the degree of security.

10.2 Role of Cyanobacteria in Nanocarrier Synthesis

There is requirement for environmental-friendly and safe synthetic methods. Various natural methods are considered safe and non-toxic and supply environmental-friendly nanocarrier system. The exploitation of biomass in the form of aqueous extracts or whole cells is based on these emerging and expanding bottom-up methodologies. Particular attention has been paid to microalgae, as they are involved in toxic metal bioremediation, amongst the biological systems used for NP synthesis. It has been shown that microalgae generate silver, gold, cadmium and nanocarrier platinum (Brayner et al. 2007; Parial et al. 2012a, b). Cyanobacteria have unique characteristics amongst the microalgae, such as high biomass yield, oxygen photosynthesis, diversity of ecosystems (contaminated and degraded environments) and growth on non-arable lands, useful by-products and production of biofuel, improving soil fertility and reducing emissions of greenhouse gases. A characteristic feature of the conversion of carbon dioxide to other types of sunlight-catalysed carbon has consequences for the growth medium's reduced costs. Microalgae have been shown to contain silver, gold, cadmium and nanocarrier platinum (Brayner et al. 2007; Parial et al. 2012a, b). As a result of negatively charged groups on the membrane and in extracellular polymeric substances (EPS), the rapid binding of metals to the cyanobacterial cell surface occurs while intracellular levels are preserved through (1) phytochelatin and metallothionein metal chelation and (2) polyphosphate body sequestration (Hudek et al. 2012). The delicate balance between metal requirements and toxic overexposure is closely regulated by exopolysaccharides (EPS) layers and cell membrane uptake and trafficking (Hudek et al. 2012; Mehta and Gaur 2005). Cyanobacteria have developed special metal delivery mechanisms for metals to be internalized (imported) (Ma et al. 2009). Outer membrane porins promote the non-selective passive diffusion of metal ions across the outer membrane, while both the outer and inner membranes of active transport systems work (Ma et al. 2009; Yee et al. 2004) demonstrated that metal-binding sites on its surface have been heterogeneously distributed by *Calothrix*, a filamentous cyanobacterium. Proton-active functional groups of surface carboxyl, phosphoryl, hydroxyl and amine found on the cell wall and exopolymer sheath communicate with metal ions (Yee et al. 2004). Cyanobacteria are regarded as the best natural systems for NP synthesis both intracellularly and extracellularly. However, the reports on the synthesis of noble metal-based nanocarriers are lacking (El-Naggar et al. 2017). *Spirulina* has been shown to synthesize gold, silver and Au/Ag bimetallic nanocarrier by extracellular method (Chakraborty et al. 2009). *Anabaena*, *Calothrix* and *Leptolyngbya* are reported to form Au, Ag, Pd and Pt nanocarrier intracellularly (Brayner et al. 2007). The study suggested that the size of the particles as well as the yield depends on the genus of cyanobacteria. The studies on NP formation studies showed that the "nitrogenase" intracellular enzyme is responsible for metal reduction, but that the cellular environment is involved in the colloid growth process (Brayner et al. 2007). *Nostoc linckia* was used to synthesize selenium nanocarrier from selenite, amorphous nanocarrier formations, both extracellularly and intracellularly (Cepoi et al. 2015). In the so-called "paintertype" reaction, selenite can be reduced to elemental selenium by reaction with active thiol groups of proteins (Zinicovscaia and Cepoi 2016) (Table 10.1).

In general, chemicals (oxybenzone, octyl methoxycinnamate, luteolinone andoctocrylene) and physical ingredients (zinc oxide, titanium dioxide) used in the sunscreens are in the form of nanoparticles (nanocarrier), which can apparently optimize the delivery of materials into the outer layers of the skin (Jain and Chakravarty 2014). In different instances, nanoformulated sunscreen with efficient non-toxic nanocarriers is considered advantageous by health care professionals (Diffey 2009). In recent years, cases of non-melanoma skin cancer (NMSC) have increased (Halpern and Kopp 2005), and health care professionals agree that the use of nanoformulated sunscreen with useful non-toxic nanocarriers in these cases is advantageous (Halpern and Kopp 2005; Diffey 2009).

Table 10.1 Cyanobacterial nanocarriers and their activity

Species and strains	Size (nm)	Application of nanocarrier	References
Anabaena flos-aquae (ALCPB24)	40, 25.2	Anti-bacterial activity	Brayner et al. (2007)
Anabaena ambigua (NCCU-1)	48	NM	Husain et al. (2015)
Anabaena variabilis (NCCU-441)	50	NM	Husain et al. (2015)
Anabaena sp.	9	NM	Rosken et al. (2014)
Anabaena doliolum	10–50	Anti-bacterial activity Anti-tumour activity	Singh et al. (2014)
Anabaena sp. (PCC 7120)	NM	NM	Gulin et al. (2016)
Anabaena sp. (SAG 12.82)	9	NM	Rosken et al. (2014)
Anabaena cylindrica (SAG 1403.2)	10	NM	Rosken et al. (2016)
Anabaena laxa	3–5	NM	Rosken et al. (2014)
Anabaena sphaerica	1–50	NM	Roychoudhury et al. (2016a, b, c)
Anabaena flos-aquae (ALCP B24)	12, 25	NM	Dahoumane et al. (2010)
Anabaena strain (L31)	70–100	UV-protecting activity	Singh et al. (2014)
Anabaena flos-aquae	100	NM	Brayner et al. (2009)
Nostoc sp.	100	Anti-bacterial activity	Al Rashed et al. (2018)
Nostoc sp. Bahar M	8.5–26.44	Anti-cancer activity Anti-bacterial activity	Bin-Meferij and Hamida (2019)
Nostoc linckia	9.39–25.89	Anti-bacterial activity Anti-cancer activity	El-Naggar et al. (2017)
Nostoc carneum	7.1–26.68	Anti-cancer activity Anti-bacterial activity Anti-haemolytic activity	El-Naggar et al. (2018)
Nostoc muscorum NCCU-442	42	NM	Husain et al. (2015)
Nostoc sp. strain HKAR-2	51–100	Anti-cancer activity Anti-fungal activity Anti-bacterial activity	Sonker et al. (2017)
Nostoc commune	15–45	Anti-fungal activity Anti-bacterial activity	Morsy (2014)
Nostoc sp.	20.3	Anti-bacterial activity	Ahmed et al. (2015)
Nostoc sp. (PCC 7422)	10–30	Anti-bacterial activity	Radtsig et al. (2016)
Nostoc sp. (PCC 9229)	10–30	NM	Radtsig et al. (2016)
Nostoc sp. (PCC 6310)	10–30	NM	Radtsig et al. (2016)
Nostoc sp. (PCC 9305)	10–30	NM	Radtsig et al. (2016)
Nostoc ellipsosporum	8–42, 60 × 18.5	Anti-cancer activity Anti-bacterial activity	Parial et al. (2016)

(continued)

Table 10.1 (continued)

Species and strains	Size (nm)	Application of nanocarrier	References
Nostoc ellipsosporum	137–209 length and 33–69 diameters	NM	Parial et al. (2012a, b)
Nostoc sp. (EA03)	50–80	Anti-bacterial activity Anti-biofilm activity Anti-cancer activity	Ebadi et al. (2019)
Spirulina platensis	30–40	Anti-bacterial activity	Saran et al. (2017)
Spirulina platensis	17–25	NM	Mohamed et al. (2017)
Spirulina (CFTRI)	47	NM	Husain et al. (2015)
Spirulina NCCU-477	49	NM	Husain et al. (2015)
Spirulina (NCCU-479)	52	NM	Husain et al. (2015)
Spirulina-481	64	NM	Husain et al. (2015)
Spirulina (NCCU-482)	42	NM	Husain et al. (2015)
Spirulina (NCCU-483)	51	NM	Husain et al. (2015)
Spirulina platensis (NCCU-S5)	46	NM	Husain et al. (2015)
Spirulina platensis	˂ 100	Anti-algal activity	El-Sheekh et al. (2014)
Spirulina platensis	10–200	Anti-cancer activity	Govindaraju et al. (2008)
Spirulina platensis	12	NM	Kaliamurthi et al. (2016)
Spirulina platensis	5–20	NM	Tsibakhashvili et al. (2011)
Spirulina platensis	12.77	NM	Sugandha et al. (2014)
Spirulina platensis	11.5	Anti-bacterial activity	Ahmed et al. (2015)
Spirulina platensis	11.6	NM	Mahdieh et al. (2012)
Spirulina subsalsa	5–30	NM	Parial and Pal (2014)
Spirulina platensis	6–10	NM	Govindaraju et al. (2008)
Spirulina platensis	2–8	Anti-bacterial activity	Suganya et al. (2015)
Spirulina platensis	15–60	Pharmaceutical	Kalabegishvili et al. (2012)
Lyngbya majuscula (CUH/Al/MW-150)	20–50	Anti-leukemic activity Anti-bacterial activity	Roychoudhury et al. (2016a, b, c)
Lyngbya (NCCU-102)	54	NM	Husain et al. (2015)
Lyngbya majuscula	2–25	NM	Parial and Pal (2014)
Lyngbya majuscula	˂20	NM	Chakraborty et al. (2009)
Lyngbya sp.	NM	Anti-bacterial activity	Patel et al. (2014)
Lyngbya majuscula	41.7	Anti-myocardial infarction properties	Bakir et al. (2018)
Leptolyngbya	6	NM	Brayner et al. (2007)
Leptolyngbya tenuis (BDU 20,391)	NM	Anti-cancer activity Anti-bacterial activity	Parial et al. (2016)
Lyngbya majuscula	5–25	NM	Roychoudhury et al. (2016a, b, c)
Synechococcus sp.	NM	Anti-bacterial activity	Patel et al. (2014)

(continued)

Table 10.1 (continued)

Species and strains	Size (nm)	Application of nanocarrier	References
Synechococcus sp.	15.2–266.7	Photocatalytic activity Anti-bacterial activity	
Synechococcus elongatus	>100–77.3	NM	Keskin et al. (2016)
Synechocystis sp. (PCC 6803)	13	NM	Focsan et al. (2011)
Synechococcus elongatus	70–85	NM	Asmathunisha and Kathiresan (2018)
Calothrix pulvinate	100	NM	Brayner et al. (2007)
Calothrix pulvinate	3.5 3.2	NM	Brayner et al. (2009)
Phormidium cyanobacterium	10–40	NM	Rahman et al. (2009)
Phormidium valderianum	15	NM	Parial et al. (2012a, b)
Plectonema boryanum (UTEX 485)	30 nm- 0.3 µm	NM	Lengke et al. (2006a, b)
Plectonema boryanum (UTEX 485)	>30	NM	Lengke et al. (2007)
Dolichospermum sp.	23	NM	Kratosovs et al. (2016)
Gleocapsa sp.	<100	Anti-tumour agent	Geetha et al. (2014)
Oscillatoria sp.	80	NM	Husain et al. (2015)
Phormidium tenue	5.1	Biolabelling activity	MubarakAli et al. (2012)
Plectonema boryanum	<10–25	NM	Lengke et al. (2006a, b)
Phormidium willei	25	Antioxidant agent	MubarakAli et al. (2013)

NM not mentioned

10.3 Types of Nanoformulations

10.3.1 Carbon-Based Nanocarrier

For industrial applications such as fillers, efficient gas adsorbents for environmental remediation (Ngoy et al. 2014), and as a support medium for various inorganic and organic catalysts, fullerenes and carbon nanotubes are two key groups of nanocarrier used (Mabena et al. 2011). Electrical conductivity, heat conductivity and mechanical properties of the carbon nanocarrier are important. The carbon-based nanocarriers are highly stabile, conductive, less toxic and environment-friendly. Carbon-based nanocarriers have been planned to be used in drug delivery, bioimaging and energy storage (Ray and Jana 2017; Fig. 10.2).

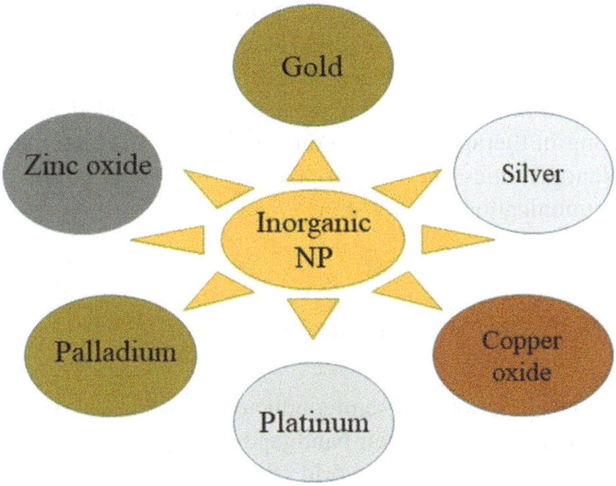

Fig. 10.2 Different metals used to develop nanoformulations loaded with cyanobacterial compounds. (Oscar et al. 2016)

10.3.2 Metallic Nanocarrier

Submicron-scale entities made of pure metals (e.g. gold, platinum, silver, titanium, zinc, cerium, iron, thallium) or their compounds are metal nanocarriers (e.g. oxides, hydroxides, sulphides, phosphates, fluorides, chlorides). The unique properties, particularly optical properties such as dispersion colour, are determined by the size and aspect ratio of the metallic nanocarrier (such as the gold nanocarrier). Because of their metallic existence, a wide range of nanoscale oxides is also known as the metal oxide nanocarrier. The electronic properties of these metal and metal oxide nanocarriers are an important result of the reduction in size. Owing to their small scale, the heavy effect of conductivity and chemical reactivity never affects the bandgap (Hoffmann 1988). In molecular diagnostics, electronics, catalysis, drug delivery and sensing, metallic nanocarriers are used (Castro and Lannes 2014). It is possible to synthesize and alter metallic nanocarrier with various chemical functional groups, which enables them to be combined with antibodies, ligands and drugs of interest and thus opens up a wide range of potential applications in biotechnology, magnetic separation, target analyte pre-concentration, targeted drug delivery, gene and drug delivery vehicles and, more importantly, diagnostics (Mody et al. 2010).

10.3.3 Ceramic Nanocarriers

These nanoceramics (also known as nanoceramics or nanopowders) are classified as inorganic, heat-resistant, non-metallic solids formed from metallic and non-metallic compounds of oxides, carbides, phosphates and carbonates. The substance has distinctive characteristics, such as high heat resistance and chemical inertness (Thomas and Gupta 2016). The biomedical field is the ceramic nanocarrier field that has been most studied. As drug carriers, nanoceramics have great potential to deliver and target the active pharmaceutical ingredient in a controlled way to the targeted location, resulting in therapeutic drug concentration at the target site (Ben-Nissan et al. 2004). Nanoceramics are also proposed in areas such as energy supply and storage and communication and transport systems.

10.3.4 Semiconductor Nanocarrier

A large bandgap is a function of a semiconductor nanocarrier, so these nanocarriers showed a major improvement in their bandgap tuning properties. Electronic devices have gained importance in today's world (Sun et al. 2000). In the conversion of sun's energy, optoelectronic instruments, molecular and cellular imaging, and in the ultrafine diagnostic, semiconductor nanocrystals have manyfold applications (Smith and Nie 2010).

10.3.5 Polymeric Nanocarrier

The particles prepared from polymers are the polymeric nanocarrier (Pnanocarrier). A polymer is a class of macromolecules consisting of natural or synthetic substances that are multiples of monomers. Waterborne paints, adhesives and coatings are the most traditional areas of use. They are mostly formed like nanospheres or nanocapsules (Mansha et al. 2017). Their applications in photonics, sensors, pollution control and environmental technology have been reported. Lately applications in biomedical areas including bioimaging, drug delivery and diagnostics have been reported (Mallakpour and Behranvand 2016).

Nannochloropsis gaditana, *Tetraselmis impellucida* and *Arthrospira* (*Spirulina*) *maxima* were studied with algae soluble protein isolates (ASPIs, 66–77 percent w/w protein), using commercially available WPI as a comparison (93 percent w/w protein). All protein isolates can form cream-stable emulsions (d3.2 0.2–0.3 μm) at pH 8.0.0. In order to shape these stable emulsions, the amount of each ASPI required (Ccr; protein-based) differed between isolates, but was within the range of proteins from both similar (photosynthetic) sources (algae and sugar beet leaves) and other sources of protein (dairy, legume and egg). Amongst the ASPI stabilized emulsions, minor differences were observed in the pH dependency of flocculation. The expected correlation between interfacial and molecular properties (constant and potential adsorption rate) and emulsion behaviour (Ccr and droplet size as a function of pH) was missing for the ASPIs (Teuling et al. 2019).

10.3.6 Lipid-Based Nanocarrier

The nanocarrier includes lipid moieties and is used efficiently in many biomedical applications. Like nanocarrier, the use of lipid nanocarrier in the pharmaceutical industry enhances the bioavailability of medicines whose solubility is poor in water. The lipid nanocarrier has a solid lipid centre and soluble lipophilic molecules are found in a matrix. In biomedical fields such as bioimaging, drug delivery and diagnostics, lipid-based nanocarriers are now commonly used (Mallakpour and Behranvand 2016; Puri et al. 2009). The lipid nanocarrier increases the absorption of drug in the gastrointestinal tract because of increased mucosal adhesion and residence time. They can also protect drugs from chemical and enzymatic degradation. Our outcomes therefore show that the typical lipid extract inhibits both SREBP-1 and -2 maturation, resulting in a decrease in the expression of genes involved in the metabolism of cholesterol and fatty acids. Using various concentrations of surfactant (SDS) in water and cyanobacteria algal oil, nanoemulsions were developed using ultrasonic techniques. The nanoemulsion's physical presence has shown that there is no self-surfactant property in algal oil. Turbidity was dominant and rendered the mechanism uncertain due to the high concentration of oil in water. There is an optimum absorption of nanoemulsions as measured by UV-Vis spectrophotometer. The UV-Vis study showed that they can be applied to minimize problems of UV radiation on human skin as the absorption of nanoemulsions is high. In case of nanoemulsions, the particle diameter of the oil sample has been efficiently reduced. It was found that all samples with constant sonication time were thermodynamically stable. Finally, studies of optical microscopic images show that nanoemulsions with a particle diameter of less than 100 nm can be optically observed. It was inferred from these experiments that the nanoemulsions were metastable and had a good particle size, making them usable for use in many useful food applications. These nanoemulsions of algae oil can be processed and used as milk enriched with algae oil, which stays stable for long time. Nanoemulsions of the oil in yoghurt as a food medium may improve the bioavailability of algae oils. In lathe machines, nanoemulsions can be used as cutting fluid. Some other nanoemulsion applications include cosmetics, such as mucosal vaccines, cell culture technology, sunscreen formulations, anti-ageing drugs and drug delivery systems. Better results were obtained by nanoemulsions with different SDS concentrations and a constant sonic time of 10 min. A crucial method used in the study to perceive the meta-stable existence of the nanoemulsions and the size of the particle was analysed by dynamic light scattering (DLS). Different ASPI was required in order to shape these stable emulsions on the basis of the type of isolate, but was within the range of proteins from both similar (photosynthetic) sources (algae and sugar beet leaves) and other sources of protein (dairy, legume and egg). Amongst the ASPI stabilized emulsions, minor differences were observed in the pH dependency of flocculation. The predicted association between interfacial and molecular properties (constant and potential adsorption rate) and emulsion behaviour (Ccr and droplet size as a function of pH) was absent for the ASPIs (Teuling et al. 2019).

10.4 Methods for the Synthesis of Solid Lipid Nanoformulations

10.4.1 High Shear Homogenization

High shear homogenization technique was initially used for the processing of solid lipid nanoemulsions (Olbrich et al. 2002; Domb 1993). Both strategies are widespread and simple to administer. However, the existence of microparticles also compromises dispersion efficiency. The process of high-speed homogenization is used to generate SLN by melt emulsification (Ahlin et al. 1998). The effect on particle size and zeta potential of various process parameters, including emulsification time, stirring rate and cooling state, was investigated by Olbrich et al. (2002). Trimyristin, tripalmitin, an amalgamation of mono-, di- and triglycerides (Witepsol W35, Witepsol H35) with glycerol behenate and poloxamer 188 were used as steric stabilizers (0.5 percent w/w). The best quality SLN was obtained for Witepsol W35 dispersions (after stirring for 8 min at 20,000 rpm) followed by cooling (10 min) and stirring (5000 rpm) at room temperature. In comparison, to emulsification (10 minutes) at (25,000 rpm) followed by cooling (5 minutes) and stirring (5000 rpm) in cool water (16 °C) were the best conditions for Dynasan dispersions (Speiser 1990). Higher stirring rates did not alter the particle size substantially, but the polydispersity index increased marginally.

10.4.2 Hot Homogenization

At temperatures above the melting point of the lipid, hot homogenization is carried out and is close to emulsion homogenization. High-shear mixing devices obtain a pre-emulsion of the drug-loaded lipid melt and the aqueous emulsifier step (the same temperature) (like silversion-type homogenizer). The consistency of the pre-emulsion has a significant effect on the quality of the final product and it is ideal to produce droplets of a few micrometres in the size range. Above the lipid melting point, high pressure homogenization of the pre-emulsion is achieved. Typically, smaller particle sizes are obtained due to lower lipid phase viscosity at higher processing temperatures (Lander et al. 2000), but this may also increase the degradation of the drug and carrier. Multiple passes (3–5 passes) through the high-pressure homogenizer gives better products. The temperature of the sample (10 °C at 500 bar) was usually enhanced by high pressure processing (Jahnke 1998); 3–5 homogenization cycles at 500–1500 bars are appropriate in most of the cases. Due to the coalescence of the particle, increasing homogenization results into an increase in size of the particles due to the high kinetic energy of the particles.

10.4.3 Cold Homogenization

The method of cold homogenization with the solid lipid is carried out and is therefore identical to milling a suspension at elevated pressure. Effective temperature control is required to ensure the solid state of the lipid during homogenization (Jahnke 1998). To overcome the following problems of the hot homogenization technique, cold homogenization has been developed, such as temperature induced accelerated degradation of the drug payload, partitioning and therefore loss of drug during homogenization into the aqueous phase, uncertain polymorphic lipid transitions due to the complexity of the nanoemulsion crystallization stage leading to cooled temperature.

The first preparatory step is the same as in the procedure of hot homogenization and involves the lipid melt solubilization or dispersion of the drug. The melt-containing drug is cooled easily (using dry ice or liquid nitrogen) to help the uniform distribution of drugs in the lipid matrix. As a result, the drug that contains solid lipid is pulverized into microparticles by ball/mortar milling. The average particle sizes obtained were in the range of 50–100 microns by enhancing the lipid fragility, and chilled processing furthermore helped particle milling. In a cold emulsifier solution, the SLNs are dispersed. The dispersion is subjected to high pressure homogenization with sufficient temperature regulation at or below room temperature, bearing in mind the normal increase in temperature during processing at high pressure. However, larger particle sizes and a wider size range are characteristic of cold homogenized samples relative to hot homogenization (Mullen zur 1996). The cold homogenization method minimizes the sample's thermal exposure, but because of the melting of the lipid/drug mixture in the initial stage, it does not eliminate it.

10.4.4 Ultrasonication or High Speed Homogenization

High-speed stirring or sonication was also developed by SLN (Eldem et al. 1991). A range of benefits are that in any laboratory, instruments that are used here are very common. The problem with this technique is the distribution of larger particle sizes in the micrometre scale. Physical instability leads to particle growth after storage. In this technique, possible metal contamination due to ultrasonication is also a major issue. In order to establish a stable formulation, experiments have been undertaken by different research groups that combine and perform high-speed stirring and ultrasonics at high temperatures.

10.4.5 SLN Prepared by Solvent Emulsification/Evaporation

The fat-soluble material is dissolved in a water-immiscible organic solvent (e.g. cyclohexane) for the synthesis of nanoparticle dispersions by precipitation in o/w emulsions (Sjöström and Bergenståhl 1992). Dispersion of the solvent nanoparticle is obtained after evaporation by precipitation of the lipid in the aqueous medium. In case of cholesterol acetate when taken as the model drug and lecithin/sodium glycocholate blend as the emulsifier, the average diameter of the particles obtained was 25 nm. Siekmann and Westesen (1996) who developed cholesterol acetate nanoparticles with a mean size of 29 nm have confirmed the reproducibility of the result (Sjöström and Bergenståhl 1992).

10.4.6 Microemulsion-Based Solid Lipid Nanoformulation Preparations

SLN preparation techniques that are focused on microemulsion dilution have been developed (Gasco 1993). They are formed at 65–70 ° C by stirring an optically transparent mixture that usually consists of a low melting fatty acid (stearic acid), an emulsifier (polysorbate 20, polysorbate 60, phosphatidylcholine soy and taurodeoxycholate sodium), co-emulsifiers (monooctylphosphate sodium) and water. Upon stirring, the hot microemulsion was distributed in cold water (2–3 °C). The standard hot microemulsion to cold water volume ratios were in the range of 1:25 to 1:50. The composition of the microemulsion critically determines the dilution process. According to the literature (Boltri et al. 1993), the microemulsion already contains the droplet structure, and thus, no energy is needed to achieve the particle size in microns. Different strategies may be taken with respect to the similarities in the synthetic process of polymer nanoparticles as identified by French scientists. By diluting polymer solutions in water, Fessi creates polymer particles. According to De Labouret et al. (1995), the velocity of the propagation processes critically decides the particle size. Only solvents which disperse very quickly into the aqueous phase (acetone) were generated with nanoparticles, whereas larger particle sizes were obtained with more lipophilic solvents. In the formation of lipid nanoparticles, hydrophilic co-solvents of microemulsion could play a similar function as acetone for the formation of polymer nanoparticles (Cavalli et al. 1996).

10.4.7 SLN Preparation by Using Supercritical Fluid

For SLN output, this is a relatively new technique and has the advantage of solventless processing (Chen et al. 2006). This platform technology for powder and nanoparticle preparation has many variants. It is carried out by the quick expansion of supercritical carbon dioxide solutions and is known as RESS process. The good choice as a solvent for this process was carbon dioxide (99.99 percent) (Gosselin et al. 2003).

10.4.8 Spray Drying Method

In order to transform an aqueous SLN dispersion into a drug product, it is an alternative technique to lyophilization. This procedure is cheaper than lyophilization. Owing to the high temperature, shear forces and partial melting of the sample, this process induces particle aggregation. For spray drying, Freitas and Muüller (1998) proposed the use of lipids with a melting point of >70°. With an SLN concentration of 1% in a solution of trehalose in water or 20% trehalose in ethanol-water mixtures (10/90 v/v), the best result was obtained.

10.4.9 Double Emulsion Method

A new protocol of solvent emulsification-evaporation has been applied for the synthesis of hydrophilic SLN loaded with drugs (Cortesi et al. 2002). Here, during solvent evaporation in the external water phase of w/o/w double emulsion, the drug is encapsulated with a stabilizer to avoid drug division to the external water phase. The bioactivity of a nanocarrier synthesized by various strains of cyanobacteria has been shown in several studies. Anti-cancer, anti-microbial, antioxidant activities are included in these bioactivities.

10.5 Biological Activities of Cyanobacteria-Based Nanoformulations

10.5.1 Anti-tumour Activity

New approaches for stopping tumour growth and metastasis have led to advances in nanoscience. Magnetic iron oxide nanocarrier, for example, has enhanced cellular imaging, improving the efficiency and speed of the diagnostic process. Nanoscience has advanced as a result of new methods for preventing tumourigenesis. Magnetic nanocarriers, for example, have improved cellular imaging, increasing diagnostic flexibility and reliability. A diverse microbial culture, including microalgae, macroalgae (seaweeds) and plant species, contributes to marine diversity (mangroves or halophytes). They can be both a prokaryote (cyanobacteria) and a eukaryote because they develop in diverse environmental conditions and produce different metabolites (Mata et al. 2010). The microalgae, that is, *Spirulina* (*Arthrospira platensis* and *Arthrospira maxima*), are often referred to as cyanobacteria or blue-green algae (Cyanophyceae) (Barsanti and Gualtieri 2014; Sithranga Boopathy et al. 2010). With a view to the abundant biodiversity of the oceans, microflora and microalgae make up more than 90% of the oceanic biomass (Sithranga Boopathy et al. 2010). This unique selection of microalgae presents enormous opportunities for the discovery of new medicines. Importantly, 50% of marine blue-green algal species are cultivated on commercial platforms for the extraction of bioactive compounds with immense ability to damage a variety of cancer cells through induced apoptosis or by inducing the activation of cell signalling proteins, particularly members of the protein kinase-c family (Lordan et al. 2011; Wells et al. 2017). In addition, marine microalgae produce significant quantities of different forms of protein, dietary fibre, polyunsaturated fatty acids (PUFAs), vitamins and minerals, depending on the species and environmental flora (Lordan et al. 2011; Wells et al. 2017). Polysaccharides derived from microalgae play a primary role as anti-cancer agents amongst these compounds in commercialized platforms. Scytonemin, a well-known commercialized extracellular pigment derived from the cyanobacterium *Scytonema*, has demonstrated a good ability to regulate the formation of mitotic spindles and enzyme kinase, having a clear connection with fibroblast and cell control (Stevenson et al. 2002). In addition, scytonemin, from cyanobacterium *Lyngbya* sp. U2555 was characterized and proven for its induction by UV radiation, and stability under different abiotic factors (Rastogi and Incharoensakdi 2014). Calothrixins are natural products derived from quinone isolated from Calothrix cyanobacteria that have demonstrated potent therapeutic potential against a variety of cancerous cell lines. Calothrixin B was discovered to have anti-proliferative activity against the HCT-116 colon cancer cell line, with an IC50 value of 0.32 M. *Lyngbya* sp. and characterized by UV radiation and stability under various abiotic conditions for its initiation (Rastogi and Incharoensakdi 2014). Calothrixins are natural products derived from quinone isolated from *Calothrix* cyanobacteria that have demonstrated potent therapeutic potential against a variety of cancerous cell lines (Ben-Nissan et al. 2004). Calothrixin B was discovered to have anti-proliferative activity against the HCT-116 colon cancer cell line, with an IC50 value of 0.32 M (Hatae et al. 2014). Some compounds based on microalgae have shown enormous potential for a variety of colon cancer cells to be inhibited. Malyngamides, for example, are small amides formed by marine cyanobacteria; malyngamides, such as malyngamide C and 8-epi-malyngamide C isolated from *Lyngbya majuscula*, were found to be cytotoxic to colon cancer cells of HT29 with an IC50 value of 5.2 and 15.4 μM, respectively (Kwan et al. 2010). Isolated from terrestrial *Nostoc* sp., Merocyclophanes A and B (UIC 10022A) showed HT-29 cell line anti-proliferative activity with an IC50 value of 3.3 and 1.7 μM, respectively (Kang et al. 2012). In addition, hierridine B, a secondary (polyketide) metabolite, was isolated from the marine cyanobacterium, *Cyanobium* sp. Selective cytotoxicity against colon cancer cell line HT-29 with an IC50 value of 0.1 mM was demonstrated by LEGE 06113, tested in a panel of eight human cancer cell lines (Leao et al. 2013). In addition, a phase II-approved cyanobacterial peptide glembatumumab vedotin demonstrated potent efficacy for the treatment of breast and melanoma cancers. Malyngamides, for example, are small amides produced by marine cyanobacteria; malyngamides, such as malyngamide C and 8-epi-malyngamide C (Luesch et al. 2006). It was demonstrated that apratoxin A induces pronounced G cell cycle arrest and apoptosis. In addition, by preventing co-translational translocation, apratoxin A was shown to reversibly inhibit the secretory pathway (Konickova et al. 2014) using an experimental pancreatic cancer model, previously documented a few bilirubin-like compounds from microalgal sources such as *Spirulina platensis* and *S. platensis*-derived tetrapyrroles. In addition, largazole isolated from *Symploca* sp. with a special chemical structure has demonstrated significant anti-proliferative activity (Taori et al. 2008). Apratoxins are another class of cyanobacterial compounds which, at nanomolar concentration, inhibit a range of cancer cell lines. Apratoxin A (parental compound) was found to exhibit cytotoxicity against adenocarcinoma cells isolated from a cyanobacterial strain of *Lyngbya boulloni*, and Coibamide-A compound was isolated from *Leptolyngbya* sp. It was stated to be toxic against NCIH460 lung and mouse neuro-2a cells (Medina et al. 2008). A common mechanism of action of many cyanobacterial compounds is cytotoxicity. In this sense, borophycin, cryptophycin 1 and 8 and cyanovirin are the most important discoveries. Borophycin is a boron-containing metabolite isolated from the *Nostoc linckia* and *N. marine* cyanobacterial species. The compound exhibits strong cytotoxicity against cell lines of human epidermoid carcinoma (LoVo) and human colorectal adenocarcinoma (KB) (Davidson 1995). In addition, *Chlorella vulgaris* extracts, the key constituent of which was lutein, demonstrated anti-proliferative effects on the human colon cancer cell line (HCT-116) (Cha et al. 2008). More recently, in an animal model, lutein has demonstrated chemoprotective efficacy against colorectal cancer by modulating K-ras, protein kinase B (PKB) and β-cateninase B (PKB) proliferative function (Reynoso-Camacho et al. 2011). In addition, extracts of the marine microalgae *Chlorella ellipsoidea*, which has violaxanthin as its main component, have shown potent anti-proliferative effects by inducing apoptosis on a human colon cancer cell line (HCT-116) (Cha et al. 2008). In order to show pro-apoptotic and anti-proliferative activities against human cancer cell lines, violaxanthin was also found (Pasquet et al. 2011). Figure 10.3 displays the chemical structures of a few selected compounds isolated from microalgae for efficient nanoformulation. Relevant cyanobacterial genera such as *Nostoc*, *Nodularia* and *Anabaena* have been described as having a unique biodiversity of

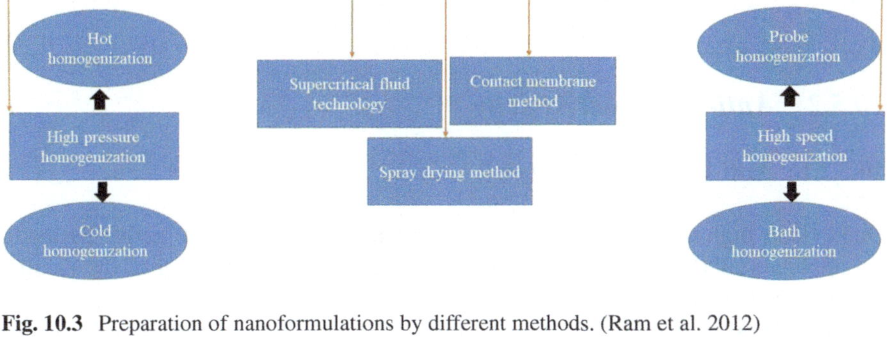

Fig. 10.3 Preparation of nanoformulations by different methods. (Ram et al. 2012)

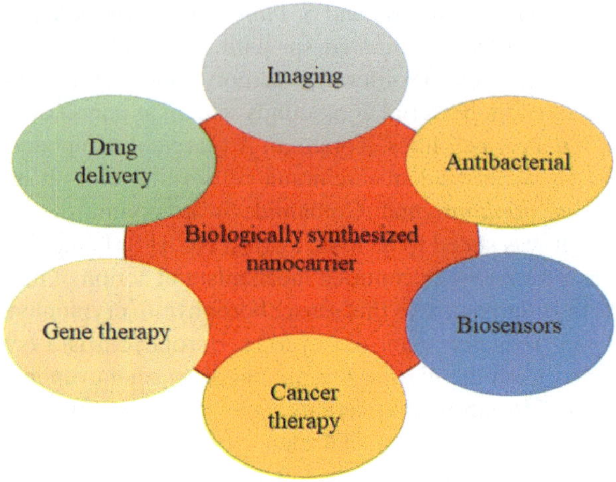

Fig. 10.4 Different biological activities of nanoformulations. (Oscar et al. 2016)

natural products, as evidenced by the isolation of natural oxadiazine nocuolin A (NoA) from some of the organisms belonging to these genera with a unique structure of 1,2,3-oxadiazine and the detailing in their genomes of the clusters of putative genes for NoA biosynthesis (Voraáččovaá et al. 2017). NoA-induced cell death and caspase-dependent apoptosis have been shown to have the same pattern. It is important to note that NoA was seen to show anti-proliferative effects against a variety of human cancer cells, more specifically against p53-mutated cell lines within an IC50 range of 0.7–4.5 μM (Voraáččovaá et al. 2017). Based on the above evidence, attachment of nanoformulation loaded with a microalgal-based molecule to the cell membrane will result in damage to the integrity of the membrane, thereby triggering damage to the membrane integrity. As a result, p53, a known pro-apoptotic gene activator, activates the proteins Bax, Bad and Bak that are known to induce leakage of the mitochondrial membrane and release Cyt-C, thus activating caspase-3 during the cascade reaction. Eventually, to cause DNA fragmentation, caspase-3 cleaves the nuclear membrane. At the same time, up-regulation of C-myc, an established apoptosis inducer, leads to further amplification of apoptotic signals and down-regulation of anti-apoptotic genes, corroborating apoptosis manifestation and consequent cell blebbing by Bcl-2 and Bcl-XL (Fig. 10.4).

10.5.2 Anti-microbial Activity

10.5.2.1 Anti-bacterial Activity

Due to their high surface area to volume ratio, the nanocarrier is attractive for medical applications (Pathak et al. 2019). There is a broad surface area of the smaller nanocarrier that enhances the contact between the nanocarrier and the target bacterial cells (Keskin et al. 2016; Aziz et al. 2014, 2015). The silver nanocarrier that

Synechococcus sp. synthesized has anti-bacterial properties against different strains of bacteria, including *Bacillus subtilis*, *Staphylococcus aureus* and *Escherichia coli*. The most vulnerable to biogenic Ag-nanocarrier was *B. subtilis* (Hamida et al. 2020). 1.54 mg/mL synthesized silver-nanocarrier from *Desertifilum* sp. (D-Snanocarrier) had substantial anti-bacterial activity against five Gram-negative and Gram-positive bacteria strains: *Pseudomonas aeruginosa*, *Bacillus cereus*, *Shigella flexneri*, *B. subtilis* and *Salmonella enterica* (clinical isolates). *Shigella flexneri* showed the greatest response to the nanocarrier showing the large inhibition zone (22.7 mm). In one more recent analysis, *E. coli*, *Salmonella typhimurium*, *Klebsiella pneumoniae* (drug-resistant) and *Streptococcus mutans* methicillin-resistant, *S. aureus* (MRSA) (clinical isolate) was treated with 1.5 mg/mL of silver-nanocarrier aqueous solution synthesized by *Desertifilum* sp. The anti-bacterial activity produced by Ag-nanocarrier of the *Oscillatoria* sp. against pathogens *E. coli* was observed and also against *Klebsiella* sp. of *Salmonella* and sp. of *Pseudomonas* using the principle of disc diffusion. Maximum anti-bacterial effects against *Pseudomonas* sp. have been seen (15 mm), and a minor anti-bacterial effect against *E. coli* was reported (10 mm). Synthesized Ag-nanocarrier from *Anabaena* sp. showed good anti-bacterial activity, which was tested against three multidrug-resistant bacteria, *K. pneumoniae*, *E. coli* and *S. aureus*, isolated from diabetic patients' foot ulcers. Sonker et al. (2017) examined silver-nanocarrier synthesized using *Nostoc* sp. cyanobacterial extract, which showed inhibitory activities against two plant-pathogenic bacteria, *Ralstonia solanacearum* and *Xanthomonas campestris*. As a result, Ag-nanocarrier (5, 10, 15 μg/mL) resulted in the growth suppression of two bacterial strains. The researchers found that Ag-nanocarrier synthesized from different cyanobacteria species had bactericidal activity against *Bacillus megaterium*, *E. coli*, *B. subtilis*, *P. aeruginosa*, *S. aureus* and *Micrococcus luteus* (Patel et al. 2015). They also stated that resistance to Ag-nanocarrier produced by *Limnothrix* sp. was shown by all bacteria studied, while SNP was synthesized by *Anabaena* sp. with an IZ of 22 mm, it was the most appropriate one. The researchers announced that *Spirulina* sp. synthesized Ag-nanocarrier and the bacterial growth of the tested bacteria (including *S. aureus*, *Staphylococcus epididymis*, *K. pneumoniae*, *E. coli* and *P. aeruginosa*) was suppressed more strongly than silver nitrate and Ag-nanocarrier synthesized by two isolates was resisted by *P. aeruginosa* (Ahmed et al. 2015). *Lyngbya majuscula*-synthesized non-toxic, highly stable Ag-nanocarrier demonstrated substantial biocidal activity against *P. aeruginosa* (Roychoudhury et al. 2015). The authors subjected *P. aeruginosa* to various Ag-nanocarrier concentrations, including 1, 0.5 and 0.1 mg/mL at 37 °C for 24 hours. The *P. aeruginosa* treatment with Ag-nanocarrier resulted in a maximum IZ. The Ag-nanocarrier synthesized by *Nostoc* sp., *Scytonema* sp., was identified in another study. *Phormidium* sp. and strong anti-bacterial agents have been used against different pathogens, like *S. aureus*, *P. aeruginosa*, MRSA, *E. coli* from Ag-nanocarrier, synthesized by *Nostoc* sp., *Scytonema* sp., *Phormidium* sp. and bacterial growth was inhibited by all previous microbes. The antibiotic activity of the *Nostoc*-mediated ZnO nanocarrier was shown against *P. aeruginosa* and *S. aureus*. From atomic force microscopic analysis, it was revealed that the biofilms of the examined bacteria were damaged by the ZnO nanocarrier so that they changed their morphology.

10.5.2.2 Anti-algal Activity

The literature investigated the inhibitory function of the *S. platensis*-synthesized Ag-nanocarrier against *Microcystis aeruginosa*, *Chlorella vulgaris* and *Scenedesmus obliquus*. Ag-nanocarrier has been used as anti-algal agents to decrease the harmful impact of *P. aeruginosa*. The viability of *Microcystis* cells decreased as the concentration of Ag-nanocarrier increased; at 10 mg/L, Ag-nanocarrier induced a 92 percent reduction of viable cells, compared to a 30 percent reduction at 1 mg/L (El-Sheekh et al. 2014).

10.5.2.3 Anti-fungal Activity

A recent research has shown that *Nostoc* sp. produces Ag-nanocarrier. Ag-antifungal nanocarrier's activity synthesized by EPSs of *N. commune* against various fungal strains infecting seeds of sorghum, maize, soybean and sesame was effective. With Ag-nanocarrier, they sterilized the grain surface and then, by the direct plate process, the seeds in Dichloran chloramphenicol peptone agar grew at 25 °C. On sorghum grains and soybean seeds with Ag-NP sterilization, complete inhibition of *Fusarium nygamai* was observed (Morsy 2014).

10.5.3 Photocatalytic Activity

Previous studies have documented the development of Ag-nanocarrier from *Synechococcus* sp. The operation was photocatalytic. These Ag nanocarriers induced methylene blue photodegradation; after 4 h, for 20 mg L^{-1} of aqueous dye solution, the decolourization of the organic pigment reached 18 ± 0.3 percent. Likewise, the Ag-nanocarrier provided by *Microchaete* sp. NCCU-342 degraded methyl red dye (84.60%) more effectively than cyanobacterial extract (49.80%) within 2 hours (Keskin et al. 2016).

10.5.4 Other Applications

The investigator stated that *L. majuscula* synthesized the Au-nanocarrier which has anti-myocardial infarction characteristics. They showed that the isoproterenol-induced changes in serum cardiac injury indicators, ECG, arterial pressure indices and antioxidant abilities of the heart are inhibited by Au-nanocarrier alone or with cyanobacterial extract. Ag-nanocarrier is also synthesized by phycobiliproteins isolated from *Nostoc* sp. Ag-nanocarrier inhibits haemolytic activity that was close to those induced by vitamin C. The scholars also reported that the possible mechanism of Ag-nanocarrier anti-haemolytic activity is due to the capped Ag-nanocarrier of *Nostoc* content (such as amino compounds (-NH2), phenols (Ar-OH)) with anti-haemolytic and antioxidant activity (Bakir et al. 2018; Younis et al. 2019).

The other studies showed that C-phycoerythrin-CdS nanocarrier could be used as biolabelling fluorescence samples for ultrasensitive DNA detection. Similarly, Au-NP synthesized by *Phormidium willei* and *Coelastrella* sp. has been shown to be used as biolabeling by Mubarak Ali et al. (2012). The researchers showed that the green Au-nanocarrier is synthesized by *Coelastrella* sp. only at higher concentration.

10.6 Strategies to Address Difficulties: Development of High Potency Drug Delivery Formula

10.6.1 Possibility of Peptide-Based Microalgal Drug Development

A peptide conjugation strategy was introduced in the pharmaceutical industry before the creation of a marketed platform to address the difficulties associated with natural compound-based drug production, such as stability and solubility (Chuang et al. 2009). Natural peptides and their synthetic analogues from various sources have been the basis for a variety of experiments to find new therapies for the treatment of malignant cells (Thundimadathil 2012). Different types of cationic anti-tumour peptides have been considered as possible therapy for anti-tumour treatment because of many favourable qualities like low molecular weight, comparatively simple structure, high cytotoxicity to tumours, little side effect, good absorption and many routes of administration (Mader and Hoskin 2006; Hoskin and Ramamoorthy 2008; Schweizer 2009; Riedl et al. 2011; Liu et al. 2013). This can contribute to increased drug effectiveness of the treatment agent when used in combination with other conventional therapeutic agents (Chuang et al. 2009). Even, with peptide-based conjugational anti-cancer drug production, many cancer drugs are in scientific progress.

Many cancer drugs are being produced using peptide-based conjugational anticancer drug. Overall, the combinatorial effect with peptide formulation could be a better alternative to a single moiety of the microalgal isolated compound in pharmaceutical commercialization due to the advantages of the low-cost scenario of marine-based drug development. These peptide-based technologies have a range of disadvantages, including poor bioavailability, instability and natural compound insolubility, as well as host cell toxicity (Reichert et al. 2010).

10.6.2 Possibilities and Facts Related to Nanoformulated Microalgal Drug Development

Organic (carbon-based nanoparticles) and inorganic nanoparticles are categorized into nanoparticles. Inorganic nanoparticles include magnetic nanoparticles (iron), noble metal nanoparticles (silver, gold and platinum) and semiconductor nanoparticles (cadmium, zinc oxide, zinc sulphite, silica and titanium oxide). Indeed, the specific properties of nanoparticles, such as biocompatibility, large-scale processing, easy use and functional performance, make them efficient in the delivery of drugs (Ramasubburayan et al. 2017). In the field of cancer drug delivery and diagnostics, nanoparticles have a broad range of applications (Ramkumar and Kirubakaran 2016; Ahila et al. 2016). In order to overcome the disadvantages of the use of peptide-based drugs, due to their low solubility and stability, nanonization of natural product-based medicines has attracted considerable interest. Nanotechnology has been reported to be one of the fastest platforms for developing nanoformulations (Ahila et al. 2016). The data on the marine-based synthesis of nanoparticles, their biomedical applications and related mechanisms were reviewed by Singh et al. (2015). A number of marine tools have recently been used extensively in the fields of nanoscience and nanotechnology (Singh et al. 2015). Marine krill lipid-based liposome nanoparticles known as "marinosomes" have recently been developed by Ibrahim et al., which encapsulate curcumin into krill lipid-based liposome nanoparticles, and have developed a safe anti-cancer nanoformulation from low-cost and readily accessible nutraceuticals (Ibrahim and Lin 2018). These "marinosomes" are based on marine krill lipid containing liposomes with a large marine ratio (Moussaoui et al. 2002). There are two subgroups of marine PUFAs: omega-6 (n-6) and omega-3 (n-3), which are metabolized and contained in phospholipids of the cell membrane. They regulate membrane fluidity, control a wide range of body functions and correct brain and nervous system development and functioning (Moussaoui et al. 2018). A significant portion of n-3 PUFAs are bound to phospholipids in marine lipids isolated from marine species such as krill, which have a higher content of n-3 long-chain PUFAs such as eicosapentaenoic acid (EPA; 20:5 n-3) and docosahexaenoic acid (DHA; 22:6 n-3). The properties of n-3 PUFAs have been documented (Alaarg et al. 2016). Various mechanisms of cancer prevention with n-3 PUFAs have been documented in other studies, including apoptosis, angiogenesis inhibition, metastasis inhibition and others (Yin et al. 2017). Previously, Anand et al. synthesized and reported anti-cancer effects of monodispersed and spherical silver nanoparticles of 20–60 nm from marine fungal species (*Aspergillus flavus* SP-3, *Trichoderma gamsii*, *Talaromyces flavus* and *Aspergillus oryzae*) (Anand et al. 2015).

10.6.3 Interest in Albumin-Based Nanoparticles: Useful Tool for Nanoformulation

The main problem in the clinical translation of medicine is biosafety. The efficacy of albumin has been confirmed as a valuable material for the manufacture of nanoparticles in recent years, especially in the field of bioimaging and delivery of drugs (An and Zhang 2017). Effective diagnosis and effective drug delivery in cancer treatment have piqued the interest of biomedical professionals (Cheng et al. 2012). Furthermore, nanosystems with effective drug delivery results showed tremendous applications in cancer treatment by focusing on increased blood circulation period (Wang et al. 2015). In addition, nanocarrier materials with adverse side effects have inspired researchers to explore relatively more biocompatible nanoparticle building materials (Sun et al. 2015). The active use of albumin in clinical nanomedicine for the use of abraxane has shown its considerable potential to be used as the most powerful nanocarrier for the production of novel nanomedicines (Chen et al. 2015). For the construction of nanoparticles, two forms of albumin are often used: human serum albumin (HSA) and bovine serum albumin (BSA). Both forms of albumins are human and cow serum albumin proteins and share many of the same characteristics, including elevated water solubility, long blood half-life, similar molecular weight (65–70 kDa) and similar amino acid residue numbers (585-HSA and 583-BSA) (Lee and Youn 2016; Karimi et al. 2016). The specific multiproperties of albumin can offer substantial opportunities with the expected properties to create multifunctional nanoparticles. By enabling albumin binding to hydrophobic or amphiphilic small molecules, albumin hydrophobic pockets play an important role in successful drug delivery (Lee et al. 2015). Direct interactions between albumin and drugs/peptides or interactions between albumin and prodrugs/peptide-conjugate may achieve binding between albumin and drugs/peptides. If there are strong interactions between them, the albumin and drugs/peptide bind together (Van Leeuwen et al. 2011). A stable nanocomplex with albumin might form the resulting prodrug/peptide-conjugate. This nanocomplex can achieve excellent stability when in a body-fluid environment, thereby confirming its practical application in a difficult in vivo system (Tessa Kurukulasuriya 2010). Although there is a very high commercial value for albumin, the study of site-selective albumin modification is still uncommon. We assume that the accumulated site-selective antibody modification information would promote the production of site-selective albumin modification and provide greater reliable quality of functionalized albumin nanomedicine. In the case of marine microalgal nanoformulations for improved commercialization, this interest may also be relevant.

10.6.4 Microalgal Polysaccharide-Based Nanoformulation

Macroalgae are excellent source for the production of sulphated polysaccharides which has applications in the nutraceutical, skincare and pharmaceutical companies (Manivasagan and Oh 2016). Furthermore, nanobiotechnologists are paying close attention to polysaccharide-based nanomaterials as novel nanocarriers for bioimaging and therapeutic applications due to its excellent bioactivity and safety, high biodegradability, low-cost effectiveness, non-toxic character and unique physicochemical properties (Lee et al. 2000). These properties are of particular interest in the nanotechnology sector and, as biomaterials, make marine polysaccharides promising. In addition, numerous studies have examined nanomaterials based on polysaccharides for biomedical applications, such as anti-microbial action, drug delivery, gene delivery, tissue engineering, cancer therapy and wound healing (Cardoso et al. 2016; Manivasagan et al. 2017; Prasad et al. 2017). Research has shown that polysaccharides released by cyanobacteria into culture media have multiple biological activities and anti-viral effects in the human body (Kanekiyo et al. 2005). In order to minimize the formation of cancer in the human body, free radical scavenging compounds such as sulfated polysaccharide derived from cyanobacteria can also be used. In addition, for medicinal use, polysaccharides derived from *Spirulina* inhibit tumour cell proliferation (Akao et al. 2009). Furthermore, several species of *Chaetoceros, Isochrysis, Nannochlropsis, Pavlova, Phaeodactylum, Skeletonema, Thalassiosira* and *Tetraselmis* have shown potential use in animal nutrition due to the abundance of biologically active substances in essential lipids such as omega-3 and omega-6 fatty acids (Muller-Feuga 2000). Although microalgae differ in composition, when formulated with other nanostructures, components derived from them may have a better application. Based on a literature survey, several reports have advocated that polysaccharides from microalgae, especially sulphated polysaccharides, have the potential to inhibit tumour growth by preventing tumourigenesis by oral consumption (Sinha 2021). For a few microalgal sulfated polysaccharide-containing nanomaterials against cancer and other synergistic microalgal nanoformulation innovations and their nanoform commercialization, anti-cancer strategies can be designed based on the above-mentioned facts.

10.7 Conclusion

Biologically active compounds that are obtained from a diversified group of algae and cyanobacteria have been commonly used in a number of industrial and biomedical aspects since the beginning of civilization. Several biomolecules, including primary metabolites, pigments, secondary metabolites, antioxidants and halogenated compounds, are present in good amount in in algae. Due to their multifunctional uses, these compounds are commonly used in the nutraceuticals, pharmaceuticals and biomedical fields. Undoubtedly, our understanding of algal metabolites has improved dramatically over the past decade, but its application in the medicine field and its nanoformulation and commercialization still face many challenges. The development of nanomedicine and marine drugs would lead to the uncovering of new enhanced anti-cancerous mechanisms of nanoformulations consisting of algal secondary metabolites. The lack of commercialized drug production utilizing microalgae and their structurally complex, biologically active compounds such as terpenes, alkaloids, steroids, polysaccharides, lipids and polyphenolics has been highlighted in this chapter. Furthermore, advanced technologies are essential to find out the efficiency of these compounds against cancer in the form of nanoformulations. To exploit these cyanobacterial and microalgae anti-cancer drug nanoformulations, nanofusion technology and green technology should be combined. A definition was drawn from beginning to end to capture the attention of the reader for more future studies to achieve high creativity in commercialized platforms. A major challenge for cheap marine microalgae-based nanoformulations is the large production for commercialization and the technology growth. Research must be undertaken in future to explore the unidentified functions of secondary metabolites of microalgae.

References

Ahila NK, Ramkumar VS, Prakash S et al (2016) Synthesis of stable nanosilver particles (AgNPs) by the proteins of seagrass *Syringodium isoetifolium* and its biomedicinal properties. Biomed Pharmacother 84:60–70

Ahlin P, Kristl J, Smid-Korbar J (1998) Optimization of procedure parameters and physical stability of solid lipid nanoparticles in dispersions. Acta Pharmaceutica (Zagreb) 48:259–267

Ahmed EA, Hafez A, Ismail F et al (2015) Biosynthesis of silver nanoparticles by *Spirulina platensis* and *Nostoc* sp. Global Adv Res J Microbiol 4:036–049

Akao Y, Ebihara T, Masuda H et al (2009) Enhancement of antitumor natural killer cell activation by orally administered *Spirulina* extract in mice. Cancer Sci 100:1494–1501

Al Rashed S, Al Shehri S, Moubayed N (2018) Extracellular biosynthesis of silver nanoparticles from Cyanobacteria. Bio Res 29

Alaarg A, Jordan NY, Verhoef JJ et al (2016) Docosahexaenoic acid liposomes for targeting chronic inflammatory diseases and cancer: an *in vitro* assessment. Int J Nanomed 11:5027

Ali M, Gopinath D, Rameshbabu V et al (2012) Synthesis and characterization of CdS nanoparticles using C-phycoerythrin from the marine cyanobacteria. Mater Lett 74:8–11

An FF, Zhang XH (2017) Strategies for preparing albumin-based nanoparticles for multifunctional bioimaging and drug delivery. Theranostics 15:3667

Anand BG, Thomas CN, Prakash S et al (2015) Biosynthesis of silver nano-particles by marine sediment fungi for a dose dependent cytotoxicity against HEp2 Hep2 cell lines. Biocatal Agric Biotechnol 4:150–157

Asmathunisha NAR, Kathiresan K (2018) Synthesis of silver and gold nanoparticles by mangrove-derived cyanobacteria. J Advan Nanomater 3:1–7

Aziz N, Fatma T, Varma A, Prasad R (2014) Biogenic synthesis of silver nanoparticles using *Scenedesmus abundans* and evaluation of their antibacterial activity. Journal of Nanoparticles, Article ID 689419, 2014. https://doi.org/10.1155/2014/689419

Aziz N, Faraz M, Pandey R, Sakir M, Fatma T, Varma A, Barman I, Prasad R (2015) Facile algae-derived route to biogenic silver nanoparticles: Synthesis, antibacterial and photocatalytic properties. Langmuir 31:11605–11612. https://doi.org/10.1021/acs.langmuir.5b03081

Bais AF, Lucas RM, Bornman JF et al (2018) Environmental effects of ozone depletion, UV radiation and interactions with climate change: UNEP Environmental Effects Assessment Panel, update. Photochem Photobiol Sci 17:127–179

Bajpai VK, Shukla S, Kang et al (2018) Developments of cyanobacteria for nano-marine drugs: relevance of nanoformulations in cancer therapies. Mar Drugs 16:179

Bakir EM, Younis NS, Mohamed ME et al (2018) Cyanobacteria as nanogold factories: chemical and anti-myocardial infarction properties of gold nanoparticles synthesized by *Lyngbya majuscula*. Mar Drugs 16:217

Barsanti L, Gualtieri P (2014) Algae: anatomy, biochemistry, and biotechnology, 2nd edn. CRC Press, Florida

Ben-Nissan B, Milev A et al (2004) Morphology of sol–gel derived nano-coated coralline hydroxyapatite. Biomaterials 25:4971–4975

Bernerd F, Compan D, Bernard B et al (2003) Modulation of gene expression induced in human epidermis by environmental stress in vivo. J Invest Dermatol 121:1447–1458

Bin-Meferij MM, Hamida RS (2019) Biofabrication and antitumor activity of silver nanoparticles utilizing novel *Nostoc* sp. Bahar M. Int J Nanomed 14:9019–9029

Boltri L, Canal T, Esposito PA et al (1993) Lipid nanoparticles: evaluation of some critical formulation parameters. In Proc Int Symp Control Release Bioact Mater 20:346–347

Booth BC (1988) Size classes and major taxonomic groups of phytoplankton at two locations in the subarctic Pacific Ocean in May and August, 1984. Marine Biology 97:275–286

Brayner R, Barberousse H, Hemadi M et al (2007) Cyanobacteria as bioreactors for the synthesis of Au, Ag, Pd, and Pt nanoparticles via an enzyme-mediated route. J Nanosci Nanotechnol 7:2696–2708

Brayner R, Yepremian C, Djediat C et al (2009) Photosynthetic microorganism-mediated synthesis of akaganeite (β-FeOOH) nanorods. Langmuir 25:10062–10067

Bultel-Poncé V, Félix-Théodose F, Sarthou C et al (2004) New pigments from the terrestrial cyanobacterium *Scytonema* sp. collected on the Mitaraka Inselberg, French Guyana. J Nat Prod 67:678–681

Cardoso MJ, Costa RR, Mano JF (2016) Marine origin polysaccharides in drug delivery systems. Mar Drugs 14:34

Castro A, Lannes D (2014) Fully nonlinear long-waves models in presence of vorticity. arXiv preprint arXiv*arXiv preprint arXiv* 1406:4096

Cavalli R, Marengo E, Rodriguez L et al (1996) Effects of some experimental factors on the production process of solid lipid nanoparticles. Eur J Pharm Biopharm 42:110–115

Cepoi L, Rudi L, Chiriac T et al (2015) Biochemical changes in cyanobacteria during the synthesis of silver nanoparticles. Can J Microbiol 61:13–21

Cha KH, Koo SY, Lee DU (2008) Antiproliferative effects of carotenoids extracted from *Chlorella ellipsoidea* and *Chlorella vulgaris* on human colon cancer cells. J Agric Food Chem 56:0521–10526

Chakraborty N, Banerjee A, Lahiri S et al (2009) Biorecovery of gold using cyanobacteria and ana eukaryotic alga with special reference to nanogold formation – a novel phenomenon. J Appl Phycol 21:145–152

Chen Q, Liang C, Wang C et al (2015) An imaginable and photothermal "Abraxane-like" nanodrug for combination cancer therapy to treat subcutaneous and metastatic breast tumors. Adv Mater 27:903–910

Chen YJ, Jin RX, Zhou YQ et al (2006) Preparation of solid lipid nanoparticles loaded with Xionggui powder-supercritical carbon dioxide fluid extraction and their evaluation *in vitro* release. Zhongguo Zhongyao Zazhi 31:376–379

Cheng A, Zaki A, Hui JZ et al (2012) Multifunctional nanoparticles: cost versus benefit of adding targeting and imaging capabilities. Science 338:903–910

Chuang CM, Monie A, Wu A et al (2009) Treatment with LL-37 peptide enhances antitumor effects induced by CpG oligodeoxynucleotides against ovarian cancer. Hum Gene Ther 20:303–313

Cortesi R, Esposito E, Luca G et al (2002) Production of lipospheres as carriers for bioactive compounds. Biomaterials 23:2283–2294

Dahoumane SA, Djediat C, Yéprémian C et al (2010) Design of magnetic akaganeite-cyanobacteria hybrid biofilms. Thin Solid Films 518:5432–5436

Davidson BS (1995) New dimensions in natural products research: cultured marine microorganisms. Curr Opin Biotechnol 6:284–291

De Labouret A, Thioune O, Fessi H et al (1995) Application of an original process for obtaining colloidal dispersions of some coating polymers. Preparation, characterization, industrial scale-up. Drug Dev Ind Pharm 21:229–241

Diffey B (2009) Sunscreens: expectation and realization. Photodermatol Photoimmunol Photomed 25:233–236

Domb AJ (1993) Lipospheres for controlled delivery of substances. US Patent, US 188837

Dreher TW, Collart LP et al (2019) *Anabaena/Dolichospermum* as the source of lethal microcystin levels responsible for a large cattle toxicosis event. Toxicon 1:100003

Ebadi M, Zolfaghari MR, Aghaei SS et al (2019) A bio-inspired strategy for the synthesis of zinc oxide nanoparticles (ZnO NPs) using the cell extract of cyanobacterium *Nostoc* sp. EA03: from biological function to toxicity evaluation. RSC Adv 9:23508–23525

Edlich R., Winters K L, Lim H W et al (2004) Photoprotection by sunscreens with topical antioxidants and systemic antioxidants to reduce sun exposure. J Long Term Eff Med Implants 14(4):317–340

Eldem T, Speiser P, Hincal A (1991) Optimization of spray-dried and-congealed lipid micropellets and characterization of their surface morphology by scanning electron microscopy. Pharma Res 8:47–54

El-Naggar NEA, Hussein MH et al (2017) Bio-fabrication of silver nanoparticles by phycocyanin, characterization, in vitro anticancer activity against breast cancer cell line and *in vivo* cytotxicity. Sci Rep 7:1–20

El-Naggar NE, Hussein MH, El-Sawah AA (2018) Phycobiliprotein-mediated synthesis of biogenic silver nanoparticles, characterization, *in vitro* and *in vivo* assessment of anticancer activities. Sci Rep 8:8925

El-Sheekh MM, Daboor SM, Swelim MA et al (2014) Production and characterization of antimicrobial active substance from *Spirulina platensis*. Iran J Microbiol 6:112

Fleming ED, Castenholz RW (2007) Effects of periodic desiccation on the synthesis of the UV-screening compound, scytonemin, in cyanobacteria. Environ Microbiol 9:1448–1455

Focsan M, Ardelean II, Craciun C et al (2011) Interplay between gold nanoparticle biosynthesis and metabolic activity of cyanobacterium *Synechocystis* sp. PCC 6803. Nanotechnology 22:485101

Freitas C, Muüller RH (1998) Spray-drying of solid lipid nanoparticles (SLNTM). Eur J Pharm Biopharm 46:145–151

Garcia-Pichel F, Pichel F, Castenholz RW (1993) Occurrence of UV-absorbing, mycosporine-like compounds among cyanobacterial isolates and an estimate of their screening capacity. Appl Environ Microbiol 59:63–169

Gasco MR (1993) Method for producing solid lipid microspheres having a narrow size distribution. U.S. Patent 5,250,236, 5 October 1993

Geetha S, Sathakkathul Z, Aarthi R et al (2014) Green synthesis of gold nanoparticle using marine cyanobacteria *Gloeocapsa* sp. and the antitumor potential. J Chem Pharm Sci 4:172–174

Geoffrey K, Mwangi AN, Maru SM (2019) Sunscreen products: rationale for use, formulation development and regulatory considerations. Saudi Pharm J 27:1009–1018

Gosselin PM, Thibert R, Preda R, Preda M et al (2003) Polymorphic properties of micronized carbamazepine produced by RESS. Int J Pharm 252:225–233

Govindaraju K, Basha SK, Kumar VG et al (2008) Gold and bimetallic nanoparticles production using single-cell protein (*Spirulina platensis*) Geitler. J Mater Sci 43:5115–5122

Gulin A, Koksharova O, Popova A et al (2016) Visualization of the spatial distribution of Ag ions in cyanobacteria *Anabaena* sp. PCC 7120 by time-of-flight secondary ion mass spectrometry and two-photon luminescence microscopy. Nanotechnol Russia 11:361–363

Halpern AC, Kopp LJ (2005) Awareness, knowledge and attitudes to non-melanoma skin cancer and actinic keratosis among the general public. Int J Dermatol 44:107–111

Hamida RS, Albasher G, Bin-Meferij MM (2020) Oxidative stress and apoptotic responses elicited by *Nostoc*-synthesized silver nanoparticles against different cancer cell lines. Cancers 12:2099

Hatae N, Satoh R, Chiba H et al (2014) N-Substituted calothrixin B derivatives inhibited the proliferation of HL-60 promyelocytic leukemia cells. Med Chem Res 23:4956–4961

Hoffmann RA (1988) Chemical and theoretical way to look at bonding on surfaces. Re Mod Phys 60:601

Hoskin DW, Ramamoorthy A (2008) Studies on anticancer activities of antimicrobial peptide. Biochimica et Biophysica Acta (BBA)-Biomembranes 1778:357–375

Hudek L, Rai S, Michalczyk A et al (2012) Physiological metal uptake by *Nostoc punctiforme*. Biometals 25:893–903

Husain S, Sardar M, Fatma T (2015) Screening of cyanobacterial extracts for synthesis of silver nanoparticles. World J Microbiol Biotechnol 31:1279–1283

Ibrahim AF, Lin YS (2018) Synthesis of graphene oxide membranes on polyester substrate by spray coating for gas separation. Chem Eng Sci 190:312–319

Jain RP, Chakravarty S (2014) Medivation Technologies LLC.Pyrido [4, 3-B] indoles and methods of use. U.S. Patent No. 8,859,561, 14 Oct 2014.U.S. Patent 8,741:919

Jahnke S (1998) The theory of high-pressure homogenization. In: Muller RH, Benita S, Bohm BHL (eds) Emulsions and nanosuspensions for the formulation of poorly soluble drugs. Medpharm Scientific Publishers, Stuttgart, pp 177–200

Kalabegishvili T, Kirkesali E, Rcheulishvili A (2012) Synthesis of gold nanoparticles by blue-green algae *spirulina platensis*.

Kale A, Kale S, Yadav P et al (2011) Magnetite/CdTe magnetic–fluorescent composite nanosystem for magnetic separation and bio-imaging. Nanotechnology 22:225101

Kaliamurthi S, Selvaraj G, Çakmak ZE et al (2016) Production and characterization of spherical thermostable silver nanoparticles from *Spirulina platensis* (*Cyanophyceae*). Phycologia 55:568–576

Kanekiyo K, Lee JB, Hayashi K et al (2005) Isolation of an antiviral polysaccharide, nostoflan, from a terrestrial cyanobacterium, *Nostoc flagelliforme*. J Nat Prod 68:1037–1041

Kang HS, Santarsiero BD, Kim H et al (2012) Merocyclophanes A and B, antiproliferative cyclophanes from the cultured terrestrial cyanobacterium *Nostoc* sp. Phytochemistry 79:109–115

Karimi M, Bahrami S, Ravari SB (2016) Albumin nanostructures as advanced drug delivery systems. Expert Opin Drug Deliv 13:1609–1623

Karsten U (2008) Defense strategies of algae and cyanobacteria against solar ultraviolet radiation. In: Algal chemical ecology. Springer, Berlin, Heidelberg, pp 273–296

Keskin S, Oya N, Koçberber Kılıc N et al (2016) Green synthesis of silver nanoparticles using cyanobacteria and evaluation of their photocatalytic and antimicrobial activity. J Nano Res 40:120–127. https://doi.org/10.4066/biomedicalresearch.29-17-3209

Kwan JC, Teplitski M, Gunasekera SP et al (2010) Isolation and biological evaluation of 8-epi-malyngamide C from the Floridian marine cyanobacterium *Lyngbya majuscula*. J Nat Prod 73:463–466

Lander R, Manger W, Scouloudis M et al (2000) Gaulin homogenization: a mechanistic study. Biotechnol Prog 16:80–85

Leao PN, Costa M, Ramos V et al (2013) Antitumor activity of hierridin B, a cyanobacterial secondary metabolite found in both filamentous and unicellular marine strains. PLoS One 8:e69562

Lee ES, Youn YS (2016) Albumin-based potential drugs: focus on half-life extension and nanoparticle preparation. Int J Pharm Investig 4:305–315

Lee JW, Park JH, Robinson JR (2000) Bioadhesive-based dosage forms: the next generation. J Pharm Sci 89:850–866

Lee SSY, Li J, Tai JN, Ratliff TL et al (2015) Avasimibe encapsulated in human serum albumin blocks cholesterol esterification for selective cancer treatment. ACS Nano 9:2420–2432

Lengke MF, Fleet ME, Southam G (2006a) Morphology of gold nanoparticles synthesized by filamentous cyanobacteria from Gold(I)–Thiosulfate and Gold (III)–Chloride complexes. Langmuir 22:2780–2787

Lengke MF, Fleet ME, Southam G (2006b) Synthesis of platinum nanoparticles by reaction of filamentous cyanobacteria with platinum (IV)-chloride complex. Langmuir 22(17):7318–7323

Lengke MF, Fleet ME, Southam G (2007) Synthesis of palladium nanoparticles by reaction of filamentous cyanobacterial biomass with a palladium (II) chloride complex. Langmuir 23:8982–8987

Liu S, Yang H, Wan L et al (2013) Penetratin-mediated delivery enhances the antitumor activity of the cationic antimicrobial peptide magainin II. Cancer Biother Radiopharm 28:289–297

Lordan S, Ross RP, Stanton C (2011) Marine bioactives as functional food ingredients: potential to reduce the incidence of chronic diseases. Mar Drugs 9:1056–1100

Los DA, Murata N (1994) Low-temperature induced accumulation of the desaturase gene transcript in Synechocystis PCC 6803 results from both acceleration of transcription and increase in mRNA stability. Russ J Plant Physiol 41:146–151

Luesch H, Chanda SK, Raya RM et al (2006) A functional genomics approach to the mode of action of apratoxin A. Nat Chem Biol 2:158–167

Ma Z, Jacobsen FE, Giedroc DP (2009) Coordination chemistry of bacterial metal transport and sensing. Chem Rev 109:4644–4681

Mabena LF, Ray SS, Coville NJ (2011) Microwave irradiation of ruthenium on nitrogen-doped carbon nanotubes. Nanostruct Mater Nanotechnol 549:33

Maddela NR, Chakraborty S, Prasad R (2021) Nanotechnology for Advances in Medical Microbiology. Springer Singapore (ISBN 978-981-15-9915-6) https://www.springer.com/gp/book/9789811599156

Mader JS, Hoskin DW (2006) Cationic antimicrobial peptides as novel cytotoxic agents for cancer treatment. Expert Opin Investig Drugs 15:933–946

Mahdieh M, Zolanvari A, Azimee A (2012) Green biosynthesis of silver nanoparticles by *Spirulina platensis*. Scientia Iranica 19:926–929

Mallakpour S, Behranvand V (2016) Nanocomposites based on biosafe nano ZnO and different polymeric matrixes for antibacterial, optical, thermal and mechanical applications. Eur Polym J 84:377–403

Manivasagan P, Bharathiraj S, Moorthy MS et al (2017) Marine biopolymer-based nanomaterials as a novel platform for theranostic applications. Polym Rev 57:631–667

Manivasagan P, Oh J (2016) Marine polysaccharide-based nanomaterials as a novel source of nanobiotechnological applications. Int J Biol Macromol 82:315–327

Mansha M, Khan I, Ullah N et al (2017) Synthesis, characterization and visible-light-driven photoelectrochemical hydrogen evolution reaction of carbazole-containing conjugated polymers. Int J Hydrog 42:10952–10961

Mata TM, Martins AA, Caetano NS (2010) Microalgae for biodiesel production and other applications: a review. Renew Sust Energ Rev 14:217–232

Medina RA, Goeger DE, Hills P et al (2008) Coibamide A, a potent antiproliferative cyclic depsipeptide from the Panamanian marine cyanobacterium *Leptolyngbya* sp. J Am Chem Soc 130:6324–6325

Mehta SK, Gaur JP (2005) Use of algae for removing heavy metal ions from wastewater: progress and prospects. Crit Rev Biotechnol 25:113–152

Mody VV, Siwale R, Singh A et al (2010) Introduction to metallic nanoparticles. J Pharm Bioallied Sci 2:282

Mohamed A, Fouda A, Elgamal M et al (2017) Enhancing of cotton fabric antibacterial properties by silver nanoparticles synthesized by new Egyptian strain *Fusarium keratoplasticum* A1-3. Egypt J Chem 60:63–71

Morsy SM (2014) Role of surfactants in nanotechnology and their applications. Int J Curr Microbiol App Sci 3:237–260

Moussaoui N, Cansell M, Denizot A (2002) Marinosomes®, marine lipid-based liposomes: physical characterization and potential application in cosmetics. Int J Pharm 242:361–365

Moussaoui S, Hauksdottir H et al (2018). Curcumin marinosomes as promising nano-drug delivery system for lung cancer Int. J Pharm 540:40-49

MubarakAli D, Arunkumar J, Nag KH et al (2013) Gold nanoparticles from Pro and eukaryotic photosynthetic microorganisms—comparative studies on synthesis and its application on biolabelling. Colloids Surf B Biointerfaces 103:166–173

MubarakAli D, Gopinath V, Rameshbabu N et al (2012) Synthesis and characterization of CdS nanoparticles using C-phycoerythrin from the marine cyanobacteria. Mater Lett 74:8–11

Mullen zur A (1996) Feste Lipid-Nanopartikel mit prolongierter Wirkstoffliberation: Herstellung, Langzeitstabilitat, Charakterisierung, Freisetzungsverhalten und mechanismen (Doctoral dissertation, Ph. D. Thesis, Free University of Berlin)

Muller-Feuga A (2000) The role of microalgae in aquaculture: situation and trends. J Appl Phycol 12:527–534

Ngoy JM, Wagner N, Riboldi L et al (2014) CO_2 capture technology using multi-walled carbon nanotubes with polyaspartamide surfactant. Energy Procedia 63:2230–2248

Olbrich C, Gessner A, Kayser O et al (2002) Lipid-drug-conjugate (LDC) nanoparticles as novel carrier system for the hydrophilic antitrypanosomal drug diminazenediaceturate. J Drug Target 10:387–396

Oscar FL, Vismaya S, Arunkumar M et al (2016) Algal nanoparticles: synthesis and biotechnological potentials. In: Thajuddin N, Dhanasekaran D (eds) Algae-organisms for imminent biotechnology. IntechOpen, London, UK

Ostiguy C, Brigitte R, Catherine W, Soucy B (2010) Engineered nanoparticles: current knowledge about OHS risks and prevention measures. Research and Studies Series, report # R-455, IRSST

Pareek V, Bhargava A, Gupta R, Jain NE et al (2017) Synthesis and applications of noble metal nanoparticles: a review. Adv Sci Eng Med 9:527–544

Parial D, Gopal PK, Paul S et al (2016) (III) bioreduction by cyanobacteria with special reference to in vitro biosafety assay of gold nanoparticles. J Appl Phycol 28:3395–3406

Parial D, Pal R (2014) Green synthesis of gold nanoparticles using cyanobacteria and their characterization. Indian J Appl Res 4:69–72

Parial D, Patra HK, Dasgupta AK et al (2012a) Screening of different algae for green synthesis of gold nanoparticles. Eur J Phycol 47:22–29

Parial D, Patra HK, Roychoudhury P et al (2012b) Gold nanorod production by cyanobacteria—a green chemistry approach. J Appl Phycol 24:55–60

Pasquet V, Morisset P, Ihammouine S et al (2011) Antiproliferative activity of violaxanthin isolated from bioguided fractionation of *Dunaliella tertiolecta* extracts. Mar Drugs 9:819–831

Patel VK, Maji D, Singh A K et al (2014) A natural plant growth promoter, calliterpenone, enhances growth and biomass, carbohydrate, and lipid production in cyanobacterium *Synechocystis* PCC 6803. J. Appl Phycol 26:279–286

Patel V, Berthold D, Puranik P et al (2015) Screening of cyanobacteria and microalgae for their ability to synthesize silver nanoparticles with antibacterial activity. Biotechnol Rep (Amst) 5:112–119

Pathak J, Ahmed H, Singh DK et al (2019) Recent developments in green synthesis of metal nanoparticles utilizing cyanobacterial cell factories. In: Nanomaterials in plants, algae and microorganisms. Academic Press, pp 237–265

Prasad R, Pandey R, Varma A, Barman I (2017) Polymer based nanoparticles for drug delivery systems and cancer therapeutics. In: Natural Polymers for Drug Delivery (eds. Kharkwal H and Janaswamy S), CAB International, UK 53–70

Prasad R, Kumar V, Kumar M, Choudhary D (2019) Nanobiotechnology in Bioformulations. Springer International Publishing (ISBN 978-3-030-17061-5) https://www.springer.com/gp/book/9783030170608

Prasad R, Siddhardha B, Dyavaiah M (2020) Nanostructures for Antimicrobial and Antibiofilm Applications. Springer International Publishing (ISBN 978-3-030-40336-2) https://www.springer.com/gp/book/9783030403362

Proteau PJ, Gerwick WH, Garcia-Pichel F et al (1993) The structure of scytonemin, an ultraviolet sunscreen pigment from the sheaths of cyanobacteria. Experientia 49:825–829

Puri A, Loomis K, Smith B et al (2009) Lipid-based nanoparticles as pharmaceutical drug carriers: from concepts to clinic. Crit Rev Ther Drug Carr Sys 26:523–580

Radtsig M, Koksharova O, Nadtochenko V (2016) Production of gold nanoparticles by biogenesis using bacteria. Microbiology 85:63–70

Rahman A, Ismail A et al (2009) Synthesis of copper oxide nano particles by using Phormidium cyanobacterium. Indones J Chem 9:355–360

Ram DT, Debnath S, Babu MN et al (2012) A review on solid lipid nanoparticles. Res J Pharm Technol 11:2

Ramasubburayan R, Sumathi S, Prakash S et al (2017) Synthesis of nano silver by a marine epibiotic bacterium *Bacillus vallismortis* and its potent ecofriendly antifouling properties. Environ Nanotechnol Monit Manag 8:112–120

Ramkumar S, Kirubakaran V (2016) Biodiesel from vegetable oil as alternate fuel for CI engine and feasibility study of thermal cracking: A critical review. Energy Convers. Manag 118:155–169

Rastogi RP, Incharoensakdi A (2014) Characterization of UV-screening compounds, mycosporine-like amino acids, and scytonemin in the cyanobacterium *Lyngbya* sp. CU2555. FEMS Microbiol Ecol 87:244–256

Ray SC, Jana NR (2017) Application of carbon-based nanomaterials as drug and gene delivery carrier. In: Carbon nanomaterials for biological and medical Applications. Elsevier, pp 163–203

Reichert J, Pecho P, Tartar A et al (2010) Report summary: development trends for peptide therapeutics. Peptide Therapeutics Foundation, San Diego, CA, pp 1–11

Reynoso-Camacho R, Gonzalez-Jasso E, Ferriz-Martínez R et al (2011) Dietary supplementation of lutein reduces colon carcinogenesis in DMH-treated rats by modulating K-ras, PKB, and β-catenin proteins. Nutr Cancer 63:39–45

Riedl S, Zweytick D, Lohner K (2011) Membrane-active host defense peptides–challenges and perspectives for the development of novel anticancer drugs. Chem Phys Lipids 164:766–781

Rosken LM, Cappel F, Korsten S et al (2016) Time-dependent growth of crystalline Au(0)-nanoparticles in cyanobacteria as self-reproducing bioreactors: 2. *Anabaena cylindrica*. Beilstein J Nanotechnol 7:312–327

Rosken LM, Korsten S, Fischer CB et al (2014) Time-dependent growth of crystalline Au 0-nanoparticles in cyanobacteria as self-reproducing bioreactors: 1. *Anabaena* sp. J Nanoparticle Res 16:2370

Roychoudhury A, Banerjee A, Lahiri V (2015) Metabolic and molecular-genetic regulation of proline signaling and itscross-talk with major effectors mediates abiotic stress tolerance in plants. Turk J Botany 39:887–910

Roychoudhury P, Bhattacharya A, Dasgupta A et al (2016a) Biogenic synthesis of gold nanoparticle using fractioned cellular components from eukaryotic algae and cyanobacteria. Phycol Res 64:133–140

Roychoudhury P, Ghosh S, Pal R (2016b) Cyanobacteria mediated green synthesis of gold-silver nanoalloy. J Plant Biochem Biotechnol 25:73–78

Roychoudhury P, Gopal PK, Paul S et al (2016c) Cyanobacteria assisted biosynthesis of silver nanoparticles—a potential antileukemic agent. J Appl Phycol 28:3387–3394

Saglam N, Korkusuz, F, Prasad R (2021) Nanotechnology Applications in Health and Environmental Sciences. Springer International Publishing (ISBN: 978-3-030-64410-9) https://www.springer.com/gp/book/9783030644093

Salata OV (2004) Applications of nanoparticles in biology and medicine. Journal of Nanobiotechnology 2:1–6

Saran T, Sharma G, Kumar M et al (2017) Biosynthesis of copper oxide nanoparticles using cyanobacteria *spirulina platensis* and its antibacterial activity. Int J Pharm Sci Res 8:3887

Schweizer F (2009) Cationic amphiphilic peptides with cancer-selective toxicity. Eur J Pharmacol 625:190–194

Sebastian V, Manuel A et al (2014) Reaction engineering strategies for the production of inorganic nanomaterials. Small 10:835–853

Serpone N, Dondi D, Albini A (2007) Inorganic and organic UV filters: their role and efficacy in sunscreens and suncare products. Inorg Chim Acta 360:794–802

Siekmann B, Westesen K (1996) Investigations on solid lipid nanoparticles prepared by precipitation in o/w emulsions. Eur J Pharm Biopharm 42:104–109

Singh CR, Kathiresan K, Anandhan SA (2015) A review on marine based nanoparticles and their potential applications. Afr J Biotechnol 14:1525–1532

Singh G, Babele PK, Kumar A, Srivastava A et al (2014) Synthesis of ZnO nanoparticles using the cell extract of the cyanobacterium, *Anabaena* strain L31 and its conjugation with UV-B absorbing compound shinorine. J Photochem Photobiol B 138:55–62

Sinha RP, Rastogi RP, Ambasht NK et al (2008) Life of wetland cyanobacteria under enhancing solar UV-B radiation. Proc Natl Acad Sci India B 78:53–65

Sinha R (2021) Metabolic engineering of Cyanobacteria and micro algae for enhanced production of bio fuels and high value products. Ann Romanian Soc Cell Biol 21160–21179

Sithranga Boopathy N and Kathiresan K (2010) Anticancer drugs from marine flora: an overview. J Oncol

Sjöström B, Bergenståhl B (1992) Preparation of submicron drug particles in lecithin-stabilized ow emulsions: I. Model studies of the precipitation of cholesteryl acetate. Int J Pharm 84:107–116

Smijs TG, Pavel S (2011) Titanium dioxide and zinc oxide nanoparticles in sunscreens: focus on their safety and effectiveness. Nanotechnol Sci Appl 4:95

Smith AM, Nie S (2010) Semiconductor nanocrystals: structure, properties, and band gap engineering. Acc Chem Res 43:190–200

Sonker AS, Pathak J, Kannaujiya VK et al (2017) Characterization and *in vitro* antitumor, antibacterial and antifungal activities of green synthesized silver nanoparticles using cell extract of *Nostoc* sp. strain HKAR-2. Can J Biotech 1:26

Speiser P (1990) Lipidnanopellets als Trägersystem für Arzneimittel zur peroralen Anwendung. European Patent EP 167825:0167825

Stevenson CS, Capper EA, Roshak AK et al (2002) Scytonemin-a marine natural product inhibitor of kinases key in hyperproliferative inflammatory diseases. Inflamm Res 51:112

Sugandha V, Babita K, Shrivastava J (2014) Green synthesis of silver nanoparticles using single cell protein of *Spirulina platensis*. Int J Pharm Bio Sci 5:2

Suganya KU, Govindaraju K, Kumar VG et al (2015) Blue green alga mediated synthesis of gold nanoparticles and its antibacterial efficacy against Gram positive organisms. Mat Sci Eng C, 47:351–356

Sun S, Murray CB, Weller D, Folks L, Moser A (2000) Monodisperse FePt nanoparticles and ferromagnetic FePt nanocrystal superlattices. Sci, 287:1989–1992

Sun X, Wang C, Gao M et al (2015) Remotely controlled red blood cell carriers for cancer targeting and near-infrared light-triggered drug release in combined Photothermal–chemotherapy. Adv Funct Mater 25:2386–2394

Taori K, Paul VJ, Luesch H (2008) Structure and activity of largazole, a potent antiproliferative agent from the Floridian marine cyanobacterium *Symploca* sp. J Am Chem Soc 130:806–1807

Tessa B, Kurukulasuriya P (2010) Technologies for climate change adaptation: Emerging lessons from developing countries supported by UNDP. J. Int Aff 17–31

Teuling E, Wierenga PA, Agboola JO et al (2019) Cell wall disruption increases bioavailability of Nannochloropsis gaditana nutrients for juvenile Nile tilapia (*Oreochromis niloticus*). Aquaculture 499:269–282

Thomas BS, Gupta RC (2016) Properties of high strength concrete containing scrap tire rubber. J Clean Prod 113:86–92

Thundimadathil J (2012) Cancer treatment using peptides: current therapies and future prospects. J Amino Acids 2012

Tsibakhashvili NY, Kirkesali EI, Pataraya DT et al (2011) Microbial synthesis of silver nanoparticles by *Streptomyces glaucus* and *Spirulina platensis*. Adv Sci Lett 4:3408–3417

Van Leeuwen AC, Buckle T, Bendle G et al (2011) Tracer-cocktail injections for combined pre- and intraoperative multimodal imaging of lymph nodes in a spontaneous mouse prostate tumor model. J Biomed Opt 16:016004

Voraáččovaá K et al (2017) The cyanobacterial metabolite nocuolin A is a natural oxadiazine that triggers apoptosis in human cancer cells. PLoS One 12:e0172850

Wang H, Xie H, Wang J et al (2015) Self-assembling prodrugs by precise programming of molecular structures that contribute distinct stability, pharmacokinetics, and antitumor efficacy. Adv Funct Mater 25:4956–4965

Wells ML, Potin P, Craigie JS et al (2017) Algae as nutritional and functional food sources: revisiting our understanding. J Appl Phycol 29:949–982

Yee N, Benning LG, Phoenix VR et al (2004) Characterization of metal– cyanobacteria sorption reactions: a combined macroscopic and infrared spectroscopic investigation. Environ Sci Technol 38:775–782

Yin Y, Sui C, Meng F et al (2017) The omega-3 polyunsaturated fatty acid docosahexaenoic acid inhibits proliferation and progression of non-small cell lung cancer cells through the reactive oxygen species-mediated inactivation of the PI3K/Akt pathway. Lipids Health Dis 16:1–9

Younis NS, Bakir EM, Mohame ME et al (2019) Cyanobacteria as nanogold factories II: chemical reactivity and anti-myocardial infraction properties of customized gold nanoparticles biosynthesized by *Cyanothece* sp. Mar Drugs 17:402

Zahran HH (1999) Rhizobium-legume symbiosis and nitrogen fixation under severe conditions and in an arid climate. Microbiol Mol Biol Rev 63:968–989

Zinicovscaia I, Cepoi L (2016) Cyanobacteria for bioremediation of wastewaters. Springer International Publishing

Chapter 11
Marine Algal Products and Algal Nanoparticle Synthesis Against Cancer

Arunkumar Ramu and Bakrudeen Ali Ahmed

Contents

11.1 Introduction

Cancer considered as unrestrained proliferation of cells and able to invade, metastasize other organs in the body, thus regard as substantial death causes of humans. Various kinds of cancers cause 18.1 million new cases and 9.6 million deaths in 2018 around the world (Bray et al. 2018). Numerous difficulties and discovering more successful strategies in treating cancer patients remain, including lessening treatment-related adverse effects, poor outcomes, and deficiency of a target therapeutic and minimal treatment toxicity in patients. Hence, the improvement of new

A. Ramu
Department of Biochemistry and Biotechnology, Centre for Research and Development, PRIST Deemed University, Vallam, Tamil Nadu, India

B. A. Ahmed (✉)
Department of Biochemistry and Biotechnology, Centre for Research and Development, PRIST Deemed University, Vallam, Tamil Nadu, India

Institute of Biological Sciences, Faculty of Science, University of Malaya, Kuala Lumpur, Malaysia

D. Thangadurai et al. (eds.), *Bioprospecting Algae for Nanosized Materials*, Nanotechnology in the Life Sciences,
https://doi.org/10.1007/978-3-030-81557-8_11

therapeutic systems for cancer treatment is necessity for proceeded with progress against the malignancy.

Nanotechnology and its derived nanomaterials provide substantial effects in the research and development (R&D) field, particularly in the medical sector. However, synthetic nanomaterials cause their toxic impacts, and some recent research studies ascribed it to their more noteworthy surface permeability into cells, membrane damage, inflammation, DNA damage, and change in connections between cells to cells as well as matrix membrane (Abbasalipourkabir et al. 2015). Therefore, novel nanoparticles are discovered for drug delivery with low toxicity to improve effectiveness on cancer therapy.

Nanotechnology with natural elements has risen as a promising option to be utilized in drug delivery systems to target delivery, being considered as novel platform for malignant growth treatment with low or no toxicity to normal cell (Prasad et al. 2017). Specifically, nanoparticles from marine organisms empower environmentally friendly ways to deal with drug discovery. Marine algae are unicellular microorganisms that contribute up to 40% of worldwide efficiency such as utilized in biofuels and nourishment (Aziz et al. 2017; Rawat et al. 2021). The creation of nanoparticles from marine algae can be an intriguing way to deal with potentiate anticancer therapy which defeats some limitations, such as lack of biodistribution, organ toxicity, limited bioavailability, and weak therapeutic effects (Sun et al. 2014; Blanco et al. 2015). This chapter deals with the intriguing anticancer properties and nanoformulation interceded by algae-derived compounds, and that may arrive at the clinical trials or motivate the improvement of new formulation permitting their interpretation into clinically valuable medications in the future.

11.2 Marine Algae on Apoptosis

The cytotoxic and apoptotic impacts of the marine algae are interceded by the production of ROS. Strangely, the brown and green algae *Colpomenia sinuosa* and *Ulva fasciata* give huge apoptosis stimulation by ROS generation and stimulation of caspases-3 and caspases-9, followed by PARP-1 cleavage on HCT-116 human colon cancer cells (Ryu et al. 2013; Al Monla et al. 2020). The marine microalga *C. calcitrans* was the indicated apoptosis toward human mammary cancer MDA-MB-231 cells on IC_{50} of 60 μg/mL through downregulated Bcl-2, hence actuated caspase-4 enlistment endoplasmic reticulum destruction pathway and S- and G2/M-cell cycle arrest (Goh et al. 2014). β-1,3-Xylan separated from the green alga, *Caulerpa lentillifera*, displayed apoptotic movement by PARP on intercede chromatin degradation by means of activation of caspase 3/7 in human mammary cancer MCF-7 cells (Maeda et al. 2012). Dieckol, isolated from the consumable brown algae *Ecklonia cava*, stimulates the actuation of apoptotic particles caspase-8, caspase-9, and caspase-3 on SKOV3 human ovarian cancer cells and tumor xenograft model through restrained action of upstream signals AKT and p38 (Ahn et al. 2015). Further, two isolectins from the marine red alga *Solieria filiformis* shows antinociceptive and mitigating movement by means of trigged caspase-subordinate apoptosis on MCF-7 human mammary cancer cells (Chaves et al. 2018).

11.3 Marine Algae on Antiangiogenesis

Angiogenesis is a significant sign of cancer and it is connected to the development of metastasis, and efficient compounds from edible and natural sources that hinder angiogenesis could be an important methodology for cancer management. Marine algae being one of the prevailing creatures in nature can give the best occasion to the establishment of new anticancer agents. Blue-green alga *Arthrospira platensis* has powerful antioxidant and has conceivable antiangiogenic consequences for PA-TU-8902 pancreatic cells and tumor xenograft. The angiogenic marker VEGF-A mRNA and protein expression were conquered through upstream signals ERK and Raf. Subsequently, *Arthrospira platensis* suppressive consequences for relocation may represent the anticancer impacts of this blue-green alga (Markova et al. 2020). The brown marine alga *Chnoospora minima* bioactive fraction gives its antiangiogenic potential on HepG2, HeLa, and MCF-7 cancer cells.

Chnoospora minima-treated embryos indicated that a significant decrease of large and small blood vessels alongside the presence of thin endothelial cells exhibited the anticancer property of *Chnoospora minima* (Parveen and Nadumane 2020). The polysaccharide fucoidan found in many brown macroalgae applies antiangiogenic impact by the hindrance of MMP-2 and comparable impact in transplanted lung carcinoma in vivo by influencing various cell surface proteins including integrin, VEGF-1, and p-selection transmembrane protein (SLEP) (Senthilkumar et al. 2013). The carotenoid siphonaxanthin disengaged from various marine microalgae, brown seaweeds, and green algae applies substantial antiangiogenenic properties against the scope of various cancer cells. Ongoing investigations call attention to that antiangiogenic movement of siphonaxanthin was joined by decrease in angiogenic factors like VEGF, FGF-2, and their receptors FGFR-1, EGR-1, and VEGFR-2 which possibly uncovered siphonaxanthin utilized as antiangiogenic drugs (Sugawara et al. 2014). Likewise, Aravindan et al. (2013) showed the antiangiogenic movement of polyphenols from five distinctive earthy colored algae in human pancreatic cancer cells (MiaPaCa-2, Panc-1, BXPC-3, and Panc-3.27). The significant supportive of tumorigenic EGFR, VEGFA, AKT, hTERT, kRas, Bcl2, FGFa, and PDGFa transcription factors restraint and concealment of kRas, AurKb, and Stat3. This information recommend that kelp polyphenols fill in as likely drug to murder potentiate tumor cells in pancreatic cancer.

11.4 Marine Algae on Intracellular Signaling Pathways Inhibition

Abnormal changes in cancer are significantly associated with intracellular signaling pathways. Along these lines, the adjustment of abnormal intracellular signaling pathway of development factors is urgent for preventing the improvement of cancer. Wentilactone B (WB), a tetranorditerpenoid got from marine algae demonstrated impact on G2/M phase arrest and stimulated apoptosis through alternate Ras/Raf/

MAPK signaling pathway in human hepatoma SMMC-7721 cells (Zhang et al. 2013). Algae-determined compound fucoidan restrains proliferation and instigates apoptosis in human ovarian cancer cells SKOV-3, A2780, OVCAR-3, TOV-112D, and Caov-3. Fucoidan incited cell cycle arrest at the G0/G1 phase and repressed the intracellular signal expressions, like phosphorylation of PI3K and Akt, which could be saved partially by IGF-1 may referred fucoidan a likely drug for the therapy of cancer patients (Liu et al. 2020). Phloroglucinol disengaged from brown algae which displayed antimetastatic capacity in breast cancer. The dynamic compound restrains SLUG by downregulating the expression of PI3K/AKT and RAS/RAF-1/ERK signaling (Kim et al. 2015). Laminarin found in marine brown algae *Laminaria digitata* has been appeared to have a few biological actions and prompts apoptosis in HT-29 colon cancer cells (Fischel et al. 1994). However, these compounds stimulate cell cycle arrest sub-G1 and G2-M phase subject to intracellular signaling ErbB, which enacts c-Jun N-terminal kinase recommends the capability of laminarin connected with apoptosis and anticancer activities (Park et al. 2013).

Leathesia difformis extricate, initiated apoptosis on B16F10 murine melanoma cells by a mitochondria-intervened pathway, just as the arrest of the cell cycle at the S phase. Furthermore, this compound diminished malignant melanoma through the hindrance of the cAMP/PKA/CREB pathways (Seo et al. 2019). On A2780 human ovarian cells, bromophycolide A actuated the arrest at the G1 phase of the cell cycle and a subsequent and predictable loss of cells from the S and G2/M phases, while at the same time instigated apoptosis (Kubanek et al. 2005). Novel polysaccharide separated from the algae initiates apoptosis and cell cycle arrest in human gastric carcinoma MKN45 cells through ROS/JNK signaling pathway (Xie et al. 2016). It likewise represses cancer movement through JNK, and arresting the cells at G2/M phase driving the cells to go through apoptosis. It affected the statement of a few proteins, for example, p53, caspase-9, and caspase-3 were upregulated in MCF-7 breast cancer cells that assume significant functions in these biological cycles (Xie et al. 2018). Diphlorethohydroxycarmalol, confined by Kang et al. (2012), initiated apoptosis by the collection of the sub-G1 cell populace and nuclear consideration, exhaustion of mitochondrial layer potential ($\Delta\psi m$), and guideline of the statement favorable to endurance and supportive of apoptotic Bcl-2 relatives.

Caulerpin, a secondary metabolite from the marine intrusive green algae *Caulerpa cylindracea,* initiated apoptosis on colorectal cancer cells, which was interceded by AMPKα1 signaling and cell cycle arrest (Yu et al. 2017). Cf-PS polysaccharide (Kwon and Nam 2007) hindered the proliferation and incited apoptosis by repressing IGF-IR signaling and the PI3K/Akt pathway, which are associated with the guideline of cell development, proliferation, separation, motility, metabolism, and protein synthesis (Chen et al. 2014). Dieckol from marine algae represses non-small cell cellular breakdown in the lungs cell proliferation and invasion through hindering PI3K/AKT/mTOR signaling, actuating the tumor silencer protein E-cadherin implying that dieckol is powerful normal anticancer drug to treat NSCLC (Wang et al. 2019). Fucosterol inhibits the progression of human ovarian cancer by initiating mitochondrial membrane potential and endoplasmic reticulum stress. Moreover, fucosterol directed the proliferation-related signaling pathways, the generation of responsive oxygen species, mitochondrial action, endoplasmic reticulum stress, angiogenesis, and calcium homeostasis demonstrates that fucosterol could be utilized as a possible helpful inhibitor in ovarian cancer (Bae et al. 2020). Further, fucosterol applies antiproliferative impacts by inciting apoptosis, cell cycle arrest, and modulating Raf/MEK/ERK signaling pathway on human cellular breakdown in the lungs A549 and SK-LU-1 cells. Besides, examination uncovered fucosterol decline in the statement of Cdc2, Cyclin A, Cyclin B1, and upregulation of the tumor suppressor (p21, Cip1, and p27Kip1) may end up being a lead molecular change for the therapy of cellular breakdown in the lungs (Mao et al. 2019).

At present, one of the objectives of cancer therapy, particularly in solid tumors, is angiogenesis, which is liable for the development of new blood vessels and is a necessity for the supported development and proliferation of solid tumors. Likewise, the quest for inhibitors of this cycle has become a main line of examination in anticancer exploration, with the subsequent arrival of a novel drugs available that have significantly improved results in patients with various tumor types and antimetastatic activities (Marín-Ramos et al. 2015). Fucodiphloroethol G, compound restrains angiogenesis on human umbilical vein endothelial (ECV-304 and EA.hy926) cells when incited with VEGF, just as the transcriptional factor c-fos and its objectives AP-N and MMP-2, by the hindrance of MAPK and Akt signaling pathways (Li et al. 2011). Recently, the brown algae *Sargassum serratifolium* anticancer effects were exposed on human glioblastoma U87MG, cervical cancer HeLa, gastric cancer MKN-28 cells, and liver cancer SK-HEP 1 cells. The antiproliferative impacts of *Sargassum serratifolium* were established through recover and repair the oncogenic MAPK/ERK signaling pathway which assumes a critical part in the proliferation of glioblastoma U87MG cells. Likewise, the migration and invasion of the U87MG cells were repressed by HES through the concealment of matrix metalloproteinase (MMP)-2 and -9 activations. In this manner, HES is a potential therapeutic operator which effectively affects glioblastoma (Kang et al. 2015). Tuberatolide B (TTB) is a diastereomeric meroterpenoid isolated from marine algae *Sargassum macrocarpum* incited ROS creation, improved DNA damage by instigating γH2AX foci establishment, and the phosphorylation of DNA damage-related proteins, like Chk2 and H2AX. Besides, TTB specifically restrained STAT3 initiation, which brought about a decrease in cyclin D1, MMP-9, survivin, VEGF, and IL-6. Furthermore, TTB-initiated ROS expression caused STAT3 restraint, DNA damage, and apoptotic cell death. Hence, TTB inhibits cancer progression by promoting ROS-mediated inhibition of STAT3 signaling, recommending that TTB is helpful for the therapy of cancer (Choi et al. 2017). Lee et al. (2012) found that phlorofucofuroeckol-A, a compound secluded from the palatable brown alga *Eisenia bicyclis*, is a powerful inhibitor of the aldo-keto reductase family 1 B10 (AKR1B10), an individual from the NADPH-subordinate aldo-keto reductase (AKR) super family, viewed as a potential cancer therapeutic objective. Similarly, the compound SQDG indicated a checked restraint of the telomerase action, which is a catalyst that drives the uncontrolled division and replication of cancer cells (Eitsuka et al. 2004).

Various compounds isolated from algae can be utilized as coadjuvants to improve the effectiveness of the drugs presently utilized as therapeutics. For example, the pretreatment of HepG2 cells with fucoxanthin permitted to improve the therapeutic impact of cisplatin (Yang et al. 2013). Qi et al. (2015), indicated marine bromophenol bis (2,3-dibromo-4,5-dihydroxybenzyl) ether (BDDE), curbs angiogenesis in HUVEC cells and in zebrafish embryos through repressing the VEGF signals. Also, BDDE inactivated the VEGF downstream signaling atoms including mTOR and Src. In addition, BDDE obstructed subintestinal vessel arrangement in zebrafish embryos in vivo and demonstrated harmfulness under high centralizations of BDDE. The consequences of this current examination demonstrated that BDDE could be utilized as a potential drug possibility for cancer prevention and therapy.

11.5 Anticancer Compound Discovery from Marine Algae

This interesting accessibility of microalgae offers extraordinary potential for the preparation of novel drugs. Significantly, half of marine blue-green algal species are developed in business stages to extract bioactive compounds, which have indicated colossal potential to inhibit different types of cancer cells by means of enlistment of apoptosis, or by influencing the actuation of cell signaling chemicals, particularly the protein kinase-c families (Lordan et al. 2011). Also, contingent upon the species and climate greenery, marine microalgae contain immense measures of different kinds of proteins, dietary fiber, nutrients, and minerals (Wells et al. 2017).

11.5.1 Quinone-Based Derivatives from Marine Algae Inhibit Cancers

Triprenylated toluquinones and toluhydroquinones discovered from the *Arminacean nudibranch* and *Leminda millecra* active particles were stimulates ROS-mediated apoptosis in esophageal cancer cells (Whibley et al. 2007). Interestingly, calothrixin B has been found to show antiproliferative movement against HCT-116 colon cancer cell line with IC_{50} estimation of 0.32 μM (Hatae et al. 2014). Some microalgae-based compounds have indicated tremendous potential to hinder a breast cancer cells. Sargaquinoic corrosive disconnected from the brown algae *Sargassum heterophyllum* was discovered to be restrained to the breast cancer than the nonmalignant cell line. Besides, it actuated apoptosis by induction of caspase-3, caspase-6, caspase-8, caspase-9, and caspase-13 as well as downregulates of Bcl-2 and G(1) arrest in MDA-MB-231 cells (de la Mare et al. 2012). Marine-derived fungus, *Aspergillus fumigatus* was isolated from marine green algae-delivered isosclerone. It indicated cytotoxicity toward MCF-7 human breast cancer cells with the IC_{50} dose 63 μM after 24 h cell inhibitions and hindered the invasion protein expression such as MMP-2 and MMP-9 in MCF-7 human breast cancer cells by adjusting MAPK signaling pathway (Li et al. 2014). Hierridin B, a secondary metabolite (polyketide) secluded from the marine cyanobacterium *Cyanobium* sp. LEGE 06113, tried in a board of eight human cancer cell lines, demonstrated specific cytotoxicity against colon cancer cell line HT-29 with an IC_{50} estimation of 0.1 mM (Freitas et al. 2016).

11.5.2 Terpenoid-Based Derivatives from Marine Algae Inhibit Cancers

Tuberatolide B was disengaged from *Sargassum macrocarpum* and recognized as active terpenoid that acts as a Farnesoid X receptor (FXR) adversary (Choi et al. 2011). Recently, Choi et al. (2017) demonstrated that Tuberatolide B prompted the ROS generation prompting STAT3 inhibition, DNA damage, and apoptotic cell death. Along these lines, Tuberatolide B represses cancer progression by inducing ROS-mediated inhibition of STAT3 signaling. In addition, the treatment of HL-60 cells with 10-acetylirciformonin B for 24 h expanded the initiation of caspase-3, caspase-8, and caspase-9 which induce the cleavage of PARP.

Heteronemin is a sesterterpene extracted from marine sponge *Hyrtios erecta* that stimulates diverse biological properties and cytotoxic effects on a various cancer cell lines including leukemia, colon, breast, cervical, and renal cancer in the scope of nanomolar concentration (Wu et al. 2015). Two cytotoxic squalenoid-determined triterpenoids, laurenmariannol confined and recognized from the marine red alga *Laurencia mariannensis* have shown critical cytotoxic action against P-388 tumor cells with IC_{50} estimations of 0.6 and 6.6 mg/mL separately (Ji et al. 2008). More recently, Lee et al. (2018) detailed that heteronemin applies a supportive of apoptotic effects in LNcap prostate cancer cells. Furthermore, they additionally applied that heteronemin prompted likewise autophagy in LNcap cells as affirmed by the outflow of LC3-II, a dissolvable protein engaged with autophagic action (Tanida et al. 2008).

The red seaweed *Laurencia viridis* is a rich origin in squalene-inferred brominated triterpenes dehydrothyrsiferol, isodehydrothyrsiferol, and 10-epidehydrothyrisiferol have demonstrated powerful cytotoxic effects against various cancer cell lines (Nort et al. 1996; Pec et al. 2003). Moreover, Liu et al. (2006) demonstrated that both stellettin An and stellettin B initiated oxidative pressure and apoptosis in HL-60 human leukemia and LNCaP prostate cancer cell lines affirmed by the upregulation of caspase-8, the decrease of Bcl-2 protein articulation, and the expansion of FasL. The compounds such as 16-hydroxydehydro-thyrsiferol, thyresenol-A, and thyrsenol-B were additionally separated from *Laurencia viridis*, and these mixes have displayed critical inhibitory activity on protein phosphatase at a concentration of 10 mM. Additionally, they have indicated powerful cytotoxic action against P388 cell line (Norte et al. 1997; Souto et al. 2003). Besides, Tang et al. (2018) indicated that apoptotic effects of smenospongine in MCF-7 human breast cancer by expanding the phosphorylation level of p38 and AMPKα that have been accounted for inhibits cancer cell development through promote apoptosis.

11.6 Marine Algal Utilization for Development of Nano-Based Drug

Nanoparticles are characterized into natural (carbon-based nanoparticles) and inorganic nanoparticles. Magnetic nanoparticles (iron), respectable metal nanoparticles (silver, gold, and platinum), and semiconductor nanoparticles (cadmium, zinc oxide, zinc sulfite, silica, and titanium oxide) are assembled as inorganic nanoparticles. Without a doubt, the uncommon properties of nanoparticles, for example, biocompatibility, enormous scope creation, straightforward utilization, and utilitarian proficiency, make them successful in drug delivery (Ramasubburayana et al. 2017). Gold (Au) and silver (Ag) nanoparticles have various applications in assorted fields. Nanoparticles have a wide scope of utilizations in the field of cancer drug delivery and diagnostics. Of different metal nanoparticles (Ag, Au, Pt, Pd, and Cu), Ag nanoparticles have pulled in much consideration in the field of nanotechnology due to their particular physicochemical and biological properties (Ahila et al. 2016).

To conquer the impediments to the utilization of peptide-based drugs, nanonization of characteristic item-based prescriptions has pulled in gigantic consideration because of their helpless solubility and stability. It was accounted for that nanotechnology is one of the quickest stages for creating nanoformulations. The nanoemulsion containing algae and cinnamon oils has extensively improved its apoptotic and anti-invasion activity by induced DNA fragmentation and actuate reactive oxygen species (ROS) generation on human colon cancer cells. Thus, nanoemulsion containing algae and cinnamon oils has significantly improved its anticancer viability and wiped out the attack uncovered that the antitumor movement against opposition cancer. Specialists are blending safe nanomaterials including metals, lipids, polysaccharides, and protein- or peptide-based nanodefined drug targets. Nanoformulations have expanded particle size and surface region because of increments in bioavailability, stability, and solubility, and decrease the symptoms of common compound-based drugs (Prasad et al. 2019). They are additionally helpful for the treatment, finding, observing, and control of biological frameworks and have as of late been named as a nanomedicine.

Lately, Elghazawy et al. (2017) tried the upgraded anticancer potential and improved solubility of novel quinoline derivatives through utilizing nanoformulation technique. Accordingly, synthesized nanomicelles of new quinoline derivatives indicated noticeable critical increments in cytotoxic efficiencies, showing a positive effect of the pre-owned nanoformulations on the delivery of active compounds to their cell biological targets. These realities have intrigued specialists working with microalgal compound-based drugs.

Chlorophyta (*Ulva lactuca*) algae extract stacked on albumin nanoparticle initiated antiproliferative and favorable to apoptotic expression on liver (HepG2) and breast cancer (MCF7) cells. The proapoptotic markers Caspase-8 and 9 were essentially raised in both the cell lines. The antioxidant potential and radical scavenging property of algae extract loaded on albumin nanoparticle was additionally obvious when contrasted with the vitamin C and gallic acid proposed that it could be a promising novel target-based chemotherapeutic drug (Al-Malki 2020). Singh et al. (2015)

revealed into the data on marine-based synthesis of nanoparticles, their biomedical applications, and instruments related to them. Besides, a scope of marine assets has been broadly used in the fields of nanoscience and nanotechnology. In different marine bioresources, nanostructures of 1–100 nm size are recognized like shells, pearls, and fish bones.

Recently, Ibrahim et al. (2018) formulated marine krill lipid-based liposome nanoparticles alluded to as marinosomes, which epitomize curcumin into krill lipid-based liposome nanoparticles, and built up a steady anticancer nanoformulation from ease and promptly accessible nutraceuticals. These marinosomes are marine krill lipid-based liposomes containing polyunsaturated fatty acids (PUFAs) (Moussaoui et al. 2002). Marine PUFAs are isolated into two subgroups: omega-6 (n-6) and omega-3 (n-3), which are used and put away in cell layer phospholipids. They influence thin smoothness, manage a wide scope of capacities in the body, and right turn of events and working of the brain and nervous system (Schmitz and Ecker 2008). Moreover, green-synthesized zinc oxide nanoparticles with *Sargassum muticum* algae extraction give antiangiogenic and antiapoptotic effects for human liver cancer cell line (HepG2). The growth hindrance and apoptotic movement were assessed by trypan blue, annexin/propidium iodide, acridine orange/propidium iodide, chick chorioallantoic membrane, and ABTS assay. Results revealed that diminished development and promote apoptosis by this nanoparticle affirm that these formulation could be utilized in cancer therapy (Sanaeimehr et al. 2018). In marine lipids, which are extracted from marine living beings, for example, krill, a huge part of n-3 PUFAs is bound to phospholipids and contains a higher substance of n-3 long-chain PUFAs, for example, eicosapentaenoic corrosive (EPA; 20:5 n-3) and docosahexaenoic corrosive (DHA; 22:6 n-3) (Winther et al. 2011). A feasible component of cancer counteraction dependent on the anti-fiery properties of n-3 PUFAs has been reported. Different investigations have revealed various systems of cancer avoidance by n-3 PUFAs, including apoptosis, hindrance of angiogenesis, restraint of metastasis, etc. (Yin et al. 2017). Already, Anand et al. (2015) integrated monodispersed and circular silver nanoparticles of 20–60 nm from marine parasitic species (*Aspergillus flavus* SP-3, *Trichoderma gamsii*, *Talaromyces flavus*, and *Aspergillus oryzae*) and detailed their anticancer impacts.

Bioactive compounds in marine plants and algae have been shown to inhibit several types of cancer cell lines. Gold and AgNPs have indicated extraordinary abilities and promising outcomes in both cancer diagnostics and therapeutics. The creation of biogenic Ag-NPs utilizing fluid extracts of the marine alga *Sargassum longifolium* demonstrated the possible cytotoxic movement against the carcinoma Hep-G2 cell lines, recommending that the cytotoxic action might be because of the presence of alkaloids present in the algae. Ag-NPs blended utilizing the seaweed *Hypnea* sp. are affected by the elements of the particles cytotoxic action on human Caucasian colon adenocarcinoma.

The cytotoxic impact of biosynthesized gold NPs with an extract of the red seaweed *Corallina officinalis* on the MCF-7 human breast cancer cell line sets up that the nano-Au demonstrated strong cytotoxic action against MCF-7 cells. Khalifa et al. (2016) uncovered that silver nanoparticles (AgNPs) biosynthesized via seaweeds (*Ulva fasciata*, *Corallina lengthen*, *Gelidium crinale*, *Laurencia obtusa*, *Cystoseira myrica,* and *Turbinaria turbinata*) indicated a lessening Ehrlich Ascites

Carcinoma (EAC) cells viability by expanding AgNPs fixation. The greatest restraint level of EAC cells was (94% and 99%) with 98 μg/ml AgNPs biosynthesized by *Ulva fasciata* and *Turbinaria turbinata* individually and silver nanoparticles biosynthesized by *Turbinaria turbinata* were the most cytotoxic against EAC. Anyway the cytotoxicity impact of *Spirulina platensis* blended AgNPs against MCF-7 shows noteworthy cytotoxicity at lower fixation in 48 h. An enormous number of in vitro examinations demonstrated that the AgNPs are poisonous to the mammalian cells. Hence, biologically orchestrated AgNPs demonstrated high inhibitory impact against cancer cell lines and safe to typical encompassing cells (Rejeeth et al. 2014).

11.7 Conclusion

This study demonstrated that marine organisms uncovered to be a fascinating source of existing, unrecognized compounds and nanoparticles with the potential for giving feasible human benefits. Nevertheless, natural medicinal source among marine life forms is helpful for the creation of various metabolites that assist them with adapting to the various states of the marine climate. Majority of these compounds showed incredible potential for therapeutic applications, displaying antioxidant and apoptosis activities against various cancers. However, ongoing researches stated that marine algae bioactive compounds showed significant cytotoxic activity on various cell lines and restrained intracellular signaling pathways are deeply studied. Currently, six of nine drugs from marine organism right now in the market are utilized in cancer therapy, and several compounds are going through clinical trials with signs for oncologic therapeutics. The quantity of compounds in the market or in clinical trials is low as compared with other natural and chemically derived drug.

The developing requirements for the biosynthesis of therapeutics of biological starting point are an absolute necessity. One of the most encouraging and novel therapeutic operators is the metal NP which inhibits naturally strong and repopulated cancer stem cells to make sensitive to chemotherapy. The utilizations of nanoformulations in nanotechnology can potentiate to develop anticancer drugs from algae source from algae-derived compounds. In the field of nanotechnology, algae are rising as significant creatures in green nanotechnology as proven by the huge number of studies detailed in the last 7–8 years. This chapter also attempted to summarize the studies done so far for algae-intervened biosynthesis of metallic nanoparticles giving further information on therapeutic strategies against cancer. Algae-mediated synthesis of metallic NPs would be a better alternative over chemical techniques. The investigations assembled in the current survey clearly demonstrate that algae and algae-mediated nanoparticles have incredible potential as a source of antitumor compounds with the ability to hamper distinctive key pathway targets associated with cancer development including cell cycle guideline, apoptosis, angiogenesis, migration, and invasion process. Despite the fact that to date algae have not been fully exploited, it is expected that in a few years, their interpretation to clinical use can be a reality.

References

Abbasalipourkabir R, Moradi H, Zarei S, Asadi S et al (2015) Toxicity of zinc oxide nanoparticles on adult male Wistar rats. Food Chem Toxicol 84:154–160

Ahila NK, Sri Ramkumar V, Prakasha S, Manikandan B et al (2016) Synthesis of stable nanosilver particles (AgNPs) by the proteins of seagrass *Syringodium isoetifolium* and its biomedicinal properties. Biomed Pharmacother 84:60–70

Ahn JH, Yang YI, Lee KT, Choi JH (2015) Dieckol isolated from the edible brown algae *Ecklonia cava*, induces apoptosis of ovarian cancer cells and inhibits tumor xenograft growth. J Cancer Res Clin Oncol 141:255–268

Al Monla R, Dassouki Z, Kouzayha A, Salma Y, Gali-Muhtasib H, Mawlawi H (2020) The cytotoxic and apoptotic effects of the brown algae *Colpomenia sinuosa* are mediated by the generation of reactive oxygen species. Molecules 25:1–14

Al-Malki AL (2020) In vitro cytotoxicity and pro-apoptotic activity of phycocyanin nanoparticles from *Ulval actuca* (Chlorophyta) algae. Saudi J Biol Sci 27:894–898

Anand BG, Navin Thomas CK, Prakash S, Sathish Kumar C (2015) Biosynthesis of silver nanoparticles by marine sediment fungi for a dose dependent cytotoxicity against HEp2 cell lines. Biocat Agric Biotechnol 4:150–157

Aravindan S, Delma CR, Thirugnanasambandan SS, Herman TS, Aravindan N (2013) Anti-pancreatic cancer deliverables from sea: first-hand evidence on the efficacy, molecular targets and mode of action for multifarious polyphenols from five different brown-algae. PLoS One 16:1–15

Aziz N, Prasad R, Ibrahim AIM, Ahmed AIS (2017) Promising applications for the production of biofuels through algae. In: Microbial Biotechnology (Patra JK, Vishnuprasad CN, Das G), Springer Nature Singapore Pte Ltd. 81–103

Bae H, Lee JY, Song G, Lim W (2020) Fucosterol suppresses the progression of human ovarian cancer by inducing mitochondrial dysfunction and endoplasmic reticulum stress. Mar Drugs 18:1–19

Blanco E, Shen H, Ferrari M (2015) Principles of nanoparticle design for overcoming biological barriers to drug delivery. Nat Biotechnol 33:941–951

Bray F, Ferlay J, Soerjomataram I, Siegel RL, Torre LA, Jemal A (2018) Global cancer statistics 2018: GLOBOCAN estimates of incidence and mortality worldwide for 36 cancers in 185 countries. CA Cancer J Clin 68:394–424

Chaves RP, Silva SRD, Nascimento Neto LG, Carneiro RF et al (2018) Structural characterization of two isolectins from the marine red alga *Solieria filiformis* (Kützing) P.W. Gabrielson and their anticancer effect on MCF-7 breast cancer cells. Int J Biol Macromol 107:1320–1329

Chen Y, Gao D Y, Huang L (2014) In vivo delivery of miRNAs for cancer therapy: challenges and strategies. Adv Drug Deliv Rev. 81:128–141

Choi H, Hwang H, Chin J, Kim E et al (2011) Tuberatolides, potent FXR antagonists from the Korean marine tunicate *Botryllus tuberatus*. J Nat Prod 74:90–94

Choi YK, Ki J, Lee KM, Choi YJ et al (2017) Tuberatolide B suppresses cancer progression by promoting ROS-mediated inhibition of STAT3 signaling. Mar Drugs 55:1–12

de la Mare JA, Lawson JC, Chiwakata MT, Beukes DR et al (2012) Quinones and halogenated monoterpenes of algal origin show anti-proliferative effects against breast cancer cells in vitro. Investig New Drugs 30:2187–2200

Eitsuka T, Nakagawa K, Igarashi M, Miyazawa T (2004) Telomerase inhibition by sulfoquinovosyldiacylglycerol from edible purple laver (*Porphyra yezoensis*). Cancer Lett 212:15–20

Elghazawy NH, Hefnawy A, Sedky NK, El-Sherbiny IM et al (2017) Preparation and nanoformulation of new quinolone scaffold-based anticancer agents: enhancing solubility for better cellular delivery. Eur J Pharm Sci 105:203–211

Fischel J, Lemee R, Formento P, Caldani C et al (1994) Cell growth inhibitory effects of caulerpenyne., a sesquiterpenoid from the marine algae *Caulerpa taxifolia*. Anticancer Res 15: 2155–2160

Freitas S, Martins R, Costa M, Leao PN et al (2016) Hierridin B isolated from a marine cyanobacterium alters vdac1, mitochondrial activity, and cell cycle genes on HT-29 colon adenocarcinoma cells. Mar Drugs 14:1–13

Goh SH, Alitheen NB, Yusoff FM, Yap SK, Loh SP (2014) Crude ethyl acetate extract of marine microalga, *Chaetoceros calcitrans*, induces apoptosis in MDA-MB-231 breast cancer cells. Pharmacogn Mag 10:1–8

Hatae N, Satoh R, Chiba H, Osaki T et al (2014) N-Substituted calothrixin B derivatives inhibited the proliferation of HL-60 promyelocyticleukemia cells. Med Chem Res 23:4956–4961

Ibrahim S, Tagami T, Kishi T, Ozeki T (2018) Curcuminmarinosomes as promising nano-drug delivery system for lung cancer. Int J Pharm 540:40–49

Ji NY, Li XM, Xie H, Ding J et al (2008) Highly oxygenated triterpenoids from the marine red alga *Laurencia mariannensis* (Rhodomelaceae). Helv Chim Acta 91:1940–1946

Kang GJ, Han SC, Koh YS, Kang HK et al (2012) Diphlorethohydroxycarmalol, isolated from *Ishige okamurae*, increases prostaglandin E2 through the expression of cyclooxygenase-1 and -2 in HaCaT human keratinocytes. Biomol Ther 20:520–525

Kang CW, Park MS, Kim NH, Lee JH et al (2015) Hexane extract from *Sargassum serratifolium* inhibits the cell proliferation and metastatic ability of human glioblastoma U87MG cells. Oncol Rep 34:2602–2608

Khalifa KS, Hamouda RA, Hanafy D, Hamza A (2016) In vitro antitumor activity of silver nanoparticles biosynthesized by marine algae. Dig J Nanomater Biostruct 11:213–221

Kim RK, Suh Y, Yoo KC, Cui YH, Hwang E et al (2015) Phloroglucinol suppresses metastatic ability of breast cancer cells by inhibition of epithelial-mesenchymal cell transition. Cancer Sci 106:94–101

Kubanek J, Prusak AC, Snell TW, Giese RA et al (2005) Antineoplastic diterpene-benzoate macrolides from the Fijian red alga *Callophycus serratus*. Org Lett 7:5261–5264

Kwon MJ, Nam TJ (2007) A polysaccharide of the marine alga *Caposiphon fulvescens* induces apoptosis in AGS gastric cancer cells via an IGF-IR mediated PI3K/Akt pathway. Cell Biology international 31:768–775

Lee MS, Kim JI, Utsuki T, Park NG, Kim HR (2012) Cytoprotective effects of phlorofucofuroeckol a isolated from *Ecklonia stolonifera* against tacrine-treated HepG2 cells. Fitoterapia 83:1060–1067

Lee MG, Liu YC, Lee YL, El-Shazly M et al (2018) Heteronemin, a marine sesterterpenoid-type metabolite, induces apoptosis in prostate LNcap cells via oxidative and ER stress combined with the inhibition of Topoisomerase II and Hsp90. Mar Drugs 16:1–24

Li Y-X, Li Y, Qian Z.-J, Ryu B et al (2011) Suppression of vascular endothelial growth factor (VEGF) induced angiogenic responses by fucodiphloroethol G. Process Biochem 46:1095–1103

Li YX, Himaya SWA, Dewapriya P, Kim HJ, Kim SK (2014) Anti-proliferative effects of isosclerone isolated from marine fungus *Aspergillus fumigatus* in MCF-7 human breast cancer cells. Process Biochem 49:2292–2298

Liu WK, Cheung FW, Che CT (2006) Stellettin A induces oxidative stress and apoptosis in HL-60 human leukemia and LNCaP prostate cancer cell lines. J Nat Prod 69:934–937

Liu S, Yang J, Peng X, Li J, Zhu C (2020) The natural product fucoidan inhibits proliferation and induces apoptosis of human ovarian cancer cells: focus on the PI3K/Akt signaling pathway. Cancer Manag Res 12:6195–6207

Lordan S, Ross RP, Stanton C (2011) Marine bioactives as functional food ingredients: potential to reduce the incidence of chronic diseases. Mar Drugs 9:1056–1100

Maeda R, Ida T, Ihara H, Sakamoto T (2012) Induction of apoptosis in MCF-7 cells by β-1,3-xylooligosaccharides prepared from *Caulerpa lentillifera*. Biosci Biotechnol Biochem 76:1032–1034

Mao Z, Shen X, Dong P, Liu G et al (2019) Fucosterol exerts antiproliferative effects on human lung cancer cells by inducing apoptosis, cell cycle arrest and targeting of Raf/MEK/ERK signalling pathway. Phytomedicine 61:1–9

Markova I, Konickova R, Vaakova K, Lenicek M et al (2020) Anti-angiogenic effects of the blue-green alga *Arthrospira platensis* on pancreatic cancer. J Cell Mol Med 24:2402–2415
Marín-Ramos NI, Alonso D, Ortega-Gutiérrez S, Ortega-Nogales et al (2015) New inhibitors of angiogenesis with antitumor activity in vivo. J Med Chem 58:3757–3766
Moussaoui N, Cansell M, Denizot A (2002) Marinosomes, marine lipid-based liposomes: physical characterization and potential application in cosmetics. Int J Pharm 242:361–365
Nort M, Fernandez JJ, Souto ML, Garcia-Gravalos MD (1996) Two new antitumoral polyether squalene derivatives. Tetrahedron Lett 15:2671–2674
Norte M, Fernandez JJ, Saouto ML, Gavin JA et al (1997) Thyrsenols A and B two unusual polyether squalene derivatives. Tetrahedron 53:3173–3178
Park HK, Kim IH, Kim J, Nam TJ (2013) Induction of apoptosis and the regulation of ErbB signaling by laminarin in HT-29 human colon cancer cells. Int J Mol Med 32:291–295
Parveen S, Nadumane VK (2020) Anti-angiogenesis and apoptogenic potential of the brown marine alga, *Chnoospora minima*. Future J Pharm Sci 6:1–14
Pec MK, Aguirre A, Moser-Their K, Fernandez JJ et al (2003) Induction of apoptosis in estrogen dependent and independent breast cancer cells by the marine terpenoid dehydrothyrsiferol. Biochem Pharmacol 65:1451–1461
Prasad R, Pandey R, Varma A, Barman I (2017) Polymer based nanoparticles for drug delivery systems and cancer therapeutics. In: Natural Polymers for Drug Delivery (eds. Kharkwal H and Janaswamy S), CAB International, UK 53–70
Prasad R, Kumar V, Kumar M, Choudhary D (2019) Nanobiotechnology in Bioformulations. Springer International Publishing (ISBN 978-3-030-17061-5) https://www.springer.com/gp/book/9783030170608
Qi X, Liu G, Qiu L, Lin X, Liu M (2015) Marine bromophenolbis(2,3-dibromo-4,5-dihydroxybenzyl) ether, represses angiogenesis in HUVEC cells and in zebra fish embryos via inhibiting the VEGF signal systems. Biomed Pharmacother 75:58–66
Ramasubburayana R, Sumathi S, Prakash S, Ramkumar VS et al (2017) Synthesis of nano silver by a marine epibiotic bacterium *Bacillus vallismortis* and its potent eco friendly antifouling properties. Environ Nanotechnol Monit Manag 8:112–120
Rawat J, Gupta PK, Pandit S, Prasad R, Pande V (2021) Current perspectives on integrated approaches to enhance lipid accumulation in microalgae. 3 Biotech 11, 303. https://doi.org/10.1007/s13205-021-02851-3
Rejeeth C, Nataraj B, Vivek R, Sakthivel M (2014) Biosynthesis of silver nanoscale particles using *Spirulina platensis* induces growth-inhibitory effect on human breast cancer cell line MCF 7. Med Aromat Plants 3:1–7
Ryu MJ, Kim AD, Kang KA, Chung HS et al (2013) The green algae *Ulva fasciata* Delile extract induces apoptotic cell death in human colon cancer cells. In Vitro Cell Dev Biol Anim 49:74–81
Sanaeimehr Z, Javadi I, Namvar F (2018) Antiangiogenic and antiapoptotic effects of green-synthesized zinc oxide nanoparticles using *Sargassum muticum* algae extraction. Cancer Nanotechnol 9:1–16
Schmitz G, Ecker J (2008) The opposing effects of n-3 and n-6 fatty acids. Prog Lipid Res 47:147–155
Senthilkumar K, Manivasagan P, Venkatesan J, Kim SK (2013) Brown seaweed fucoidan: biological activity and apoptosis, growth signalling mechanism in cancer. Int J Biol Macromol 60:366–374
Seo GY, Ha Y, Park AH, Kwon OW, Kim YJ (2019) *Leathesia difformis* extract inhibits α-MSH-induced melanogenesis in B16F10 cells via down-regulation of CREB signaling pathway. Int J Mol Sci 20:1–16
Singh CR, Kathiresan K, Anandhan S (2015) A review on marine based nanoparticles and their potential applications. Afr J Biotechnol 14:1525–1532
Souto ML, Manriquez CP, Norte M, Leira F, Fernandez J (2003) The inhibitory effects of squalene derived triterpenes on protein phosphatase PP2A. Bioorg Med Chem Lett 13:1261–1264
Sugawara T, Ganesan P, Li Z, Manabe Y, Hirata T (2014) Siphonaxanthin, a green algal carotenoid, as a novel functional compound. Mar Drugs 12:3660–3668

Sun J, Fernandez-Cid A, Riera A, Tognetti S et al (2014) Structural and mechanistic insights into Mcm2-7 double-hexamer assembly and function. Genes Dev 28:2291–2303

Tang J, Wu W, Yang F, Liu L et al (2018) Marine sponge-derived smenospongine preferentially eliminates breast cancer stem-like cells via p38/AMPKα pathways. Cancer Med 7:3965–3976

Tanida I, Ueno T, Kominami E (2008) LC3 and autophagy. Methods Mol Biol 445:77–88

Wang CH, Li XF, Jin LF, Zhao Y et al (2019) Dieckol inhibits non-small-cell lung cancer cell proliferation and migration by regulating the PI3K/AKT signaling pathway. J Biochem Mol Toxicol 33:1–8

Wells ML, Potin P, Craigie JS, Ravenm JA et al (2017) Algae as nutritional and functional food sources: revisiting our understanding. J Appl Phycol 29:949–982

Whibley CE, McPhail KL, Keyzers RA, Maritz MF et al (2007) Reactive oxygen species mediated apoptosis of esophageal cancer cells induced by marine triprenyltoluquinones and toluhydroquinones. Mol Cancer Ther 6:2535–2543

Winther B, Hoem N, Berge K, Reubsaet L (2011) Elucidation of phosphatidylcholine composition in krill oil extracted from *Euphausia superba*. Lipids 46:25–36

Wu SY, Sung PJ, Chang YL, Pan SL, Teng CM (2015) Heteronemin, a spongeanses terterpene, induces cell apoptosis and autophagy in human renal carcinoma cells. Biomed Res Int 2015:1–13

Xie P, Fujii I, Zhao J, Shinohara M, Matsukura M (2016) A novel polysaccharide derived from algae extract induces apoptosis and cell cycle arrest in human gastric carcinoma MKN45 cells via ROS/JNK signaling pathway. Int J Oncol 49:1561–1568

Xie P, Horio F, Fujii I, Zhao J et al (2018) A novel polysaccharide derived from algae extract inhibits cancer progression via JNK, not via the p38 MAPK signaling pathway. Int J Oncol 52:1380–1390

Yang L, Wang P, Wang H, Li Q et al (2013) Fucoidan derived from *Undaria pinnatifida* induces apoptosis in human hepatocellular carcinoma SMMC-7721 cells via the ROS-mediated mitochondrial pathway. Mar Drugs 11:1961–1976

Yin Y, Sui C, Meng F, Ma P, Jiang Y (2017) The omega-3 polyunsaturated fatty acid docosahexaenoic acid inhibits proliferation and progression of non-small cell lung cancer cells through the reactive oxygen species-mediated inactivation of the PI3K/Akt pathway. Lipids Health Dis 16:1–9

Yu H, Zhang H, Dong M, Wu Z et al (2017) Metabolic reprogramming and AMPKα1 pathway activation by caulerpin in colorectal cancer cells. Int J Oncol 50:161–172

Zhang Z, Miao L, Lv C, Sun H, Wei S, Wang B et al (2013) Wentilactone B induces G2/M phase arrest and apoptosis via the Ras/Raf/MAPK signaling pathway in human hepatoma SMMC-7721 cells. Cell Death Dis 6:1–12

Chapter 12
Surface-Functionalized Diatoms for Drug Delivery and Tissue Engineering Applications

U. T. Uthappa, Mahesh P. Bhat, Ho-Young Jung, and Mahaveer D. Kurkuri

Contents

12.1 Introduction

It is well known that the porous material called diatom microalgae is one of the representative classical specimens to replace synthetic silica (De Stefano et al. 2011). Gordon and Drum foresaw the possibility of researching diatoms in 1994 in the field of nanotechnology (Sprynskyy et al. 2017). Exoskeleton of diatom silica acquired from using diatom cell culture route or via mining of naturally occurring diatomaceous earth mineral (fossilized diatoms). So, in the diatom cell culture which comprises only supply of inorganic salts as nutrients and sunlight (Lin et al. 2017), thus biosilica diatoms can be attained after purification. Diatoms are single-celled photosynthetic algae with a porous skeleton protected by a silica shell wall known as frustules, which are primarily self-possessed of silicon dioxide (SiO_2) and have hierarchical and systematically structured pores (Xing et al. 2017). More than 100,000 different species of diatoms have been recognized, and due to their

U. T. Uthappa · H.-Y. Jung
Department of Environment and Energy Engineering, Chonnam National University, Gwangju, Republic of Korea

M. P. Bhat · M. D. Kurkuri (✉)
Centre for Nano and Material Sciences, Jain University, Bengaluru, Karnataka, India
e-mail: mahaveer.kurkuri@jainuniversity.ac.in

D. Thangadurai et al. (eds.), *Bioprospecting Algae for Nanosized Materials*, Nanotechnology in the Life Sciences,
https://doi.org/10.1007/978-3-030-81557-8_12

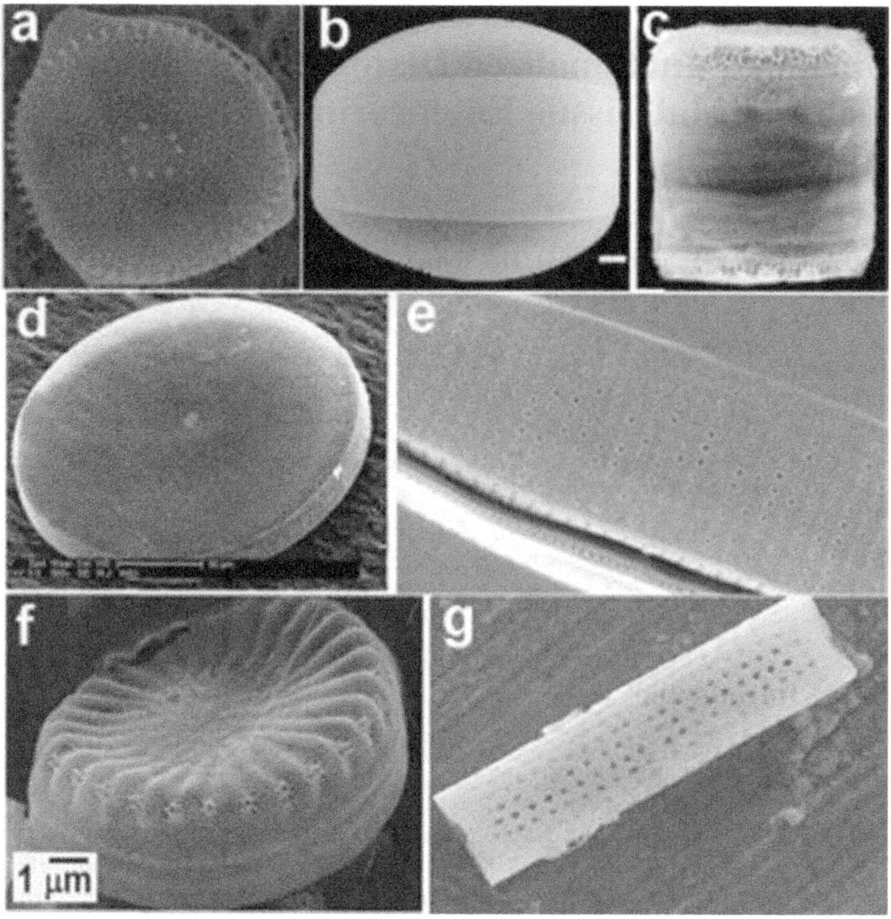

Fig. 12.1 (**a**) *Thalassiosira weissflogii* (scale bar 10 µm), (**b**) *Coscinodiscus granni* (scale bar 20 µm), (**c**) *Thalassiosira pseudonana* (scale bar 1 µm), (**d**) *Coscinodiscus wailessi* (50 µm), (**e**) *Nitzschia* sp. (5 µm), (**f**) *Cyclotella* sp., (**g**) *Aulacoseira* sp. (1 µm). (Copyright © Elsevier 2018, reprinted with permission from Uthappa et al. 2018a)

distinctive morphological existence, each species differs. Figure 12.1 displays a few selected species of diatoms and their respective SEM images.

Such nature-derived diatom silica have numerous benefits in contrast of synthetic silica with respect to low cost, biocompatibility, proven non-toxicity, abundant in nature, thermal stability, and chemical inertness. Amidst, few important diatom species such as nanoporous *Coscinodiscus concinnus* (Gnanamoorthy et al. 2014), *Thalassiosira weissflogii* (Cicco et al. 2015), *Thalassiosira pseudonana* (Delalat et al. 2015), *Nitzschia* sp. (Javalkote et al. 2015), and *Aulacoseira* have been described as future carriers for drugs because of their exceptional characteristics and structural morphology (Aw et al. 2012; Zhang et al. 2013). Additionally,

Thalassiosira pseudonana have been considered for immunogenicity as well (Delalat et al. 2015).

This chapter focuses on the modern progress of the application of diatom biosilica in the biomedical application with major emphasis on drug delivery applications. Further, various surface functionalization strategies adapted on the surface of diatoms to enhance their physicochemical properties and features for the release of drugs that are tunable, programmed, and pH-sensitive drug release will be discussed. Also, as a part of other biomedical application of diatoms with some highlights involved in the tissue engineering applications and their future prospects in the drug delivery and tissue engineering field will be discussed.

12.2 Surface-Functionalized Diatoms for Drug Delivery Applications

In the area of drug delivery, there have been several recent breakthroughs and have rapidly fast tracked the innovation and helped in the improvement of several new kind of therapeutic actions. Some of the drugs have particular disadvantages, including the prevalent ones, such as solubility problems (low or high), poor stability, many lateral effects, including deficient targeting and elevated toxicity. Consequently, the constant research and development in advanced drug delivery systems (aDDs) have shown tremendous noteworthy progress. These aDDs have been shown to enhance the aqueous solubility of poorly soluble drugs by improving their physicochemical properties, targeting distribution, and accumulating at target sites that are not accessible by traditional drug delivery systems that deliver/transport drug candidates across the entire body. All at once, the commercial attainment of the aDDs comprises biocompatibility, biodegradability, low cost, facile manufacturing, controlled release profile to offer constant plasma concentration in order to minimize the level of administration and reduce patient discomfort, for extended stages (Natarajan et al. 2014; Wang et al. 2015).

Nanotechnology showed impressive impacts on all fields of biomedical science in recent decades, such as drug delivery and tissue engineering applications (Prasad et al. 2017; Saglam et al. 2021). Innovative nanoscale drug delivery systems by exploring nanomaterials/particles provided new avenues for the delivery of varied range of therapeutics more effectively by tackling evolutionary glitches of innate conventional drug delivery systems. In this perspective, a selective range of synthetic nanomaterials like metal nanoparticles, metal oxide nanoparticles, polymers, micro/nanoparticles, and semiconductor nanoparticles and many more have been advanced and discovered as drug carriers (Wenzel et al. 2002; Nokhodchi et al. 2004; Farokhzad and Langer 2006; Russo et al. 2011; Bimbo et al. 2012; Fry et al. 2014; Natarajan et al. 2014; Wang et al. 2015; De Jesus and Zuhorn 2015; Uthappa et al. 2020a).

Silica-based micro/nanomaterials have been identified as a potential alternative material to polymers and a few other nanomaterials for drug delivery applications due to their distinct physical, chemical, and optical properties. Because of its excellent physical and chemical properties, synthetic silica derived from the sol-gel route or electrochemical etching of Si wafers has gained lot of attention as a drug carriers (Zhao et al. 2009; Muhammad et al. 2011). Nonetheless, they have some drawbacks such as high cost, time-consuming, and production of toxic wastes during synthesis (Losic et al. 2009; Sumper and Brunner 2006). So, to overcome the hindrance of these synthetic silica, it is very essential to find alternative silica materials. Thus, through millions of years by genetic evolution, nature have designed and gifted unicellular photosynthetic algae known as "Diatoms/DE" (Round et al. 2007). Due to the astonishing features of diatoms such as non-toxic, biocompatible, chemically stable, low cost and facile functionalization which grades as an effective micro/nanocarriers to substitute existing synthetic silica (Uthappa et al. 2019b). Furthermore, this DE silica material attributed to their non-toxic nature has approved by US-EPA/FDA to be employed in food industry and agriculture sectors (Chao et al. 2014). The exploration of DE silica materials for other allied industrial applications and mining of DE silica in bulk range provoked concern of attention toward long-term cytotoxicity. In this aspect, a meticulous investigation made by Mossman and Glenn systematically briefed the influence of long-term contact of large amounts of DE silica dust which may lead to breathing disorders (Mossman Glenn 2013). Despite extended contact of large amounts of DE silica is not a major and foremost issue for biomedical applications as it requires in minute extents and short-term contact via oral route and has rapid clearance rate. Although DE is presently used as insulators in medical devices since it does not cause any lung allergy when compared to former asbestos (Santos 2014). Furthermore, diatom silica materials widely explored for broad range of other applications such as nanofabrication, chromatographic separation, catalysis, pest control, solar cell, supercapacitor (Patil et al. 2017), dye removal (Sriram et al. 2019), solid-state sensors for the detection of toxins in water and so on (Bhat et al. 2019).

In view of all these considerable facts, for the very first time, diatoms (*Aulacoseira*) explored in drug delivery applications. Surface of diatoms modified with dopamine iron oxide nanoparticles, the drug release outline can be fine-tuned by using magnetically guided drug delivery systems. Magnetized diatoms were formed by electrostatically driven self-assembly of dopamine-modified iron oxide nanoparticles (NPs) onto the surface of diatoms in this study. The cationic magnetic materials accumulated into anionic diatoms due to electrostatic interaction. The drug loading potential of Indomethacin, a poorly water-soluble drug, was 0.280.05 g/g, and the drug release for DOPA/Fe_3O_4/diatom in the first stage was around 60% in the first 6 h, which can be attributed to the burst release, and the remaining 40% was released in 2 weeks (Losic et al. 2010).

From the same group in other work, surface functionalization of diatoms was carried out by using various organosilanes, to offer the preferred surface functionalities. Different organosilanes are regulated by two different forms of hydrophobic and hydrophilic properties on the surface of diatoms. The presence of various functional groups on the diatomic surface and their variance in drug-loading ability

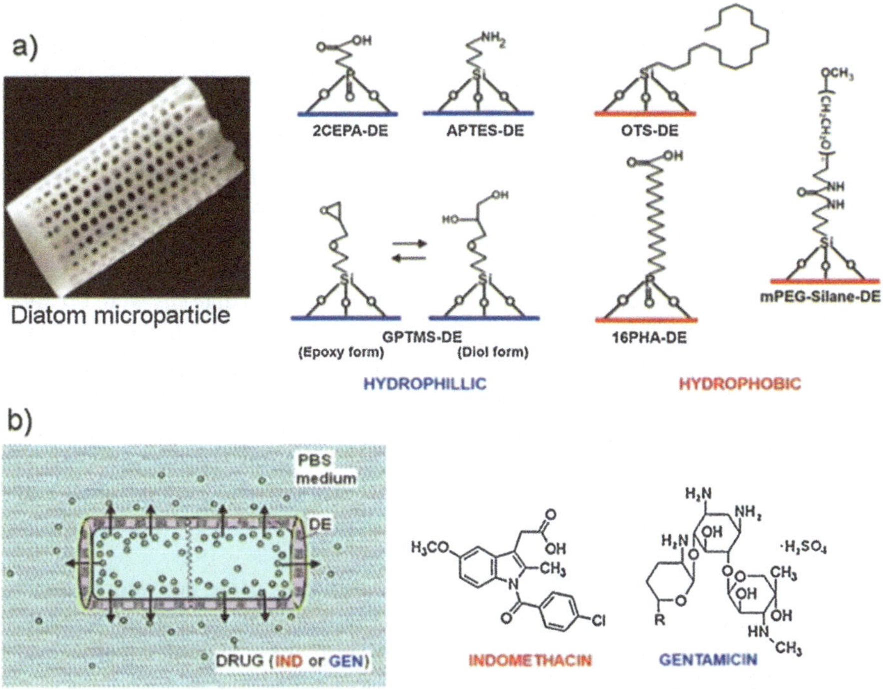

Fig. 12.2 (**a**) Surface functionalization on the diatom surfaces with hydrophobic and hydrophilic properties to regulate the rate of drug release, (**b**) drug release mechanism for two distinct drugs. (Copyright © Elsevier 2013, reprinted with permission from Bariana et al. 2013)

(15–24%) and release profile (6–15 days) have been observed (Aw et al. 2013). General surface functionalization approaches are schematically shown in Fig. 12.2.

Similar way in other work validated by using organosilanes and phosphonic acid, to generate dissimilar functionalities. Following that, two separate drugs, indomethacin (hydrophobic) and gentamicin (hydrophilic), were used to modify the surface. It was revealed that hydrophilic surfaces like amino-rich-APTES (3-aminopropyltriethoxysilane), epoxy-rich-GPTMS 3-(glycidyloxy propyl) trimethoxysilane, and carboxyl-rich-2-CEPA (2-carboxyethyl-phosphonic acid) boosted the loading of the hydrophobic drug with a substantial controlled release in contrast to pristine diatoms. Instead, slighter drug loading capacity and rapid release were witnessed for hydrophobic surface modified diatoms with 16-PHA (16-phosphonohexadecanoic acid), methoxy-poly-(ethylene-glycol)-silane (mPEG-silane), and OTS (7-octadecyltrichlorosilane). But for the gentamicin drug, the hydrophobic surface modified with 16-PHA-enriched drug loading and controlled release, while in hydrophilic modified with 2CEPA and APTES indicated a minor extent of drug loading and shorter drug release profile (Bariana et al. 2013).

A functionalization process established through covalent attachment of GO to the APTES surface via NHS-EDC coupling chemistry is described here, resulting in stable composites. The electrostatic process, on the other hand, involves interactions between GO functional groups (COOH) and amine groups (NH2) formed on diatom surfaces (APTES-DE). Indomethacin was used to study the release profile from both types of surface-functionalized diatoms, wherein drug loading for GO-DE was 28%, which is noticeably greater than that of APTES-DE (control), that is, 15.8%. The drug release profile for the control sample at pH 3.5 specified only 14 days of extended release. Surprisingly, the release of surface-functionalized GO-DE was extended up to 37 days, providing compelling evidence for the long-term release of drug through GO-modified diatoms (Kumeria et al. 2013).

In other work (Todd et al. 2014), iron oxide NPs loaded to diatoms were obtained for magnetically regulated delivery of tiny molecules to tumors. It was feasible to bear more functionalities by physical adsorption by exploring 10 μm diatoms. Two different kinds of approaches like fluorescence and magnetic resonance imaging systems aided to examine the movement of tumors in animal model. Interestingly, diatoms were focused near tumors because of magnetic properties indicated enrichment in tumor retention. Iron oxide diatoms have demonstrated minor toxicity and biodegradability and are thus becoming more prospective in the area of drug delivery.

On the surface of diatom frustules, a grafting process using activators regenerated by electron transfer-based atom transfer radical polymerization (ARGET-ATRP) regenerates thermoresponsive oligo (ethylene glycol) methacrylate copolymer. Levofloxacin was loaded onto the developed matrix that showed the temperature-responsive release of drugs above and below the polymer's lower critical solution temperature (LCST). Furthermore, antimicrobial studies were carried out using typical two different wound pathogens for the released drug from the designed carriers to ensure the activity of therapeutics (Vasani et al. 2015).

Another work depicted the prospective of diatomite nanoparticles (DNPs) for drug delivery through effective biofunctionalization. To extend the intracellular absorption of 3-aminopropyltriethoxysilane (APT)-functionalized DNPs and to proliferate the biocompatibility of 3-aminopropyltriethoxysilane (APT)-functionalized DNPs, researchers used polyethylene glycol (PEG) and a cell-penetrating peptide (CPP) conjugation. PEGylation and CPP conjugation will significantly reduce the cytotoxicity of the DNPs-APT in cytotoxicity studies with 200 g/mL incubation with MCF-7 and MDA-MB-231 breast cancer cells for about 1 day. In addition, cellular realization of the functionalized DNPs also assessed by means of the above- stated cancer cell lines, presenting CPP conjugation significantly upsurge the DNPs cellular uptake. Additionally, surface functionalization of DNPs enriched loading of a sorafenib drug, with a loading capacity of 22wt% and likewise improved the drug release profiles in aqueous solutions (Terracciano et al. 2015). The surface functionalization strategies are schematically shown in Fig. 12.3.

In this report, diatom frustule *Nitzschia* for improved chemotherapy by using curcumin for magnetically active drug delivery was achieved. Two separate processes, such as curcumin magnetic diatom microparticles ferrofluid (CMDM-F) and curcumin magnetic diatom microparticle by in situ process, were treated with iron oxide NPs (CDM-I). In addition, cell viability studies using the HeLa cell lines

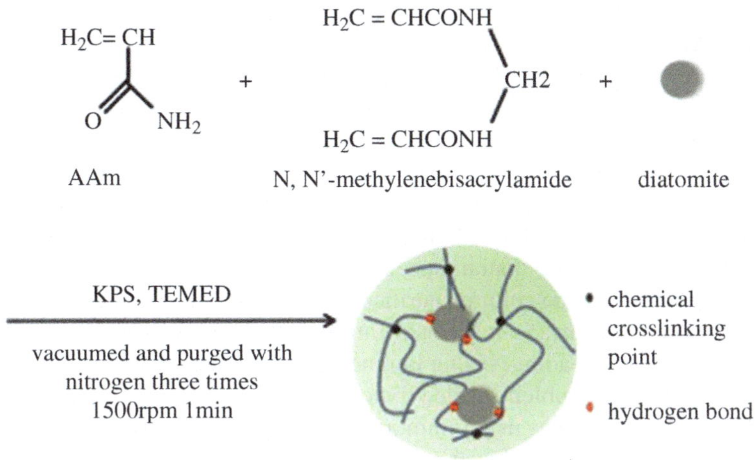

Fig. 12.3 Schematic depicts the fabrication of PAAm/diatom composite hydrogel system and their hydrogen bonding interactions between PAAm and diatoms. (Copyright © Elsevier 2019, reprinted with permission from Lv et al. 2019)

disclosed that CMDM-F showed high cytotoxicity (60%) in contrast to curcumin´ (42%), DM (1.9%), CDM (44.8%), and CMDM-I (59.9%). Importantly, CMDM-F and CMDM-I indicated enhanced cytotoxicity when linked with pristine curcumin. Finally, it was discovered that CMDM-F has a high cell viability, which is a novel route and could be a suitable stage in the path for magnetic hyperthermia applications (Javalkote et al. 2015).

In other work, DE microparticles were tagged with vitamin B12 as the tumor-directing ligand and loaded with anticancer agents' cisplatin, 5-fluorouracil, and tris-tetraethyl[2,2´-bipyridine]-4,4´-diamineruthenium(II) complex. In vitro studies demonstrated the enhanced adherences of the vitamin B12-coated diatoms close to HT-29 and MCF-7 cell lines. Adherences of the vitamin B12-tagged diatoms was almost three times greater than unmodified ones and was associated with the improved transcobalamin II (TC(II)) and transcobalamin II receptor (TC(II)-R) appearance of the targeted tissue (Delasoie et al. 2018).

Currently, in drug delivery field, various encapsulation/loading processes have been used to load the drugs to the designed carriers including dropwise/solvent evaporation techniques. Yet, reproducible approach that lessens management errors needs to be minimized. In this study, new technique known as ESCARGOT (Embedded Scaffold RemovinG Open Technology) explored to formulate microfluidic device to encapsulate hydrophobic drug into the biogenic diatom silica which belongs to *Cyclotella* Isorhamnetin drug with hydrophobic in nature and with varied concentrations: 20, 60, and 100 μg/mL, and three altered resident times in the device 0.4, 1, and 2 min were studied. Based on the statistical investigation, the ideal situations to attain an extreme encapsulation efficiency were 2 min and 20 μg/mL. Isorhamnetin drug release showed 48.26% of burst release in the first hour, whereas the total amount of drug was supplied in 3 h. However, present methodology might be used as a standard method to attain reproducible results (Mancera-Andrade et al. 2019).

In the present report, in situ free radical polymerization of PAAm/diatomite-based composite hydrogel was obtained and represented in Fig. 12.3. The release rate of carbendazim in hydrogels in the presence and absence of diatomite was found to be moderate, compared to carbendazim dissolved in water. Mechanism involved in carbendazim release from the diatomite composite hydrogel systems was completely constructed based on polymer relaxation (Lv et al. 2019).

For the very first time, diatom species *Amphora subtropica* explored for drug delivery applications. The cultures purified to get Amphora frustules (exoskeleton) (AF), monitored by surface functionalization with biopolymer chitosan. Results disclosed better drug loading, strong luminescence, biodegradable, and biocompatible nature of the doxorubicin-loaded diatoms. In comparison to pristine doxorubicin (DOX), toxicity studies using immortalized lung cancer cell line (A549) direct sustained drug delivery and lower toxicity, suggesting that AF can be used as an alternative replacement for synthetic silica nanomaterials (Sasirekha et al. 2019).

It is well known that conventional codelivery routines are regularly hindered by the various physicochemical properties among diverse kind of drugs. To overcome this issue, an innovative twofold drug delivery system using an R5 peptide-fused ferritin (R5FT) has been designed. Herein, R5 can facilitate biosilica accumulation on the ferritin surface (SiO_2/R5FT), where two types of molecules loaded into the ferritin cage and biosilica matrix. However, entrapped mRFP or PTX in the shell unveiled a short-term release, whereas DOX in the core systems exhibited a long-term sustained release. Here, twofold drug delivery systems with the supplementary release of PTX reduced the IC50 value of DOX by twofold in contrast to only DOX (Ki et al. 2020). Schematic representation of biosilica wrapped with ferritin cages by carrying two types of drugs and their temporal release profile shown in Fig. 12.4.

In other work, Kabir et al. (2020) adapted DE-embedded core shell materials by loading two chemotherapeutic drugs at perpetual molar ratios, in non-identical

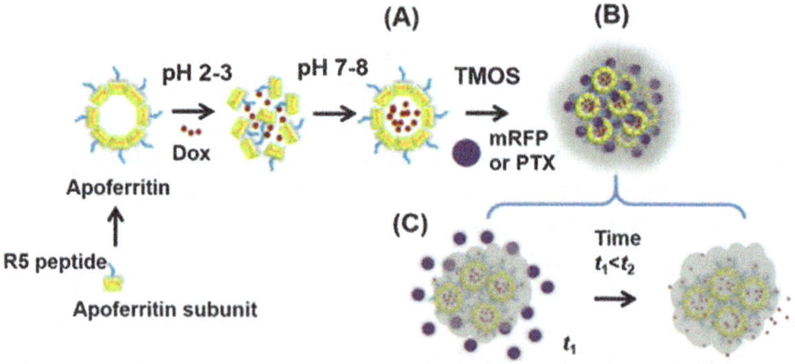

Fig. 12.4 Schematic of doxorubicin (Dox)-laden R5 ferritin [R5FT(Dox)] representation of biosilica wrapped with ferritin cages by carrying two types of drugs and their temporal release profile. (Copyright © Elsevier 2020, reprinted with permission from Ki et al. 2020)

partitions of a distinct drug delivery carriers. Herein, loaded drug in different partitions of delivery carrier (core of diatoms and shell of cyclodextrin) then controls the release percentage of both drugs in situ, and helps in preserving optimum molar ratios which is important for their synergistic outcomes through in vitro studies.

Recently, surface-functionalized DE (*Aulacoseira*) have been documented with xerogel particles of high surface area content as an efficient drug delivery system. Diclofenac sodium (DS) was used in this analysis and the loading capacity of the drug was 38.8 ± 1.3% for pristine DE. However, the loading capacity was 46.2 ± 1.6% for the surface-modified material DE-XER. The drug release for DE-XER matrices was outstandingly monitored for up to 36 days, suggesting the prominence of surface modification (Uthappa et al. 2018b).

Another latest advancement carried out by our group where curcumin (CUR) drug was loaded to DE *(Aulacoseira)* material and the polydopamine (PDA) was coated, followed by tagging with the target ligand known as folic acid (FA). In vitro drug release validated at pH 1.2 for DE-CUR 87% and 74% of drug was released for DE-CUR-PDA-FA. Similarly, at pH 7.4 for pristine DE-CUR 67.53% and 51.90% of drug release was obtained for DE-CUR-PDA-FA. It is important to mention that the developed DE-CUR-PDA-FA systems could be used as anticancer drug delivery systems for tumor therapy only after extensive in vitro/in vivo studies (Uthappa et al. 2019a).

MIL-100(Fe)-DE hybrid material was designed and investigated recently by our group for the delivery of isoniazid anti-TBC (tuberculosis) medication. The improved drug loading capacity of 9% for MIL-100(Fe)-DE which is explained by high surface area and enriched functionalities existent on the surface of MIL-100(Fe)-DE when compared with bare DE only 5% of drug loading capacity witnessed. In vitro drug release showed at pH 1.2 for DE showed only 6 and 13 days of drug release for MIL-100(Fe)-DE was observed. Notably, in pH 7.4 hybrid material, MIL-100(Fe)-DE exhibited 23 days of exceptional controlled drug release was witnessed (Uthappa et al. 2020b).

12.3 Surface-Functionalized Diatoms for Tissue Engineering Applications

Growing demand of multifunctional biomaterials with definite and specific physical, chemical, and biological properties for tissue engineering applications is frequently in conflict due to the high cost of the allied products, which encounter strong gap between the research and clinics. Biosilica derived from diatoms symbolizes as a potential substrate and alternative candidates owing to their abundance in nature, low production cost, and high reproducibility of mesoporous and nanostructural features when compared to more probed biomaterials for tissue engineering for instance: ceramics, polymers, and synthetic silica materials.

In this study, silk fibroin was combined with silicon-releasing diatom particles as a potential for bone tissue engineering applications. In this case, diatom-loaded fibroin sponges were delivered via solvent casting-particulate leaching mode. Cell adhesion, metabolic activity, and proliferation assays were used to assess the biological impact on the human osteosarcoma cell line MG63, an effective osteoblast-like model. Furthermore, alkaline phosphatase activity (ALP), osterix, and collagen type I development in the MG63 cell line were measured as early bone formation markers to validate the engineered scaffolds as pro-mineralization (Le et al. 2018).

In this study, systematically compared two dissimilar diatomite purification approaches, with and without calcination, and diverse size diatom particles resultant in terms of their composition, morphology, size, and cytotoxic effect on cell cultures. Two examined methods ensued in diatoms with unlike features, chiefly in terms of composition and stability. Subsequent diatom particles can go through degradation in aqueous environmental conditions, consequently vigorously releasing silicon ions. Obviously impact of the purification process on silicon release and a decline of silicon release from calcined samples demonstrated in the reported wok. Additionally, diatom-derived particles proved no cytotoxic effect in vitro, but refinement path might influence the cytotoxicity, particularly for microparticles (Le et al. 2016).

Going forward in other part of work, diatomite particles combined with chitosan to form composite membranes via solvent-casting technique for applications of bone tissue engineering. The existing study found that diatomite integration improved chitosan membrane surface area, roughness, swelling properties, and protein adsorption capacities, all of which are important factors in osteoblast adhesion and proliferation. Furthermore, cytotoxicity-exposed chitosan/diatomite composites had no cytotoxic effect on 3T3 and Saos-2 cell lines and had a positive impact on virtuous biocompatibility. Diatomite incorporation increased osteoblastic Saos-2 cell dispersion and proliferation on membranes in vitro cell culture. Later on, ALP activity of Saos-2 cells improved extraordinarily for 7- and 21-day incubation phases for 3wt% of diatomite encapsulated trials (Tamburaci and Tihminlioglu 2017).

Likewise, in other part of work, biosilica loaded with porous chitosan efficaciously. The research findings indicated that chitosan-biosilica composites were greatly in the form of porous morphology comprising micropores and macropores. While diatomite accumulation reduced the compressive mechanical strength of the dry composites, wet chitosan-diatomite composites revealed superior compression moduli when related to pristine chitosan. Further, cell attachment and proliferation results indicated that chitosan-diatomite composites favorable for osteoblast proliferation, ALP activity, and improved biomineralization on the developed scaffold surface (Tamburaci and Tihminlioglu 2018).

A wet co-electrospinning technique was designed to integrate PHBV/PCL and PUL fibers into the 3D fibrous scaffold. Diatoms were electrospun and integrated into fibers in their actual shape and size. Additionally, any feasible size, concentration, and shape reliant cytotoxicity of diatoms were inhibited by loading diatoms into polymer fibers subsequently declining precise interaction area with cells.

Effectively crosslinking of particular phase PUL fibers throughout and after electrospinning was established. Antibiotic encapsulating ability of diatoms while loaded into fibers was assessed with cefuroxime axetil. Controlled release of cefuroxime axetil from scaffold was attained while diatoms used as a doping agent to stimulate osteogenic activity. Diatom-loaded scaffold groups (PHBV/PCL/Diatoms and PHBV/PCL/CA: PUL/Diatoms) showed increased Saos-2 cell viability, which was directly proportional to the presence of diatoms in the structure. Since diatoms supported cell bioactivity and PUL fibers enhanced cell distribution by increasing hydrophilicity, the engineered co-electropun scaffold resulted in improved cell distribution and healthier cell spreading (Dalgic et al. 2019).

For the first time (Cicco et al. 2019), fabricated sodium alendronate in vivo integrated into nanostructured biosilica shells of cultured *Thalassiosira weissflogii* diatoms and upon an acid-oxidative procedure, Na ALE-functionalized frustules obtained without varying their structural mesoporosity. Additionally, the ALE-biosilica has been validated to be an effectual osteoconductive biomaterial substrate by triggering the tissue regeneration model represented in vitro by osteoblast-like cells (SaOS-2) and bone marrow stem cells (BMSC) and hindering osteclast-like cells (J774).

12.4 Conclusion and Future Perspectives

This chapter focused on the applications of the naturally available innumerable types of diatoms; biosilica in drug delivery and tissue engineering area. Different sorts of surface functionalization strategies have been elucidated. For the processing of few synthetic silica, laboratory procedures are costly, laborious, and require the use of noxious chemicals, and these materials have been widely used and explored in the field of drug delivery and tissue engineering to date. Diatoms, on the other hand, have become one of the most appealing candidates due to their natural accessibility, low cost, non-toxic existence, eco-friendly, biocompatibility, high surface area, systematically organized porous structure, and ease of surface functionalization. To investigate the biological realization of diatom silica as an advanced pharmaceutical drug vehicle prerequisite for clinical applications, comprehensive study must be systematically analyzed in detail. Additional experiments to assess the effectiveness and biocompatibility of diatoms for drug delivery and tissue engineering fields expending in vivo responses with animal models and preclinical evaluation are highly desirable. Although the use of DE silica by various authorities like EPA/USDA/USFDA granted to be employed in food industry and agriculture. Yet, there is no any official regulatory approval for diatoms to be used in pharmaceutical industry and medicine fields. Currently, due to their biological origin, verified biocompatibility and low cytotoxicity the diatom silica/DE showed efficacious results through in vitro/vivo studies. However, the DE/diatom, biosilica biomaterial, has resilient prospective demand for biomedical applications near future.

Acknowledgments The authors acknowledge the financial support of DST, India (DST/BDTD/EAG/2019), DST/TDT/DDP-31/2021 and Jain University. Prof. Ho-Young-Jung strongly honors the financial assistance received from Chonnam National University (Grant number: 2018-3274), South Korea.

References

Aw MS, Simovic S, Yu Y, Addai-Mensah J, Losic D et al (2012) Porous silica microshells from diatoms as biocarrier for drug delivery applications. Powder Technol 223:52–58. https://doi.org/10.1016/j.powtec.2011.04.023

Aw MS, Bariana M, Yu Y, Addai-Mensah J, Losic D et al (2013) Surface-functionalized diatom microcapsules for drug delivery of water-insoluble drugs. J Biomater Appl 28:163–174. https://doi.org/10.1177/0885328212441846

Bariana M, Aw MS, Kurkuri M, Losic D (2013) Tuning drug loading and release properties of diatom silica microparticles by surface modifications. Int J Pharm 443:230–241. https://doi.org/10.1016/j.ijpharm.2012.12.012

Bhat MP, Kigga M, Govindappa H, Patil P, Jung H-Y, Yu J, Kurkuri et al (2019) A reversible fluoride chemosensor for the development of multi-input molecular logic gates. New J Chem 43:12734–12743. https://doi.org/10.1039/C9NJ03399H

Bimbo M, Peltonen L, Hirvonen J, Santos A et al (2012) Toxicological profile of therapeutic nanodelivery systems. Curr Drug Metab 13:1068–1086. https://doi.org/10.2174/138920012802850047

Chao JT, Biggs MJ, Pandit (2014) Diatoms: a biotemplating approach to fabricating drug delivery reservoirs. Expert Opin Drug Deliv 11:1687–1695. https://doi.org/10.1517/17425247.2014.935336

Cicco SR, Vona D, De Giglio E, Cometa S, Mattioli-Belmonte M, Palumbo F, Ragni R, Farinola GM et al (2015) Chemically modified diatoms biosilica for bone cell growth with combined drug-delivery and antioxidant properties. ChemPlusChem 80:1104–1112

Cicco SR, Vona D, Leone G, De Giglio E, Bonifacio MA, Cometa S, Fiore S, Palumbo F, Ragni R, Farinola GM et al (2019) In vivo functionalization of diatom biosilica with sodium alendronate as osteoactive material. Mater Sci Eng C Mater Boil Appl 104:109897. https://doi.org/10.1016/j.msec.2019.109897

Dalgic AD, Atila D, Karatas A, Tezcaner A, Keskin D et al (2019) Diatom shell incorporated PHBV/PCL-pullulan co-electrospun scaffold for bone tissue engineering. Mater Sci Eng C Mater Biol Appl 100:735–746. https://doi.org/10.1016/j.msec.2019.03.046

De Jesus MB, Zuhorn IS (2015) Solid lipid nanoparticles as nucleic acid delivery system: properties and molecular mechanisms. J Control Release 201:1–13. https://doi.org/10.1016/j.jconrel.2015.01.010

De Stefano L, De Stefano M, De Tommasi E, Rea I, Rendina I et al (2011) A natural source of porous biosilica for nanotech applications: the diatoms microalgae. Phys Status Solidi C 8:1820–1825. https://doi.org/10.1002/pssc.201000328

Delalat B, Sheppard VC, Rasi Ghaemi S, Rao S, Prestidge CA, McPhee G, Rogers ML, Donoghue JF, Pillay V, Johns TG, Kroger N, Voelcker et al (2015) Targeted drug delivery using genetically engineered diatom biosilica. Nat Commun 6:8791. https://doi.org/10.1038/ncomms9791

Delasoie J, Rossier J, Haeni L, Rothen-Rutishauser B, Zobi F et al (2018) Slow-targeted release of a ruthenium anticancer agent from vitamin B12 functionalized marine diatom microalgae. Dalton Trans 47:17221–17232. https://doi.org/10.1039/C8DT02914H

Farokhzad OC, Langer R (2006) Nanomedicine: developing smarter therapeutic and diagnostic modalities. Adv Drug Deliv Rev 58:1456–1459. https://doi.org/10.1016/j.addr.2006.09.011

Fry NL, Boss GR, Sailor MJ (2014) Oxidation-induced trapping of drugs in porous silicon microparticles. Chem Mater 26:2758–2764. https://doi.org/10.1021/cm500797b

Gnanamoorthy P, Anandhan S, Prabu (2014) Natural nanoporous silica frustules from marine diatom as a biocarrier for drug delivery. J Porous Mater 21:789–796. https://doi.org/10.1007/s10934-014-9827-2

Javalkote VS, Pandey AP, Puranik PR, Deshmukh PK (2015) Magnetically responsive siliceous frustules for efficient chemotherapy. Mater Sci Eng C 50:107–116. https://doi.org/10.1016/j.msec.2015.01.079

Kabir A, Nazeer N, Bissessur R, Ahmed M et al (2020) Diatoms embedded, self-assembled carriers for dual delivery of chemotherapeutics in cancer cell lines. Int J Pharm 573:118887. https://doi.org/10.1016/j.ijpharm.2019.118887

Ki M-R, Kim JK, Kim SH, Nguyen TKM, Kim KH, Pack SP et al (2020) Compartment-restricted and rate-controlled dual drug delivery system using a biosilica-enveloped ferritin cage. J Ind Eng Chem 81:367–374. https://doi.org/10.1016/j.jiec.2019.09.027

Kumeria T, Bariana M, Altalhi T, Kurkuri M, Gibson CT, Yang W, Losic D et al (2013) Graphene oxide decorated diatom silica particles as new nano-hybrids: towards smart natural drug microcarriers. J Mater Chem B 1:6302. https://doi.org/10.1039/C3TB21051K

Le TDH, Bonani W, Speranza G, Sglavo V, Ceccato R, Maniglio D, Motta A, Migliaresi C et al (2016) Processing and characterization of diatom nanoparticles and microparticles as potential source of silicon for bone tissue engineering. Mater Sci Eng C 59:471–479. https://doi.org/10.1016/j.msec.2015.10.040

Le TDH, Liaudanskaya V, Bonani W, Migliaresi C, Motta A et al (2018) Enhancing bioactive properties of silk fibroin with diatom particles for bone tissue engineering applications. J Tissue Eng Regen Med 12:89–97. https://doi.org/10.1002/term.2373

Lin H-Y, Yen S-C, Kuo P-C, Chung C-Y, Yeh K-L, Huang C-H, Chang J, Lin H-J et al (2017) Alkaline phosphatase promoter as an efficient driving element for exogenic recombinant in the marine diatom *Phaeodactylum tricornutum*. Algal Res 23:58–65. https://doi.org/10.1016/j.algal.2017.01.007

Losic D, Mitchell JG, Voelcker NH (2009) Diatomaceous lessons in nanotechnology and advanced materials. Adv Mater 21:2947–2958

Losic D, Yu Y, Aw MS, Simovic S, Thierry B, Addai-Mensah et al (2010) Chem Commun 46:6323–6325. https://doi.org/10.1039/C0CC01305F

Lv J, Sun B, Jin J, Jiang W et al (2019) Mechanical and slow-released property of poly (acrylamide) hydrogel reinforced by diatomite. Mater Sci Eng C Mater Biol Appl 99:315–321. https://doi.org/10.1016/j.msec.2019.01.109

Mancera-Andrade EI, Parsaeimehr A, Ruiz-Ruiz F, Rorrer GL, González-Valdez J, Iqbal HM, Parra-Saldivar et al (2019) Isorhamnetin encapsulation into biogenic silica from *Cyclotella* sp. using a microfluidic device for drug delivery applications. Biocatal Agric Biotechnol 19:101175. https://doi.org/10.2174/1389200128028500047

Mossman BT, Glenn R (2013) Bioreactivity of the crystalline silica polymorphs, quartz and cristobalite, and implications for occupational exposure limits (OELs). Crit Rev Toxicol 43:632–660. https://doi.org/10.3109/10408444.2013.818617

Muhammad F, Guo M, Qi W, Sun F, Wang A, Guo Y, Zhu G et al (2011) pH-triggered controlled drug release from mesoporous silica nanoparticles via intracellular dissolution of ZnO nanolids. J Am Chem Soc 133:8778–8781. https://doi.org/10.1021/ja200328s

Natarajan JV, Nugraha C, Ng XW, Venkatraman S et al (2014) Sustained-release from nanocarriers: a review. J Control Release 193:122–138. https://doi.org/10.1016/j.jconrel.2014.05.029

Nokhodchi A, Tailor A et al (2004) In situ cross-linking of sodium alginate with calcium and aluminum ions to sustain the release of theophylline from polymeric matrices. Farmaco 59:999–1004. https://doi.org/10.1016/j.farmac.2004.08.006

Patil P, Bhat MP, Gatti MG, Kabiri S, Altalhi T, Jung H-Y, Losic D, Kurkuri et al (2017) Chemodosimeter functionalized diatomaceous earth particles for visual detection and removal of trace mercury ions from water. Chem Eng J 327:725–733

Prasad R, Pandey R, Varma A, Barman I (2017) Polymer based nanoparticles for drug delivery systems and cancer therapeutics. In: Natural Polymers for Drug Delivery (eds. Kharkwal H and Janaswamy S), CAB International, UK 53–70

Round FE, Crawford RM, Mann (2007) Diatoms: biology and morphology of the genera. Cambridge University Press

Russo L, Colangelo F, Cioffi R, Rea I, Stefano (2011) A mechanochemical approach to porous silicon nanoparticles fabrication. Materials 4:1023–1033. https://doi.org/10.3390/ma4061023

Saglam N, Korkusuz, F, Prasad R (2021) Nanotechnology Applications in Health and Environmental Sciences. Springer International Publishing (ISBN: 978-3-030-64410-9) https://www.springer.com/gp/book/9783030644093

Santos HA (2014) Porous silicon for biomedical applications. Elsevier/Woodhead Publishing

Sasirekha R, Sheena TS, Deepika MS, Santhanam P, Townley HE, Jeganathan K, Kumar SD, Premkumar K et al (2019) Surface engineered *Amphora subtropica* frustules using chitosan as a drug delivery platform for anticancer therapy. Mater Sci Eng C 94:56–64. https://doi.org/10.1016/j.msec.2018.09.009

Sprynskyy M, Pomastowski P, Hornowska M, Krol A, Rafinska K, Buszewski B et al (2017) Naturally organic functionalized 3D biosilica from diatom microalgae. Mater Des 132:22–29. https://doi.org/10.1016/j.matdes.2017.06.044

Sriram G, Bhat MP, Kigga M, Uthappa UT, Jung H-Y, Kumeria T, Kurkuri MD et al (2019) Amine activated diatom xerogel hybrid material for efficient removal of hazardous dye. Mater Chem Phys 235:121738. https://doi.org/10.1016/j.matchemphys.2019.121738

Sumper M, Brunner E (2006) Learning from diatoms: nature's tools for the production of nanostructured silica. Adv Funct Mater 16:17–26. https://doi.org/10.1002/adfm.200500616

Tamburaci S, Tihminlioglu F (2017) Diatomite reinforced chitosan composite membrane as potential scaffold for guided bone regeneration. Mater Sci Eng C 80:222–231. https://doi.org/10.1016/j.msec.2017.05.069

Tamburaci S, Tihminlioglu F (2018) Biosilica incorporated 3D porous scaffolds for bone tissue engineering applications. Mater Sci Eng C 91:274–291. https://doi.org/10.1016/j.msec.2018.05.040

Terracciano M, Shahbazi MA, Correia A, Rea I, Lamberti A, De Stefano L, Santos HA et al (2015) Surface bioengineering of diatomite based nanovectors for efficient intracellular uptake and drug delivery. Nanoscale 7:20063–20074. https://doi.org/10.1039/C5NR05173H

Todd T, Zhen Z, Tang W, Chen H, Wang G, Chuang YJ, Deaton K, Pan Z, Xie J et al (2014) Iron oxide nanoparticle encapsulated diatoms for magnetic delivery of small molecules to tumors. Nanoscale 6:2073–2076. https://doi.org/10.1039/c3nr05623f

Uthappa U, Brahmkhatri V, Sriram G, Jung H-Y, Yu J, Kurkuri N, Aminabhavi TM, Altalhi T, Neelgund GM, Kurkuri MD et al (2018a) Nature engineered diatom biosilica as drug delivery systems. J Control Release 281:70–83. https://doi.org/10.1016/j.jconrel.2018.05.013

Uthappa U, Sriram G, Brahmkhatri V, Kigga M, Jung H-Y, Altalhi T, Neelgund GM, Kurkuri MD et al (2018b) Xerogel modified diatomaceous earth microparticles for controlled drug release studies. New J Chem 42:11964–11971. https://doi.org/10.1039/C8NJ01238E

Uthappa U, Kigga M, Sriram G, Ajeya KV, Jung H-Y, Neelgund GM, Kurkuri MD et al (2019a) Facile green synthetic approach of bio inspired polydopamine coated diatoms as a drug vehicle for controlled drug release and active catalyst for dye degradation. Microporous Mesoporous Mater 288:109572. https://doi.org/10.1016/j.micromeso.2019.109572

Uthappa U, Kurkuri MD, Kigga M (2019b) Nanotechnology advances for the development of various drug carriers, nanobiotechnology in bioformulations. Springer, pp 187–224

Uthappa U, Arvind O, Sriram G, Losic D, Kigga M, Kurkuri MD et al (2020a) Nanodiamonds and their surface modification strategies for drug delivery applications. J Drug Deliv Sci Technol 60:101993. https://doi.org/10.1016/j.jddst.2020.101993

Uthappa U, Sriram G, Arvind O, Kumar S, Neelgund GM, Losic D, Kurkuri MD et al (2020b) Engineering MIL-100 (Fe) on 3D porous natural diatoms as a versatile high performing platform for controlled isoniazid drug release, Fenton's catalysis for malachite green dye degradation and environmental adsorbents for Pb^{2+} removal and dyes. Appl Surf Sci 528:146974. https://doi.org/10.1016/j.apsusc.2020.146974

Vasani RB, Losic D, Cavallaro A, Voelcker NH et al (2015) Fabrication of stimulus-responsive diatom biosilica microcapsules for antibiotic drug delivery. J Mater Chem B 3:4325–4329. https://doi.org/10.1039/C5TB00648A

Wang Y, Santos A, Evdokiou A, Losic D et al (2015) An overview of nanotoxicity and nanomedicine research: principles, progress and implications for cancer therapy. J Mater Chem B 3:7153–7172. https://doi.org/10.1039/C5TB00956A

Wenzel JG, Balaji KS, Koushik K, Navarre C, Duran SH, Rahe CH, Kompella UB et al (2002) Pluronic® F127 gel formulations of Deslorelin and GnRH reduce drug degradation and sustain drug release and effect in cattle. J Control Release 85:51–59. https://doi.org/10.1016/S0168-3659(02)00271-7

Xing Y, Yu L, Wang X, Jia J, Liu Y, He J, Jia Z et al (2017) Characterization and analysis of Coscinodiscus genus frustule based on FIB-SEM. Prog Nat Sci 27:391–395. https://doi.org/10.1016/j.pnsc.2017.04.019

Zhang H, Shahbazi MA, Makila EM, da Silva TH, Reis RL, Salonen JJ, Hirvonen JT, Santos HA et al (2013) Diatom silica microparticles for sustained release and permeation enhancement following oral delivery of prednisone and mesalamine. Biomaterials 34:9210–9219. https://doi.org/10.1016/j.biomaterials.2013.08.035

Zhao Y, Trewyn BG, Slowing II, Lin VS et al (2009) Mesoporous silica nanoparticle-based double drug delivery system for glucose-responsive controlled release of insulin and cyclic AMP. J Am Chem Soc 131:8398–8400. https://doi.org/10.1021/ja901831u

Part V
Industrial and Environmental Applications of Algal Nanomaterials

Chapter 13
Algae for Nanocellulose Production

Pieter Samyn, Michaela Pappa, Sanjaya Lama, and Dries Vandamme

Contents

P. Samyn (✉) · M. Pappa · S. Lama · D. Vandamme
Analytical and Circular Chemistry, Institute for Materials Research IMO-IMOMEC, University of Hasselt, Diepenbeek, Belgium
e-mail: dries.vandamme@uhasselt.be

D. Thangadurai et al. (eds.), *Bioprospecting Algae for Nanosized Materials*, Nanotechnology in the Life Sciences,
https://doi.org/10.1007/978-3-030-81557-8_13

13.1 Introduction

Macroalgae are a heterogeneous group of organisms from different phylogenetic origins grouped as photo-autotrophs of aquatic environments. Macroalgae or seaweeds are distinguished from higher plants because they lack a vascular structure. In general, the group of algae contains different ancestral roots and are categorized into four kingdoms: (i) Plantae (e.g. *Chlorophytes* green algae, *Rhodophytes* red algae); (ii) Chromista (e.g. *Phaeophytes* brown algae, *Dinoflagellates* single-celled eukaryotes, Diatoms microalgae), (iii) Protista (e.g. *Euglenophytes* eukaryotes) and (iv) Bacteria (Cyanobacteria) (Bhattacharya and Medlin 1998; Davis et al. 2003). They have common ancestors that share not only the gene of photosynthesis but also the gene of cellulose synthesis. Plant eukaryotes possibly acquired it when a cyanobacterium was incorporated as a plastid in their microbial ancestor, which then evolved to create the glaucophytes, red algae, green algae and finally land plants, all these belonging to the kingdom Plantae. Thus, the pool of photosynthetic organisms was enhanced, which changed the Earth's atmosphere and created terrestrial autotrophic organisms, in combination with the intrinsic cellulose synthesis gene, which via evolution shaped the higher plants. In contrast, all terrestrial plants have a single origin for cellulose and variations in their qualities have been induced by evolution (Nobles et al. 2001; Leliaert et al. 2012). Therefore, the algae diversity, regarding species morphology and characteristics along with the cellulose architecture, provides ultimate interesting biomass for the recovery of cellulosic materials and the production of various cellulose structures. Owing to the limited research on nanocellulose materials from cyanobacteria and microalgae, they are presently not further considered in this introductory chapter.

Driven by the current needs in resolving one of our everyday fundamental societal challenges, that is, limitations in available material resources on our planet Earth and particularly the rapid depletion of fossil oil, alternative resources should be investigated. The sustainable management of renewable biomass has, therefore, become a key issue in parallel with the development of extraction routes and valorization paths for the different ingredients of plant-based and animal-based resources. In recent decades, the valorization opportunities for residues from agricultural, forest and/or marine origin have been considered at the academic level as a pathway for the design of novel high-technological materials through the implementation of novel processing routes.

The advantage in using algae biomass over traditional lignocellulosic biomass for the production of cellulosic biopolymers can be summarized, amongst others, in the following main arguments:

- *Availability*: The extraction of cellulose from algae does not interfere with the food-chain and the short harvesting cycles compared to wood-based biomass are inherent to the rapid growth of algae. The cultivation of algae can be done in a simple and sustainable way by avoiding the occupation of large acres of land and the need for fertilizers. The unwanted phenomena of eutrophication or blooming due to abundance in local algae growth provide a wealthy excess of algae biomass for which currently novel application markets are needed. Most interestingly, the local algae sources can be included in the valorization chain, providing a broad variety in the properties of extracted cellulose. In particular, after the well-known commercial extraction of agar from *Gelidiella* red algae for food industries, the remaining side-products include a fibrous pulp fraction that can serve as feedstock for further conversion into cellulosic materials. After initial processing and extraction of valuable products, the cellulose content in residuals can be enriched and allows for the economically viable production of high volumes. Otherwise, the accumulation of *Posidonia oceanica* brown algae on the seashore after stormy weather poses an environmental problem that needs creative solutions. The associated residues from marine biomass provide an abundant and low-cost potential source for which processing is currently under development. Seaweeds offer a wide range of interesting bioactive molecules, chemicals and biofuels already conceptualized in several biorefinery schemes (Torres et al. 2019). Moreover, algae waste valorization has been recently reported, for example, for bioethanol or biomethane production (Uju et al. 2015; Barbot et al. 2016), as soil conditioner or fertilizer products or as bio-adsorbents for metal pollutants such as chromium (Kang et al. 2011).

- *Processing*: The presence and availability of native cellulose in a simple cell structure of algae with high purity allows to simplify extraction processes, for example, by the exclusion of elaborative processing needed for removal of the matrix and/or additional polysaccharide components. In particular, the presence of lignin in lignocellulosic biomass is responsible for a dark colouration and needs harsh chemical bleaching processes, eventually resulting in a polluting environment. Advantageously, the high cellulose content within algae biomass is favourable for high production yields after extraction of the pure cellulose. The easier separation of cellulose fractions from algae biomass with its simplified structure is more favourable compared to the lignocellulosic biomass, requiring extensive and expensive processing for the isolation of cellulose and nanocellulose products due to its hierarchical and embedded structure. As such, the milder processing conditions for algae-based cellulose aim at being more energy-efficient and lower the degradation of recovered cellulose. Only some impurities (minerals) and organic nutrients (proteins) in algae biomass may occur and can be removed through mild chemical processing. In general, the recovery of cellulose fractions from algae can be seen as a single portion of polysaccharides that is present in the cell wall besides the other valuable compounds such as proteins, pigments or fatty acids. Therefore, extensive research is set up for the implementation of an algae 'biorefinery concept' with the gradual extraction of various products during step-wise processing (Torres et al. 2019). However, the sustainable processing of algae biomass through a bio-refinery mainly focused on the valorization potential of subsequently recovered biofuels and commodity chemicals, starting from red algae such as *Gracilaria dura*, *Gelidiella acerosa* or *Gelidium pussilum* (Baghel et al. 2015; Baghel et al. 2017) or green algae such as *Ulva fasciata* (Trivedi et al. 2013). An integrated strategy for the macroalgal biorefinery of *Sargassum tenerrimum* brown algae was illustrated, with subsequent production of a nutrient-rich sap, alginic acid, protein concentrate, cellulose and salt (Baghel et al. 2020). Another proof-of-concept for the biomass fractionation of *Ulva ohnoi* green algae was demonstrated (Prabhu et al. 2020) by recovery of six major components including starch, mineral salts, lipids, ulvan, protein and cellulose. As an advantage in the entire processing cycle, the strong molecular structure of cellulose makes it less sensitive to prior extraction steps of the more sensitive components under gradually increasing process intensification. Therefore, the selection of processing conditions involves a combination of preliminary non-destructive steps for extraction of the sensitive molecules, followed by the more destructive methods. For example, protein removal prior to cellulose extraction was done under alkaline conditions, which allowed to separate both fractions as a supernatant and residual fraction, while filtration of the latter residue directly provided the cellulose material.

Since the first mentions for extraction of cellulose from algae (Stanford 1885; Percival and Ross 1948), algae cellulose has been considered as an alternative biomass next to the traditional woody and non-woody plants. Therefore, the algae-based cellulose becomes in competition with other non-lignocellulosic sources from either animal origin (e.g. tunicate) or bacterial origin (e.g. *Acetobacter xylinus*, *Agrobacterium*, *Pseudomonas*, *Rhizobium* or *Sarcina*). Since the quest for the design of functional materials with excellent properties and novel functionalities, unravelling of the intrinsic cellulose structure and the discovery of nanocellulose structures opened a fully new field of research. The disintegration of the hierarchical fibre structure into its entities at nanoscale level allows to isolate single microfibrils, nanofibrils and crystallites. As such, nanocellulose is acclaimed as a material with excellent mechanical properties, chemical and thermal stability, renewability, biodegradability and lightweight. The general review papers on nanocellulose provide a good overview on the transformation of cellulose into nanomaterials with new properties and industrial applications in different fields (Dufresne 2019). Depending on the applied thermal, chemical and/or mechanical processing routes, different preparation methods were proposed. On the other hand, the selection of specific processing parameters and/or given resources preliminarily determine the final morphology and properties of the nanocellulose. The production of nanocellulose relies on a broad range of available resources from either lignocellulosics or other origins, while the extraction efficiency and nanoparticle dimensions largely vary depending on the starting material, being wood (softwood, hardwood, bleached pulp, etc.), annual plants (cotton, flax, hemp, sisal, henep, ramie, palm fibres, etc.) or agricultural residues (coconut, sugar beet, carrots, bagasse, wheat, potato, etc.) (Kargarzadeh et al. 2017). In particular, the combination of a high surface area together with flexible surface chemistry of cellulose nanomaterials makes it an interesting material for tuning their properties towards different needs, whether it is bio-compatibility, anti-microbial properties, hydrophobicity, conductivity, etc. Therefore, the functional properties of nanocellulose are widely spreading in a broad range of applications such as paper-based devices (Barhoum et al. 2017), composites (De France et al. 2020) and future emerging applications (Trache et al. 2020).

In this chapter, current research on the recovery of nanocellulose materials from algae is highlighted as an overview on the current state of the art and new developments. Although the algae-based cellulose is highly attractive due to its quality and ease in processing, it is still used in much smaller quantities compared to the lignocellulosic-based nanocellulose. This can be revealed from the relatively limited number of reference papers that were found and summarized herein. Probably the inferior insight into the structure of algae cellulose and the highly multidisciplinary nature of the topic at the cutting edge of biology, chemistry and materials engineering are at the base of this misconception. It is a hope that bundled expertise in various domains could lead to a future breakthrough in the development of more sustainable processes and highly technological products from algae-based nanocellulose. Besides experience and insight into the cellulose processing conditions, however, also the better fundamental insight into the cellulose organization within algae biomass needs to be addressed in order to further optimize processing conditions. From this point of view, it seems that the one-to-one transition from wood-based nanocellulose materials towards algae-based nanocellulose materials does not allow for the full exploitation of this alternative biomass, but the review of the entire extraction processing cycles should allow for discovery of additional benefits in making processes more sustainable. Therefore, this work aims at contributing to a better understanding of important parameters determining the processing efficiency and material properties, while providing an outlook for different application potentials of algae-based nanocellulose. It is indeed stressed that all nanocellulose materials, properties and applications described in this chapter are derived from algae biomass, while there is indeed significant overlap in the nanocellulose materials obtained from traditional sources. In supplement to the abundance of nanocellulose literature, a specific review on the algae-based nanocellulose materials has however not been found to date.

13.2 Structural Organization and Biosynthesis of Cellulose in Algae

13.2.1 Molecular Structure of Cellulose

Cellulose is a linear homopolymer of the glucose monomer in its β-D-glucopyranose form with thermodynamically preferred $^4C_1$ chair conformation, which is also referred to as an anhydroglucopyranose or anhydroglucose unit (AGU). The AGUs are linked by $\beta(1 \rightarrow 4)$ glycosidic bonds, that is, acetal functions between the equatorial OH group of C4 and the C1 carbon atom, resulting in a long β-1,4-glucan chain. In order to establish a stable acetal linkage near the anomeric carbon atom (labelled C1), every second AGU is rotated at 180° with respect to its adjacent neighbour in the plane (Fig. 13.1a) (Credou and Berthelot 2014). Thus, two adjacent AGUs (disaccharide) form the repetitive unit of cellulose and are termed as

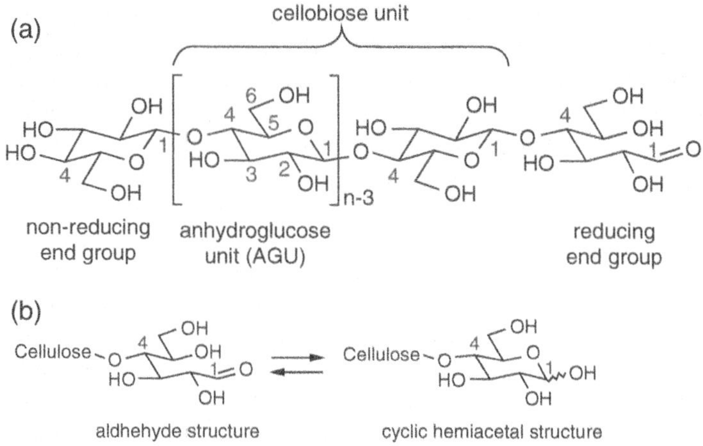

Fig. 13.1 Chemistry of cellulose, with (**a**) representation of a cellulose molecular structure showing the anhydroglucose unit (AGU), repetitive unit cellobiose (n = DP, degree of polymerization), non-reducing end and (**b**) reducing end equilibrium between two structures (Credou and Berthelot 2014)

so-called cellobiose units. Both end groups of the cellulose molecular chain are different: (i) the reducing end contains both an aldehyde structure and a cyclic hemiacetal structure (original C1–OH group) of the pyranose ring in equilibrium (Fig. 13.1b), while (ii) the non-reducing end has a closed ring of the original C4–OH group. During polymeric growth, the chain elongation reaction evidently takes place at the non-reducing end (Klemm et al. 2005). The molecular chain length denoted as the degree of polymerization (DP) is expressed as the number of AGUs in a polymer molecule. The DP value can reach up to 15000–20000 monomer units, depending on the source of the cellulose as well as extraction methods applied (O'Sullivan 1997; Moon et al. 2011).

Due to the presence of many hydroxy groups (three per AGU; C2, C3, and C6 atoms) in the cellulose chain, an extensive network of intra- and intermolecular hydrogen bonds builds up throughout the cellulose chain. The intramolecular hydrogen bonds occur as O3H–O5´ and O2H–O6´ in neighbouring AGUs (Fig. 13.2) (Credou and Berthelot 2014) and inhibit the free rotation of the monomers, creating a straight, rigid glucan chain (McNamara et al. 2015). The intermolecular hydrogen bonds and van der Waals forces are responsible for crystalline and other supramolecular (polymorphs and allomorphs) structures of cellulose (Gardner and Blackwell 1974).

Cellulose exists as a biopolymer in nature, forming a fibre of parallel-packed multiple individual molecular polymer chains. During biosynthesis, approximately 36 individual cellulose molecules are bundled together to form elementary fibrils (protofibrils), which are then packed into larger units called microfibrils. The width of microfibrils varies from approximately 25 to 30 nm in *Valonia* and other green algae to approximately 5 to10 nm in most plants (Ha et al. 1998; Somerville 2006).

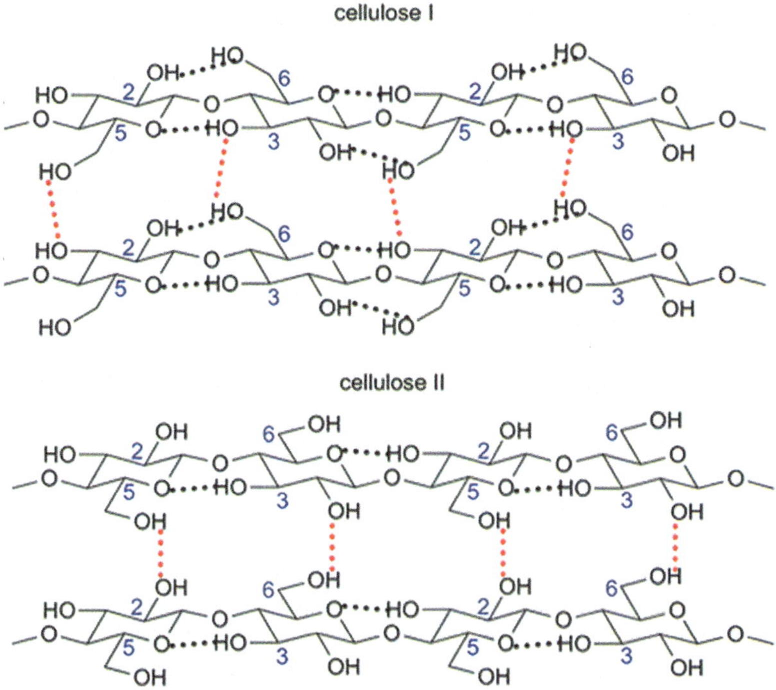

Fig. 13.2 Supramolecular distinction between cellulose I and cellulose II lies in inter- and intramolecular hydrogen bonds (Credou and Berthelot 2014)

The intra- and intermolecular hydrogen bonds (Fig. 13.2) and van der Waals forces are responsible for the aggregation/packing of these fibrils. The cellulose fibrils contain highly ordered (crystalline) and disordered (amorphous-like) regions. The ratio of crystalline to the amorphous areas depends on the source of the cellulose. Cellulose fibrils are compactly aligned together in the crystalline regions, stabilized by a vast and robust network of intra- and intermolecular hydrogen bonds. The pattern of hydrogen-bonding and molecular orientation in these crystallites can vary widely, resulting in cellulose polymorphs, depending on the source material, extraction or treatment methods (Habibi et al. 2010).

The cellulose structure can exist in several different polymorphs, referred to as Iα, Iβ, II, III_I, III_{II}, IV_I, IV_{II}. Cellulose I (with both Iα and Iβ allomorphs) with the cellulose chains in a parallel configuration is found in nature. The other allomorphs can be obtained by various chemicals or heat treatments (Jarvis 2003; Moon et al. 2011). The Iα allomorph (triclinic one-chain unit cell) is the dominant polymorph for most algae and bacteria, while the Iβ allomorph (monoclinic two-chain unit cell) is prevalent in plants and tunicates. In algae, both cellulose I allomorphs can be found: for example, the *Valonia* sp. green algae have the cellulose Iα allomorph primarily, while *Micrasterias denticulate* green algae has mainly the cellulose Iβ.

13.2.2 Biosynthesis of Cellulose in Algae

The biosynthesis of cellulose in algae species is catalysed by a family of integral membrane glycosyltransferases called cellulose synthase (CeS) complexes. Macroalgae and all eukaryotic organisms share the cellulose synthase A catalytic subunit. During synthesis, the CeS remains bound to the non-reducing end of the chain, elongating it by bonding UDP-activated glucose from the cytosol. This is done with the inversion and the anomeric configuration of the UDP-glucose from α to β. The glucose then can bond on the cellulose chain, generating a UDP as a leaving group (McNamara et al. 2015).

When CeS is visualized in freeze-fracture electron microscopy, the protein complex or the edge of the cellulose microfibril forms distinctive arrangements called 'terminal complexes'(TCs) (Brown and Montezinos 1976; Roberts and Roberts 2007). With the help of freeze-fracture of the plasma membrane, the terminal complex (TC) can be observed as a transmembrane imprint (diameter of ~30 nm) of the cellulose synthase (Giddings et al. 1980; Kimura et al. 1999; Somerville 2006). Bacteria own the most archaic form of TC, a singular solitary particle. On the other end, there exists a six-particle TC situated in a circle called 'rosettes,' which is the only TC type found in the secondary wall of higher plants. Algae are the intermediate link; hence, a wide variety of TC types can be found, with a different number of subunits arranged in distinctive geometrical shapes. The different types of TCs found in algae have been elucidated and reviewed in large detail before (Tsekos 1999).

First, the TC is responsible for the biosynthesis of the glucan chain, which has at least one catalytic site in each subunit for the polymerization of the elemental cellulose chain. Second, it manages the chain assembly into a microfibril, followed by the lateral connection of the microfibrils to bundles, creating a macrofibril. Finally, the TC movement and orientation directs the deposition pattern of the cellulose microfibrils. The TC translocates the microfibrils across the cell membrane via its membrane-embedded domain channels (McNamara et al. 2015); however, the precise mechanism has not been elucidated yet. Consequently, the TC geometry influences the cellulose architecture on many levels (Roberts and Roberts 2007).

Among all groups, the highest TC shape diversity is observed in *Chlorophyta* green algae. There is a single row of linear TCs (*Chladophorophyceae*, *Florideophyceae* and *Phaeophyceae*), linear TCs of three rows (*Chladophorophyceae* and *Chlorophyceae*, e.g. *Valonia* sp.), octagonal rosettes (*Coleochaete scutata*), rosettes of hexagonal arrays, rosettes in a single row and solitary rosettes (*Zygnematophyceae* and *Charophyceae*) (Fig. 13.3). In *Rhodophyta* red algae, single or multiple rows (2–4) of linear TCs are observed, whereas only linear TCs of a single row are reported in *Phaeophyceae* red algae (*Hetero-kontophyta*). Additionally, the blond algae *Vaucheria hamata* (*Xanthophyceae* of *Heterokontophyta*) has a linear TC geometry of multiple diagonal rows (Koyama et al. 1997). If a single rosette or a row of particles (e.g. four rows for red algae) is considered as a unit, the number of units in their arrangement determines the dimension of the microfibril bundles (the thickness of the macrofibril). Similarly, the

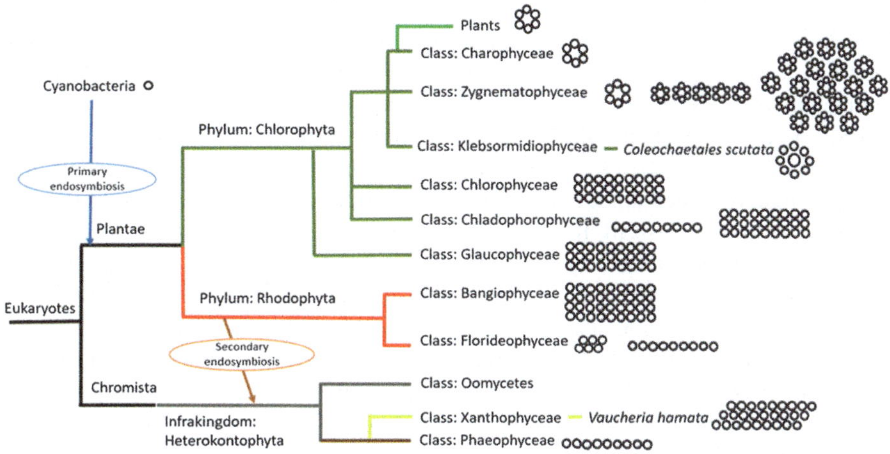

Fig. 13.3 Phylogeny of algae showing the evolution of TCs (Tsekos 1999). Not all taxa or terminal complex types are shown. Branch lengths are arbitrary (Brown and Saxena 2007; Popper et al. 2011)

subunits' proximity impacts the density of the microfibril bundle, and the catalytic sites specify the number of glucan chains in each nanofibril. The proximity of the chains is also influenced by excessive hydrogen bonding and vice versa, a characteristic of the crystalline ordered domain. Meanwhile, for a linear TC with multiple rows of particles, cellulose with a high degree of crystallinity is expected. The rosettes with a space in the centre will provide room for flexibility (Tsekos 1999).

The TC geometry correlates to the types of the cellulose-crystal allomorph. It is stated that a linear-type TC synthesizes ribbon-like celluloe Iα allomorph crystalline-rich nanofibrils. In contrast, a rosette-type TC produces flexible nanofibrils rich in the cellulose Iβ allomorph (Tsekos 1999). This corresponds with theoretical speculation that the Iα-rich microfibrils have a rigid structure with a high degree of crystallinity, while the Iβ-rich microfibrils are characterized by larger disordered domains. Nonetheless, some algae are an exception, for example, *Florideophyceae* red algae with a linear TC of six subunits (two diagonal rows of three particles, closely arranged) produces flexible elementary fibrils. This anomaly was attributed to a narrow cross-section of the macrofibril resulting in more interactions with the amorphous components and allowing twisting of the macrofibril. Therefore, generally, algae with linear TC, brown algae, most red algae and the green algae from the *Chlorophyceae*, *Chladophorophyceae* and *Glaucophyceae* families are considered to produce more rigid nanofibrils rich in metastable crystalline cellulose Iα allomorph (Tsekos 1999; Mihranyan 2011).

The TC geometry also influences the shape, orientation and deposition of cellulose fibril. If there is a single row of TC subunits, then the microfibril will have a flat ribbon-like shape. If there are multiple rows of subunits, the microfibrils will have a parallelogram or orthogonal cross-section. However, the Floridean class of red algae remains an exception due to the unique 2×3 subunit formation, which mainly produces a cylindrical-type microfibril (4–5 nm in diameter) (Martone et al. 2019). The 2 × 3 subunit formation in a diagonal assemblage creates a parallelogram cross-section closer to a rhombus – square shape. Moreover, the microfibrils will have a uniplanar orientation parallel to the surface plasma membrane. Furthermore, rosette-type TC produces thin 2–3 nm flexible cylindrical microfibrils (Tsekos 1999; Martone et al. 2019). When the rosettes are arranged in hexagonal arrays, like in *Zygnemalean* green algae, they acquire a different deposition pattern. In the first level of the algae cell wall, the microfibrils have a random orientation. In contrast, the microfibrils are oriented parallel in the second layer of the cell wall, creating bands (macrofibrils) that traverse each other and are parallel to their wall/plasma membrane surface. It has also been observed in the *Zygnematophyceae* class that solitary rosettes deposit cellulose for primary wall formation, while hexagonal arrays or linear rows of rosettes are aligned for the secondary wall formation. In plants, only solitary rosettes exist for the formation of the secondary wall. However, the primary wall in algae is suggested not to contain the typical crystalline allomorphs, which might lead back to a type of cellulose synthase with a different TC geometry (Tsekos 1999; Popper et al. 2014). In summary, algal TC can be grouped into five general categories:

- Singular rosettes: cellulose Iβ dominant, flexible cylindrical microfibril without orientation, found in evolved Charophycean green algae or the primary wall of Zygnematophycean green algae.
- The array of rosettes: usually oriented bands of microfibrils (with a larger dimension when aggregating with neighbouring microfibrils) parallel to the plasma membrane surface that intersect with each other, found in the Zygnematophycean green algae secondary cell wall.
- Linear TC in a 2 × 3 formation: flexible rosette-mimicking microfibril found in Floridean red algae.
- Linear TC with multiple rows: often oriented cellulose Iα dominant microfibril of high crystallinity, wide and broad cross-section according to the number of subunits and their proximity, mainly found in green and red algae, usually at the secondary cell wall.
- Linear TC in one row: thin ribbon-like microfibril, found in brown algae, some red algae and possibly in some green algae primary cell wall.

13.2.3 Organization of Cellulose in the Algae Cell Wall

In the previous section of cellulose biosynthesis, it was discussed how different types of TCs are accountable for the multiplicity of structures and orientations of cellulose microfibrils in various classes of algae. This section will deal more in detail with how the cellulose microfibrils are organized among other structural components in the algae cell wall. Due to the extensive diversity of algal groups, cell wall description here is mainly focused on the three major groups: *Chlorophyta* (green algae), *Rhodophyta* (red algae) and *Phaeophyceae* (brown algae). Also, most of the macroalgae species studied for nanocellulose extraction fit in one of these three main groups; thus, this consideration should rationalize the further discussion and the proposed concern of this chapter.

Chlorophyta (green algae) and *Rhodophyta* (red algae) belong to the Plant kingdom and hence similarities in the cell wall structure can be observed, especially with Charophycean green algae (evolved green algae) (Mikkelsen et al. 2014). On the other hand, *Phaeophyceae* (brown algae) have a different wall structure and composition, as they stem from the evolutionarily distant kingdom of *Chromista* (Cock et al. 2010; Popper et al. 2011). The algae cell wall is likely constituted by a two-phase system, that is, a crystalline phase (fibrous skeleton) embedded into a more amorphous phase (the matrix). A similar description specifically applies for *Charophyceae* algae (closer to plants) (Fig. 13.4a), whereas *Chlorophyceae* algae have a double layer of the fibrous rich component which encloses the amorphous matrix among them (Fig. 13.4b) (Youssef et al. 2015). Fundamentally, the algae cell wall differs from the land plant cell wall by the abundance of matrix components (e.g. lignin) with respect to the skeletal elements and by the presence of different polysaccharides (e.g. hemicelluloses) in plants (Fig. 13.5) (Kloareg and Quatrano 1988; Charrier et al. 2019). The fibrous skeleton adds mechanical strength and stiffness to the cell wall. Cellulose is the major fibrous component of brown algae and most green and red algae, while the minority is composed of xylans, mannans or even glycoproteins. Most green algae have a cellulose derivative (mercerized-like) with a low degree of crystallinity and the randomly oriented cellulose chains, while the rest are characterized by high crystallinity cellulose (Lee 2018). In many algae species, the cellulose fibre content was determined at 10% or higher (Knoshaug et al. 2013).

Depending on the season and growth period, the young green and red algae cells have a primary soft cell wall, which allows the cells to grow in size. When the cell matures, a more rigid secondary layer is gradually formed under the primary cell wall. Usually, the primary cell wall comprises microfibril bundles that are randomly deposited, whereas the secondary wall has more structure with macrofibrils with parallel orientation (Fig. 13.4d). In between walls exist the middle lamella that is enriched in pectin and connects the cells together. The red algae lamellae are lying parallel to the cell surface, with microfibrils randomly distributed in each lamella. However, there are species where the microfibrils tend to be parallel (Tsekos 1999; Nash et al. 2019). Additionally, red algae have much richer amorphous components (agar, carrageenan, etc.) than green algae and are thus primarily used to extract these compounds for various applications.

Brown algae are phylogenetically unrelated to green or red algae lineages; thus, their cell walls evolved independently with distinct composition and characteristics (Cock et al. 2010; Raimundo et al. 2017). Although the components and organization of the brown algae cell wall remain intricate, they have well-labelled representation (Deniaud-Bouët et al. 2014; Deniaud-Bouët et al. 2017; Charrier et al. 2019). On the other hand, the view of the brown algae cell wall may deviate from the typical primary-secondary layer organization and consisted of an inner layer and an

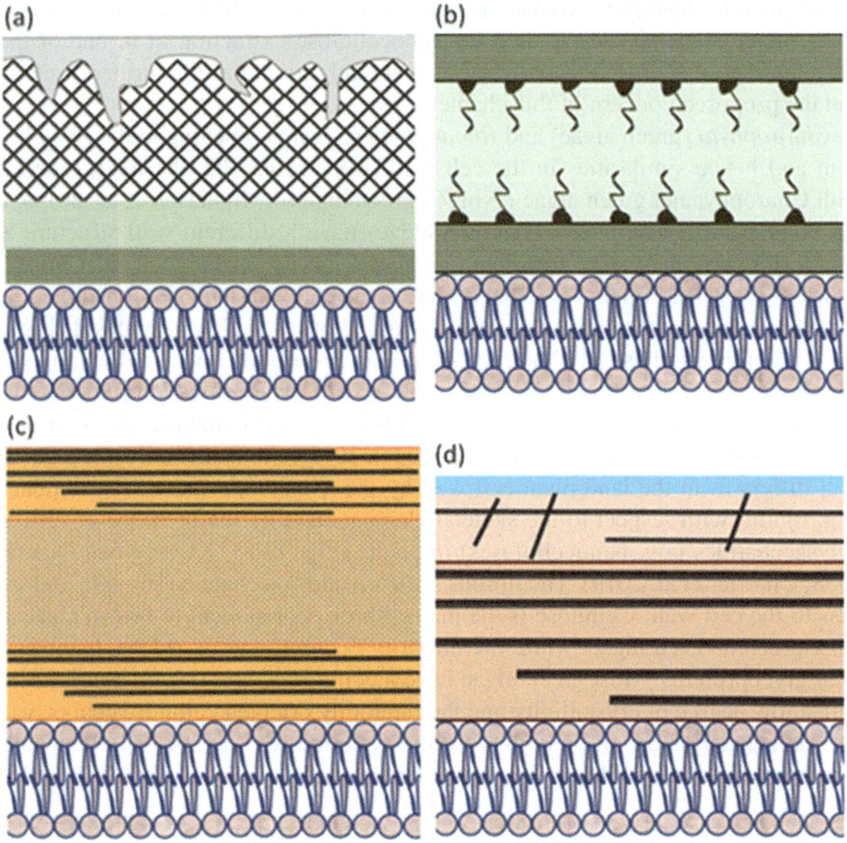

Fig. 13.4 Schematic representation of different algal cell wall structures: (**a**) Charophyceae green algae: on top of the bilayer membrane, there is an inner layer rich in fibrillous cellulose material, surrounded by a middle amorphous layer (pectin), which is covered by an outer lattice (pectin) and at the outermost part exists extracellular mucilage or substances (Youssef et al. 2015), (**b**) Chlorophyceae green algae: a double layer of fibrillous material of cellulose and hemicellulose encloses the amorphous matrix layer (sulphated polysaccharides), while glycoprotein embedded in the skeletal inner and outer layers extend towards the middle of the matrix (Youssef et al. 2015), (**c**) Phaeophyceae cell wall: inner and outer layers of a fibrillar framework of criss-crossing cellulose microfibrils parallel to the cell membrane, among them the interfibrillar matrix (alginates and fucans) (Youssef et al. 2015), (**d**) Rhodophyta cell wall: consisting from top to bottom from the middle lamella, the primary cell wall with random cellulose microfibrils and the secondary cell wall with more oriented cellulose microfibrils embedded in the amorphous matric (galactans) (Nash et al. 2019)

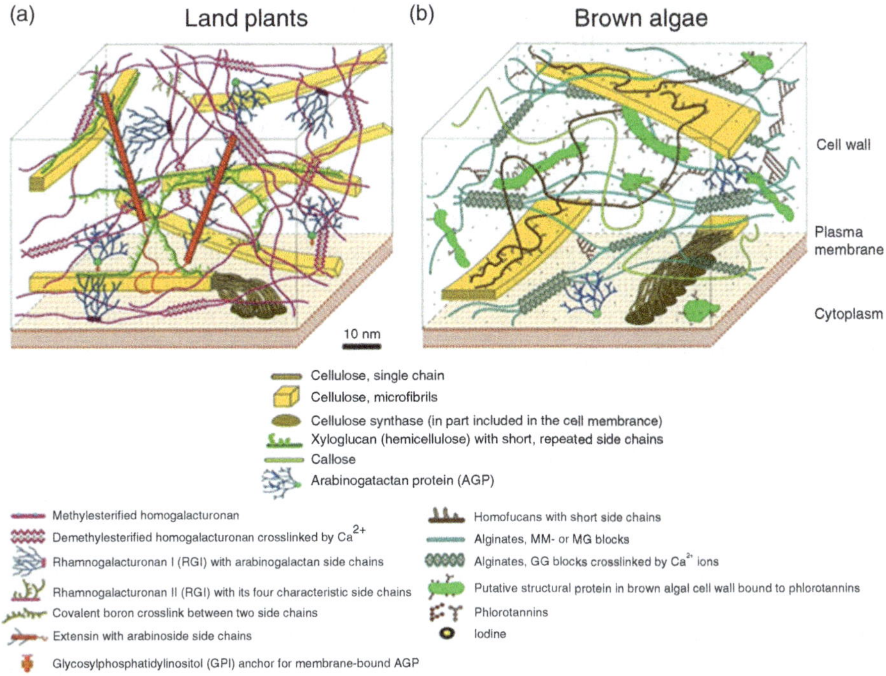

Fig. 13.5 Primary cell wall of brown algae in comparison with the wall of land plants: (**a**) land plants have a square-like cross section of random, flexible microfibril bundles, (**b**) the primary brown algae cell wall has randomly deposited microfibrils and microfibril bundles of orthogonal cross section embedded in the amorphous matrix (Charrier et al. 2019)

outer layer of random cellulose microfibrils parallel to the cell surface with a thick layer of the amorphous matrix material in between (Fig. 13.4c) (Youssef et al. 2015).The fibrous component containing the ribbon-shaped cellulose microfibrils (1–8% dry weight) are sparsely distributed in the amorphous phase with abundant alginates (~ 40%) and fucose-containing sulphated polysaccharides (FCSPs; including sulfaphed fucans, ~ 40%). The cellulose microfibrils are believed to be cross-linked by the FCSPs and proteins. Additional components such as β-1,3-glucans (callose), mixed-linkage glucans, arabinogalactan proteins (AGPs) and halide compounds are identified in the brown algae cell wall (Raimundo et al. 2017; Salmeán et al. 2017; Charrier et al. 2019). The gel-forming two main polysaccharides, alginates and FCPS (commonly referred to as fucoidans), are commercially extracted from brown algae.

The algae cell wall structures discussed here are primarily directed towards the organization of cellulose microfibrils, which could enhance the treatment methods applied in nanocellulose extraction from macroalgae, as outlined in next paragraphs. However, one should not neglect the fact that the cell wall composition can vary depending upon species, growth phase, season, habitat and other growth parameters.

13.3 Processing of Nanocellulose Structures from Algae Cellulose

After extraction of cellulose fibres from algae biomass, the hierarchical fibre structure consists of a bottom-up organization of the cellulose molecules into elementary fibrils that are intertwined into microfibrillar bundles. The different morphologies of cellulose microfibrils or microfibrillar cellulose (MFC), cellulose nanofibrils (CNFs) and cellulose nanocrystals (CNCs) can be recovered through the top-down processing routes combining chemical and mechanical treatments of the algae-based cellulose. An overview of the isolation of different forms of nanocellulose from algae is given below.

13.3.1 Microfibrillar Cellulose (MFC)

The cellulose microfibrils or microfibrillated cellulose (MFC) consists of a dense and intertwingled network structure of thin cellulose microfibrils. The cellulose microfibrils contain an aggregation of several elementary fibrils with both amorphous and crystalline regions, resulting in fibre dimensions with length of a few hundred micrometres to maximum 1 mm and diameter of 100 nm to 1 μm. It is traditionally obtained by mechanical grinding or cryo-crushing of wood fibres or pulp. However, the high energy consumption in the production of MFC from lignocellulosic materials implies an important impedance for the further industrial development of this technology. As a reference material, the MFC from softwood pulp needed additional processing in a high pressure microfluidizer at fibre consistencies of only 2 wt.-% and pressures up to 1700 bar for only one processing step (Taheri and Samyn 2016). Therefore, the processing should include several chemical or enzymatic pre-treatments to reduce the energy requirements, but they complicate the production and yield different properties to the MFC fibres. Interestingly, the MFC from algae can be extracted under purely chemical conditions and result in a comparable microfibrillar morphology without the need of severe mechanical energy input.

The inherent filamentous structure of cellulose recovered from marine green algae, such as, for example, *Cladophora*, *Chaetomorpha*, *Microdyction* and *Rhizoclonium*, provides an excellent alternative source for production of MFC. The natural arrangement of elementary fibrils in the cell wall of green algae families are characterized by a laminar outer layer with an extremely high degree of orientation and relatively thick fibril diameter in the range of 10–30 nm. The individual fibre radius obtained from *Cladophora* varied depending on the origin and processing conditions of 38 nm (Ek et al. 1998) to individual fibre width of 25–30 nm (Mihranyan et al. 2004). Comparably, the elementary fibrils in wood of lignocellulosic materials are smaller with diameter in the range of 5–10 nm, embedded in a

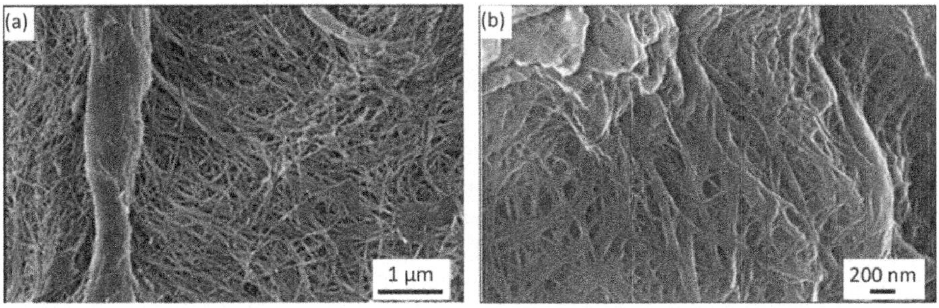

Fig. 13.6 Extraction of MFC from *Cladophora* green algae with intrinsic microfibrillar nature of its cellulose structure, after pure chemical treatment: (**a**) alkaline treatment (NaOH) and single bleaching (Na_2ClO_2), (**b**) alkaline treatment (NaOH) and two bleaching steps (Na_2ClO_2 + HCl) (Mihranyan et al. 2004)

significant amount of lignin. Therefore, green algae mostly allow for the production of MFC through a simple alkaline extraction process and bleaching, in order to expose the fibrillar nature of extracted *Cladophora* cellulose with distinct microfibrils as bundles of elementary filaments after purely chemical treatment (Fig. 13.6). As a misleading name, however, the *Cladophora* cellulose is often referred to as 'cellulose powder' or nanofibrillated cellulose, while its morphology rather resembles the produced MFC from wood pulp.

The primary treatment of green algae with methanol for several days allows for the removal of a lipid fraction, while other mineral impurities are removed during the subsequent treatment with NaOH (Siddhanta et al. 2009). The mineral salt crystals are often observed as deposits at the algae surfaces, while a smooth aspect of the fibre surface is characteristic for the presence of oils or waxes, which have first to be removed before further processing. The subsequent alkaline treatment additionally lowers the hemicellulose content, as it induces the hydrolysis of the ester bonds between lignin and carboxyl groups in the hemicellulose. In contrast, the ether bonding between the lignin and carbohydrates is not broken by the alkaline processing and the subsequent bleaching process under mild acid conditions in hydrochloric acid or acetic acid is required for the lignin removal. As an advantage, the concentration of pigments and non-cellulose components in certain algae is often lower than in lignocellulosic materials and allows for mild bleaching conditions. As a result, after the purification and bleaching of *Cladophora* green algae in sodium hypochlorite and acetic buffer, followed by several washing steps and mild acid hydrolysis, the filamentous MFC powders were produced with variable degree of crystallinity (Mihranyan et al. 2004). The mild hydrolysis under 5% HCl under boiling and subsequent freeze-drying allowed to directly individualize the fibrillar cellulose structure from the algae cell wall without the need for additional mechanical fibrillation. As a result, the extracted *Cladophora* cellulose contains a highly porous network structure of numerous intertwingled microfibrils. The direct extraction of MFC from cellulose powder from *Cladophora* green algae was further optimized by adapting the bleaching into a four-step chlorine-free process, including an alternating alkaline extraction with hydrogen peroxide (2% H_2O_2 + NaOH) and mild acidic treatment (HCl) at slightly elevated temperature of 110 °C (Liu et al. 2017). In particular, the step-wise purification allows for significant reduction in impurities and heavy metal content. As such, the amounts of leachables, extractables and microbial contaminants were significantly reduced and extraction yield of the bleached *Cladophora* cellulose increased. Indeed, the natural presence of heavy metal contaminants such as Cd, Pb, Hg and As were detected as toxic substances in the aquatic body of algae (Messyasz et al. 2015), and need severe purification treatment to reduce below the critical concentration. The *Chaetomorpha* green algae were used as an alternative source for extraction of cellulose fibrils through bleaching with hydrogen peroxide, followed by alkaline treatment in NaOH and mild acidic heating (HCl). The final product was part of a progressive step-wise extraction yielding the valorization of a lipid fraction, water-soluble sulphated polysaccharides and cellulose (Freile-Pelegrín et al. 2020). Interestingly, the final cellulose yield after either direct or sequential extraction was about 30% to 35% and was therefore higher than reported for other types of green algae typically ranging between 1.5% and 10%. Contrarily, the extraction of cellulose from brown algae did not provide the required fibrillated structure but resulted in irregular powdery particles owing to the different organization of the elementary cellulose fibrils.

The surface modification of MFC from *Cladophora* green algae was done by a traditional TEMPO oxidation reaction in order to introduce carboxyl groups with anionic properties or modification with hydroxypropyl trimethylammonium in order to provide cationic surface groups (Hua et al. 2015). As a result, the morphology of the fibres and structure of the fibrillar network significantly changed with distinct properties depending on the surface modification: the anionic species introduce an aggregated and co-axial alignment of somewhat thicker fibres (80–100 nm) and the cationic groups rather provide a random orientation of somewhat thinner fibres (20–30 nm) into a network structure (Hua et al. 2014). As such, the films produced by vacuum filtration produced an oriented fibre network with smaller pores and higher specific surface for the anionic compared to cationic MFC (Fig. 13.7). Besides the changes in morphology and surface charge, the typical large specific surface area did not change after TEMPO modification of *Cladophora* cellulose. After spray drying of modified MFC from algae, the fibres could interestingly be re-dispersed more easily than the unmodified fibres through ultrasonication treatment in aqueous media. The final structures of processed films were also partly due to the processing of the modified MFC into the formation of a sheet structure by vacuum filtration, using a high-energy ultrasonic treatment. Alternatively, the carboxylation of MFC extracted from *Cladophora* by TEMPO oxidation created reactive groups for subsequent deposition of a metal-organic framework (Fig. 13.8). The functionalization of the MFC through Al^{3+} deposits was done by an ion exchange mechanism forming intermediate complexes of COO^-/Al^{3+} (Zhou et al. 2019). The interfacial synthesis could be performed according to the in-situ deposition and growth onto carboxylated MFC starting from an aluminium-nitrate solution added into the fibrous suspension. As such, an aluminium-containing nanolayer was deposited through interfacial synthesis and adhered to the cellulose fibrils through

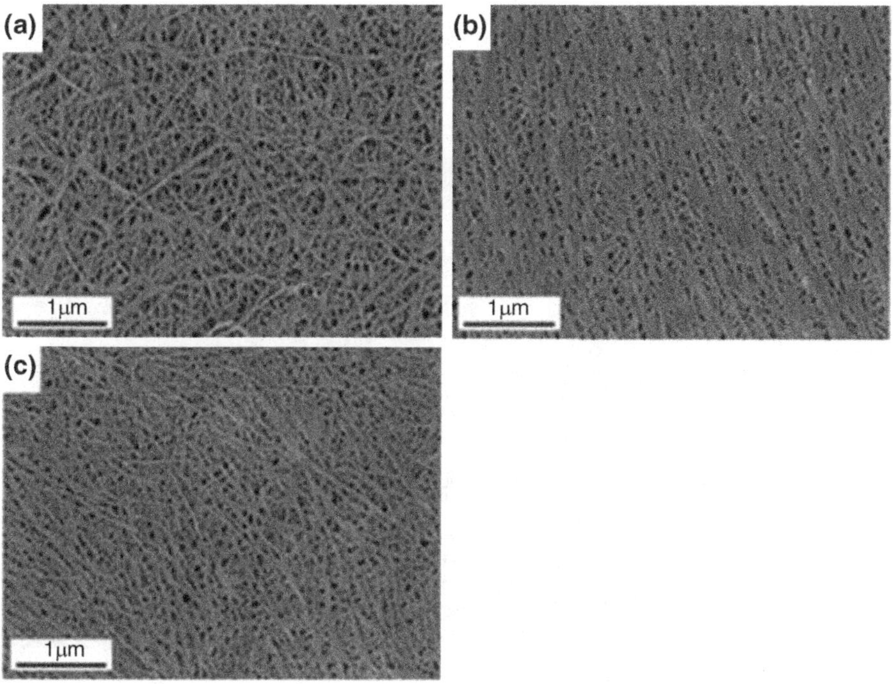

Fig. 13.7 Formation of porous fibrous films by organization of MFC from green algae in different conditions: (**a**) non-modified, (**b**) anionic MFC and (**c**) cationic MFC (Hua et al. 2014)

coordination bonding in presence of a linker and surfactant. After freeze-drying into a porous solid structure, the surface-modified fibres evidently allow for interfibrillar cross-linking at their cross-over points and form a more coherent interconnected fibrillar network.

13.3.2 Cellulose Nanofibrils (CNFs)

The individual cellulose nanofibrils (CNFs) – also referred to as nanofibrillated cellulose (NFC) – are smaller in diameter and length than the microfibrillar structure, forming a denser network. Hence, it requires the destruction of the microfibrils through extremely high mechanical shear forces to individualize the nanofibrils from the hierarchical fibre structure. The intensive mechanical processing of CNF traditionally happens in an ultra-grinder or supercollider, homogenizer or high-pressure microfluidizer. The high shear in a fibre suspension creates an impact of the colliding streams against the channel walls of the processing chamber, resulting in the de-agglomeration, fibrillation and dispersion of the fibres into the aqueous processing phase. The fibrillation process typically takes place within an interaction chamber at average pressures of 300–1300 bar and multiple cycles from 15 to 30.

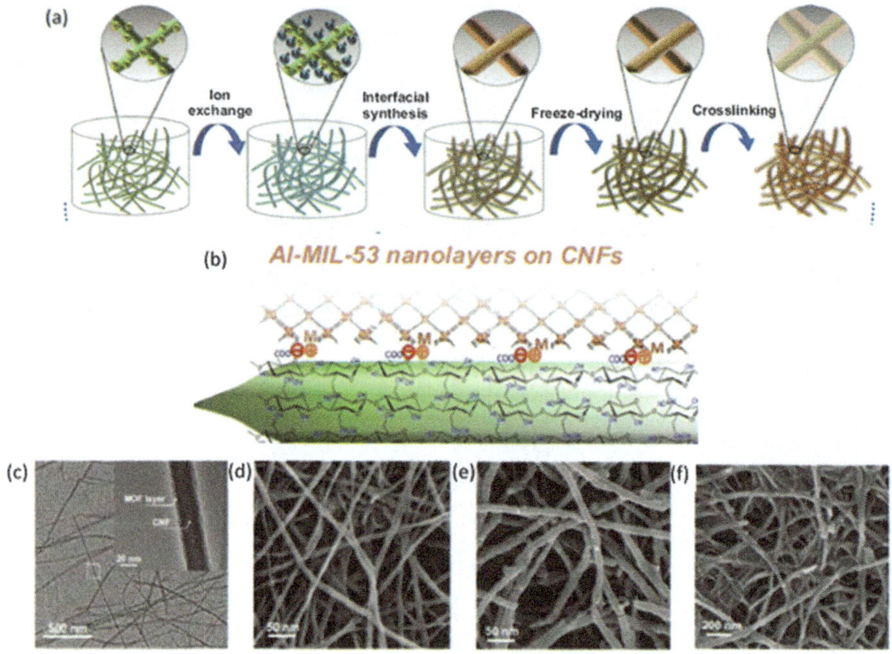

Fig. 13.8 Surface functionalization of the MFC network extracted from green algae, through interfacial synthesis and deposition of Al^{3+}: (**a**) schematic production procedure, (**b**) schematic surfaced modification with coating and cross-linking of a continuous Al-containing monolayer, and morphology of extracted and modified fibres, including (**c**) TEM image of modified fibres, (**d**) SEM of pure fibres, (**e**) SEM of modified fibres, (**f**) SEM of fibrillar network and fibre interactions (Zhou et al. 2019)

The interaction chamber within a homogenizer can essentially be considered as a fixed continuous micro-reactor geometry that provides a turbulent mixing to create a uniform pressure profile to obtain accurate and repeatable size distributions. Inside these chambers the dilute slurries of eventually pre-treated fibres are exposed to extremely high flow velocities and shear rates to yield individualized nanofibrils. The mechanical pre-treatments to decrease the size of the fibres may include milling and homogenization (to increase swelling), as well as refining (to raise the binding potential of cellulose fibres) in association with enzymatic conditions for wood pulp. Alternatively, the CNF is in majority produced from different sources of green algae (*Cladophora*, *Ulva*, *Pithora*, *Aegagropila*) and brown algae (*Posidonia oceanica*).

The homogenization of cellulose from *Cladophora* green algae could be done at a solid content of 2 wt.-%, after soaking in NaOH and chlorite bleaching. A comparative study of bleached softwood pulp fibres, bacterial cellulose under similar ultratourax homogenizing conditions was done (Fig. 13.9) (Štefelová et al. 2017): interestingly, the algae-based CNF presented straight and more uniform nanofibrils, while the others rather have short and partially bend nanofibrillated structures likely

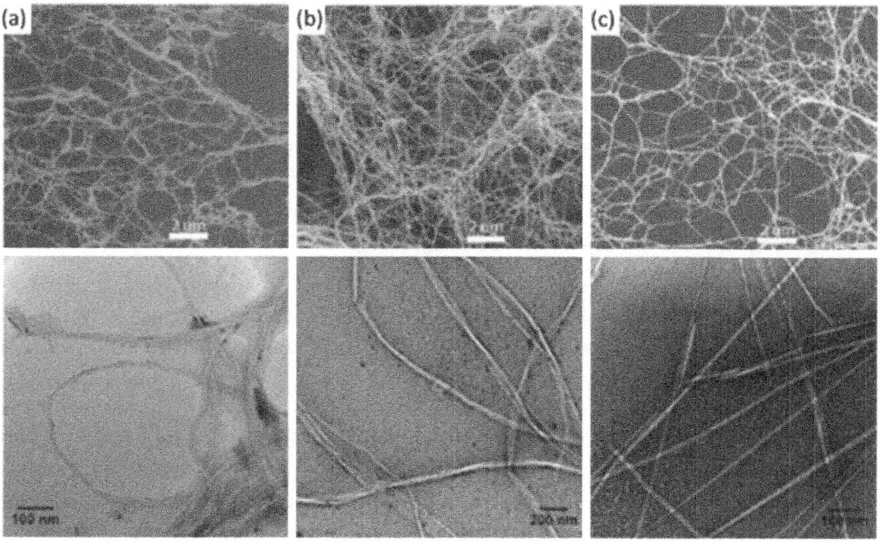

Fig. 13.9 Morphology of CNF obtained after bleaching and homogenization of different sources, including (**a**) softwood, (**b**) bacterial cellulose and (**c**) *Cladophora* cellulose (Štefelová et al. 2017)

consisting of multiple fibrils. The diameter size of CNF from algae ranged between 5 and 20 nm, while the diameter of wood-based CNF is generally larger in the range of 10–110 nm. Owing to the differences in fibre morphology, the highly porous structures and identification of single nanofibrils in the fibre mats from algae-based CNF can be obtained following conventional freeze-drying procedures, as the high stiffness of the algae fibrils prevents regular problems attributed to agglomeration. In contrast, the wood-based nanofibrils tend to easily agglomerate due to their higher mobility and packing density.

The pre-treatment of algae cellulose by TEMPO-oxidation was performed to produce CNF from green algae by homogenization, however, the high selectivity of the modification towards the local cellulose structure should be taken into account (Ono et al. 2019). For *Cladophora* cellulose, the carboxylated groups after oxidation were mainly present in the low-molecular mass fraction, while they were more almost homogeneously distributed in wood-based and cotton-based CNF. These results are in agreement with the core-clad chain packaging model of individual microfibrils from algae, where the high-molecular mass cellulose molecules with almost no carboxylated functionalities are situated in the inner core and the low-molecular species are located at the outer surface of the fibrils (Fig. 13.10). The latter are likely caused by partial depolymerization during oxidation treatment of the cellulose molecules, and are observed at the outer surface by peeling. Consequently, other pre-treatments with a likely more homogeneous modification of the fibre structure were applied. The enzymatic treatment of cellulose from *Cladophora* algae in combination with microfluidization was done with a given charge of cellulase per glucan unit in sodium acetate buffer (Xiang et al. 2016).

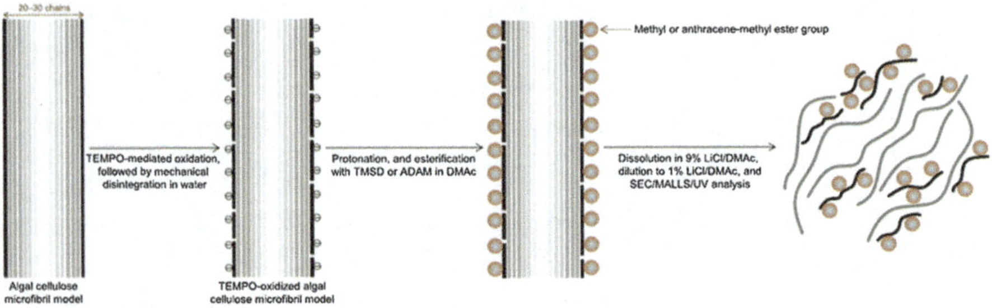

Fig. 13.10 Effect of pre-treatment of *Cladophora* microfibrils by TEMPO-oxidation, resulting in the modification of low-molecular-weight fractions at the surface and intact fibrils in the core (Ono et al. 2019)

However, the enzymatic efficiency highly depends on the accessibility and interaction of the cellulase with the cellulose structure. As a result of enzymatic hydrolysis, higher solid yields are obtained from algae cellulose compared to wood pulp because of the high crystallinity and consequently low reactivity of the algae cellulose. As a function of treatment times up to 50 h, there is only a small decrease in cellulose yield from 100% to 85% for algae cellulose and from 100% to 60% for wood cellulose. Indeed, the enzyme directly attacks the crystalline domains of algae cellulose at a slow rate, as compared to the interaction with the more amorphous domains in wood cellulose. In parallel, the differences in diameter distribution between algae and wood-based nanofibrils remain, existing after enzymatic hydrolysis at around 10–40 nm for algae-based CNF and 40–80 nm for eucalyptus CNF. Therefore, it was concluded that the enzymatic hydrolysis did not significantly affect the diameter distribution of algae-based CNF, but the intrinsically higher diameter of CNF from algae was favourable in reducing the agglomeration. Instead, the pure microfluidization of *Cladophora* cellulose without pre-treatment reactions is able to produce the CNF at the intrinsic microfibrillar level.

The alternative cellulose sources from *Ulva prolifera* or *Ulva lactuca* green algae were employed to less extend for production of CNF by grinding (Zhao et al. 2014) or mechanical disintegration through a microfluidization process (Wahlström et al. 2020). As an advantage, the fibrillation of CNF (in combination with CNC production) could be incorporated into a full extraction protocol without the need of intermediate drying and re-dispersion of the extracted cellulose or the need of any chemical or enzymatic pre-treatment (Fig. 13.11). In subsequent steps, the extraction was done under milder conditions with ethanol (for removal of pigments and fatty acids) in combination with hydrogen peroxide (for removal of residual pigments and impurities), offering a more sustainable approach as compared with the traditional hypochlorite and methanol. The homogenization process in a microfluidizer could remove all inorganic impurities from the extracted cellulose, but a remaining fraction of xylose monosaccharides (e.g. xyloglucan) was detected in the extracted cellulose and CNF. Therefore, the xylose-glucose polysaccharide was observed as a residual fraction with an aggregated sheet-like morphology that was

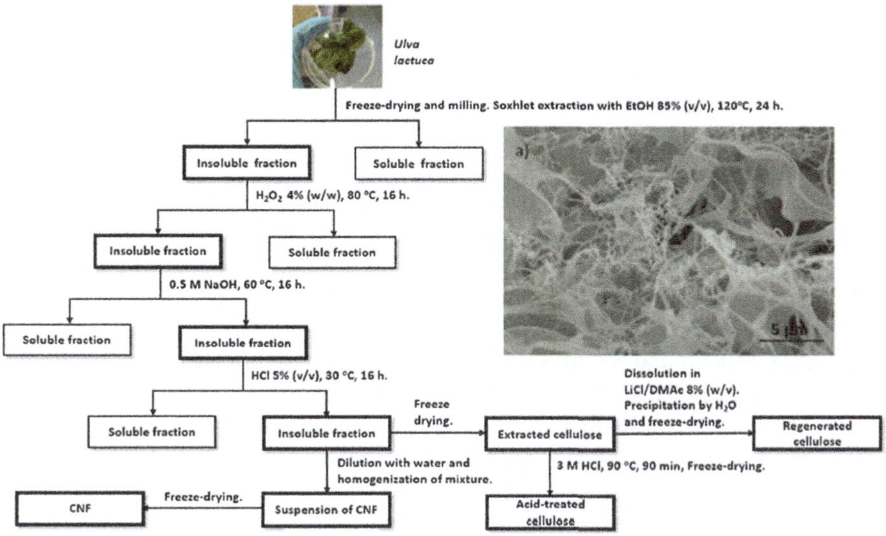

Fig. 13.11 Full biorefinery extraction concept for the extraction of cellulose and isolation of CNF and CNC from green algae, using mild conditions (inset shows the freeze-dried CNF) (Wahlström et al. 2020)

not completely fibrillated. Due to the presence of long side branches within the molecular structure of hemicelluloses, they are not able to completely fibrillate the cellulose and form strict nanofibrillar structures. Alternatively, the post-treatment with severe acidic hydrolysis in HCl or dissolution and regeneration in LiCl/DMAc could partially remove the xylose fraction from CNF.

Depending on the local availability of very specific algae species, the *Pithophora* green algae served as an alternative source for the production of CNF by subsequent alkaline extraction (NaOH), acid treatment (HCl), alkaline bleaching (H_2O_2) as pre-treatment steps for cellulose recovery. The resulting cellulose was further passed through a microfluidizer as a suspension with only 0.2 wt.-% solid content and three steps (Gustafsson et al. 2019). While the number of processing steps to obtain CNF is significantly lower compared to reference wood pulp, the dilute suspensions are still problematic. In parallel, a better homogeneity in fibre quality was observed after a low number of processing steps, due to the high purity and lack of lignin in the green algae. The other types of green algae species, such as *Aegagropila linnaei*, were explicitly considered because of the natural presence of proteins that may be beneficial in the development of mechanical properties of CNF in composite materials. Therefore, the CNF obtained by microfluidization was compared after applying different pre-treatment steps to the algae biomass, which include either a maleic acid hydrolysis, or sodium hypochlorite oxidation (Guo et al. 2017). The introduction of maleic acid allowed for a facile destruction of the algae cell wall towards more extensive fibrillation. While the maleic acid was more selective for removal of the hemicellulose, the acid treatment only induced a very minor reduction in protein concentration compared to the oxidation reactions. However, denaturation and hydrolysis of the proteins may happen through maleic acid treatment and a small reduction in molecular weight due to depolymerization of the cellulose was reported. The high degree of carboxylation of CNF in the presence of maleic acid is favourable for the increase in suspension stability and reactivity of the fibres for further surface modification.

Among the types of brown algae, both the balls and leaves from *Posidonia oceanica* were used for the grinding or microfluidization after different pre-treatment steps. First, the delignification and bleaching of brown algae happened under alkali (NaOH) and sodium chlorite conditions, followed by the TEMPO-mediated oxidation (Bettaieb et al. 2015d). However, extraction of cellulose from brown algae is often more difficult and requires additional processing steps. Therefore, the brown algae are alternatively treated by multi-step dewaxing in a toluene/ethanol mixture during Soxhlet extraction, followed by two bleaching steps for cellulose extraction. The relatively high portion of lignin and formation of the strong liquor fraction reduce the efficiency of the bleaching. The first bleaching solution of sodium chlorite/acetic acid often becomes fully saturated in lignin and a second bleaching treatment in sodium bisulphate was added for the full whitening and removal of lignin. The alkali and bleaching pre-treatments of brown algae already induce favourable exposure of micron-sized filaments with 5–10 μm diameter with a fibrillated surface aspect. While grinding in an ultrafine friction equipment (supermass collider), the concentration of sodium chlorite oxidant during TEMPO-oxidation highly influenced the morphology and degree of fibrillation. In particular, the fibrillation yield increased at a higher degree of oxidation and highest carboxyl content was obtained at the higher chlorite concentration, both for the leaves and the balls. Hence, finer and stronger fibrillar networks were formed by varying the intensity of oxidation. The easier fibrillation is evidently induced by the generation of negative charges at the fibril surface that weaken the hydrogen bonding and reduce cohesion between the fibrils within the cell wall. On the other hand, the oxidized fibrils also underwent swelling and hydration to become more flexible. After high-pressurized microfluidization of the TEMPO-oxidized *Posidonia oceanica* balls and leaves, the fibrillation of the balls was the strongest in parallel with variations in porosity and chemical composition of both parts (Bettaieb et al. 2015b). The leaves indeed present a highly porous structure of their fibres with large lumen and thin cell wall (2–5 μm), compared to the denser structure of the balls with thick cell wall (10 μm) (Fig. 13.12). On the other hand, the easier fibrillation of the balls was attributed to higher hemicellulose content in the balls. It was assumed that the presence of hemicellulose acts as a physical barrier that prevents close contact between the microfibrils and consequently reduces the hydrogen bonding. However, it is known that the bleaching reaction with sodium chlorite is highly specific for removal of the lignin and a high fraction of hemicellulose remains intact (only a small fraction of amorphous hemicellulose is partially removed), together with a significant amount of mineral compounds. Thus, the quest for easy processing of algae highly depends on the selection of appropriate biomass and properties of the resulting fibres highly depend on the type of pre-treatment. Therefore, an easier processing route for production of CNF

Fig. 13.12 Variations in CNF structures obtained by high-pressure homogenization of *Posidonia oceanica* brown algae after TEMPO-oxidation of extracted cellulose, including (**a**) balls and (**b**) leaves (Bettaieb et al. 2015a)

with high aspect ratio was developed from residual *Laminaria hyperborea* brown algae, applying mild conditions without the need for any mechanical, chemical or biological pre-treatment (Onyianta et al. 2020). Indeed, after extraction of the alginate during an initial acid and alkali treatment, the residual dry fraction containing cellulose can be immediately used as a never-dried side stream. The homogeneous CNF could be obtained up to 3 wt.-% solid content with only one single processing step through a microfluidizer, resulting in a factor 10 lower energy consumption compared to traditional CNF from wood. The favourable processing can be attributed to the intrinsic highly porous cellulose structure and high cellulose content in the dry matter of the *Laminaria* algae species.

13.3.3 Cellulose Nanocrystals (CNCs)

The cellulose nanocrystals (CNCs), cellulose nanowhiskers (CNWs) or nanocrystalline cellulose (NCC) are similar terminologies referring to the short rod-like cellulose fibres with diameters of about 5–20 nm and length below 100 nm. They are consequently characterized by low aspect ratio and high degree of crystallinity corresponding to the remaining ordering of cellulose crystal regions. The common route for fabrication of CNC involves the acid hydrolysis of cellulose sources, during which the chemically less-resistant amorphous zones are dissolved and the remaining cellulose crystallites have outperforming properties due to the high crystallinity. The resulting aqueous suspensions of CNC are obtained after subsequent centrifugation, dialysis and ultrasonic treatment. The selection of a specific acid medium for hydrolysis (e.g. sulphuric acid, phosphoric acid or hydrochloric acid) in combination with the processing conditions (e.g. temperature, time, concentration and sonication) importantly influences the final properties of the CNC in terms of process yield, mechanical properties, surface properties and rheological features. The traditional lignocellulosic species were most frequently applied either starting from the original wood structure or by treatment of several extracted cellulose fractions including microcrystalline cellulose and previously mentioned MFC or CNF.

Starting from the extraction and purification of cellulose from freshwater *Cladophora* green algae by alkaline treatment in NaOH and oxidative bleaching in acetic acid, the CNC was isolated by further treatment in hydrobromic acid at 80 °C for several hours in combination with ultrasonication (Sucaldito and Camacho 2017). The high acid concentrations and elevated temperatures (2.5 M HBr, 80 °C) allow for the penetration of hydronium ions in the amorphous cellulose domains and hydrolytic cleavage of the glycosidic bonds. While performing the traditional acid hydrolysis of algae biomass with sulphuric acid (Mondal et al. 2019), the sulphate groups additionally induce negative surface charges and reduce the thermal stability. Otherwise, the HBr reduces the formation of labile surface groups to be further used as reactive centres for further chemical modification. As an alternative, also the hydrochloric acid hydrolysis (2.5 M HCl, 70 °C) of *Cladophora* cellulose was applied to yield functional carbonyl groups at the CNC surface with relatively low surface charge density (Kalashnikova et al. 2013). From a comparative study in the dimensions of CNC from cotton, bacterial cellulose and green algae, the exceptionally high length, diameter and aspect ratio of CNC from *Cladophora* green algae were observed (Fig. 13.13). Depending on the progressive increase in concentration of H_2SO_4 from 60% to 80% for 5 h, the fibre residues with large length became systematically smaller and the yield increased up to 75%, being much higher than from lignocellulosic materials. Due to the large surface area of the CNC from *Chaetomorpha* green algae, however, the particles tend to aggregate easily in clusters and spontaneously form a nanoscale ordered open network structure rather than separated particles. In addition, the CNC production by sulphuric acid

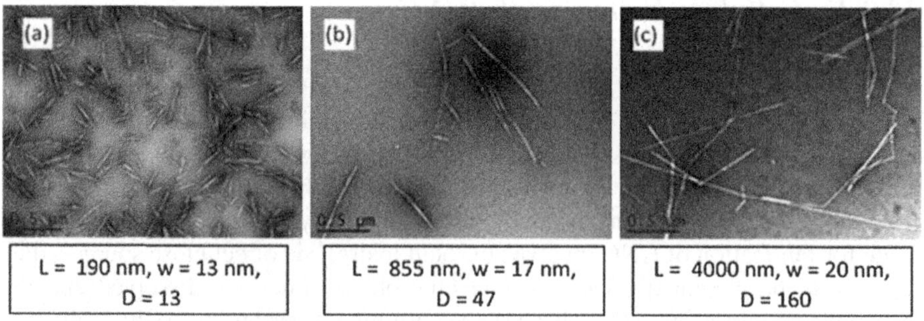

Fig. 13.13 Morphology of CNC obtained after acid hydrolysis in HCl of different sources resulting in different length (L), width (w) and aspect ratio (D), including (**a**) cotton cellulose, (**b**) bacterial cellulose and (**c**) *Cladophora* cellulose (Štefelová et al. 2017)

hydrolysis of cellulose microfibrils from *Chaetomorpha* green algae was followed by a precipitation reaction in ethanol or sodium hydroxide in order to reduce the negative effects attributed to presence of excess sulphate groups (Saritha et al. 2013).

The red algae are mainly composed of cellulose and agar with a high content of carbohydrates. Therefore, they form an interesting source and/or side-stream material after agar production for the further processing into cellulose materials because of their large abundance and availability. The huge amounts of solid fibrous residue originating from *Gelidium elegans* red algae after extraction of agar were considered as a most interesting biomasss for recovery of CNC nanomaterials through an acid hydrolysis process (Chen et al. 2016). The presence of proteins, oil, waxes and lignin in the red algae could be fully eliminated by the chemical peroxide bleaching step, in combination with the partial fibrillation and depolymerization of the fibre bundles. The fibres were already partially swollen and their surface was partially cracked after NaOH alkalization, while the bleaching products could subsequently penetrate more in depth of the inner fibre parts. Interestingly, the same conditions could be applied for the extraction and dissolution of agar components and the never-dried residues could be directly applied in CNC production. An extraction scheme was presented for *Gelidium sesquipedale*, in order to progressively isolate the MFC and CNC (El Achaby et al. 2018). The final sulphuric acid hydrolysis with increasing treatment times introduced progressive decrease in the zeta potential, as the good dispersion stability is due to elimination of polar cellulose components and the introduction of polar sulphate groups on CNC. On the other hand, additional washing of isolated CNC with NaOH until neutral pH favourably improved the thermal stability compared to the other chemically treated fibres (Chen et al. 2016). The diameter and length of the CNC progressively decrease after longer hydrolysis times in parallel with the increase in aspect ratio, but the CNC yield evidently decreases after longer time while the cellulose content becomes higher. Due to the high length and aspect ratio of CNC from red algae compared to traditional wood sources, they have a stronger tendency for self-assembly due to their chiral nematic structure.

While discovering more sustainable processing and extraction of CNC from *Gelidiella aceroso* red algae, a microwave alkaline pre-treatment was employed in combination with hydrolysis in ionic liquids (Singh et al. 2017). As such, the microwave heating could fully eliminate the need for a separate dewaxing process in toluene/ethanol and reduce the heating time. The hydrolysis under a combination of sulphuric acid and 1-ethyl-3-methylimidazolium chloride and ultrasonication has been employed as an alternative with partial regeneration of the used solvents.

The different parts of balls and leaves of *Posidonia oceanica* brown algae were equally used for CNC production by traditional hydrolysis in concentrated sulphuric acid. Due to the significant presence of lignin, however, the obtained holocellulose after removal of lignin in two bleaching steps with sodium chlorite and sodium bisulphate was applied as starting material for CNC isolation. The characterization of bleached and unbleached leaves indicates indeed that hemicellulose parts are removed at mild conditions and lignin is only removed under the more intensive bleaching process. Finally, the holocellulose part (cellulose +

hemicellulose) was used for hydrolysis with sulphuric acid at 45 °C or was additionally washed with sodium hydroxide. In parallel, also the hydrolysis time had to be prolonged from 30 to 60 min as the numerous unreacted residues showed very low yield (8%) after short processing times (Luzi et al. 2016). By comparing CNC qualities from holocellulose and cellulose fractions of the *Posidonia oceanica*, the higher yields were evidently obtained from the cellulose fractions due to its higher purity (Bettaieb et al. 2015c). In most of the research, the lignocellulosic residues from balls (Fortunati et al. 2015; Luzi et al. 2015) were used for the isolation of CNC using sulphuric acid hydrolysis as it more frequently accumulates as waste fraction compared to the original leaves. Finally, the yield was comparable to that of other annual plants while the aspect ratio of CNC from both the leaves and balls were higher than those originating from lignocellulosic sources. The yield for CNC from leaves (25%) was comparatively higher than for balls (20%), in parallel with the earlier studies on fibrillated cellulose. The more porous structure of the leaves resulted in the very long CNC (up to 520 nm) with higher aspect ratio (up to 75), compared to the denser ball structure having CNC with smaller length (up to 290 nm) and lower aspect ratio (up to 35) (Bettaieb et al. 2015a). In contrast with the strong hydrolysis conditions, the mild hydrolysis of extracted and bleached cellulose from the *Posidonia oceanica* in low-concentrated hydrochloric acid and/or enzymatic hydrolysis in combination with mild mechanical stirring rather than sonication provided a route for the synthesis of microcrystalline cellulose (MCC) (Tarchoun et al. 2019). The latter have a compact granular or rod-like shape with the arrangement of micron-fibrils with diameter at around 8 μm, whereas the mild acid did not result in a shortening of the fibril length.

Although the surface modification of CNC is commonly performed by established procedures applied on lignocellulosic-based nanocrystals, it has been relatively less used for the algae-based nanomaterials while the same techniques are transferable. The surface modification with a surfactant including an acid phosphate ester of ethoxylated nonyl phenol (Fortunati et al. 2015) resulted in a hydrophobization of the algae-based CNC with better dispersibility into a polymer matrix.

In summary, a comparison of dimensions and aspect ratio for CNC from different algae sources is given in an overview, together with an indication of the different applied hydrolysis conditions (Table 13.1). The selection of algae type clearly determines the final yield and dimensions of isolated CNC. Although final yield depends on conditions, it was mostly proportional to the amount of alpha cellulose in the different cellulose sources. Following a straightforward step-wise processing protocol of depolymerization, bleaching and acid hydrolysis, the largest size values were observed for the CNC from brown algae (Doh et al. 2020). The two bleaching steps included both sodium hypochlorite and hydrogen peroxide treatments, followed by the hydrolysis in H_2SO_4 and final washing in sodium hydroxide for reduction of the remaining sulphate groups. In particular, the chemical composition with carbohydrate, lipid, protein and mineral content of algae is highly different, even if they belong to the same group of green, red, or brown species. Therefore, the extraction yield was highest for brown algae and there was only a slight tendency that the yield for red algae might be slightly lower compared to the green algae. Although the

Table 13.1 Comparative study on CNC morphology of different algae families and types versus traditional sources

CNC source		Conditions				Length (nm)	Width (nm)	Aspect ratio	CNC yield (%)	References
Reference materials	Hardwood bleached kraft pulp	H_2SO_4	64%	50 °C	50 min	171	15	11.5	20	Van Hai et al. (2015)
	Softwood bleached kraft pulp	H_2SO_4	64%	50 °C	50 min	179	17	10.6	20.5	Van Hai et al. (2015)
	Cotton	H_2SO_4	64%	50 °C	50 min	278	33	8.4	33.4	Van Hai et al. (2015)
	Cattail	H_2SO_4	64%	50 °C	50 min	249	19	13.3	17	Van Hai et al. (2015)
Green algae	*Ulva lactuca*	H_2SO_4	51%	45 °C	30 min	53	17	3.16	16.3	Doh et al. (2020)
	Arthrospira maxima	H_2SO_4	51%	45 °C	30 min	51	15	3.51	20.3	Doh et al. (2020)
	Cladophora	HCl	2.5 M	70 °C	60 min	4000	20	160	–	Kalashnikova et al. (2013)
	Chaetomorpha antennina	H_2SO_4	60–80%	60 °C	60 min	1000	60–70	12.5	30	Saritha et al. (2013)

(continued)

Table 13.1 (continued)

CNC source		Conditions				Length (nm)	Width (nm)	Aspect ratio	CNC yield (%)	References
Red algae	*Palmaria palmata*	H_2SO_4	51%	45 °C	30 min	64	17	3.85	13.3	Doh et al. (2020)
	Porphyra umbilicalis	H_2SO_4	51%	45 °C	30 min	28	11	2.79	17.4	Doh et al. (2020)
	Gelidium elegans	H_2SO_4	64%	45 °C	45 min	547	22	25	15.5	Chen et al. (2016)
	Gelidiella aceroso	H_2SO_4 + ionic liquid	64%	95 °C	30 min	305–512	14–50	10–21	30	Singh et al. (2017)
	Fibre residues	H_2SO_4	64%	50 °C	50 min	432	29	15.1	20.5	Van Hai et al. (2015)
	Fibre waste	H_2SO_4	64%	50 °C	30 min	315	9	35	13.8	El Achaby et al. (2018)
	Fibre waste	H_2SO_4	64%	50 °C	40 min	294	7	42	12.9	El Achaby et al. (2018)
	Fibre waste	H_2SO_4	64%	50 °C	80 min	285	5	57	11.2	El Achaby et al. (2018)
Brown algae	*Laminaria japonica*	H_2SO_4	51%	45 °C	30 min	239	22	11.5	25.8	Doh et al. (2020)
	Sargassum fluitans	H_2SO_4	51%	45 °C	30 min	44	9	5.05	26.1	Doh et al. (2020)
	Posidonia oceanica leaves	H_2SO_4	64%	45 °C	30 min	500–650	15–20	40	<10 (agglomeration)	Luzi et al. (2016)
		H_2SO_4	64%	45 °C	60 min	200–450 nm	5–10	40–45	15	Luzi et al. (2016)
		H_2SO_4	64%	55 °C	40 min	338	7	48.3	30.1	Bettaieb et al. (2015b)
	Posidonia oceanica balls	H_2SO_4	64%	45 °C	60 min	180	4.9	36.7	–	Luzi et al. (2016)
		H_2SO_4	64%	55 °C	40 min	276	8	34.6	27.9	Bettaieb et al. (2015b)

sizes and aspect ratio of CNC from different species are highly different, they also strongly depend on the processing conditions and intensity of the ultrasonication treatment which may likely result in the smaller particle sizes. However, it reflects the variability in obtained morphology under similar processing conditions, which may be related to the original sizes and organization of the cellulose microfibrils in different species, which cannot seemingly not be categorized according to the respective groups of green, red and brown algae.

13.3.4 Other Nanocellulose Structures from Algae Cellulose

The nanocellulose extracted from *Cladophora* green algae could be transformed into solid beads and/or porous spheres with an internal aerogel structure. The green algae nanofibrils were oxidized into 2,3-dialdehyde cellulose (DAH) via a periodate oxidation reaction, resulting in a one-pot synthesis route for the formation of microparticles and microspheres (Fig. 13.14a). Depending on the reaction conditions, the 2,3-hydroxyl groups of *Cladophora* the cellulose could either be partly oxidized to 30%, or fully oxidized while retaining the nanofibrillar structure at high degree of oxidation (Kim et al. 2000). As such, the porosity of the morphological structures could be tailored depending on the reaction conditions for the same cellulose source. The periodate oxidation of cellulose mostly generates a more compact structure, but the solid spherical beads with diameter of 1–20 μm are consequently formed by almost fully oxidized nanofibrils. The morphologies can form through the arrangement of the nanofibrils that become less stiff because of the ring-opening of the glucose and changes in fibre dimensions. The cross-links are simultaneously formed through the inter- and intrafibrillar hemiacetal bonds contributing to the stabilization of the ordered bead shape. Surprisingly, the formation specifically occurred for the highly crystalline cellulose from algae, while it did not appear for the less crystalline microfibrillated cellulose types. In addition, the particle shape can be further adapted into the formation of porous nanocellulose beads with aerogel-type core

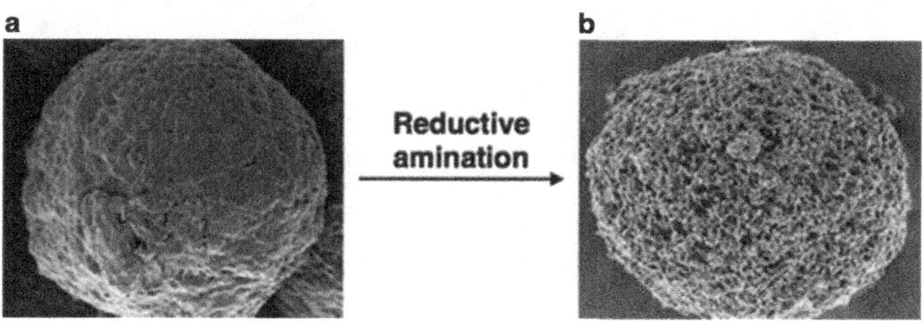

Fig. 13.14 Periodate oxidization of *Cladophora* cellulose forming (**a**) spherical microparticles (Kim et al. 2000) and (**b**) porous beads (Lindh et al. 2016)

structure (Fig. 13.14b), by introducing a diamine spacer on the oxidized cellulose from the green algae *Cladophora* (Lindh et al. 2016). The reductive amination resulted in the formation of open spheres with 10–100 nm diameter and up to 30 times higher surface area. In particular, the covalent cross-linking of the DAH-modified nanofibrils with alkyl- or aryl-tethered diamines caused the reaction between the aldehyde groups of DAH and the amine groups, which was less pronounced while using the mono-amines. In addition, the aminated mesoporous spheres possess good potential for the further functionalization due to the presence of remaining aldehyde groups at the surface.

The nanocellulose aerogels were produced from *Cytosphaera jacquinotii* brown algae, based on the freeze-drying of isolated cellulose nanofibres obtained after chemical extraction by alkaline treatment (NaOH) and bleaching in sodium hypochlorite, following a short and straightforward reaction process (Fig. 13.15a) (Paniz et al. 2020). The absence of sulphuric acid treatment eliminates the presence of negative charges at the fibre surface, with consequently low zeta potential values and formation of an almost unstable colloidal suspension of cellulose fibrils. A porous, sponge-like structure is consequently formed through favourable interaction of the nanofibrils during freeze-drying, where agglomerated nanofibres accumulated in the cell walls of the aerogel with a thickness of around 32 nm

(a)
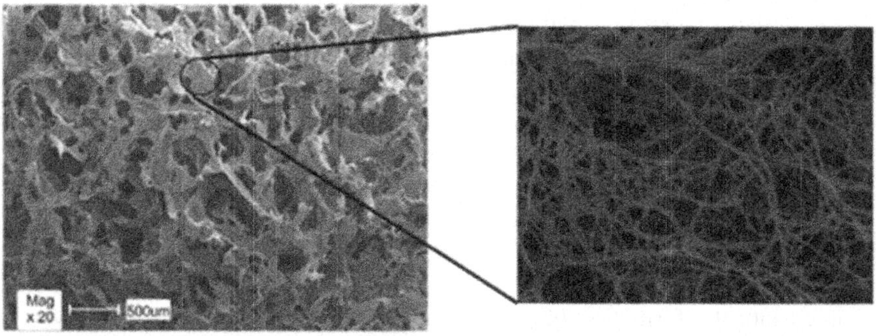

(b)
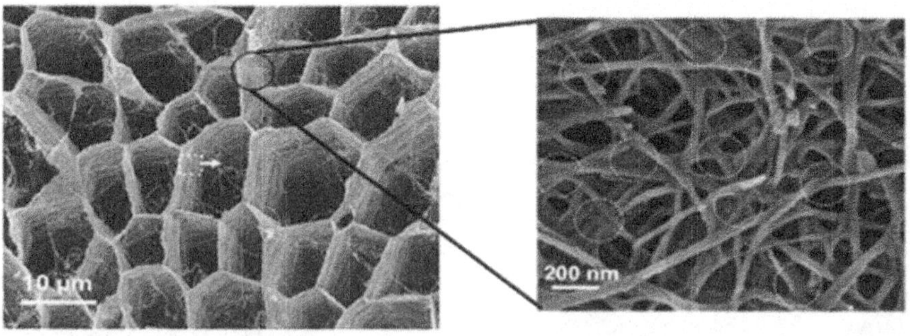

Fig. 13.15 Porous aerogel structures obtained after freeze-drying of extracted cellulose from algae, (**a**) irregular structure of non-surface-modified CNF (Paniz et al. 2020) and (**b**) regular structure and reinforcement of the aerogel walls after surface modification of the CNF (Zhou et al. 2019)

corresponding to the nanofibril diameter. The method allowed to recover almost all of the cellulose content from the algae into the nanofibrous aerogel structure due to the elimination of harsh processing conditions. The assembly of CNF from *Ulva prolifera* into an aerogel by freeze-drying was further facilitated after t-BuOH modification (Zhao et al. 2014), forming a flexible foam with better interaction between the fibres, resulting in higher performance. Interestingly, the freeze-drying of modified CNF fibres with deposition of an Al^{3+} nanolayer formed an aerogel structure that could be stabilized through a cross-linking reaction between the fibrils (Zhou et al. 2019). The skeleton of a cross-linked aerogel consisted of a 'welded' structure with interconnections between the entangled fibrils (Fig. 13.15b), which form the side walls of the porous network structure of the aerogel and enhance the elasticity of the structure.

13.4 Properties of Nanocellulose from Algae

The exceptional properties of algae-based nanocellulose can mainly be related to the high quality and purity of obtained cellulose and consequent high degree of crystallinity. As the latter is considered as the main parameter, it directly determines the superior properties of algae-based nanocellulose over traditional nanocellulose from lignocellulosic origin in many aspects, including mechanical strength, stiffness and thermal stability. The unique geometries with long nanofibres and nanofibrils also influence the processing conditions, mechanical reinforcement efficiency and rheological properties. Some of the more fundamental properties are highlighted below in relation to their wood-based counterparts.

13.4.1 Crystallinity

Both the extracted cellulose from algae biomass and the isolated nanocellulose materials are characterized by extremely high crystallinity compared to other lignocellulosic materials or wood, except for the cellulose extracted from cotton with crystallinity around 90%. As a natural material, the biosynthesis route and native occurrence of cellulose in algae allow for the creation of a high-molecular-weight polymer with high degree of orientation. The high degree of crystallinity for algae cellulose is mainly attributed to the occurrence of relatively thick microfibrils (10–30 nm) and high degree of polymerization, in contrast to the occurrence of smaller microfibrils (5–10 nm) in wood. In particular, the thick microfibrils of marine algae cellulose favour a high degree of orientation of the cellulose molecules and elementary fibrils, whereas the smaller microfibrils may allow for easier twisting of the fibrillar structure and reduced their orientation in plant or wood species. As such, the cellulose extracted from green algae (e.g. *Valonia* or *Cladophora*) may have exceptionally high degree of crystallinity, which may be up to around 95%.

The crystalline cellulose may occur in two composite forms, that is, Iα cellulose with a highly crystalline ordering and the Iβ cellulose with lower crystallinity. The two allomorphs Iα and Iβ crystalline structures are present in algae cellulose, with predominance of the tri-clinic Iα structure in algae cellulose and predominance of the monoclinic Iβ structure in plant or wood cellulose. For example, the composition of *Chaetomorpha gracilis* green algae includes 26.5% Iα cellulose and only 0.4% Iβ cellulose (Freile-Pelegrín et al. 2020), with overall crystallinity of 93.5%. The transition zones between the two alternating allomorph phases are formed by the interface between hydrogen-bonded molecular sheets.

The MFC extracted from green algae is characterized by the intrinsic high crystallinity of the microfibrils with cellulose Iα structure. The changes in crystallinity may be induced during subsequent treatment by grinding, swelling or extraction. The effects of extraction and bleaching results in the preservation of the original crystalline, even the removal of the lignin and amorphous hemicellulose parts during alkali treatment may contribute to a further increase in crystallinity (e.g. 93.8% for unbleached to 94.5% for beached MFC from *Cladophora*) (Liu et al. 2017). Comparatively, the MFC from green algae (e.g. *Cladophora*) resulted in a higher crystallinity than brown algae (e.g. *Algiflor*) (Mihranyan et al. 2004), which might be further related to differences in density. In particular, the high crystallinity of MFC from *Cladophora* is highly desired for reduction in moisture or water sorption properties, being non-hygroscopic at low to moderate relative vapour pressure, which is in contrast with the highly hygroscopic properties of wood-based CNF (Le Bras et al. 2015).

The CNF after processing of algae-based cellulose by microfluidization often exposes higher crystallinity under suitable processing conditions. The mild homogenization conditions may not significantly change the crystallinity, but the associated mild pre-treatment conditions may be responsible for small variations in crystallinity (Wahlström et al. 2020). Indeed, the increase in crystallinity might be related to the removal of amorphous phases or disordered cellulose, impurities and exposure of the large crystallite size for algae cellulose. It is evident that any remaining mineral fraction, residue or hemicellulose fraction hinders the crystallization of the cellulose. On the other hand, the use of an enzymatic hydrolysis step as pre-treatment does not necessarily contribute to a better separation of the algae microfibrils into their nanofibrillar structure, while instead a small decrease in crystallinity may occur in parallel with a partial depolymerization after long treatment time (Xiang et al. 2016). The mechanical damage of the crystalline structure under extensive shear could reduce the crystallinity of *Cladophora* green algae cellulose from 93% to 85% after 10 passes and 80% after 20 passes. Also, the pre-treatment of cellulose from *Aegagropila* green algae by sodium hypochlorite oxidation resulted in a lower crystallinity after the mechanical fibrillation, in contrast with the mild pre-treatment conditions for maleic acid that is more selective in removing the amorphous hemicellulose fractions (Guo et al. 2017). In general, the number of passes during the mechanical processing can cause the breaking-up of certain small crystalline domains, inducing the small variations in the preferred crystalline orientation of the cellulose Iα structures and slight tendency towards the properties of a randomly oriented structure of the cellulose crystallites. Therefore, a slight increase in crystallinity degree upon mechanical processing in a microfluidizer is often observed after 1 or 2 passes (Taheri and Samyn 2016), in parallel with the better homogenization under parallel shear forces and slight separation or delamination of the crystalline domains that allow for a better exposure and quantification of the single domains. The reduction in crystallinity after more intense processing, however, can result from mechanical damage and depolymerization. Advantageously, the lowering in crystalline content may enhance the reactivity of the algae CNF or later chemical modification.

In a comparative XRD study for wood-based, bacterial and *Cladophora* CNC (Fig. 13.16a), the spectra of CNF from algae cellulose are characterized by relatively sharp peaks and calculated crystallinity index C.I. = 96% (Štefelová et al. 2017). The smaller FWHM for diffraction peaks of algae cellulose compared to wood may agree with the larger crystalline index and higher width of the crystalline sizes in algae cellulose, for example, calculated at 3.6 nm for wood-based up to 11.3 nm for algae CNF. The XRD spectra typically show diffraction peaks at $2\theta = 16°, 22°, 34°$, respectively, corresponding to the crystal planes [110], [002] and [004] of the cellulose I structure. Mostly in wood-based CNF, the diffraction peaks at $2\theta = 14.3°$ and 16.5° corresponding to the [100] and [010] crystallographic planes overlap and form a single broad diffraction peak at around $2\theta = 16°$. The presence

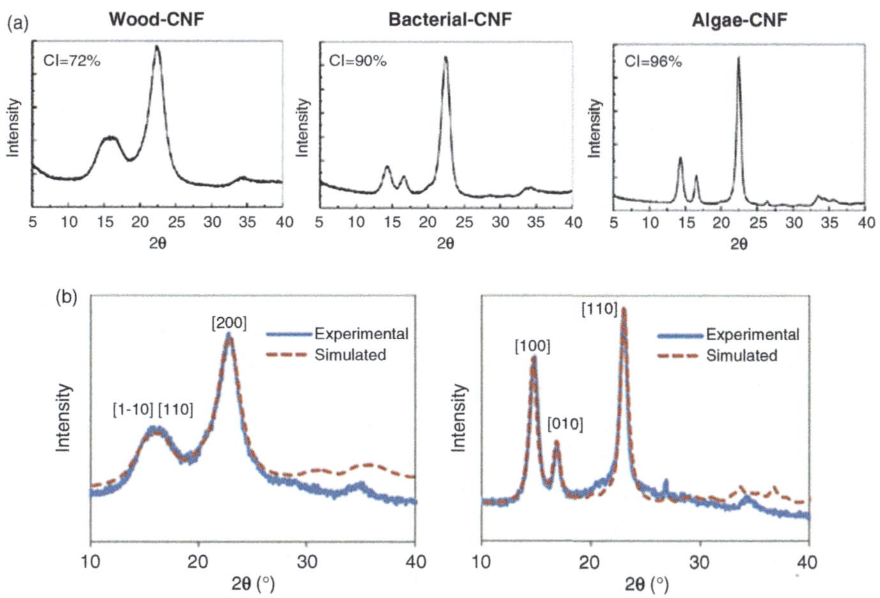

Fig. 13.16 XRD spectra for CNF originating from different sources according to experimental and simulated data: (**a**) experimental spectra for W-CNF, B-CNF and A-CNF, respectively, obtained by microfluidization (Štefelová et al. 2017) and (**b**) simulation of XRD data for wood-based CNF (left) with cellulose Iα and algae-based CNF with cellulose Iβ (right) (Xiang et al. 2016)

of an amorphous fraction corresponding to the lowest intensity I_{AM} at $2\theta = 18°$ can be used for calculation of the crystallinity index C.I. according to the empirical method of Segal, with C.I. = $100 \times (I_{002}-I_{AM})/I_{002}$ (Segal et al. 1959). However, the separate contributions of crystalline orientation are better expressed in the algae-based CNF. Therefore, the more detailed studies on the orientation of the crystalline planes in cellulose microfibrils need to be implemented for better correctness. The existence of both cellulose Iα and Iβ polymorphs in simulation experiments of XRD spectra were indeed considered (Xiang et al. 2016). Based on the simulated spectra of CNF from bleached eucalyptus pulp and *Cladophora* green algae, the correct assignment of cellulose Iα or cellulose Iβ structures was made in order to determine the degree of orientation in CNF crystalline structures (Fig. 13.16b). The algae CNF showed preferred orientation along the [100] and [010] axis, assuming that it has a hybrid cellulose Iα structure with about 75% and 25% of the respective crystal phases. The wood-based CNF, in contrast, showed preferred orientation with overlap in the peaks at [1–10] and [110] with a characteristic reference peak at [200] corresponding to the cellulose Iβ structure.

The crystallinity of CNC from algae-based cellulose may gradually increase up to 87% with treatment time during hydrolysis in sulphuric acid. In parallel with a slight removal of the amorphous zones during a first bleaching experiment in sodium hypochlorite, the crystallinity of 66% after bleaching was already higher than from the virgin algae cellulose and followed a more intense rise in crystallinity during acid hydrolysis (El Achaby et al. 2018). Depending on the source, however, there was not always a clear correlation between the crystallinity of extracted cellulose and crystallinity of CNC from brown, red and green algae (Doh et al. 2020). Compared to the extracted cellulose, it is clear that the acid hydrolysis provides an increase in crystallinity. As such, it attains a value for crystallinity index in comparable ranges of that for cellulose Iα in microcrystalline cellulose (MCC) (Chen et al. 2016). However, the penetration of hydronium ions may obviously also lead to the release of eventual individual crystallite structures, while the creation of other semi-ordered structures besides the cellulose I were eventually suggested. In particular, the longer treatment times with strong acids may induce some degradation in crystallinity during the isolation of CNC depending on the acid concentration and temperature. In general, however, the Segal crystallinity index for CNC from wood is comparable to slightly lower within ranges between 60 and 65%n while it approaches the high crystallinity of around 71% for CNC from cotton (Van Hai et al. 2015). The higher values for Segal crystallinity index for CNC from wood species could be obtained at around 75–80%, but only after extensive process optimization studies (Kandhola et al. 2020). The relatively higher values for algae-based CNC compared to wood- and annual-plant-derived CNC is in line with the higher contents of cellulose Iα with better quality in algae biomass. In summary, an overview of the crystallinity for different nanocellulose materials including MFC, CNF and CNC from green, red and brown algae biomass is compared in Table 13.2.

Table 13.2 Comparative study on MFC, CNF and CNC crystallinity from different algae families and types versus traditional sources

Nanocellulose source		Conditions	Crystallinity				References
			Extracted cellulose	MFC	CNF	CNC	
Reference materials	MFC wood	Homogenization	36	36	–	–	Hua et al. (2014)
		TEMPO oxidation	36	32	–	–	Hua et al. (2014)
	CNF wood	Microfluidization	36	–	66	–	Le Bras et al. (2015)
	CNC hardwood	H_2SO_4 hydrolysis	55	–	–	62	Van Hai et al. (2015)
	CNC softwood	H_2SO_4 hydrolysis	58	–	–	65	Van Hai et al. (2015)
	CNC Cotton	H_2SO_4 hydrolysis	65	–	–	72	Van Hai et al. (2015)
	CNC Cattail	H_2SO_4 hydrolysis	54	–	–	69	Van Hai et al. (2015)
Green	*Chaetomorpha gracilis*	Alkali + H_2O_2 bleaching	93	93	–	–	Freile-Pelegrín et al. 2020
	Cladophora	Alkali + $NaClO_2$ bleaching	>95	93	–	–	Le Bras et al. (2015)
		Alkali + $NaClO_2$ bleaching	>95	92	–	–	Hua et al. (2014)
		Alkali + $NaClO_2$ bleaching + mild HCl acid	>95	94.7	–	–	Mihranyan et al. (2004)
		TEMPO oxidation	92	93	–	–	Hua et al. (2014)
		Chlorine-free bleaching	93	95	–	–	Liu et al. (2017)
		Alkali + $NaClO_2$ bleaching + homogenizer	95	–	96	–	Štefelová et al. (2017)
		TEMPO oxidation + homogenizer	98	–	92	–	Ono et al. (2019)
		Enzymatic hydrolysis + microfluidizer 10 passes	93	–	85	–	Xiang et al. (2016)
		Enzymatic hydrolysis + microfluidizer 20 passes	93	–	80	–	Xiang et al. (2016)

(continued)

Table 13.2 (continued)

Nanocellulose source		Conditions	Crystallinity				References
			Extracted cellulose	MFC	CNF	CNC	
	Aegagropila linnaei	Maleic acid hydrolysis + homogenizer	89	–	95	–	Guo et al. (2017)
		$NaClO_2$ oxidation+ homogenizer	89	–	94	–	Guo et al. (2017)
	Pithophora	Alkaline + H_2O_2 bleaching + microfluidizer	88	–	96.6	–	Gustafsson et al. (2019)
	Ulva lactuca	Alkali + H_2O_2 bleaching + microfluidizer	48	–	48	–	Wahlström et al. (2020)
		Alkali + H_2O_2 bleaching + HCl acid hydrolysis	48	–	–	63	Wahlström et al. (2020)
		HCl + $NaClO_2$ bleaching + H_2SO_4 acid hydrolysis	46.7	–	–	67.0	Doh et al. (2020)
	Arthrospira maxima	HCl + $NaClO_2$ bleaching + H_2SO_4 acid hydrolysis	67.2	–	–	89.7	Doh et al. (2020)
	Chaetomorpha antennina	H_2SO_4 hydrolysis	70	–	–	95	Saritha et al. (2013)
Red	*Palmaria palmata*	HCl + $NaClO_2$ bleaching + H_2SO_4 acid hydrolysis	61.0	–	–	84.9	Doh et al. (2020)
	Porphyra umbilicalis	HCl + $NaClO_2$ bleaching + H_2SO_4 acid hydrolysis	65.9	–	–	87.5	Doh et al. (2020)
	Gelidium elegans	Alkali + H_2O_2 bleaching + H_2SO_4 hydrolysis	33	–	–	73	Chen et al. (2016)
	Gelidiella aceroso	Alkali microwave + H_2SO_4 hydrolysis	27	–	–	60	Singh et al. (2017)
	Fibre residues	H_2SO_4 hydrolysis	63	–	–	70	Van Hai et al. (2015)
	Fibre waste	Alkali + $NaClO_2$ bleaching + H_2SO_4 hydrolysis (30 min)	39	–	–	81	El Achaby et al. (2018)
	Fibre waste	Alkali + $NaClO_2$ bleaching + H_2SO_4 hydrolysis (40 min)	39	–	–	83	El Achaby et al. (2018)
	Fibre waste	Alkali + $NaClO_2$ bleaching + H_2SO_4 hydrolysis (60 min)	39	–	–	87	El Achaby et al. (2018)

(continued)

Table 13.2 (continued)

Nanocellulose source		Conditions	Crystallinity				References
			Extracted cellulose	MFC	CNF	CNC	
Brown	*Algiflor* residue	Alkali + bleaching + mild HCl acid	81.7	81.7	–	–	Mihranyan et al. (2004)
	Laminaria hyperborea	Microfluidizer 1 pass	55	–	68	–	Onyianta et al. (2020)
		Microfluidizer 2 passes	55	–	65	–	Onyianta et al. (2020)
		Microfluidizer 5 passes	55	–	64	–	Onyianta et al. (2020)
	Posidonia oceanica leaves	Alkali + $NaOCl_2$ bleaching + TEMPO oxidation + microfluidizer	53	–	52.9	–	Bettaieb et al. (2015b)
		Alkali + $NaClO_2$ bleaching + H_2SO_4 acid hydrolysis	46	–	–	62	Bettaieb et al. (2015b)
	Posidonia oceanica balls	Alkali + $NaOCl_2$ bleaching + TEMPO oxidation + microfluidizer	58	–	63.5		Bettaieb et al. (2015b)
		Alkali + $NaClO_2$ bleaching + H_2SO_4 acid hydrolysis	60	–	–	64	Bettaieb et al. (2015b)
	Laminaria japonica	HCl + $NaClO_2$ bleaching + H_2SO_4 acid hydrolysis	75.7	–	–	98.9	Doh et al. (2020)
	Sargassum fluitans	HCl + $NaClO_2$ bleaching + H_2SO_4 acid hydrolysis	65.6	–	–	83.1	Doh et al. (2020)

13.4.2 Thermal Stability and Degradation

The thermal decomposition and stability of cellulose nanomaterials are critical for the processing of the composite materials, while a more detailed study on the gradual degradation mechanisms provide better insights in the qualities of CNF and CNC. The superior thermal stability of algae-based nanocellulose compared to wood-based nanocellulose particularly reflects the high degree of crystallinity and purity of algae-based cellulose. In particular, the alkali-treated cellulose extracted from algae may possess slightly higher thermal degradation temperature owing to the presence of remaining lignin fractions (Singh et al. 2017). The lignin has a strongly aromatic structure and may protect the cellulose fibres against degradation, with main degradation temperature at around 450 °C compared to a degradation temperature of around 300–350 °C for cellulose. Alternatively, the alkali-treated cellulose from *Gelidium* red algae caused the solubilization of the agar components and consequently shifted the degradation temperature to a higher value of 317 °C compared to the untreated cellulose fibres at 308 °C. The further increase in degradation temperature after bleaching of the algae cellulose to values of 319 °C is in parallel with the removal of non-cellulosic impurities and inorganic mineral salts (Chen et al. 2016). The remaining hemicellulose parts form a weak amorphous fraction with lower thermal stability owing to the presence of the acetyl groups.

By comparing the step-wise thermal degradation of wood-based and algae-based CNF (Xiang et al. 2016), the latter only shows a single major degradation step at 390 °C corresponding to the thermal decomposition of the cellulose material. In contrast, the thermal degradation of the wood-based CNF follows a step-wise process with evaporation of the free physiosorbed water or moisture (100 °C) and the chemically bound water (180 °C), followed by the decomposition of hemi-cellulose (240–250 °C) and final degradation of the cellulose (326 °C). The higher degradation temperature of the algae-based cellulose is a direct consequence of the higher degree of crystallinity and/or higher degree of polymerization. The inertness of algae-based CNF and absence of water through easy removal under drying are beneficial in improving the thermal stability in respect to wood-based CNF, although it is also strongly influenced by the drying procedure and effectiveness (Štefelová et al. 2017). The thermal degradation is indeed strongly initiated by surface effects and preferential degradation of the amorphous zones, which both control the accessibility of the cellulose structure and sensitivity to degradation of the polymer chains. After processing, the mild mechanical shear exerted on CNF from *Laminaria hyperborea* brown algae had no significant detrimental effect on a reduction in thermal stability of the CNF after one or two processing steps (Onyianta et al. 2020). However, the presence of eventual residual fraction of proteins and/or low-molecular-weight fractions of hemicellulose might have detrimental effects on the thermal stability at low temperatures. Therefore, the bleaching and oxidative pre-treatments may have benefits for the improved thermal stability both through the removal of impurities and/or the enhancement of the network through stronger binding between the single CNF fibrils at the surface level (Guo et al. 2017). On the other hand, the TEMPO-oxidation reaction for isolation CNF from brown algae residues induced simultaneous depolymerization and reduction in molecular weight of the cellulose polymer, which obviously results in a lowering of the thermal degradation temperature for CNF (230 °C) compared to the native cellulose from brown algae (350 °C) (Gao et al. 2018). In comparing the main thermal degradation temperature of CNF from *Ulva lactuca* green algae (i.e. 342 °C) and from *Cladophora* green algae (i.e. 326 °C), differences were noticed depending on the source and algae family (Wahlström et al. 2020). The latter correlation could be mainly explained through the variations in crystallinity between the different CNF depending on the species, with increasing thermal stability at the higher crystallinity index. In parallel, also small variations in sizes of the micro- and nanofibrillar structures might influence the thermal stability, as the total surface area increases at the higher degree of fibrillation and becomes more prone to thermal degradation.

The thermal stability of CNC from algae biomass strongly depends on the hydrolysis conditions. While the expected high crystallinity would imply high thermal stability, however, surface effects may play a role in the premature thermal decomposition of CNC. After acid hydrolysis, the densification into a compact cellulose structure and removal of the non-cellulosic components should be in favour of a better thermal stability, while the presence of remaining active sulphate groups after hydrolysis with sulphuric acid generally induces premature degradation. In a comparative study after sulphuric acid hydrolysis, the algae-based CNC had a degradation temperature (220 °C) that is relatively higher compared to hardwood-based CNC (185 °C) and softwood-based CNC (192 °C), but which was lower than the CNC from cattail (213 °C). Although the differences in thermal stability of the extracted cellulose from different sources could relate to variations in crystallinity among species, the thermal stability of the isolated CNC did not directly correspond after sulphuric acid hydrolysis. Therefore, the neutralization of CNC from either green, red or brown algae should be done by excessive washing for the removal of excess sulphuric acid, followed by a final washing treatment in sodium hydroxide towards adjusted neutral pH, where the sulphate groups are reduced (Doh et al. 2020). In that case, the crystalline structure of algae-based CNC prevail the thermal performance and exhibit lower weight loss over the entire temperature domain.

13.4.3 Rheological Properties

The knowledge on shear viscosity and rheological properties of algae-based nanocellulose suspensions is an important instrument in better understanding the nanocellulose structure, quality and its processing conditions. While the rheology of wood-based nanocellulose was intensively characterized and reviewed for both the fibrillated MFC and CNF materials (Taheri and Samyn 2015), or the crystalline CNC materials (Taheri and Samyn 2017), the specific information on rheological properties for the algae-based nanocellulose is not yet described in detail. Therefore, some studies only focused on the viscoelastic properties of CNF from *Laminaria hyperborea* brown algae (Onyianta et al. 2020), and the rheological properties of CNF from *Posidonia oceanica* brown algae based on both the oscillatory measurements and continuous shear rotational data (Bettaieb et al. 2015d). The materials usually indicate gel-like properties due to the strong network formation of the entangled fibrillated cellulose fibres. Although general trends in rheological properties of algae-based CNF agree to the evolutions observed for wood-based CNF, the peculiar morphology of CNF from algae with higher length and aspect ratio may induce benefits in terms of elastic properties, hysteresis versus recovery and consequent energy consumption during processing. The gelling properties for suspensions with fibres of high length and aspect ratio can be favoured and make them more favourable as a thickening agent, fluently inducing an increase in viscosity. Besides the gel formation through fibrillar entanglements, also the water-binding capacity of algae-based nanofibres is higher due to the higher surface area with presentation of hydroxyl groups.

The rheological properties of CNF from brown algae were studied in the linear viscoelastic region as a function of the number of processing steps in a microfluidizer. A huge increase in storage modulus has been observed after one single pass, while later remaining constant for the additional number of passes (Onyianta et al. 2020). This is in parallel with the easy fibrillation of the algae cellulose in a limited number of processing steps and the lower need for additional energy consumption at a higher number of passes. When compared with the storage modulus of wood-pulp based CNF, the value for algae-based CNF is about 50% higher after one single pass. In parallel, the suspension properties of algae-based CNF reflect a higher portion of elastic response compared to wood-based CNF with a storage modulus that is larger than the loss modulus and relatively independent of the angular velocity. The higher contribution of elastic effects compared to viscous effects are generally reflected in a lower dampening factor and consequently lower energy loss through accumulation in the suspension. The latter is observed as a better recovery of the deformation in the suspension upon ascending and descending frequency cycles, known as hysteresis loss. The high values of the moduli are characteristic for representation of stronger interactions in the fibrillar network, which further increase after the surface-modification of CNF due to additional fibre interactions and entanglement. The high dynamic moduli (G', G'') for algae-based CNF from *Posidonia oceanica* balls relatively to the low dynamic moduli for algae-based CNF from *Posidonia oceanica* leaves could be attributed to the formation of a stronger CNF network from balls, mainly as a result from its different morphologies. In particular, the increase in CNF concentration evokes the transition from solution state ($G' \approx G''$) into the formation of a weak gel state ($G' > G''$) at concentrations above 1 wt.-%. Under shear conditions, the network of CNF fibrils is able to retain its shape rather than it starting to flow depending on the frequency and strain ranges. As such, the deformation of the fibrillar network allows for a shear thinning effect and the chances for reversible or irreversible flocculation only happen at extremely high frequencies outside the linear visco-elastic range. As such, the fibrillar network structure in the suspension breaks down and is further characterized by non-linear dynamic properties after exceeding a critical strain. Indeed, the CNF suspensions from brown algae possess a yield stress above which the material behaves fluid-like, while it has more solid-like performance at the lower shear stresses below yield stress (Gao et al. 2018). An example for the flow properties with apparent viscosity and hysteresis effects of algae-cellulose from the *Posidonia oceanica* balls and leaves is illustrated in Fig. 13.17, indicating the progressive plateaus that may relate to the timely formation and break-up of floccule (Bettaieb et al. 2015d).

Interestingly, the evolution of storage moduli with increase in fibrillation could be related to the higher water retention capacity of the fibrillated cellulose, as a measure for the degree of fibrillation (Onyianta et al. 2020). The CNF from algae, however, was more prone to swelling than their wood-based counterparts and could be processed into stable suspensions with a solid content up to 3 wt.-%, in contrast with the lower solid contents for wood-based CNF suspensions that are usually

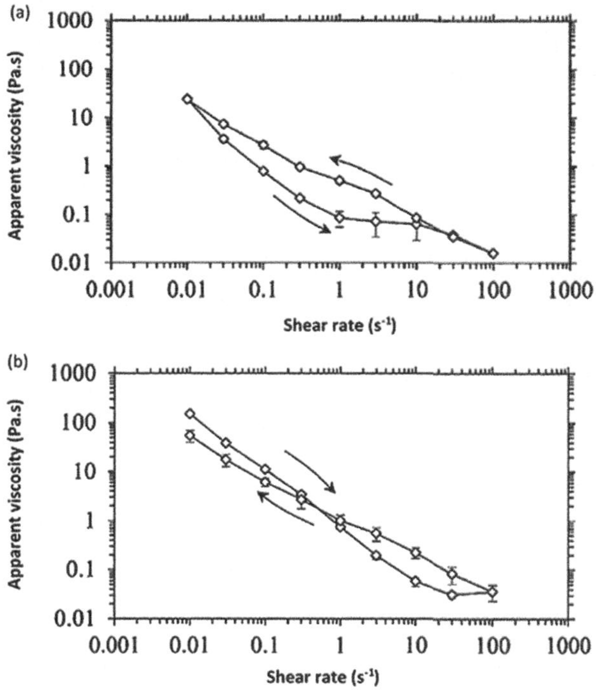

Fig. 13.17 Continuous flow measurements with ascending and descending shear rates during rotational testing of CNF suspensions containing TEMPO-oxidized CNF (comparable carboxyl content) from two different algae parts (**a**) *Possidonia oceanica* leaves and (**b**) *Possidonia oceanica* balls (Bettaieb et al. 2015d)

processed at 1 wt.-%. As such, the intrinsic properties and high quality of CNF from algae allow for the better processing conditions with significant enhancement in viscoelastic shear properties.

13.5 Application Potential of Algae-Based Nanocellulose

The general application domains of nanocellulose have been rapidly expanding in the last decades, with main applications in materials sciences, environmental sciences, energy storage, electronics and healthcare. The application potentials for general nanocellulose materials have been reviewed lately in some excellent works (Klemm et al. 2018). However, the works considered in this chapter only focus on some actual application reports of nanocellulose isolated from algae cellulose. In particular, the algae-based nanocellulose has specifically been applied in several industrial domains such as biomedicine, material sciences and water treatment, as summarized in Table 13.3.

Table 13.3 General overview of various applications based on algae nanocellulose

Domain	Application	Strain	Nanocellulose properties	References
Biomedicine	Virus removal in recombinant protein production	*Cladophora* sp.	CNF paper: 17 nm pore size; crystallinity index: 97%	Asper et al. (2015)
	Dietary fibre for patients with inflammatory bowel disease	*Sargassum* sp.	CNF 0.1 w% diluted; hydrothermal pre-treatment and extruded (100um nozzle; 245 MPa)	Azuma et al. (2013)
	Haemodialysis	*Cladophora* sp.	CNF polypyrole functionalized membranes	Ferraz et al. (2013)
	Virus-retentive size exclusion filtration	*Cladophora* sp.	CNF filters with average pore size distribution of 100 nm	Metreveli et al. (2014)
	Virus-retentive size exclusion filtration	*Cladophora* sp.	CNF filters: pore size distribution 10–25 nm	Gustafsson and Mihranyan (2016)
	Virus-retentive size exclusion filtration	*Cladophora* sp.	CNF drained over support membrane: pore size distribution 22 nm	Manukyan et al. (2019)
	Cytocompatilibiy mono/macrophages	*Cladophora* sp.	CNF with TEMPO-oxidation and film casting	Hua et al. (2015)
	Matrix material in immunosorbent devices	*Cladophora* sp.	Porous sulphonated CNF beads (15 μm) with surface area of 100 $m^2 g^{-1}$ and pore sizes between 20 and 60 nm	Rocha et al. (2018)
	Removal of protein aggregates and viruses in human serum albumin	*Cladophora* sp.	CNF filter with average pore sizes from 2 to 80 nm	Wu et al. (2020)
Material sciences	Concrete enforcement	*Cladophora* sp.	CNF 1 w% in matrix of sand, water and cement	Cengiz et al. (2017)
	Thermal insulators and fire retardants	*Cladophora* sp.	CNF aerogel with metal-organic framework	Zhou et al. (2019)
	Polypyrrole/ nanocellulose based supercapacitors	*Cladophora* sp.	CNF + polypyrrole as positive electrode	Tammela et al. (2015)
	Adsorbents/Filtration	*Cladophora* sp.	CNF charred at 500 °C for 1 h with Sbet of >300 $m^2 g^{-1}$	Štefelová et al. (2017)
	Electrical insulation	*Cladophora* sp.	CNF paper sheets	Le Bras et al. (2015)
	Nanocomposite materials	*Gelidium elegans*	CNC acid hydrolysis H_2SO_4	Chen et al. (2016)
	Nanocomposite materials	*Gelidium sesquipedale*	CNC acid hydrolysis H_2SO_4	El Achaby et al. (2018)
	Nanocomposite materials	*Ulva lactuca*	CNF	Wahlström et al. (2020)

(continued)

Table 13.3 (continued)

Domain	Application	Strain	Nanocellulose properties	References
Water treatment	Drinking water purification	*Pithopora* sp.	CNF (20–30 nm thick), filters with open porous structure	Gustafsson et al. (2019)
	Removal of mercury ions	*Cystoseria* sp.	CNC with Fe_3O_4 functionalization	Zarei et al. (2018)
	Red dye absorption	*Cladophora* sp.	CNF porous beads	Ruan et al. (2018)
Emulsions and food	Stabilizers for oil-water emulsions	*Cladophora* sp.	CNC + bed ion exchange resin	Kalashnikova et al. (2013)
	Milk thickeners	Waste brown algae	CNF + TEMPO-oxidation	Gao et al. (2018)

As a first priority, the majority of reported studies on algal cellulose nanofibres have their application in biomedicine such as size-exclusion filtration for retention of bacteria and viruses and in electroactive membranes in haemodialysis (Ferraz et al. 2013). Filter papers composed of 100% naturally derived cellulose were validated and it was shown that algal nanocellulose paper was useful for the removal of endogenous retroviruses and retro-virus like particles during the production of recombinant proteins (Asper et al. 2015; Gustafsson and Mihranyan 2016). Other biomedical applications are centred on the effects of algal nanocellulose on colon inflammation using the mouse model of inflammatory bowel disease: it was demonstrated that algal cellulose nanofibres may be an interesting source for dietary fibre (Azuma et al. 2013). The porous beads prepared from *Cladophora* green algae nanocellulose were sulphonated and showed great potential as immunosorbents (Rocha et al. 2018). Recently, the potential leachables of nanocellulose from *Cladophora* green algae were characterized and showed to be acceptable for future biomedical applications (Liu et al. 2017).

The algae cellulose nanofibres gain a lot of interest to be applied in materials (Bacakova et al. 2019), owing to their excellent properties, with high mechanical strength, chemical inertness and tailorable morphological, physical, chemical, electrical, thermal and optical properties. It can act as a filler and binder in pharmaceutical capsules (Samiee et al. 2019), as concrete reinforcement (Cengiz et al. 2017), thermal insulator and fire retardants (Zhou et al. 2019), or electrode material in supercapacitors (Tammela et al. 2015) or be applied in several nanocomposite materials (Chen et al. 2016; El Achaby et al. 2018; Wahlström et al. 2020).

A third major domain of application currently is water treatment (Shak et al. 2018). Filters fabricated from *Phitophora* green algae were applied for drinking water purification (Gustafsson et al. 2019) and filters containing cellulose from *Cladophora* green algae were tested and proven for trapping swine influenza virus particles with retention efficiency that matches industrial filters (Metreveli et al. 2014). In addition, adsorbent beads have been fabricated using *Cladophora* cellulose for fast adsorption of palladium (II) ions of up to 80% maximum capacity in 2 h (Ruan et al. 2018). The initiation of this study could develop an alternative for electronic waste management by recovering and recycling palladium through an efficient and low-cost technology (El-Safty et al. 2015). The nanocellulose from *Cystoseria* brown algae and Fe_3O_4 nanocellulose were utilized successfully for mercury removal (Zarei et al. 2018). The 2,3-dialdehyde cellulose beads crosslinked with chitosan have been reported to adsorb effectively red dye (Ruan et al. 2018). A recent innovative approach used *Cladophora* nanofibres-based biochar for dye and oil adsorption (Štefelová et al. 2017).

Other recent applications for algal cellulose nanofibres were reported, for example, as stabilizers for water/oil emulsion or in food as milk thickeners (Kalashnikova et al. 2013; Gao et al. 2018). It is expected that continued research effort will lead to further exploration of algae-based nanocellulose materials to be applied in different domains such as food in, for example, packaging materials or in biomedicine for skin tissue engineering and repair or wound healing.

13.6 Conclusion and Outlook

The exploitation of cellulose from macroalgae as a source for the production of nanocellulose materials provides an excellent alternative for the traditional sources of nanocellulose from lignocellulosic origin. As such, the availability of cellulose sources and sustainability in terms of resource exploitation and production technology may be significantly broadened. From an industrial point of view, the extremely high quality of algae-based nanocellulose and reduction in complexity of processing with associated energy-savings are the driving forces for a transition towards the valorization of algae-based cellulose and/or the development of comprehensive biorefinery schemes for algae biomass. The use of algae-cellulose as the starting material may therefore provide a solution for main obstacles and future industrial breakthrough of nanocellulose. However, the selection of algae types and families and the optimization of processing parameters are critical in the control of the final nanocellulose properties.

Starting from fundamental understanding on the arrangement of cellulose synthase complexes (i.e. multiple-row linear types, consolidated rosette types or isolated rosette types), the organization of relatively thick cellulose microfibrils in the algae cell wall with a high degree of crystallinity offers a natural morphology for extraction of microfibrillar cellulose (MFC) following purely chemical extraction steps without harsh conditions for extensive removal of the circumfering matrix components. The further isolation of cellulose nanofibrils (CNFs) and cellulose nanowhiskers (CNCs) may be performed according to procedures that are well known from the processing of wood-based biomass, but the properties of the individual microfibrils from algae-cellulose remain predominant in determination of the unique nanocellulose properties, with superior length of fibrils, extremely high aspect ratio, crystallinity, thermal stability and consequent mechanical reinforcement capability. Based on the presented recent works in the field, either general algae types or locally harvested algae (or algae residues) have been selected for nanocellulose production, while they seem to be preferentially processed towards a specific type of nanocellulose component (at least in the overall view, although different combinations have been applied), for example, production of MFC from green algae, production of CNF from green or brown algae and production of CNC from red algae. Interestingly, the variability in morphologies and properties between different algae species agree relatively well among various studies, but of course the selection of attractive processing conditions remains a key role. In this respect, first steps have been undertaken in environmentally friendly processing through peroxide bleaching and ionic liquid hydrolysis.

In future, the additional role of biotechnology in further tuning the nanocellulose properties and functionalization of nanocellulose materials remains an open field of research. In parallel with the production of bacterial nanocellulose, the steering on properties of algae-based nanocellulose through explicit cultivation under controlled conditions and/or modification of algae growth media have not at all been considered. The highly interdisciplinary research field requires better understanding of the structures and properties of nanocellulose in relation to the genetics of the algae biomass. As such, a vast breakthrough could be realized towards more independence of nanocellulose products and one-step biological production of functionalized materials with high economic and technological value.

References

Asper M, Hanrieder T, Quellmalz A, Mihranyan A (2015) Removal of xenotropic murine leukemia virus by nanocellulose based filter paper. Biologicals 43(6):452–456

Azuma K, Osaki T, Ifuku S, Maeda H, Morimoto M, Takashima O, Tsuka T, Imagawa T, Okamoto Y, Saimoto H, Minami S (2013) Suppressive effects of cellulose nanofibers—made from adlay and seaweed - on colon inflammation in an inflammatory bowel-disease model. Bioact Carbohydr Diet Fibre 2(1):65–72

Bacakova L, Pajorova J, Bacakova M, Skogberg A, Kallio P, Kolarova K, Svorcik V (2019) Versatile application of nanocellulose: from industry to skin tissue engineering and wound healing. Nano 9(2):164

Baghel RS, Trivedi N, Gupta V, Neori A, Reddy CRK, Lali A, Jha B (2015) Biorefining of marine macroalgal biomass for production of biofuel and commodity chemicals. Green Chem 17(4):2436–2443

Baghel RS, Mantri VA, Reddy CRK (2017) A new wave of research interest in marine macroalgae for chemicals and fuels: challenges and potentials. In: Kerton FM, Yan N (eds) Fuels, chemicals and materials from the oceans and aquatic sources, 1st edn. Wiley and Sons, New York, pp 43–63

Baghel RS, Suthar P, Gajaria TK, Bhattacharya S, Anil A, Reddy CRK (2020) Seaweed biorefinery: a sustainable process for valorising the biomass of brown seaweed. J Clean Prod 263:121359

Barbot YN, Al-Ghaili H, Benz R (2016) A review on the valorization of macroalgal wastes for biomethane production. Mar Drugs 14(6):120

Barhoum A, Samyn P, Öhlund T, Dufresne A (2017) Review of recent research on flexible multifunctional nanopapers. Nanoscale 9(40):15181–15205

Bettaieb F, Khiari R, Dufresne A, Mhenni MF, Belgacem MN (2015a) Mechanical and thermal properties of *Posidonia oceanica* cellulose nanocrystal reinforced polymer. Carbohydr Polym 123:99–104

Bettaieb F, Khiari R, Dufresne A, Mhenni MF, Putaux JL, Boufi S (2015b) Nanofibrillar cellulose from *Posidonia oceanica*: properties and morphological features. Ind Crop Prod 72:97–106

Bettaieb F, Khiari R, Hassan ML, Belgacem MN, Bras J, Dufresne A, Mhenni MF (2015c) Preparation and characterization of new cellulose nanocrystals from marine biomass *Posidonia oceanica*. Ind Crop Prod 72:175–182

Bettaieb F, Nechyporchuk O, Khiari R, Mhenni MF, Dufresne A, Belgacem MN (2015d) Effect of the oxidation treatment on the production of cellulose nanofiber suspensions from *Posidonia oceanica*: the rheological aspect. Carbohydr Polym 134:664–672

Bhattacharya D, Medlin L (1998) Algal phylogeny and the origin of land plants. Plant Physiol 116(1):9–15

Brown RM, Montezinos D (1976) Cellulose microfibrils: visualization of biosynthetic and orienting complexes in association with the plasma membrane. PNAS 73(1):143–147

Brown RM, Jr, Saxena IM (2007) Cellulose: Molecular and structural biology - Selected articles on the synthesis, structure, and applications of cellulose: Springer Netherlands.

Cengiz A, Kaya M, Bayramgil NP (2017) Flexural stress enhancement of concrete by incorporation of algal cellulose nanofibers. Constr Build Mater 149:289–295

Charrier B, Rabillé H, Billoud B (2019) Gazing at cell wall expansion under a golden light. Trends Plant Sci 24(2):130–141

Chen YW, Lee HV, Juan JC, Phang SM (2016) Production of new cellulose nanomaterial from red algae marine biomass *Gelidium elegans*. Carbohydr Polym 151:1210–1219

Cock JM, Sterck L, Rouzé P, Scornet D, Allen AE, Amoutzias G, Anthouard V, Artiguenave F, Aury JM, Badger JH, Beszteri B, Billiau L, Bonnet E, Bothwell JH, Bowler C, Boyen C, Brownlee C, Carrano CJ, Charrier B, Cho GY, Coelho SM, Collén J, Corre E, Da Silva C, Delage L, Delaroque N, Dittami SM, Doulbeau S, Elias M, Farnham G, Gachon CMM, Gschloessl B, Heesch S, Jabbari K, Jubin C, Kawai H, Kimura K, Kloareg B, Küpper FC, Lang D, Le Bail A, Leblanc C, Lerouge P, Lohr M, Lopez PJ, Martens C, Maumus F, Michel G, Miranda-Saavedra D, Morales J, Moreau H, Motomura T, Nagasato C, Napoli CA, Nelson DR, Nyvall-Collén P, Peters AF, Pommier C, Potin P, Poulain J, Quesneville H, Read B, Rensing SA, Ritter A, Rousvoal S, Samanta M, Samson G, Schroeder DA, Ségurens B, Strittmatter M, Tonon T, Tregear JW, Valentin K, von Dassow P, Yamagishi T, Van de Peer Y, Wincker P (2010) The *Ectocarpus* genome and the independent evolution of multicellularity in brown algae. Nature 465:617–621

Credou J, Berthelot T (2014) Cellulose: from biocompatible to bioactive material. J Mater Chem B 2(30):4767–4788

Davis TA, Volesky B, Mucci A (2003) A review of the biochemistry of heavy metal biosorption by brown algae. Water Res 37(18):4311–4330

De France K, Zeng A, Wu T, Nyström G (2020) Functional materials from nanocellulose: Utilizing structure–property relationships in bottom-up fabrication. Adv Mater 33:2000657

Deniaud-Bouët E, Kervarec N, Michel G, Tonon T, Kloareg B, Hervé C (2014) Chemical and enzymatic fractionation of cell walls from Fucales: insights into the structure of the extracellular matrix of brown algae. Ann Bot 114(6):1203–1216

Deniaud-Bouët E, Hardouin K, Potin P, Kloareg B, Hervé C (2017) A review about brown algal cell walls and fucose-containing sulfated polysaccharides: Cell wall context, biomedical properties and key research challenges. Carbohydr Polym 175:395–408

Doh H, Lee MH, Whiteside WS (2020) Physicochemical characteristics of cellulose nanocrystals isolated from seaweed biomass. Food Hydrocol 102:105542

Dufresne A (2019) Nanocellulose processing properties and potential applications. Curr Forestry Rep 5(2):76–89

Ek R, Gustafsson C, Nutt A, Iversen T, Nyström C (1998) Cellulose powder from *Cladophora* sp. algae. J Mol Recogn 11(1–6):263–265

El Achaby M, Kassab Z, Aboulkas A, Gaillard C, Barakat A (2018) Reuse of red algae waste for the production of cellulose nanocrystals and its application in polymer nanocomposites. Int J Biol Macromol 106:681–691

El-Safty SA, Shenashen MA, Sakai M, Elshehy A, Halada K (2015) Detection and recovery of palladium, gold and cobalt metals from the urban mine using novel sensors/adsorbents designated with nanoscale wagon-wheel-shaped pores. J Vis Exp 106:e53044

Ferraz N, Leschinskaya A, Toomadj F, Fellström B, Strømme M, Mihranyan A (2013) Membrane characterization and solute diffusion in porous composite nanocellulose membranes for hemodialysis. Cellulose 20(6):2959–2970

Fortunati E, Luzi F, Puglia D, Petrucci R, Kenny JM, Torre L (2015) Processing of PLA nanocomposites with cellulose nanocrystals extracted from *Posidonia oceanica* waste: Innovative reuse of coastal plant. Ind Crop Prod 67:439–447

Freile-Pelegrín Y, Chávez-Quintal C, Caamal-Fuentes E, Vázquez-Delfín E, Madera-Santana T, Robledo D (2020) Valorization of the filamentous seaweed *Chaetomorpha gracilis* (Cladophoraceae, Chlorophyta) from an IMTA system. J Appl Phycol 32(4):2295–2306

Gao H, Duan B, Lu A, Deng H, Du Y, Shi X, Zhang L (2018) Fabrication of cellulose nanofibers from waste brown algae and their potential application as milk thickeners. Food Hydrocoll 79:473–481

Gardner KH, Blackwell J (1974) The hydrogen bonding in native cellulose. Biochim Biophys Acta 343(1):232–237

Giddings TH, Brower DK, Staehelin LA (1980) Visualization of particle complexes in the plasma membrane of *Micrasterias denticulata* associated with the formation of cellulose fibrils in primary and secondary cell walls. J Cell Biol 84(2):327–339

Guo J, Uddin KMA, Mihhels K, Fang W, Laaksonen P, Zhu JY, Rojas OJ (2017) Contribution of residual proteins to the thermomechanical performance of cellulosic nanofibrils isolated from green macroalgae. ACS Sustain Chem Eng 5(8):6978–6985

Gustafsson S, Mihranyan A (2016) Strategies for tailoring the pore-size distribution of virus retention filter papers. ACS Appl Mater Interfaces 8(22):13759–13767

Gustafsson O, Manukyan L, Gustafsson S, Tummala GK, Zaman S, Begum A, Alfasane MA, Siddique-e-Rabbani K, Mihranyan A (2019) Scalable and sustainable total pathogen removal filter paper for point-of-use drinking water purification in Bangladesh. ACS Sustain Chem Eng 7(17):14373–14383

Ha MA, Apperley DC, Evans BW, Huxham IM, Jardine WG, Viëtor RJ, Reis D, Vian B, Jarvis MC (1998) Fine structure in cellulose microfibrils: NMR evidence from onion and quince. Plant J 16(2):183–190

Habibi Y, Lucia LA, Rojas OJ (2010) Cellulose nanocrystals: chemistry, self-assembly, and applications. Chem Rev 110(6):3479–3500

Hua K, Carlsson DO, Ålander E, Lindström T, Strømme M, Mihranyan A, Ferraz N (2014) Translational study between structure and biological response of nanocellulose from wood and green algae. RSC Adv 4(6):2892–2903

Hua K, Strømme M, Mihranyan A, Ferraz N (2015) Nanocellulose from green algae modulates the in vitro inflammatory response of monocytes/macrophages. Cellulose 22(6):3673–3688

Jarvis M (2003) Cellulose stacks up? Nature 426(6967):611–612

Kalashnikova I, Bizot H, Bertoncini P, Cathala B, Capron I (2013) Cellulosic nanorods of various aspect ratios for oil in water Pickering emulsions. Soft Matter 9(3):952–959

Kandhola G, Djioleu A, Rajan K, Labbé N, Sakon J, Carrier DJ, Kim JW (2020) Maximizing production of cellulose nanocrystals and nanofibers from pre-extracted loblolly pine kraft pulp: a response surface approach. Bioresour Bioprocess 7(1):19

Kang OL, Ramli N, Said M, Ahmad M, Yasir SM, Ariff A (2011) *Kappaphycus alvarezii* waste biomass: a potential biosorbent for chromium ions removal. J Environ Sci 23(6):918–922

Kargarzadeh H, Ioelovich M, Ahmad I, Thomas S, Dufresne A (2017) Methods for extraction of nanocellulose from various sources. In: Kargarzadeh H, Ahmad I, Thomas S, Dufresne A (eds) Handbook of nanocellulose and cellulose nanocomposites, 1st edn. Wiley and Sons, pp 1–49

Kim UJ, Kuga S, Wada M, Okano T, Kondo T (2000) Periodate oxidation of crystalline cellulose. Biomacromolecules 1(3):488–492

Kimura S, Laosinchai W, Itoh T, Cui X, Linder CR, Brown RM (1999) Immunogold labeling of rosette terminal cellulose-synthesizing complexes in the vascular plant *Vigna angularis*. Plant Cell 11(11):2075–2085
Klemm D, Heublein B, Fink HP, Bohn A (2005) Cellulose: fascinating biopolymer and sustainable raw material. Angew Chem Int Ed 44(22):3358–3339
Klemm D, Cranston ED, Fischer D, Gama M, Kedzior SA, Kralisch D, Kramer F, Kondo T, Lindström T, Nietzsche S, Petzold-Welcke K, Rauchfuß F (2018) Nanocellulose as a natural source for groundbreaking applications in materials science: today's state. Mater Today 21(7):720–748
Kloareg B, Quatrano RD (1988) Structure of the cell walls of marine algae and ecophysiological functions of the matrix polysaccharides. Oceanogr Mar Biol Annu Rev 26:259–315
Knoshaug EP, Shi B, Shannon TG, Mleziva MM, Pienkos PT (2013) The potential of photosynthetic aquatic species as sources of useful cellulose fibers - a review. J Appl Phycol 25(4):1123–1134
Koyama M, Sugiyama J, Itoh T (1997) Systematic survey on crystalline features of algal celluloses. Cellulose 4(2):147–160
Le Bras D, Strømme M, Mihranyan A (2015) Characterization of dielectric properties of nanocellulose from wood and algae for electrical insulator applications. J Phys Chem B 119(18):5911–5917
Lee RE (2018) Phycology, 5th edn. Cambridge University Press
Leliaert F, Smith DR, Moreau H, Herron MD, Verbruggen H, Delwiche CF, De Clerck O (2012) Phylogeny and molecular evolution of the green algae. Crit Rev Plant Sci 31(1):1–46
Lindh J, Ruan C, Strømme M, Mihranyan A (2016) Preparation of porous cellulose beads via introduction of diamine spacers. Langmuir 32(22):5600–5607
Liu J, Willför S, Mihranyan A (2017) On importance of impurities, potential leachables and extractables in algal nanocellulose for biomedical use. Carbohydr Polym 172:11–19
Luzi F, Fortunati E, Puglia D, Petrucci R, Kenny JM, Torre L (2015) Study of disintegrability in compost and enzymatic degradation of PLA and PLA nanocomposites reinforced with cellulose nanocrystals extracted from *Posidonia oceanica*. Polym Degrad Stab 121:105–115
Luzi F, Fortunati E, Puglia D, Petrucci R, Kenny JM, Torre L (2016) Modulation of acid hydrolysis reaction time for the extraction of cellulose nanocrystals from *Posidonia oceanica* leaves. J Renew Mater 4(3):190–198
Manukyan L, Padova J, Mihranyan A (2019). Virus removal filtration of chemically defined Chinese Hamster Ovary cells medium with nanocellulose-based size exclusion filter. Biologicals 59:62–67
Martone PT, Janot K, Fujita M, Wasteneys G, Ruel K, Joseleau JP, Estevez JM (2019) Cellulose-rich secondary walls in wave-swept red macroalgae fortify flexible tissues. Planta 250(6):1867–1879
McNamara JT, Morgan JLW, Zimmer J (2015) A molecular description of cellulose biosynthesis. Annu Rev Biochem 84(1):895–921
Messyasz B, Leska B, Fabrowska J, Pikosz M, Roj E, Cieslak A, Schroeder G (2015) Biomass of freshwater *Cladophora* as a raw material for agriculture and the cosmetic industry. Open Chem 13(1):1108–1118
Metreveli G, Wågberg L, Emmoth E, Belák S, Strømme M, Mihranyan A (2014) A size-exclusion nanocellulose filter paper for virus removal. Adv Healthc Mater 3(10):1546–1550
Mihranyan A (2011) Cellulose from Cladophorales green algae: From environmental problem to high-tech composite materials. J Appl Polym Sci 119(4):2449–2460
Mihranyan A, Llagostera AP, Karmhag R, Strømme M, Ek R (2004) Moisture sorption by cellulose powders of varying crystallinity. Int J Pharm 269(2):433–442
Mikkelsen MD, Harholt J, Ulvskov P, Johansen IE, Fangel JU, Doblin MS, Bacic A, Willats WGT (2014) Evidence for land plant cell wall biosynthetic mechanisms in charophyte green algae. Ann Bot 114(6):1217–1236
Mondal K, Sakurai S, Bhagabati P, Goud VV, Katiyar V (2019) Fabrication of nano cellulose from marine green algae residue and their effects as filler in developing biodegradable polymer com-

posite. In Proceedings of international symposium on a new era in food science and technology, Japan, p 28

Moon RJ, Martini A, Nairn J, Simonsen J, Youngblood J (2011) Cellulose nanomaterials review: structure, properties and nanocomposites. Chem Soc Rev 40(7):3941–3994

Nash MC, Diaz-Pulido G, Harvey AS, Adey W (2019) Coralline algal calcification: a morphological and process-based understanding. PLoS One 14(9):e0221396

Nobles DR, Romanovicz DK, Brown RM (2001) Cellulose in cyanobacteria. Origin of vascular plant cellulose synthase? Plant Physiol 127(2):529

Ono Y, Fukui S, Funahashi R, Isogai A (2019) Relationship of distribution of carboxy groups to molar mass distribution of TEMPO-oxidized algal, cotton, and wood cellulose nanofibrils. Biomacromolecules 20(10):4026–4034

Onyianta AJ, O'Rourke D, Sun D, Popescu CM, Dorris M (2020) High aspect ratio cellulose nanofibrils from macroalgae *Laminaria hyperborea* cellulose extract via a zero-waste low energy process. Cellulose 27(14):7997–8010

O'Sullivan AC (1997) Cellulose: the structure slowly unravels. Cellulose 4(3):173–207

Paniz OG, Pereira CMP, Pacheco BS, Wolke SI, Maron GK, Mansilla A, Colepicolo P, Orlandi MO, Osorio AG, Carreño NLV (2020) Cellulosic material obtained from Antarctic algae biomass. Cellulose 27(1):113–126

Percival EGV, Ross AG (1948) The cellulose of marine algae. Nature 162(4127):895–896

Popper ZA, Michel G, Hervé C, Domozych DS, Willats WGT, Tuohy MG, Kloareg B, Stengel DB (2011) Evolution and diversity of plant cell walls: From algae to flowering plants. Annu Rev Plant Biol 62(1):567–590

Popper ZA, Ralet MC, Domozych DS (2014) Plant and algal cell walls: diversity and functionality. Ann Bot 114(6):1043–1048

Prabhu MS, Israel A, Palatnik RR, Zilberman D, Golberg A (2020) Integrated biorefinery process for sustainable fractionation of *Ulva ohnoi* (Chlorophyta): process optimization and revenue analysis. J Appl Phycol 32(4):2271–2282

Raimundo SC, Pattathil S, Eberhard S, Hahn MG, Popper ZA (2017) β-1,3-Glucans are components of brown seaweed (Phaeophyceae) cell walls. Protoplasma 254(2):997–1016

Roberts AW, Roberts E (2007) Evolution of the cellulose synthase (cesa) gene family: insights from green algae and seedless plants. In: Brown RM, Saxena IM (eds) Cellulose: molecular and structural biology – selected articles on the synthesis, structure, and applications of cellulose, 1st edn. Springer, Amsterdam, pp 17–34

Rocha I, Ferraz N, Mihranyan A, Strømme M, Lindh J (2018) Sulfonated nanocellulose beads as potential immunosorbents. Cellulose 25(3):1899–1910

Ruan CQ, Strømme M, Lindh J (2018) Preparation of porous 2,3-dialdehyde cellulose beads crosslinked with chitosan and their application in adsorption of Congo red dye. Carbohydr Polym 181:200–207

Salmeán AA, Duffieux D, Harholt J, Qin F, Michel G, Czjzek M, Willats WGT, Hervé C (2017) Insoluble $(1 \rightarrow 3)$, $(1 \rightarrow 4)$-β-D-glucan is a component of cell walls in brown algae (Phaeophyceae) and is masked by alginates in tissues. Sci Rep 7(1):2880

Samiee S, Ahmadzadeh H, Hosseini M, Lyon S (2019) Algae as a source of microcrystalline cellulose. In: Hosseini M (ed) Advanced bioprocessing for alternative fuels, biobased chemicals, and bioproducts:technologies and approaches for scale-up and commercialization. Woodhead Publishing, Sawston, pp 331–350

Saritha S, Nair SM, Kumar NC (2013) Nano-ordered cellulose containing I_α crystalline domains derived from the algae *Chaetomorpha antennina*. BioNanoScience 3(4):423–427

Segal L, Creely JJ, Martin AE, Conrad CM (1959) An empirical method for estimating the degree of crystallinity of native cellulose using the X-ray diffractometer. Text Res J 29(10):786–794

Shak K, Yee P, Pang YL, Mah SK (2018) Nanocellulose: recent advances and its prospects in environmental remediation. Beilstein J Nanotechnol 9:2479–2498

Siddhanta AK, Prasad K, Meena R, Prasad G, Mehta GK, Chhatbar MU, Oza MD, Kumar S, Sanandiya ND (2009) Profiling of cellulose content in Indian seaweed species. Bioresour Technol 100(24):6669–6673

Singh S, Gaikwad KK, Park SI, Lee YS (2017) Microwave-assisted step reduced extraction of seaweed (*Gelidiella aceroso*) cellulose nanocrystals. Int J Biol Macromol 99:506–510

Somerville C (2006) Cellulose synthesis in higher plants. Annu Rev Cell Dev Biol 22(1):53–78

Stanford EC (1885) New method of treating seaweed products at inventions exhibition. Jour Soc Chem Ind 4:518–520

Štefelová J, Slovák V, Siqueira G, Olsson RT, Tingaut P, Zimmermann T, Sehaqui H (2017) Drying and pyrolysis of cellulose nanofibers from wood, bacteria, and algae for char application in oil absorption and dye adsorption. ACS Sustain Chem Eng 5(3):2679–2692

Sucaldito MR, Camacho DH (2017) Characteristics of unique HBr-hydrolyzed cellulose nanocrystals from freshwater green algae (*Cladophora rupestris*) and its reinforcement in starch-based film. Carbohydr Polym 169:315–323

Taheri H, Samyn P (2015) Rheological properties and processing of polymer blends with micro- and nanofibrillated cellulose. In: Hakeem KR, Jawaid M, Alothman OY (eds) Agricultural biomass based potential materials. Springer, Heidelberg, pp 259–291

Taheri H, Samyn P (2016) Effect of homogenization (microfluidization) process parameters in mechanical production of micro- and nanofibrillated cellulose on its rheological and morphological properties. Cellulose 23(2):1221–1238

Taheri H, Samyn P (2017) Rheological properties and self-assembly of cellulose nanowhiskers in renewable polymer biocomposites. In: Srinivasan A, Bandyopadhyay S (eds) Advances in polymer materials and technology. CRC Press, Boca Raton, pp 623–662

Tammela P, Wang Z, Frykstrand S, Zhang P, Sintorn IM, Nyholm L, Strømme M (2015) Asymmetric supercapacitors based on carbon nanofibre and polypyrrole/nanocellulose composite electrodes. RSC Adv 5(21):16405–16413

Tarchoun AF, Trache D, Klapötke TM (2019) Microcrystalline cellulose from *Posidonia oceanica* brown algae: extraction and characterization. Int J Biol Macromol 138:837–845

Torres MD, Kraan S, Domínguez H (2019) Seaweed biorefinery. Rev Environ Sci Biotechnol 18(2):335–388

Trache D, Tarchoun AF, Derradji M, Hamidon TS, Masruchin N, Brosse N, Hussin NH (2020) Nanocellulose: From fundamentals to advanced applications. Front Chem 8:392

Trivedi N, Gupta V, Reddy CRK, Jha B (2013) Enzymatic hydrolysis and production of bioethanol from common macrophytic green alga *Ulva fasciata* Delile. Bioresour Technol 150:106–112

Tsekos I (1999) The sites of cellulose synthesis in algae: diversity and evolution of cellulose-synthesizing enzyme complexes. J Phycol 35(4):635–655

Uju WAT, Goto M, Kamiya N (2015) Great potency of seaweed waste biomass from the carrageenan industry for bioethanol production by peracetic acid–ionic liquid pretreatment. Biomass Bioenergy 81:63–69

Van Hai L, Son HN, Seo YB (2015) Physical and bio-composite properties of nanocrystalline cellulose from wood, cotton linters, cattail, and red algae. Cellulose 22(3):1789–1798

Wahlström N, Edlund U, Pavia H, Toth G, Jaworski A, Pell AJ, Choong FX, Shirani H, Nilsson KPR, Richter-Dahlfors A (2020) Cellulose from the green macroalgae *Ulva lactuca*: isolation, characterization, optotracing, and production of cellulose nanofibrils. Cellulose 27(7):3707–3725

Wu L, Mantas A, Gustafsson S, Manukyan L, Mihranyan A. (2020) Aggregate removal nanofiltration of human serum albumin solution using nanocellulose-based filter paper. Biomedicines 8:209.

Xiang Z, Gao W, Chen L, Lan W, Zhu JY, Runge T (2016) A comparison of cellulose nanofibrils produced from Cladophora glomerata algae and bleached eucalyptus pulp. Cellulose 23(1):493–503

Youssef NH, Farag IF, Rinke C, Hallam SJ, Woyke T, Elshahed MS (2015) In silico analysis of the metabolic potential and niche specialization of candidate phylum "Latescibacteria" (WS3). PLoS One 10(6):e0127499

Zarei S, Niad M, Raanaei H (2018) The removal of mercury ion pollution by using Fe_3O_4-nanocellulose: synthesis, characterizations and DFT studies. J Hazard Mater 344:258–273

Zhao W, Lu Y, Ouyang H, Yan X, Pan R, Yang D, She X, Li J, Xia Y (2014) Algal cellulose aerogel: from environmental problem (green tide) to novel high-tech materials. Sci Technol Rev 32(4–5):34–39

Zhou S, Apostolopoulou-Kalkavoura V, Tavares da Costa MV, Bergström L, Strømme M, Xu C (2019) Elastic aerogels of cellulose nanofibers@metal–organic frameworks for thermal insulation and fire retardancy. Nano-Micro Lett 12:9

Chapter 14
Pelagic *Sargassum* as a Source of Micro- and Nanocellulose for Environmentally Sustainable Plastics

Héctor Uriel Castañeda-Serna, Georgina Calderón-Domínguez, Ma. De la Paz Salgado-Cruz, Arturo García-Bórquez, and Reynold Ramón Farrera-Rebollo

Contents

H. U. Castañeda-Serna · G. Calderón-Domínguez (✉) · R. R. Farrera-Rebollo
Instituto Politécnico Nacional, Escuela Nacional de Ciencias Biológicas, Departamento de Ingeniería Bioquímica, Ciudad de México, Mexico
e-mail: gcalderon@ipn.mx

M. De la Paz Salgado-Cruz
Instituto Politécnico Nacional, Escuela Nacional de Ciencias Biológicas, Departamento de Ingeniería Bioquímica, Ciudad de México, Mexico

Consejo Nacional de Ciencia y Tecnología (CONACYT), Ciudad de México, Mexico

A. García-Bórquez
Instituto Politécnico Nacional, Escuela Superior de Físico Matemáticas, Ciudad de México, Mexico

D. Thangadurai et al. (eds.), *Bioprospecting Algae for Nanosized Materials*, Nanotechnology in the Life Sciences,
https://doi.org/10.1007/978-3-030-81557-8_14

14.1 Introduction

One of the most critical worldwide environmental problems is plastic disposal, mainly into the sea, affecting marine life. One solution proposed in the European Union is improving recycling and limiting single-use plastics (SUPs) (Norton 2020). However, this solution is not entirely effective because people have given this material much importance. Thus, it is necessary to find other possibilities, such as using environmentally friendly plastics or producing materials using biodegradable polymers. It is important to mention that the best way to solve this problem is the conjunction of several alternatives (Shen et al. 2020).

Biodegradable plastics are materials that are decomposed by microbes and integrated into the natural ecosystem without ecotoxic effect or residual by-products. They are formed by biopolymers such as alginate, cellulose, chitosan, or starch, obtained from different sources, such as plants, cells, or algae (Tabassum and Khan 2020).

From these biopolymers, alginate and cellulose can be obtained from pelagic *Sargassum*, in concentrations varying between 19% and 34.6% (Davis et al. 2004; Rhein-Knudsen et al. 2017; Mohammed et al. 2018, 2020) and between 4.8% and 10% (Siddhanta et al. 2009; Kumar and Sahoo 2017), respectively.

Alginate is composed of mannuronic acid and guluronic acid, and the FDA recognized it as a GRAS material (Rhein-Knudsen et al. 2017; Shukla et al. 2019; Puscaselu et al. 2020). Concerning cellulose, it is one of the most abundant biopolymers on the earth, and nowadays, it is used to remove molecules or to strengthen biodegradable films or plastics (Rathod et al. 2015; De Oliveira et al. 2018; Doh and Whiteside 2020; Doh et al. 2020), with a high tendency to use nanocellulose.

In this respect, some authors (Doh et al. 2019; Doh and Whiteside 2020) reported nanocellulose yield values higher than 25.8–42.7% when using *S. fluitans*, while with other seaweed sources, the cited yields were smaller (16.3–26.1%). From these studies can be extrapolated the possible use of pelagic *Sargassum* species to obtain nanocellulose and alginate (Doh et al. 2019; Doh and Whiteside 2020).

Another compound of interest, extracted from the pelagic *Sargassum* species, is the fucoidan, which is a sulfated polysaccharide composed by repeating disaccharide units of α-1,3- and α-1,4-linked fucose residues with branchings attached at C2 positions (Rodríguez-Jasso et al. 2010a, b; Bachelet-Violette et al. 2014), which has been found in a concentration between 2.61% and 11.67% (Chale-Dzul et al. 2017, 2020; Fernando et al. 2020b). Additionally, fucoidan is reported to have different types of biological activities, such as an anticoagulant, anti-inflammatory, contraceptive, antibacterial, and antiviral, which could be applied to make intelligent plastics (Rodríguez-Jasso et al. 2010b; Lee and Lim 2014; Kimura et al. 2013; Ashayerizadeh et al. 2019).

In this chapter, recent information about the extraction of compounds from pelagic *Sargassum* as alginate, nanocellulose, and fucoidan, for applying in the elaboration of bioplastics with functional activities, will be discussed, as a way to decrease the contamination generated by the unusual arrival of this seaweed, as well as the nonbiodegradable plastics.

14.2 Environmental Problem with Plastics

The production of plastics at a large scale started in the 1940s and 1950s. Recently, plastics' annual production has doubled, reaching about 348 million tons in 2017 (Hahladakis et al. 2018; Shen et al. 2020). These substances have been widely used due to their properties of stability, low cost, durability, and lightness. However, plastics generate an environmental problem due to their resistance to natural degradation, which becomes a significant challenge in the waste management process, especially in sustainability. This situation is recognized as a white pollution problem (Rameshkumar et al. 2020; Shen et al. 2020).

Large volumes of plastics are generated due to their short lifespan (about less than 1 month), but their resistance to degradation makes this situation a potential environmental problem. Furthermore, during plastic decomposition, microplastics and nanoplastics can be formed, causing toxic effects on humans and ecological systems (Sheavly and Register 2007; Hahladakis et al. 2018; Shen et al. 2020). For example, the microplastics can alter the phytoplankton photosynthesis, affecting organic matter and oxygen production in the seas and even finding these small materials in the sea animals' stomachs (Shen et al. 2019; Norton 2020). However, they provide many benefits. For example, they are used as barrier materials, protecting food against decomposition promoted by ambient or microbiological factors. They are also utilized as containers due to their lightweight, which is reflected in less fuel consumption and decreased nocive gas emissions. Other reports have cited their use in textile and medical fields (Andrady and Neal 2009; Hahladakis et al. 2018). Hence, even though policies around the world have directed to ban the single-use plastics (Norton 2020), they have not been entirely adequate to mitigate this problem. One approach has been combining other types of actions, for example, the employment of bioplastics (Shen et al. 2020).

14.3 Bioplastics

The need to replace, partially or totally, conventional plastic material with natural polymers like cellulose derivatives, thermoplastic starch, polylactic acid, or alginate has become a worldwide subject of interest (Puscaselu et al. 2020; Rameshkumar et al. 2020).

In this regard, the term "bioplastic" is generic and can refer to bio-based plastics synthesized from biomass and renewable sources and the plastics (including petrochemical ones) that are biodegradable (Emadian et al. 2017; Folino et al. 2020; Hemalakshita et al. 2021). The possibility of making bioplastics using the chemicals compounds in waste or invasive species might combat two serious environmental problems that occur nowadays (Doh et al. 2020; Folino et al. 2020).

Among candidates to make bioplastics are the brown seaweed species. They are rich in carbohydrates, mainly alginate, which can build the backbone of biopolymer film (Gomaa et al. 2018; Doh et al. 2020). It has been reported that alginate has a linear structure that can form a strong polymer matrix and films, with fibrous structures and characteristics comparable to conventional plastic materials (Kumar and Sahoo 2017; Doh et al. 2020). The main disadvantage of alginate films is the hydrophilicity, but the crosslinking with ions Ca^{2+}, nanocellulose, or fucoidan might improve the film (Gomaa et al. 2018).

14.4 Alginate

The alginate is a biopolymer extracted from brown algae composed of (1–4) linked β-D-Mannuronic acid (M) and α-L-Guluronic acid (G) residues (Bhowmik et al. 2006; Rhein-Knudsen et al. 2017). Several authors reported alginate extraction from different brown seaweed species (Table 14.1), finding dissimilar yields (17–48.1%), as well as M/G ratios smaller than one. The M/G ratios give information about the gelling properties of alginate. For example, higher ratios have been related to the formation of soft and elastic gels, while smaller ratios to hard and brittle gels (Mohammed et al. 2018). These ratio differences could indicate that the amount of mannuronic and guluronic acids was alga dependent, as acids' concentration varied with the species. However, the largest differences in values could be related to the method employed to determine the composition. In this regard, Rosado-Espinosa et al. (2020) used FTIR, while other authors obtained the relation using the ^{1}H NMR method.

The pelagic *Sargassum* species have an alginate yield between 19% and 34.6%, while the M/G ratio indicates applications that require stronger gels (Davis et al.

Table 14.1 Alginate yield and M/G ratio for brown seaweed species

Seaweed source	Alginate yield (%)	M/G ratio	References
S. fluitans	21	0.52	Davis et al. (2004)
L. digitata	34.6	–	Schiener et al. (2014)
A. esculenta	37.4	–	Schiener et al. (2014)
S. muticum	25.6	1.04	El Atouani et al. (2016)
S. natans	23	0.6	Rhein-Knudsen et al. (2017)
S. vulgare	17	0.6	Rhein-Knudsen et al. (2017)
S. natans	19	0.51	Mohammed et al. (2018)
S. natans	28	0.45	Mohammed et al. (2020)
S. fluitans	34.6	51.6	Rosado-Espinosa et al. (2020)
S. hystrix	48.1	62.3	Rosado-Espinosa et al. (2020)

M mannuronic acid, *G* guluronic acid

2004; Rhein-Knudsen et al. 2017; Mohammed et al. 2018, 2020). Usually, pelagic *Sargassum* species have an M/G ratio less to 1; however, in the study of Rosado-Espinosa et al. (2020) for *S. fluitans*, this quantity was larger than the other authors, but the method to obtain this relation is not the most used.

14.5 Seaweed Alginate Extraction

The extraction of alginate from seaweed usually is carried out through chemical processes. Most of the methods are based on seaweed bleach, using a formaldehyde or sodium hypochlorite solutions. The product obtained from this reaction is calcium alginate, which is converted to alginic acid through the addition of hydrochloric or sulfuric acids, and finally extracted again as sodium alginate by the reaction of the alginic acid with sodium carbonate or sodium hydroxide and recovered by centrifuging or adding alcohol (Table 14.2).

Usually, the bleaching process on *Sargassum* species is carried out to produce colorless alginate by eliminating the phenolic compounds (Davis et al. 2004; Andriamanantoanina and Rinaudo 2010; Rhein-Knudsen et al. 2017; Mohammed et al. 2018, 2020). The second stage of conversion to alginic acid is fundamental to obtain alginate with higher purity. This is due to the presence of fucoidan inside the seaweed, which by this treatment is dissolved. The elimination of this carbohydrate is important to have a better sodium alginate yield (Davis et al. 2004; Rosado-Espinosa et al. 2020). Finally, the alkaline reaction is carried out to produce the salt; this treatment, to be more efficient, is recommended to be done at high temperatures (Mohammed et al. 2018). The posttreatment by centrifuging serves to remove the insoluble compounds, while the addition of ethanol precipitates the alginate (Davis et al. 2004; Rhein-Knudsen et al. 2017; Mohammed et al. 2018). Another posttreatment is through the calcium chloride route, which transforms the sodium alginate to calcium alginate (Gomez et al. 2009). Although the pelagic *Sargassum* species has not better yields of alginate than other seaweeds, this alga can be considered a good raw material due to its high disponibility.

14.6 Alginate Films

Different authors have reported the use of alginate from seaweed as a filmogenic material with the potential to form bioplastics; for example, Santana and Kieckbusch (2013) made films of alginate prepared from *Macrocystis pyrifera* and using different plasticizers as glycerol, xylitol, and mannitol, having the first two similar functional attributes for food wrappings. Gomaa et al. (2018) elaborated films using *Sargassum latifolium*, adding ion Ca^{2+} and fucoidan. They demonstrated that both substances enhanced the thickness of the films, and specifically, the fucoidan increased the permeability to water vapor and oxygen. All these films showed

Table 14.2 Types of treatment and posttreatment to get alginate from seaweeds

Seaweed	Pretreatment	Treatment	Posttreatment	Alginate yield (%)	References
S. fluitans	Formaldehyde solution	HCl 3 times, and extraction with NaOH at 22 °C or Na_2CO_3 at 80 °C	Precipitation by ethanol	21.1–24.5	Davis et al. (2004)
M. pyrifera	None	HCl 3 times during 15 min at room temperature, extraction with Na_2CO_3 2 h at 60 °C	Reconversions to calcium alginate with $CaCl_2$, alginic acid with HCl, and sodium alginate with Na_2CO_3, precipitation and washing with ethanol by Soxhlet	25–34	Gomez et al. (2009)
Spatoglossum sp. *Sargassum* sp.	Bleaching with NaClO	HCl pH = 4, extraction with Na_2CO_3 2 h at 60–70 °C	Precipitation with centrifugation (16,200 × *g* 2 min), filtration and precipitation by ethanol	17–34	Andriamanantoanina and Rinaudo (2010)
L. digitata *L. hyperborea* *S. latissima* *A. esculenta*	None	Extraction with Na_2CO_3 at 60 °C	Precipitation with centrifugation (16,200 × *g* 2 min)	28.5–37.4	Schiener et al. (2014)
S. muticum	Formalin during 1 day	HCl 1 day, extraction with Na_2CO_3 1 day	Precipitation by centrifugation (4500 × *g* 20 min), washing with ethanol and acetone	25.6	El Atouani et al. (2016)

(continued)

Table 14.2 (continued)

Seaweed	Pretreatment	Treatment	Posttreatment	Alginate yield (%)	References
S. natans *S. vulgare*	Formaldehyde 1 night	HCl 1 night, extraction with Na_2CO_3 2 h at 99 °C	Pressured filtration, mix with diatomaceous earth and precipitated with isopropanol	17–23	Rhein-Knudsen et al. (2017)
S. natans	Formaldehyde 1 night	H_2SO_4 at 25–60 °C, extraction with Na_2CO_3 0.5–6 h at 25–75 °C	Precipitation by centrifugation (8000 × *g* 10 min), filtered, reconversion with H_2SO_4, centrifugation (8000 × *g* 10 min), extraction with Na_2CO_3	17	Mohammed et al. (2018)
S. natans	Formaldehyde	H_2SO_4, alkali extraction with Na_2CO_3	Bleaching with NaClO, precipitation by centrifugation (8000 × *g* 10 min)	24.37	Mohammed et al. (2020)
S. filipendula *S. buxifolium* *S. hystrix* *S. fluitans* *S. ramifolium*	Formaldehyde 30 min	HCl 15 min, extraction with Na_2CO_3 2 h at 80 °C	Precipitation by ethanol	34.6–48.1	Rosado-Espinosa et al. (2020)

barrier properties against UV light, delaying the oxidative degradation of foods. More recently, Marismandani and Husni (2020) reported an edible film using alginate prepared from *Laminaria japonica* and mixed to glycerol/virgin coconut oil, finding that alginate concentration influences the characteristics of edible film (thickness, tensile strength, elongation, water transmission rate, and solubility). In other work, Kaczmarek (2020) obtained films using sodium alginate prepared from *Laminaria, Macrocystis,* and *Ascophyllum* species, where the films based only on this material were not stable in wet conditions. However, the addition of tannic acid was useful in promoting antioxidant properties in those films. In all these cases, the addition of different substances to improve the film properties was necessary, and in this regard, the use of nanocellulose, which could be obtained from the same seaweed, represents a suitable alternative (Draget et al. 1991; Doh et al. 2020).

14.7 Green Extraction of Nanocellulose

Cellulose is a complex carbohydrate and one of the most abundant biomaterials on the earth. It is the main structural polymer of plant cell walls, and it is also synthesized by seaweed. In algae, the cellulose is found in two types of crystalline forms: α and β (Brown 2003; Siddhanta et al. 2013). The α-cellulose has a one-chain triclinic structure, and it is the most abundant in *Valonia* and bacterial celluloses. The β-cellulose has a two-chain monoclinic structure, and it is richer in ramie and cotton celluloses (Sugiyama et al. 1991; Siddhanta et al. 2009; Le Bras et al. 2015). Usually, all seaweed classes have a higher amount of α-cellulose than of β type, as shown in Table 14.3 (Siddhanta et al. 2009, 2013), which differences affect the elastic moduli (Newman 1999). In *Sargassum* species (*Phaeophyceae*), the content of cellulose is between 4.8% and 10%. This concentration is smaller than the other seaweeds species (Table 14.3) and related to their higher alginate proportion (Kumar and Sahoo 2017). However, there are no studies about cellulose content in pelagic *Sargassum* species.

The chemical modification of cellulose has generated a broad range of novel forms such as nanocrystals, whiskers, nanofibrils, and nanofibers, which are more related to their shape. Parallel to, other categories have been proposed, more related to their sizes, such as microcrystalline cellulose (MCC), microfibrillated cellulose (MFC), nanofibrillated cellulose (NFC), and cellulose nanocrystals (CNC) (Mishra et al. 2018). All these materials have unique characteristics (Klemm et al. 2011).

Microcrystalline cellulose (MCC) is a powder with a size between 10 and 50 μm in diameter, with a crystalline structure, and multi-size cellulose microfibrils (Oyeniyi and Itola 2012). The microfibrillated cellulose (MFC) is a fine cellulose form with multiple elementary fibers, including a crystalline-amorphous region, with dimensions about a few microns of diameter (Ding et al. 2014). Nanofibrillated cellulose (NFC) is fibers with diameters in the nanometric scale, formed by individually (3–5 nm) and aggregated fibers (20–50 nm), and with lengths around 500–1000 nm (Liu et al. 2014). Cellulose nanocrystals (CNC) describe a rod-like configuration, with diameters (3–30 nm) and lengths (100–2000 nm) dependent on the source (Habibi et al. 2010).

14.8 Nanocellulose Synthesis

The isolation of fibers and nanoparticles from cellulose is performed with strong acid solutions or hydroxides. This allows the dissolution of the amorphous domains and the cut in longitudinal microfibers. The type of crystals obtained are called generally as nanocrystals of cellulose, and it is obtained in aqueous solution. During the acid treatment, the hydronium ions penetrate the cellulose chains, and only individual crystals are obtained after mechanical treatment; their geometric dimension depends on the origin of the cellulose substrate and the hydrolysis conditions. In

Table 14.3 Cellulose extraction from seaweeds of different sources and methods

Seaweed class	Species	Cellulose (%)	α-Cellulose (%)	β-Cellulose (%)	References
Chlorophyceae	*Chaetomorpha antennina*	7.9	7.5	0.4	Siddhanta et al. (2009)
	Chamaedoris auriculata	10	8.7	0.8	
	Ulva fasciata	5.3	2.2	1.6	Siddhanta et al. (2013)
	Chaetomorpha aerea	20.0	18.5	0.8	
	Enteromorpha prolifera	5.0	4.1	0.8	
	Bryopsis hypnoides	4.3	1.1	2.8	
	Caulerpa imbricata	7.4	4.0	3.1	
	Boodlea composita	5.2	4.8	0.1	
Phaeophyceae	*Dictyota dichotoma*	9.5	5.0	2.5	Siddhanta et al. (2009)
	Dictyota bartayresiana	9.5	5.0	2.5	
	Sargassum tenerrimum	10.0	8.4	1.25	
	Sargassum subrepandum	7.0	–	–	Abou-El-Wafa et al. (2011)
	Sargassum naozhouense	4.8	–	–	Peng et al. (2012)
	Sargassum horneri	5.5	–	–	Rabemanolontsoa and Saka (2013)
	Sargassum wightii	9.9	–	–	Kumar and Sahoo (2017)
Rhodophyceae	*Gracilaria edulis*	5.5	3.75	1.0	Siddhanta et al. (2009)
	Gelidiella acerosa	13.8	8.5	3.5	
	Kappaphycus alvarezii	2.5	2.0	0.3	
	Gracilaria textorii	3.95	1.75	2.0	
	Gracilaria debilis	4.5	1.75	2.5	
	Champia indica	2.5	2.2	0.3	
	Sarconema scinaioides	2.5	1.25	0.2	

this sense, for example, the crystal sizes along the longitudinal direction of cellulose chains present in the original cellulose of kelp are tremendous compared to terrestrial plants (Dufresne 2013; Doh et al. 2019). Some researchers report that the isolation of cellulose nanoparticles from seaweed is difficult due to the alginate presence (Visanko et al. 2014).

Nanocellulose is obtained from seaweed through five stages. The first one consists of removing the wax from the algae using ethanol/toluene mixtures. This step is recommended when the algae are rich in lipids (Taherdanak and Zilouei 2014; De Oliveira et al. 2018). The second stage is the depolymerization with hydrochloric acid and sodium hydroxide. This treatment is similar to alginate extraction in brown seaweed, converting the calcium alginate to alginic acid and then to sodium alginate that it is soluble in water (Tarchoun et al. 2019; Doh et al. 2019; Doh and Whiteside 2020). The third stage is a bleaching process to remove the color of the remaining material; it is usually carried out in two steps. The first one uses potassium hydroxide, sodium hypochlorite, and glacial acetic acid to adjust pH, and the second, activated hydrogen peroxide. These combinations of solvents can remove all types of pigments to obtain a colorless product (De Oliveira et al. 2018; Doh et al. 2019; Tarchoun et al. 2019; Doh and Whiteside 2020). At this stage, the cellulose has been isolated, and a hydrolysis protocol with sulfuric acid is usually carried out to obtain nanoparticles, stopping the reaction with ice distilled water (De Oliveira et al. 2018; Gao et al. 2018; Doh et al. 2019; Doh and Whiteside 2020). Finally, to avoid the particles' aggregation, ultrasonication of the samples is necessary (Doh et al. 2019; Doh and Whiteside 2020), keeping them in refrigeration or being lyophilized.

As shown in Table 14.4, most of these steps are common in the different presented methods, being the main difference in the dewaxing process, only used by De Oliveira et al. (2018) when extracting nanocellulose from *Gelidium sesquipedale* (red algae).

It was also observed from the reports of Doh et al. (2019) and Doh and Whiteside (2020) that the place where seaweed was collected could affect the nanocellulose yields, reporting both of them significantly different results, but following the same methodology.

Another method to obtain cellulose nanoparticles is through mechanical actions, breaking the cellulose fibers. This process consists of high-pressure homogenization or grinding, but this route requires high energy (Visanko et al. 2014). This situation is avoided by applying different pretreatments to the samples; for example, an alkaline or acid hydrolysis, an enzyme treatment, or the oxidation of 2,2,6,6-tetramethylpiperidine-1-oxyl (TEMPO) (Hu and Ragauskas 2012; Dufresne 2013) can be used. These pretreatments decrease the mechanical processes energy requirements from 30,000 to 1000 kWh/ton (Mishra et al. 2018).

The alkaline hydrolysis is more efficient to solubilize lignin, hemicellulose, and pectin (Taherdanak and Zilouei 2014), while the acid hydrolysis breaks the rigid structure of the hemicellulose and cellulose (Pedersen and Meyer 2010). Both methods are efficient in isolating the cellulose structure. However, the acid hydrolysis process is considered toxic and corrosive, besides requiring the use of high temperatures, while the alkaline process, although less severe, is more expensive (Pedersen

Table 14.4 Nanocellulose synthesis from seaweeds of different sources and methods

Treatment	Chemical or apparatus	Cellulose source	Length (nm)	Width (nm)	Yield (%)	References
Dewaxing	Toluene/ethanol	*G. sesquipedale*	493	12.43	–	De Oliveira et al. (2018)
Bleaching and lignin removal	Sodium chlorite and potassium hydroxide					
Acid treatment	Sulfuric acid					
De-polymerization	Hydrochloric acid and sodium hydroxide	*L. japonica*	239.4	22.45	26.1	Doh et al. (2019)
Bleaching	Potassium hydroxide, sodium hypochlorite, glacial acetic acid, and active hydrogen peroxide	*S. fluitans*	43.7	8.81	25.8	
Acid hydrolysis	Sulfuric acid	*P. palmate*	64.2	17.38	13.3	
		P. umbilicalis	27.6	10.89	17.4	
Mechanical dispersion	Ultrasonicator	*U. lactuca*	53.2	17.30	16.3	
De-polymerization	Hydrochloric acid and sodium hydroxide	*S. fluitans*	43.06	–	42.7	Doh and Whiteside (2020)
Bleaching	Potassium hydroxide, sodium hypochlorite, Glacial acetic acid, and active hydrogen peroxide					
Acid hydrolysis	Sulfuric acid					
Mechanical dispersion	Ultrasonicator					

and Meyer 2010; Hu and Ragauskas 2012). To overcome those disadvantages, other methods have been proposed, like the enzymatic ones, that degrades the content of lignin and hemicellulose, by using cellobiohydrolases or endoglucanases (Kumar and Sharma 2017) or via the oxidation of TEMPO, that consists of chemical pre-treatment, carried out in aqueous solution and mild environmental conditions, that helps to prevent the postaggregation of nanoparticles throughout a drying stage (Saito et al. 2007).

14.9 Nanocellulose Applications

The principal applications of nanocellulose considered its use as a barrier for liquid and gaseous materials (Syverud and Stenius 2009), as reinforcement in polymer matrices, in water purification (Mishra et al. 2018), and as drug delivery for biomedical application. In these last applications, the cell attachment is enhanced by the cellulose modifications that is carried out through the union with proteins and plasma and also by the introduction of surface charges (Lin and Dufresne 2014).

The advantages of nanocellulose are related to their cheap production costs, accessibility, sustainability, excellent mechanical properties, high superficial area, the possibility of making a chemical modification, and the biocompatibility with no toxic effects (Abitbol et al. 2016).

There are several uses for nanocellulose isolated from seaweeds (Table 14.5). For example, the NCN obtained from *Ulva lactuca* has been applied in the removal of tetracycline, and other antibiotics in aqueous solution, with a dependence of pH (Rathod et al. 2015), as well as in mefenamic acid photodegradation in aqueous media (Rathod et al. 2018). In other work, Liu et al. (2017a) extracted nanocellulose crystals from *Cladophora* sp. for biomedical applications, demonstrating that impurities are insignificant to have cytotoxic effects, and Wu et al. (2020) used CNF extracted from *L. japonica* to obtain Pickering emulsion in sunflower oil, that are stable during 30 days and have a high emulsification effect (50%). NCN from red and brown seaweed (*G. sesquipedale*, *L. japonica, S. fluitans*, and *S. natans*) has also been applied in films or aerogels to obtain better mechanical and optical properties (Doh and Whiteside 2020; Doh et al. 2020) and to enhance the release of bioactive compounds from food packing material (De Oliveira et al. 2018).

Regarding pelagic *Sargassum* species as a source of nanocellulose, recent publications reported that cellulose nanocrystals from these species increase the tensile strength of films, reducing their permeability to water vapor and oxygen (Doh and

Table 14.5 Seaweed nanocellulose applications

Nanocellulose source	Application	References
U. lactuca	Particles of NCN for removal of tetracycline and other antibiotics in aqueous solution	Rathod et al. (2015)
Cladophora sp.	Isolation of ultrapure nanocellulose for biomedical applications	Liu et al. (2017a)
G. sesquipedale	Aerogels with PVA and agar and nanocellulose composites	De Oliveira et al. (2018)
U. lactuca	Nanocomposites with TiO_2 for photodegradation of mefenamic acid in aqueous media	Rathod et al. (2018)
S. fluitans	Alginate nanocomposite film with CNCs	Doh and Whiteside (2020)
L. japonica	Synthetized CNF particles for applications in Pickering emulsion	Wu et al. (2020)
L. japonica and *S. natans*	Nanocomposite films with cellulose nanocrystals	Doh et al. (2020)

Whiteside 2020; Doh et al. 2020). As described before, the nanocellulose could be applied in bioplastics improving the final product mechanical, microbial, and nutraceutical properties.

14.10 Fucoidan

Fucoidans are a sulfated polysaccharide and constitute up to 25–30% of the alga dry weight. These molecules are composed of α-1,3-backbones or repeating disaccharide units of α-1,3- and α-1,4-linked fucose residues with branchings attached at C2 positions (Rodríguez-Jasso et al. 2010a, b; Bachelet-Violette et al. 2014).

Recently, it has been established that the fucoidan possesses several biologically active substances with anticoagulant, antithrombotic, anti-inflammatory, antitumor, contraceptive, and antiviral properties (Rodríguez-Jasso et al. 2010b; Ale et al. 2011; Kimura et al. 2013; Lee and Lim 2014; Chollet et al. 2016).

Several authors reported the extraction of fucoidan from some *Sargassum* species, as shown in Table 14.6. These species have a fucoidan contents between 0.25% and 6.07%; for the composition of carbohydrates, some authors report that the content of neutral carbohydrates and sulfated groups can vary depending on the extraction method, harvest location, season, and algae species (Chale-Dzul et al. 2020). The fucoidan content in pelagic *Sargassum* is between 2.61% and 11.67% for *S. fluitans,* meanwhile, it is 5.25% for *S. natans*, relating these differences to the species, season, year, and place of the collection (Chale-Dzul et al. 2017, 2020; Fernando et al. 2020b; Rosado-Espinosa et al. 2020).

Table 14.6 Fucoidan yield for several *Sargassum* species

Seaweed	Fucoidan yield (%)	Neutral carbohydrates (%)	Sulfated groups (%)	References
S. fluitans	2.16	62.02	3.78	Chale-Dzul et al. (2017)
S. fluitans	11.67	25.80	7.56	Chale-Dzul et al. (2020)
Sargassum sp.	0.25	–	–	Rafi and Veerichetty (2020)
S. natans	5.25	45.96–72.02	13.85–28.01	Fernando et al. (2020b)
S. filipendula	4.86–6.07	–	–	Laeliocattleya et al. (2020)
S. fluitans	4.4	–	–	Rosado-Espinosa et al. (2020)
S. buxifolium	3.2	–	–	Rosado-Espinosa et al. (2020)
S. hystrix	2.7	–	–	Rosado-Espinosa et al. (2020)

14.11 Fucoidan Extraction from *Sargassum*

The isolation of fucoidan from *Sargassum* species includes the elimination of phenolic compounds with ethanol, formaldehyde, chloroform, or methanol (Chale-Dzul et al. 2017, 2020; Fernando et al. 2020b; Laeliocattleya et al. 2020; Rosado-Espinosa et al. 2020), followed by the extraction of fucoidan carbohydrates in distilled water or acid solutions and finally, the purification with calcium salts or acid to remove the alginate. The fucoidan is precipitated with ethanol (Chale-Dzul et al. 2017, 2020; Rafi and Veerichetty 2020) (Table 14.7). A novel method to extract this carbohydrate is through enzymes, using a commercial cellulase (CelluclastR) and Alcalase to degrade peptides that could contaminate the fucoidan (Fernando et al. 2020a).

Another way to improve the extraction yields of fucoidan is by microwave irradiation, which has generated better results than those obtained in conventional ovens without affecting the carbohydrate's antioxidant activity (Kurnialahi et al. 2020, Liu et al. 2020).

Any method elected for fucoidan extraction had promising results with pelagic *Sargassum* species, as shown in Table 14.7, but apparently, the better results were with distilled water treatment because the yield is 11.67% (Chale-Dzul et al. 2020). Although the fucoidan content is less of any other carbohydrate previously described, the extraction could link up with other carbohydrates isolation as cellulose or alginate, to have an integral use of pelagic *Sargassum*.

14.12 Fucoidan Functional Activities

Some authors reported antioxidant (Yuan and Macquarrie 2015; Kurnialahi et al. 2020; Liu et al. 2020) and antimicrobial activities (Marudhupandi and Kumar 2013; Liu et al. 2017b; Ashayerizadeh et al. 2019; Kim et al. 2020) for fucoidan. In this regard, these researchers proved that fucoidan from algae (*L. japonica, S. terrenium*) inhibited several human bacterial pathogens, such as *E. coli* and *S. aureus* or norovirus, and showed protective effects for a bacterial infection when the carbohydrate had more sulfated fucose amount than galactose or glucose.

Furthermore, fucodian has also been reported to improve the mechanical properties of films prepared with alginate. Hence, it is possible to add this substance to films conserving their functional characteristics (Gomaa et al. 2018). As described before, the use of this substance in bioplastics has functional properties for the protection of oxidative degradation or bacterial degradation, without having toxic effects on the environment.

Table 14.7 Fucoidan treatment and posttreatment for *Sargassum* species

Seaweed	Pretreatment	Extraction	Posttreatment	Fucoidan yield (%)	References
S. fluitans	Ethanol 80%, 12 h room temperature, and 70 °C	Distilled water at room temperature 7 h, 70 °C 7 h, and 70 °C 4 h	Treatment with $CaCl_2$, precipitation with centrifugation (10,000 rpm 30 min), dialysis during 48 h	2.16–11.67	Chale-Dzul et al. (2017); Chale-Dzul et al. (2020)
Sargassum sp.	None	Distilled water at 85 °C 2 h	Adjusting pH = 7 with TCA over night at 4 °C, precipitation with centrifugation, the supernatant, adding ethanol in proportion 1:3, during 1 night at 4 °C	0.25	Rafi and Veerichetty (2020)
S. natans	Ethanol and formaldehyde 8 h 37 °C	Enzymatic extraction with cellulase at 50 °C 24 h, pH = 4.5 and Alcalase at 50 °C 8 h, pH = 8.0	Filtration and precipitation with centrifugation (6000 × *g*), treatment with $CaCl_2$, adding ethanol in proportion 1:4, during 1 night at 4 °C and dialysis	5.25	Fernando et al. (2020b)
S. filipendula	Chloroform methanol solution (2:1) at room temperature 20 min	HCl in ultrasonic bath at 70 °C 10–20 min	Filtration, treatment with $CaCl_2$ at 4 °C 8 h, precipitation with centrifugation (8000 rpm), treatment with ethanol 4 °C 8 h	4.86–6.07	Laeliocattleya et al. (2020)
S. filipendula *S. buxifolium* *S. hystrix* *S. fluitans* *S. ramifolium*	Formaldehyde 30 min	HCl at room temperature 3 h, pH = 4, 15 min, and water during 15 min	Neutralization with NaOH, precipitation with ethanol, concentration in rotatory evaporator at 40 °C 50 rpm, dialysis 24 h	2.1–4.4	Rosado-Espinosa et al. (2020)

14.13 Future Perspectives

The accumulation and unusual affluence of pelagic *Sargassum* is a dangerous problem for ecosystems in countries of the Atlantic Ocean and seashores of the Caribbean Sea. Similarly, it is necessary to solve the plastic pollution problem by changing traditional plastics to bioplastics. The use of pelagic *Sargassum* could be an appropriate solution for both situations, as the concentrations of carbohydrates and fiber that these seaweeds contain could be the raw materials to synthesize nanoparticles and other functional compounds, helping in the reduction of environmental contamination. Moreover, it is possible with the integral use of pelagic *Sargassum*, overcoming their adverse effects on the coasts, by applying a sequential extraction methodology that allows the obtention of the different functional components, turning the problem into an economically attractive solution.

Acknowledgment Héctor Uriel Castañeda Serna thanks to CONACyT and BEIFI-IPN for the scholarships provided and the financial support for this work. This research was funded through the projects: 20195500, 20210024, 20211381 and 20210918 from the Instituto Politécnico Nacional (IPN, Mexico) and 1668 from CONACyT.

References

Abitbol T, Rivkin A, Cao Y, Nevo Y, Abraham E, Ben-Shalom T, Lapidot S, Shoseyov O (2016) Nanocellulose, a tiny fiber with huge applications. Curr Opin Biotechnol 39:76–88. https://doi.org/10.1016/j.copbio.2016.01.002

Abou-El-Wafa GSE, Shaaban KA, El-Naggar MEE, Shaaban M (2011) Bioactive constituents and biochemical composition of the Egyptian brown alga *Sargassum Subrepandum* (Forsk). Rev Latinoam Quím 39(1–2):62–74

Ale MT, Mikkelsen JD, Meyer AS (2011) Important determinants for fucoidan bioactivity: a critical review of structure-function relations and extraction methods for fucose-containing sulfated polysaccharides from brown seaweeds. Mar Drugs 9(10):2106–2130. https://doi.org/10.3390/md9102106

Andrady AL, Neal MA (2009) Applications and societal benefits of plastics. Philos Trans R Soc Lond B Biol Sci 364(1526):1977–1984. https://doi.org/10.1098/rstb.2008.0304

Andriamanantoanina H, Rinaudo M (2010) Relationship between the molecular structure of alginates and their gelation in acidic conditions. Polym Int 59(11):1531–1541. https://doi.org/10.1002/pi.2943

Ashayerizadeh O, Dastar B, Pourashouri P (2019) Study of antioxidant and antibacterial activities of depolymerized fucoidans extracted from *Sargassum tenerrimum*. Int J Biol Macromol 151:1259–1266. https://doi.org/10.1016/j.ijbiomac.2019.10.172

Bachelet-Violette L, Silva AKA, Maire M, Michel A, Brinza O, Ou P, Ollivier V, Nicoletti Am Wilheim C, Letourneur D, Ménager C, Chaubet F (2014) Strong and specific interaction of ultra small superparamagnetic iron oxide nanoparticles and human activated platelets mediated by fucoidan coating. RSC Adv 4(10):4864–4871. https://doi.org/10.1039/c3ra46757k

Bhowmik BB, Sa B, Mukherjee A (2006) Preparation and in vitro characterization of slow release testosterone nanocapsules in alginates. Acta Pharma 56:417–429

Brown RM (2003) Cellulose structure and biosynthesis: what is in store for the 21st century? J Polym Sci A Polym Chem 42:487–495. https://doi.org/10.1002/pola.10877

Chale-Dzul J, Freile-Pelegrín Y, Robledo D, Moo-Puc R (2017) Protective effect of fucoidans from tropical seaweeds against oxidative stress in HepG2 cells. J Appl Phycol 29(5):2229–2238. https://doi.org/10.1007/s10811-017-1194-3

Chale-Dzul J, Pérez-Cabeza de Vaca R, Quintal-Novelo C, Olivera-Castillo L, Moo-Puc R (2020) Hepatoprotective effect of a fucoidan extract from *Sargassum fluitans* Borgesen against CCl_4-induced toxicity in rats. Int J Biol Macromol 145:500–509. https://doi.org/10.1016/j.ijbiomac.2019.12.183

Chollet L, Saboural P, Chauvierre C, Villemin JN, Letourneur D, Chaubet F (2016) Fucoidans in Nanomedicine. Mar Drugs 14(8):145. https://doi.org/10.3390/md14080145

Davis TA, Ramirez M, Mucci A, Larsen B (2004) Extraction, isolation and cadmium binding of alginate from *Sargassum* spp. J Appl Phycol 16:275–284. https://doi.org/10.1023/b:japh.0000047779.31105.ec

De Oliveira JP, Bruni GP, Fabra MJ, da Rosa ZE, Lopez-Rubio A, Martínez-Sanz M (2018) Development of food packaging bioactive aerogels through the valorization of *Gelidium sesquipedale* seaweed. Food Hydrocoll 89:337–350. https://doi.org/10.1016/j.foodhyd.2018.10.047

Ding SY, Zhao S, Zeng Y (2014) Size, shape, and arrangement of native cellulose fibrils in maize cell walls. Cellulose 21(2):863–871. https://doi.org/10.1007/s10570-013-0147-5

Doh H, Whiteside WS (2020) Isolation of cellulose nanocrystals from brown seaweed, *Sargassum fluitans*, for development of alginate nanocomposite film. Polym Cryst 3(4):e10133. https://doi.org/10.1002/pcr2.10133

Doh H, Lee MH, Whiteside WS (2019) Physicochemical characteristics of cellulose nanocrystals isolated from seaweed biomass. Food Hydrocoll 102:105542. https://doi.org/10.1016/j.foodhyd.2019.105542

Doh H, Dunno KD, Whiteside WS (2020) Preparation of novel seaweed nanocomposite film from brown seaweeds *Laminaria japonica* and *Sargassum natans*. Food Hydrocoll 105:105744. https://doi.org/10.1016/j.foodhyd.2020.105744

Draget KI, Østgaard K, Smidsrød O (1991) Homogeneous alginate gels: a technical approach. Carbohydr Polym 14(2):159–178. https://doi.org/10.1016/0144-8617(90)90028-q

Dufresne A (2013) Nanocellulose: a new ageless bionanomaterial. Mater Today 16(6):220–227. https://doi.org/10.1016/j.mattod.2013.06.004

El Atouani S, Bentiss F, Reani A, Zrid R, Belattmania Z, Pereira L, Mortadi A, Cherkaoui O, Sabour B (2016) The invasive brown seaweed *Sargassum muticum* as new resource for alginate in Morocco: spectroscopic and rheological characterization. Phycol Res 64(3):185–193. https://doi.org/10.1111/pre.12135

Emadian SM, Onay TT, Demirel B (2017) Biodegradation of bioplastics in natural environments. Waste Manag 59:526–536. https://doi.org/10.1016/j.wasman.2016.10.006

Fernando IPS, Dias MKHM, Madusanka DMD, Han EJ, Kim MJ, Jeon YJ, Lee K, Cheong SH, Han YS, Park SR, Ahn G (2020a) Human keratinocyte UVb-protective effects of a low molecular weight fucoidan from *Sargassum horneri* purified by step gradient ethanol precipitation. Antioxidants 9(4):340. https://doi.org/10.3390/antiox9040340

Fernando IPS, Sanjeewa KKA, Lee HG, Kim HS, Vaas APJP, De Silva HIC, Nanayakkara CM, Abeytunga DTU, Lee WW, Lee DS, Jeon YJ (2020b) Characterization and cytoprotective properties of *Sargassum natans* fucoidan against urban aerosol-induced keratinocyte damage. Int J Biol Macromol 159(15):773–781. https://doi.org/10.1016/j.ijbiomac.2020.05.132

Folino A, Karageorgiou A, Calabrò PS, Komilis D (2020) Biodegradation of wasted bioplastics in natural and industrial environments: a review. Sustainability 12(15):6030. https://doi.org/10.3390/su12156030

Gao H, Duan B, Lu A, Deng H, Du Y, Shi X, Zhang L (2018) Fabrication of cellulose nanofibers from waste brown algae and their potential application as milk thickeners. Food Hydrocoll 79:473–481. https://doi.org/10.1016/j.foodhyd.2018.01.023

Gomaa M, Fawzy MA, Hifney AF, Abdel-Gawad KM (2018) Use of the brown seaweed *Sargassum latifolium* in the design of alginate-fucoidan based films with natural antioxidant properties and kinetic modeling of moisture sorption and polyphenolic release. Food Hydrocoll 82:64–72. https://doi.org/10.1016/j.foodhyd.2018.03.053

Gomez CG, Pérez Lambrecht MV, Lozano JE, Rinaudo M, Villar MA (2009) Influence of the extraction–purification conditions on final properties of alginates obtained from brown algae (*Macrocystis pyrifera*). Int J Biol Macromol 44(4):365–371. https://doi.org/10.1016/j.ijbiomac.2009.02.005

Habibi Y, Lucia LA, Rojas OJ (2010) Cellulose nanocrystals: chemistry, self-assembly, and applications. Chem Rev 110(6):3479–3500. https://doi.org/10.1021/cr900339w

Hahladakis JN, Velis CA, Weber R, Iacovidou E, Purnell P (2018) An overview of chemical additives present in plastics: migration, release, fate and environmental impact during their use, disposal and recycling. J Hazard Mater 344:179–199. https://doi.org/10.1016/j.jhazmat.2017.10.014

Hemalakshita T, Dubey AR, Pande A, Wagh NS, Prasad R, Gupta AS, Pandit S (2021) Bioplastics: Fundamentals to Application. In: Prasad R, Singh J, Kumar V, Upadhyay CP (eds) Recent Developments in Microbial Technologies. Springer Nature Singapore, 301–321

Hu F, Ragauskas A (2012) Pretreatment and lignocellulosic chemistry. Bioenergy Res 5(4):1043–1066. https://doi.org/10.1007/s12155-012-9208-0

Kaczmarek B (2020) Improving sodium alginate films properties by phenolic acid addition. Materials 13(13):2895. https://doi.org/10.3390/ma13132895

Kim H, Lim CY, Lee DB, Seok JH, Kim KH, Chung MS (2020) Inhibitory effects of *Laminaria japonica* fucoidans against noroviruses. Viruses 12(9):997. https://doi.org/10.3390/v12090997

Kimura R, Rokkaku T, Takeda S, Senba M, Mori N (2013) Cytotoxic effects of fucoidan nanoparticles against osteosarcoma. Mar Drugs 11(11):4267–4278. https://doi.org/10.3390/md11114267

Klemm D, Kramer F, Moritz S, Lindström T, Ankerfors M, Gray D, Dorris A (2011) Nanocelluloses: a new family of nature-based materials. Angew Chem Int Ed 50(24):5438–5466. https://doi.org/10.1002/anie.201001273

Kumar S, Sahoo D (2017) A comprehensive analysis of alginate content and biochemical composition of leftover pulp from brown seaweed *Sargassum wightii*. Algal Res 23:233–239. https://doi.org/10.1016/j.algal.2017.02.003

Kumar AK, Sharma S (2017) Recent updates on different methods of pretreatment of lignocellulosic feedstocks: a review. Bioresour Bioprocess 4(1):7. https://doi.org/10.1186/s40643-017-0137-9

Kurnialahi ID, Husni A, Sinurat E, Isnansetyo A (2020) Antioxidant activity of tropical seaweed *Sargassum muticum* fucoidan. Aquac Aquar Conserv Legis 13(1):230–240

Laeliocattleya RA, Yunianta A, Suloi AF, Gayatri PP, Putri NA, Anggraeni YC (2020) Fucoidan content from brown seaweed (*Sargassum filipendula*) and its potential as radical scavenger. J Phys Conf Ser 1430:012023. https://doi.org/10.1088/1742-6596/1430/1/012023

Le Bras D, Strømme M, Mihranyan A (2015) Characterization of dielectric properties of nanocellulose from wood and algae for electrical insulator applications. J Phys Chem B 119(18):5911–5917. https://doi.org/10.1021/acs.jpcb.5b00715

Lee EJ, Lim K (2014) Polyelectrolyte complexes of chitosan self-assembled with fucoidan: an optimum condition to prepare their nanoparticles and their characteristics. Korean J Chem Eng 31:664–675. https://doi.org/10.1007/s11814-013-0243-0

Lin N, Dufresne A (2014) Nanocellulose in biomedicine: current status and future prospect. Eur Polym J 59:302–325. https://doi.org/10.1016/j.eurpolymj.2014.07.025

Liu J, Korpinen R, Mikkonen KS, Willför S, Xu C (2014) Nanofibrillated cellulose originated from birch sawdust after sequential extractions: a promising polymeric material from waste to films. Cellulose 21(4):2587–2598. https://doi.org/10.1007/s10570-014-0321-4

Liu J, Willför S, Mihranyan A (2017a) On importance of impurities, potential leachables and extractables in algal nanocellulose for biomedical use. Carbohydr Polym 172:11–19. https://doi.org/10.1016/j.carbpol.2017.05.002

Liu M, Liu Y, Cao MJ, Liu GM, Chen Q, Sun L, Chen H (2017b) Antibacterial activity and mechanisms of depolymerized fucoidans isolated from *Laminaria japonica*. Carbohydr Polym 172:294–305. https://doi.org/10.1016/j.carbpol.2017.05.060

Liu J, Wu SY, Chen L, Li QJ, Shen YZ, Jin L, Zhang X, Chen PC, Wu MJ, Choi JI, Tong HB (2020) Different extraction methods bring about distinct physicochemical properties and antioxidant activities of *Sargassum fusiforme* fucoidans. Int J Biol Macromol 155:1385–1392. https://doi.org/10.1016/j.ijbiomac.2019.11.113

Marismandani ADP, Husni A (2020) Development and characterization of biobased alginate/glycerol/virgin coconut oil as biodegradable packaging. E3S Web Conf 147:03016. https://doi.org/10.1051/e3sconf/202014703016

Marudhupandi T, Kumar TTA (2013) Antibacterial effect of fucoidan from *Sargassum wightii* against the chosen human bacterial pathogens. Int Curr Pharm J 2(10):156–158. https://doi.org/10.3329/icpj.v2i10.16408

Mishra RK, Sabu A, Tiwari SK (2018) Materials chemistry and the futurist eco-friendly applications of nanocellulose: status and prospect. J Saudi Chem Soc 22(8):949–978. https://doi.org/10.1016/j.jscs.2018.02.005

Mohammed A, Bissoon R, Bajnath E, Mohammed K, Lee T, Bissram M, John N, Jalsa NK, Lee KY, Ward K (2018) Multistage extraction and purification of waste *Sargassum natans* to produce sodium alginate: an optimization approach. Carbohydr Polym 198:109–118. https://doi.org/10.1016/j.carbpol.2018.06.067

Mohammed A, Rivers A, Stuckey DC, Ward K (2020) Alginate extraction from *Sargassum* seaweed in the Caribbean region: optimization using response surface methodology. Carbohydr Polym 245(1):116419. https://doi.org/10.1016/j.carbpol.2020.116419

Newman RH (1999) Estimation of the relative proportions of cellulose I alpha and I beta in wood by carbon-13 NMR spectroscopy. Holzforschung 53(4):335–340. https://doi.org/10.1515/hf.1999.055

Norton M (2020) Tackling the challenge of packaging plastic in the environment. Chem Eur J 26:1–4. https://doi.org/10.1002/chem.202001890

Oyeniyi YJ, Itiola OA (2012) The physicochemical characteristic of microcrystalline cellulose, derived from sawdust, agricultural waste products. Int J Pharm Pharm Sci 4:197–200

Pedersen M, Meyer AS (2010) Lignocellulose pretreatment severity – relating pH to biomatrix opening. New Biotechnol 27(6):739–750. https://doi.org/10.1016/j.nbt.2010.05.003

Peng Y, Han Y, Gardner DJ (2012) Spray-drying cellulose nanofibrils: effect of drying process parameters on particle morphology and size distribution. Wood Fiber Sci 44(4):448–461

Puscaselu RG, Gutt G, Amariei S (2020) The use of edible films based on sodium alginate in meat product packaging: an eco-friendly alternative to conventional plastic materials. Coatings 10(2):166. https://doi.org/10.3390/coatings10020166

Rabemanolontsoa H, Saka S (2013) Comparative study on chemical composition of various biomass species. RSC Adv 3(12):3946–3956. https://doi.org/10.1039/c3ra22958k

Rafi ASM, Veerichetty V (2020) In vitro antioxidant and antiproliferative activity of crude fucoidan from *Sargassum* species. J Algal Biomass Util 11(1):54–67

Rameshkumar S, Shaiju P, O'Connor KE, Babu PR (2020) Bio-based and biodegradable polymers - state-of-the-art, challenges and emerging trends. Curr Opin Green Sustain Chem 21:75–81. https://doi.org/10.1016/j.cogsc.2019.12.005

Rathod M, Haldar S, Basha S (2015) Nanocrystalline cellulose for removal of tetracycline hydrochloride from water via biosorption: equilibrium, kinetic and thermodynamic studies. Ecol Eng 84:240–249. https://doi.org/10.1016/j.ecoleng.2015.09.031

Rathod M, Moradeeya PG, Haldar S, Basha S (2018) Nanocellulose/TiO_2 composites: preparation, characterization and application in the photocatalytic degradation of a potential endocrine disruptor, mefenamic acid, in aqueous media. Photochem Photobiol Sci 17:1301–1309. https://doi.org/10.1039/C8PP00156A

Rhein-Knudsen N, Ale MT, Ajalloueian F, Meyer AS (2017) Characterization of alginates from Ghanaian brown seaweeds: *Sargassum* spp. and *Padina* spp. Food Hydrocoll 71:236–244. https://doi.org/10.1016/j.foodhyd.2017.05.016

Rodríguez-Jasso RM, Mussatto SI, Pastrana L, Aguilar CN, Teixeira JA (2010a) Fucoidan-degrading fungal strains: screening, morphometric evaluation, and influence of medium composition. Appl Biochem Biotechnol 162(8):2177–2188

Rodríguez-Jasso RM, Mussatto SI, Pastrana L, Aguilar CN, Teixeira JA (2010b) Microwave-assisted extraction of sulfated polysaccharides (fucoidan) from brown seaweed. Carbohydr Polym 86(3):1137–1144. https://doi.org/10.1007/s12010-010-8992-2

Rosado-Espinosa LA, Freile-Pelegrín Y, Hernández-Nuñez E, Robledo D (2020) A comparative study of *Sargassum* species from the Yucatan peninsula coast: morphological and chemical characterisation. Phycologia 59(3):261–271. https://doi.org/10.1080/00318884.2020.1738194

Saito T, Kimura S, Nishiyama Y, Isogai A (2007) Cellulose nanofibers prepared by TEMPO-mediated oxidation of native cellulose. Biomacromolecules 8:2485–2491. https://doi.org/10.1021/bm0703970

Santana AA, Kieckbusch TG (2013) Physical evaluation of biodegradable films of calcium alginate plasticized with polyols. Braz J Chem Eng 30(4):835–845. https://doi.org/10.1590/s0104-66322013000400015

Schiener P, Black KD, Stanley MS, Green DH (2014) The seasonal variation in the chemical composition of the kelp species *Laminaria digitata*, *Laminaria hyperborea*, *Saccharina latissima* and *Alaria esculenta*. J Appl Phycol 27(1):363–373. https://doi.org/10.1007/s10811-014-0327-1

Sheavly SB, Register KM (2007) Marine debris and plastics: environmental concerns, sources, impacts and solutions. J Polym Environ 15:301–305. https://doi.org/10.1007/s10924-007-0074-3

Shen M, Ye S, Zeng G, Zhang Y, Xing L, Tang W, Wen X, Liu S (2019) Can microplastics pose a threat to ocean carbon sequestration? Mar Pollut Bull 150:110712. https://doi.org/10.1016/j.marpolbul.2019.110712

Shen M, Song B, Zeng G, Zhang Y, Huang W, Wen X, Tang W (2020) Are biodegradable plastics a promising solution to solve the global plastic pollution? Environ Pollut 263:114469. https://doi.org/10.1016/j.envpol.2020.114469

Shukla S, Park JH, Kim M (2019) Efficient, safe, renewable, and industrially feasible strategy employing *Bacillus subtilis* with alginate bead composite for the reduction of ochratoxin-A from wine. J Clean Prod 242:118344. https://doi.org/10.1016/j.jclepro.2019.118344

Siddhanta AK, Prasad K, Meena R, Prasad G, Mehta GK, Chhatbar MU, Oza MD, Kumar S, Sanandiya ND (2009) Profiling of cellulose content in Indian seaweed species. Bioresour Technol 100(24):6669–6673. https://doi.org/10.1016/j.biortech.2009.07.047

Siddhanta AK, Kumar S, Mehta GK, Chhatbar MU, Oza MD, Sanandiya ND, Chejara DR, Godiya CB, Kondaveeti S (2013) Cellulose contents of some abundant Indian seaweed species. Nat Prod Commun 8(4):497–500. https://doi.org/10.1177/1934578x1300800423

Sugiyama J, Vuong R, Chanzy H (1991) Electron diffraction study on the two crystalline phases occurring in native cellulose from an algal cell wall. Macromolecules 24(14):4168–4175. https://doi.org/10.1021/ma00014a033

Syverud K, Stenius P (2009) Strength and barrier properties of MFC films. Cellulose 16:75–85. https://doi.org/10.1007/s10570-008-9244-2

Tabassum N, Khan MA (2020) Modified atmosphere packaging of fresh-cut papaya using alginate based edible coating: quality evaluation and shelf life study. Sci Hortic 259:108853. https://doi.org/10.1016/j.scienta.2019.108853

Taherdanak M, Zilouei H (2014) Improving biogas production from wheat plant using alkaline pretreatment. Fuel 115:714–719. https://doi.org/10.1016/j.fuel.2013.07.094

Tarchoun AF, Trache D, Klapötke TM (2019) Microcrystalline cellulose from *Posidonia oceanica* brown algae: extraction and characterization. Int J Biol Macromol 138:837–845. https://doi.org/10.1016/j.ijbiomac.2019.07.176

Visanko M, Liimatainen H, Sirviö JA, Heiskanen JP, Niinimäki J, Hormi O (2014) Amphiphilic cellulose nanocrystals from acid-free oxidative treatment: physicochemical characteristics and use as an oil–water stabilizer. Biomacromolecules 15(7):2769–2775

Wu J, Zhu W, Shi X, Li Q, Huang C, Tian Y, Wang S (2020) Acid-free preparation and characterization of kelp (*Laminaria japonica*) nanocelluloses and their application in Pickering emulsions. Carbohydr Polym 236:115999. https://doi.org/10.1021/bm500628g

Yuan Y, Macquarrie D (2015) Microwave assisted extraction of sulfated polysaccharides (fucoidan) from *Ascophyllum nodosum* and its antioxidant activity. Carbohydr Polym 129:101–107. https://doi.org/10.1016/j.carbpol.2015.04.057

Chapter 15
Algal Nanotechnology for Wastewater Treatment

Mrinal Kanti Adak and Ajit Das

Contents

15.1 Introduction

Uncontrolled domestic, industrial, agricultural, and hospital wastes accumulated in drinking water sources pose a threat toward the living bodies through water pollution, which require developing advanced technology for wastewater treatment. Rapid industrialization and urbanization in the modern era lead wastewater problem a global issue to balance the ecosystem. In wastewater treatment, the main goal is the removal of suspended solid or liquid, nutrients, biological oxygen demand (BOD), microbial pathogens, toxic metals, and coliform bacteria (Hwang et al. 2016; Agarwal et al. 2019). World Health Organization (WHO) has reported 58% of diarrheal deaths could be prevented through safe drinking water, sanitation, and

M. K. Adak (✉)
Department of Chemistry, Sidho-Kanho-Birsha University, Purulia, West Bengal, India

Department of Chemical Sciences, Indian Institute of Science Education and Research, Mohali, India

A. Das
Department of Chemistry, Sidho-Kanho-Birsha University, Purulia, West Bengal, India

Departmentof Chemistry, Balarampur College, Balarampur, West Bengal, India

D. Thangadurai et al. (eds.), *Bioprospecting Algae for Nanosized Materials*, Nanotechnology in the Life Sciences,
https://doi.org/10.1007/978-3-030-81557-8_15

hygiene, and 2.1 billion people are deprived of safe drinking water in 2015 (WHO 2014).

Dyes from different industries and personalized product, especially from textile industries, are considered the main sources of polluted water. Visible dyes as azo dyes are dangerous to the environment and ecosystem due to their affective role in changing biological cycle; they destroy the photosynthesis process and create disorder in immunity system. Pharmaceutical wastes from hospital, industries, animal farm, and irrigation strongly affect ecosystems through water pollution. These types of pharmaceutical waste are responsible for antibiotic-resistant microbial strains (Bhattacharyya et al. 2022). Wastewater from petroleum industries is also a big challenge to us. In a conventional wastewater treatment process, combination of preliminary treatment, secondary treatment, and tertiary treatment cannot remove 100% waste (Agarwal et al. 2019). Different technologies like membrane filtration, precipitation, coagulation, reverse osmosis, ion exchange, ozonation, and electrochemical are involved for wastewater treatments (Ali 2012), but these leads to limitations like sludge production problem, high cost, energy requirement, clearance problem, managing, and also creation of secondary toxic waste. Again, any single method is not sufficient to remove all pollutants. In this situation, a sustainable management of wastewater is required to resolve this problem. As a result, specific restrictions and strict legislation are applied at different industrial fields to save the future. Most of the countries follow American and European model (Hessel et al. 2007). In this respect, Clean Water Act in 1972, Water (Prevention and Control of Pollution) Act in 1974, and Safe Drinking Water Act (SDWA) of 1974 are most important.

To solve the wastewater problem, algal pond system is very significant since 1950 (Delanka-Pedige et al. 2019) due to greener, cheap, environment friendly pathway. But microalgal culture system process requires long time, tedious task, and huge land area to purify wastewater. To address this issue, technology with small area, cost effective, and greener process requirement is necessary. Nanotechnology provides this alternative path for wastewater treatment due to its pioneering role for design and development of new product with enhanced performance (Leach et al. 2002; Dhingra et al. 2010). Nanomaterials have high surface-to-volume ratio, high surface reactivity, and tunable characteristics that are more applicable than bulk materials. Microbial platform has been used in nanotechnology as a cleaner, green, and sustainable route to focus on hazardous chemical-free synthesis route with more economical and less energy intensive (Aziz et al. 2015). Literature study shows that interactions between microorganism and metal salt lead to the formation of metal nanoparticle through viable biotechnological route like bioleaching and bioremediation process (Gericke and Pinches 2006; Prasad et al. 2016). In the synthesis of nanoparticle, algae act as "bio-nanofactories" through green path where live or died biomass has been used. The size and shape of the nanoparticle depend on the medium pH, temperature, processing time, and substrate concentration (Gericke and Pinches 2006; Aziz et al. 2014, 2015; Khalid 2020).

Different physical or chemical methods like mechanical milling, ball milling, laser ablation, sputtering, precipitation, vapor deposition, and sol-gel process are mainly used for nanoparticle synthesis. However, these types of processes lead to several negative effects on environment with high cost production and high-energy consumption (Li et al. 1999). Microbes like fungi, bacteria, and yeast are reported for the biosynthesis of metal nanoparticle (Prasad 2016, 2017, 2019a, b). These types of biosynthesis have several advantages: (i) reduce the formation of byproduct and waste, (ii) avoid toxic solvent and reagent, (iii) reaction done at ambient temperature and pressure, and (iv) process use natural and renewable sources (Anastas and Kirchhoff 2002). In this respect, *Pseudomonas stutzeri* (Klaus et al. 1999), *Schizosaccharomyces pombe* (Kowshik et al. 2002), *Bacillus* sp. (Beveridge and Murray 1980), and *Rhodococcus* (Ahmad et al. 2003) are indicated in literature. Microalgae have diverse advantage than other in synthesis purpose due to its high ability of metal intake, reducing metal ion for superior contender, low manufacturing cost, easy handling, availability, and environment friendly. Again, microalgae are grown rapidly than any higher plants that give importance for large-scale nanoparticle synthesis. Chlorophyceae, diatoms, Phaeophyceae, euglenoids, cyanobacteria (blue algae), Cyanophyceae, and Rhodophyceae algae species are generally used in nanoparticle synthesis (Brayner et al. 2007; Borowitzka 2013; Singh and Singh 2019). Microalgae are also now used as valuable chemicals in different fields especially in biofuel, natural dyes, and cosmetics. However, full multiplicity of microalgae is still unknown and searchable (Borowitzka 2013). Each cell of microalgae acts as a nano-bioreactor where total microalgal culture behaves as a photobioreactor.

Microalgal culture for wastewater treatment faces problem not only wide space with large amount of water but also biomass separation from water for further application (Xu et al. 2011; Eroglu et al. 2012). Again, its low resistance to chemicals and heat limits its application. The above problem can be concluded through immobilization of microorganism through nanomaterial fabricating, electrospun nanofibers (Eroglu et al. 2012, 2013a), laminar nanomaterials (Eroglu et al. 2013b; Wahid et al. 2013), or magnetic nanoparticle impregnated in polymer (Eroglu et al. 2013b). In these methods, microorganisms are separated from their environment through immobilization or by entrapment where cells are captured by nanomaterials or nano-support or involvement of magnetic nanoparticles into microalgal cells (Eroglu et al. 2013b). These types of nanotechnology application lead to smaller space, smaller volume of growth medium, easy handling, and reuse of microalgae for further application (Hall-Stoodley et al. 2004; Liu et al. 2009). In this technique, nanomaterials are best choice as their large surface enhances algal viability.

In this chapter, the advantages of nanotechnology has been discussed in two different ways. In the first way, microalgae are considered as main chemicals for nanoparticle synthesis and the application of synthesized nanoparticle in wastewater treatment. In another way, nanotechnology has been used as supported medium for immobilization of microorganism. In each case, the mechanism of the treatment process, involvement of metal, nanoparticle formation, and its application in diverse field have been discussed stepwise.

15.2 Biosynthesis of Nanoparticle by Algae

Nanoparticle means the particle which size should be between 1 and 100 nm in any dimension. Commonly "top-down" and "bottom-up" approaches were used for nanoparticle synthesis, where first one follow physical method and second one follow chemical method. These physical and chemical methods are harmful to environment. An alternative biological path is very important to sustain environment with the production of nontoxic and stable nanoparticle. Nanoparticle synthesis by the use of microalgae is much attractive and simpler way in terms of economic and environmental consideration. Microalgae can live in diverse atmospheric conditions, such as alkaline, acidic, and hyper saline medium with different temperatures. In recent decade, algae-mediated metal reduction reaches at nearly 84% during the year of 2011–2015 (Sharma et al. 2016). There are several process and methodologies used to synthesize metallic nanoparticle from their aqueous metal salt solution. Planned bioreactor is proposed to consider nanoparticle recovery without microalgal cell death, immobilization of hosts, optimization of reaction yield, and avoid toxic effect of inorganic salts (Dahoumane et al. 2012a). "Living cultures" of microalgae are now become much interesting toward scientific community due to its simplicity, one-step process, enzyme-mediated route, and finally recover of live cell after the process. In "living cultures," algal cells are not destroyed and washed, but the cells continue for photosynthesis, metabolism, and division (Dahoumane et al. 2017). This type of cultures can be used in different ways for nanosynthesis: (i) cell culture along with metal salt aqueous solution, (ii) addition of cells and its cell-free exponent at middle stage of its growth to aqueous metal salt solution, and (iii) sonication and centrifugation or microwave irradiation of cells before the addition of metal salt aqueous solution (Dahoumane et al. 2017).

15.2.1 Mechanistic Approach

The general process in algae-mediated biosynthesis of nanoparticle follows three steps: (i) preparation of algal cell culture or algal biomass, (ii) preparation of a fixed molar metal salt solution, and (iii) incubation of mixed algal and metal solutions for a certain time in controlled condition. Gold and silver salt solutions are mostly used for synthesis of corresponding nanomaterials. The collection of algal biomass follows different procedures. Algal cells are collected from their culture at a fixed time during their growth phase and dispersed these in aqueous solution with metallic salts, cell free aqueous extract, aqueous solution of ground, fresh or dried algae (Dahoumane et al. 2017). Microalgae have ability to form nanoparticles from metal salts through intracellular or extracellular reduction path. In intracellular path, metals are reduced intracellularly into corresponding metal nanoparticle and then released by the cell, which is recovered from culture medium. The positively charged metal ions are trapped on the surface of cell wall or in the cytoplasm through negatively charged enzymes or proteins. Here, siderophores are also produced by microorganism to intake the metal ions into the cell for reduction (Brayner et al. 2009). The trapped metal ions are then reduced to corresponding small nuclei which give specified morphological characteristic nanomaterials. In extracellular nanoparticle synthesized path, microalgae use extracellular materials like proteins, amino acids, and enzymes (Patil and Kim 2018). In this respect, the role of microbial protein and enzymes secretions are very important and differ for different species. The morphology of nanoparticle and its yield with kinetic rate depend on the type of microalgae species used and the medium of culture. Most of the algae follow second-type process. Extracellular process is more preferable for nanoparticle synthesis due to easy steps, and there is no necessary for repeated centrifugation to remove unnecessary elements like cellular components. The polysaccharide matrices at the surface of the cell and intracellular nanoparticle formed very stable colloid where nanoparticles were stabilized with same size and shape. Here, organic matrices act as capping agent (Dahoumane et al. 2014). In biological entities, different biomolecules like flavonoids, polyols, terpenoids, and organic acids control the topography of nanomaterials. Different carboxyl groups from acid, aldehyde, ketone of biomolecules, and proteins actively take part to bind the metal for the metal reduction. Here, keto groups are useful for electron donation for metal reduction, where proteins act as capping agent and prevent oxidation of metal (Subramaniyam et al. 2015). The organic moieties on the surface of nanomaterials reduce aggregation forces. Capping agent is used to reduce nanoparticle aggregation, stabilize nanocrystal with fine size, and shape distribution. Again, the formation of extracellular substances by microalgae depends on nitrogen sources of the algal growth medium and changes according to algal growth stage. In this process, metal toxicity toward microalgal cell is very important as excess amount of metal causes the cell death. Above the optimum value of metal concentration causes the cell death, which leads to the loss of photosynthetic activity (Dahoumane et al. 2012a). Therefore, the process should be balanced between metal salt uptake and nanoparticle release without cell damage and death.

Scenedesmus acutiformis and *Scenedesmus acumitis* algae precipitated copper at vacuoles and nuclei when grown in higher copper concentrate medium (Ferstenberg et al. 1975). *Tetraselmis cochinensis* has been used for the synthesis of gold nanoparticle intracellularly. The purple color formations of algal cells indicate the formation of desired nanoparticle. Here, synthesized particles are more concentrated on cell wall than on cytoplasmic membrane (Senapati et al. 2012). *Chlorella vulgaris* used extracellular pathway for the synthesis of silver nanoparticle from $AgNO_3$ solution. In this synthesis path, solution color of silver salt changes colorless to yellow-brown which indicates the formation of nanoparticle. This color change is attributed to the excitation of surface plasmon in the silver-associated nanoparticle (da Silva Ferreira et al. 2017). *Chlorella vulgaris* is also effective for the formation of gold nanoparticle through intercellular path, but lesser effect is found for palladium, ruthenium, and rhodium salt (Luangpipat et al. 2011). Eroglu et al. (2013a) used 25 mg L^{-1} of $Na_2[PdCl_4]$ salt as optimum amount for the synthesis of palladium nanoparticle by *Chlorella vulgaris*. In mechanistic study, NADH and NADPH act as reducing agent for the reduction of Pd(II) to Pd(0) nanoparticle during photosynthesis. Soil microalga, *Chlorococcum* sp. MM11, is used in 48-h exposure time at dark condition for the formation of iron nanoparticle where reddish yellow color of iron chloride solution turns yellowish brown. This type of color change is due to surface plasmon resonance (SPR) (Subramaniyam et al. 2015). Single SPR peak in UV-vis analysis indicates monodisperse isotropic nanoparticle formation where multiple SPR bands or bands with shoulder indicate poly-dispersed or nonuniform characteristics of nanoparticle (Ebrahiminezhad et al. 2016). Fresh water microalgae *Chlamydomonas reinhardtii* follow extracellular path for silver nanoparticle production, where nanoparticle formation shows a linear relationship with the input of $AgNO_3$ solution in the presence of exopolysaccharide (EPS)-containing cell culture (Rahman et al. 2019). *Scenedesmus obliquus* green microalgae are cultivated at different nitrogen sources for nanoparticle synthesis. Study shows that $NaNO_3$ and KNO_3 media are much effective for nanoparticle synthesis, where urea and ammonium nitrate increase the biomass of algae. These types of differences are due to the production of extracellular enzymes in the presence of different nitrogen sources that control the reduction of metal cation (Darwesh et al. 2019). *Klebsormidium flaccidum* algal cells are encapsulated with silica gel to form living bio-hybrid material for the synthesis of gold nanoparticle (Sicard et al. 2010). Diatom *Nitzschia* has been used for gold nanoparticle formation, where protein and polysaccharides are used for reduction and stabilization of nanoparticle. In the culture medium, greenish diatom becomes ruby red in the presence of gold chloride solution which indicates the formation of nanoparticle (Borase et al. 2017). A comparative study for Ag nanoparticle formation using *Chlamydomonas reinhardtii* shows that the collected whole cell rapidly forms nanoparticle rather than cell-free extract. This type of fast formation of nanoparticle is due to the presence of active biomolecules in first case (Barwal et al. 2011). Magnetotactic unicellular microalgae *Chlamydomonas reinhardtii* and *Nannochloropsis salina* can produce magnetite nanomaterials within their cell, and this type of microalgae can be easily separable from culture media by a permanent magnet (Dahoumane et al. 2017). Highly stable Au-Ag alloy nanoparticle formation is done by the living cells of *C. reinhardtii*. The reduction of metal is done intracellularly to form nanoparticle which is stabilized by the extracellular matrix of microalgae (Dahoumane et al. 2014). The cell viability is very important in the presence of different metal ions for nanoparticle formation. Generally, two procedures are used to measure viability of cell. In one process, pulse amplitude modulation fluorometry is used to measure chlorophyll a fluorescence, and in another process, extraction of chlorophyll a is done by acetone. First one follows noninvasive and nondestructive path, while second one follows destructive path (Dahoumane et al. 2017). Literature survey shows that most of the biosynthesis of nanoparticle is gold nanoparticle due to its easy reduction of Au(III) to Au(0), specific optical properties at nanoscale, and a wide range of application (Dahoumane et al. 2012b).

The microalgal organism also simultaneously takes part to reduce the toxic effect of high concentration metal ions. In this respect, it takes high duty in polluted ecosystem. Some metal ions are necessary for growth, but excess use of them leads to denature of protein molecules (Gadd and Griffiths 1977). The respiratory system of *Chlorella vulgaris* is largely affected in the presence of fluoride and copper ions jointly, but this effect is reduced for single metal (Hassall 1969). The affectivity of toxic metal on micro algal cells depends on its bioavailability, binding sites on cell membrane, differs for different algal cells and in different wastewaters. Among the several detoxification mechanisms by microalgae, two pathways are most important. These are association of polyphosphate and conversion of metal nanoparticle in thylakoid. These two types of detoxification mechanism follow deactivation of cationic form by inorganic anions or reduction of higher oxidation state to lower one. Due to these reasons, microalgal cells are directly used for wastewater treatment to remove metal or organic toxicity (Table 15.1).

15.2.2 Factors Influencing Nanoparticle Synthesis

The size and shape of nanoparticle formation depend on several factors, which reflect their potential activity in different applications. In biosynthesis process, different factors are considered during the nucleation, construction, and stabilization of nano-size particles. In this respect, pH of cultural medium, concentration of metal salt, concentration of algal biomass, contact time, and reaction temperature

Table 15.1 List of algae for nanoparticle formation

Algae	Reduction path	Metal	References
Klebsormidium flaccidum	Intracellular	Akaganeite (β-FeOOH) nanorods	Brayner et al. (2009)
Klebsormidium flaccidum	Intracellular	Gold nanoparticle	Dahoumane et al. (2012a)
Sargassum wightii	Extracellular	Gold nanoparticle	Singaravelu et al. (2007)
Chlorella vulgaris	Intracellular	Gold nanoparticle	Luangpipat et al. (2011)
Tetraselmis cochinensis	Intracellular	Gold nanoparticle	Senapati et al. (2012)
Chlorella vulgaris	Extracellular	Silver chloride nanoparticle	da Silva Ferreira et al. (2017)
Chlorella vulgaris	–	Palladium nanoparticle	Eroglu et al. (2013a)
Chlorococcum sp. MM11	–	Iron nanoparticle	Subramaniyam et al. (2015)
Chlamydomonas reinhardtii	Extracellular	Silver nanoparticle	Rahman et al. (2019)
Scenedesmus obliquus	Extracellular	Silver nanoparticle	Darwesh et al. (2019)
Nitzschia	Extracellular	Gold nanoparticle	Borase et al. (2017)
Chlamydomonas reinhardtii	–	Silver nanoparticle	Barwal et al. (2011)
Eolimna minima	Intracellular	Gold nanoparticle	Feurtet-Mazel et al. (2016)
Chlamydomonas reinhardtii	Intracellular	Ag–Au alloy nanoparticles	Dahoumane et al. (2014)

decide the fate of nanoparticle. Variation of these parameters leads to variation in shape and size, agglomeration of nanoparticle, and its activity.

15.2.2.1 Concentration

The concentration of biomolecules and metal ions decides the reaction rate as well as the yield of corresponding nanoparticle. High concentration of metal salt shows toxic effect and destroys the cell in culture medium. The size and shape of nanoparticle formation depend on the concentration of reactant. As example, addition of 5–20% *C. serrulata* extract to 10^{-3} (M) $AgNO_3$ solution leads to blue shift, which indicates reduction of diameter of nanoparticle. But 25% extract causes the aggregation of nanoparticle that reduces the intensity of SPR bands (Aboelfetoh et al. 2017). In silver nanoparticle formation by *N. oculata* and *C. vulgaris*, microalgae show effective effort for low concentrated $AgNO_3$ solution than higher concentration (Mohseniazar et al. 2011).

15.2.2.2 pH

pH is an important parameter for the formation of nanoparticle due to its variation control the morphology of nanoparticle which reflected in SPR band peak intensity. Silver nanoparticle by the use of *Caulerpa serrulata* green microalga extract shows high stability at alkaline medium where aggregation is observed at low pH value (Aboelfetoh et al. 2017). *Shewanella* algae form small size gold nanoparticle at pH 7, where at lower pH (2.0) leads to aggregation of gold nanoparticle extracellularly (Konishi et al. 2007).

15.2.2.3 Contact Time

Contact between metal salt and microalgae cell culture or extract decides the fate of nanoparticle formation rate, morphology, and its stability. *Spirulina platensis* extracellularly forms gold nanoparticle, where increasing contact time leads to the increase of particle size (Govindaraju et al. 2008), but *Caulerpa serrulata* shows reverse effect for silver nanoparticle and stabilized after 5 months (Aboelfetoh et al. 2017) (Table 15.2).

15.2.2.4 Temperature

Plectonema boryanum has been used for the formation of silver nanomaterials with the varying of temperature from 25 °C to 100 °C. Study shows that particle size of materials gradually increases with increasing temperature as at high temperature,

Table 15.2 List of synthesized nanoparticles from algae for pollutant degradation in wastewater

Algae	Nanoparticle	Pollutant	Kinetics	References
Chlorella sp. and *Spirulina*	Au	4-nitrophenol	Pseudo first order	Zayadi and Bakar (2020)
Turbinaria conoides and *Sargassum tenerrimum*		Rhodamine B and Sulforhodamine 101	–	Ramakrishna et al. (2016)
Padina tetrastromatica		Eosin yellow and Congo red	Pseudo first order	Princy and Gopinath (2018)
Chlamydomonas reinhardtii	CdS	Methylene blue	Pseudo first order	Rao and Pennathur (2017)
Spirulina (Arthrospira) platensis		Malachite green	–	Mandal et al. (2016)
Spirulina platensis	Pd	Lead	Pseudo second order	Sayadi et al. (2018)
Chlorella	ZnO	Dibenzothiophene	–	Khalafi et al. (2019)
Chlorosarcinopsis sp. MAS04		Nutrient	–	Vasistha et al. (2020)
Anabaena cylindrica	CuO	Detoxification of drinking water	–	Bhattacharya et al. (2019)
Caulerpa racemosa	Ag	Methylene blue	Pseudo first order	Edison et al. (2016)
Chlorella pyrenoidosa		Methylene blue	Pseudo first order	Aziz et al. (2015)
Chlorella ellipsoidea		Methylene blue and methyl Orange	Pseudo first order	Borah et al. (2020)
Sargassum myriocystum		Methylene blue	–	Balaraman et al. (2020)
Hypnea musciformis		Methyl orange	–	Selvam and Sivakumar (2015)
Chlorella pyrenoidosa	TiO_2	Crystal violet	Pseudo first order	Sharma et al. (2018)
Padina gymnospora	CdO-ZnO	Reactive blue 198 dye	–	Rajaboopathi and Thambidurai (2017)
Sargassum wightii	MgO	Methylene blue	–	Pugazhendhi et al. (2019)
Chlorella vulgaris	Fe_3O_4	PO_4^{3-}, NH_4^+	–	Govarthanan et al. (2020)

nucleation of new silver crystal formation is prohibited by the deposition of silver atoms (Lengke et al. 2007).

15.2.3 Microalgal Nanoparticle for Wastewater Treatment

Biosynthesized nanoparticles have a great role for environmental remediation. These types of nanomaterials can be used to remove organic pollutants from wastewater. 4-nitrophenol is reduced to less toxic 4-aminophenol by bio-stabilized colloidal Au nanoparticle in presence of $NaBH_4$. The reduction of 4-nitrophenol is indicated by color change from yellow to colorless solution. Wastewater from industries especially from medicines, pesticides, plastics, and leather should be treated by nanocatalyst to reduce the toxicity of 4-nitrophenol. In the detoxification process, Au nanoparticle acts as a catalyst that helps electron relay process between electron donor BH_4^- and organic acceptor. Higher chlorophyll content *Spirulina*-stabilized gold nanoparticle shows much higher activity than *Chlorella*-stabilized gold nanoparticle (Zayadi and Bakar 2020). ZnO nanoparticle synthesis from microalgae *Chlorella* shows photocatalytic activity for the degradation of pollutant dibenzothiophene under neutral pH condition. This type of algal ZnO nanoparticle also shows that high degradation efficiency for dibenzothiophene contains real wastewater (Khalafi et al. 2019). CuO nanoparticle synthesized from *Anabaena cylindrica* shows potential application for drinking water purpose for disinfection. This CuO nanomaterial also shows good antimicrobial activity against gram-negative bacteria, *Escherichia coli* (Bhattacharya et al. 2019).

Organic dyes from paper and textile industries are very dangerous to aquatic ecosystem due to its limit to the sunlight for water bodies. The toxic nature of these dyes also creates toxicity for human health through drinking water which causes disorder in central nervous, respiratory, and immune system. CdS nanoparticle from single-cell green alga *Chlamydomonas reinhardtii* shows its activity to degrade methylene blue through advanced oxidation process. The bio-synthesized CdS nanoparticle shows higher stability and activity for the degradation of dye than other process. In the degradation mechanism, under UV radiation, CdS produces a large number of electron-hole pair. Hole reacts with water to form hydroxyl radicals, and electrons are taken up by oxygen to form anionic superoxide radicals. These reactive species mineralize methylene blue to carbon dioxide and water (Rao and Pennathur 2017).

$$CdS + h\nu \rightarrow h^+ + e^- \tag{15.1}$$

$$h^+ + \mathrm{H_2O} \rightarrow HO^\cdot + \mathrm{H}^+ \tag{15.2}$$

$$MB + \mathrm{HO^\cdot \, or \, O_2^{\cdot-}} \rightarrow \text{Harmless products} \left(\mathrm{CO_2 + H_2O + NH_4^+ + NO_3^- + SO_4^{2-}} + Cl^-\right) \tag{15.3}$$

Marine green alga *Caulerpa racemosa* has been used as reducing and capping agent for the synthesis of highly stable silver nanoparticle. The stabilized silver nanoparticle shows photocatalytic effect in the presence of $NaBH_4$ during the degradation of methylene blue. In the degradation mechanism, Ag nanoparticle acts as electron transfer mediator between BH_4^- ion and dye, where $NaBH_4$ acts as both electron donor and hydrogen provider (Edison et al. 2016). Silver nanoparticle from *Chlorella pyrenoidosa* shows high photocatalytic activity and recyclability for the degradation of methylene blue under visible light irradiation. In the degradation mechanism, the photogenerated electron-hole pair acts as powerful oxidizing and reducing agent, respectively. The high band gap facilitates to reduce the recombination of electron-hole pair and enhances photocatalytic activity by the formation of hydroxyl and superoxide radicals. Again, superoxide radicals combine with proton to give $HOO^{\bullet}$ that also form H_2O_2 for highly reactive hydroxyl radical's source. These $HO\bullet$ and $O_2^{\bullet-}$ degrade the aromatic ring and form CO_2, H_2O, SO_4^{2-}, NO_3^-, and NH_4^+ ions (Aziz et al. 2015). Silver nanoparticle is also effective for the degradation of methylene blue and methyl orange organic dyes in pseudo first-order kinetic pathway (Borah et al. 2020). Single crystalline nature silver nanoparticle has been synthesized by the use of red alga *Hypnea musciformis* at room temperature and capable for methyl orange dye degradation under sunlight irradiation (Selvam and Sivakumar 2015). Hexagonal-shaped silver nanoparticle synthesis from *Sargassum myriocystum* shows significant antibacterial activity against clinical pathogens like *Staphylococcus aureus*, *Proteus vulgaris*, *Escherichia coli*, *Pseudomonas aeruginosa*, *S. epidermidis*, and *Klebsiella pneumoniae*. The synthesized nanoparticle is also capable to degrade 98.2% methylene blue in 60 min under high-pressure halogen light source (Balaraman et al. 2020). *Chlorella pyrenoidosa* has been used to prepare TiO_2 nanoparticle, which shows photocatalytic activity for the degradation of crystal violet under visible light irradiation. Influence of graphene oxide to this nanomaterial exhibits two times better performance than only TiO_2 due to photosensitizer behavior of graphene oxide (Sharma et al. 2018). *Spirulina (Arthrospira) platensis* organism has been used to prepare CdS nanoparticle that shows photocatalytic activity for the degradation of malachite green dye under UV radiation. In this nano-synthesis, toxic Cd(II) is transformed to less toxic CdS nanoparticle without any external addition of sulfur source (Mandal et al. 2016). CdO-ZnO nanoparticle from seaweed-macro algae degrades Reactive Blue 198 dye under visible light, UV light, and natural sunlight irradiation. The band gap energy of ZnO decreases in the presence of CdO, which facilitates electron transfer from conduction band to valence band and creates electron-hole pair. Again electron of dye molecule becomes excited and transferred from HOMO to LUMO. The holes react with water molecule to produce hydroxyl radical, and photogenerated electron reacts with oxygen and produces superoxide radicals. These reactive species are responsible for dye degradation (Rajaboopathi and Thambidurai 2017). Biosynthesized gold nanoparticle from brown algae *Turbinaria conoides* and *Sargassum tenerrimum* shows photocatalytic activity for the degradation of Rhodamine B and Sulforhodamine 101 in UV-vis irradiation (Ramakrishna et al. 2016). Marine microalgae *Padina tetrastromatica* is used for the preparation of mono dispersed gold nanoparticle. Hazardous organic pollutant dyes eosin yellow and Congo red are degraded by the gold nanoparticle in the presence of sodium borohydrate (Princy and Gopinath 2018). *Sargassum wightii* brown algae form MgO nanoparticle which show anticancer, antimicrobial, and photocatalytic activities. Methylene blue dye has been successfully degraded by the use of MgO nanoparticle under both UV and sunlight irradiation (Pugazhendhi et al. 2019). ZnO nanoparticle contains microalga *Chlorosarcinopsis* sp. MAS04 has been cultivated in primary- and secondary-treated wastewater for nutrient reduction as well as biodiesel production. This bio-nano approach shows 97.5%, 87.20%, and 82.21% removal efficiency for total organic carbon, total nitrogen, and total phosphorous in primary-treated wastewater (Vasistha et al. 2020). Exopolysaccharides (EPSs) derived from the microalga *Chlorella vulgaris* are composed with Fe_3O_4 nanoparticle to form Fe_3O_4@EPS nanocomposite. This technique is able to remove 91% PO_4^{3-} and 85% of NH_4^+ under optimum condition of 3.5 g/L of Fe_3O_4@EPS, pH 7.0, and 13 h of incubation in wastewater treatment plants (Govarthanan et al. 2020). Among the toxic metals, lead is most important which is able to damage brain, kidney, etc. through the contamination from the wastewater of battery, food packaging, dyeing, etc. Biosynthesized palladium nanomaterials show highest ability to absorb lead within 60 min under pH 6.0, 10 mg/L lead concentration, and adsorbent dose 0.5 g/L (Sayadi et al. 2018).

Combination of microalgae and nano-photocatalyst has been used for the removal of hydrocarbon from wastewater. In such study, nanoparticles are absorbed by the algal cell and the growth of cell is retarded by both of the nanoparticle and pollutant. According to the investigation of Salehi and coworkers, *Chlorella vulgaris*-ZnO system is more superior and effective than *Dunaliella tertiolecta*-ZnO system for the removal of hydrocarbon from oily water sample. Although this type of removal efficiency depends on algal type, hydrocarbon content, and nanoparticle concentration (Salehi et al. 2019). Microalga *Chlorosarcinopsis* sp. MAS04 cultured in primary-treated wastewater integrated with ZnO nanoparticle in secondary-treated wastewater has been investigated for the treatment of sewage wastewater with quality biodiesel production (Vasistha et al. 2020). *Sargassum* species supported nano-ZnO and Co–ZnO composite show rapid degradation of malachite green dye under visible light irradiation. Here, algae enhance dye uptake properties and increase probabilities for interaction between photons and catalyst and also prevent agglomeration of nanoparticles. Again, presence of cobalt reduces the electron-hole recombination rate which favors the production of oxidizing radicals (Rabie et al. 2020).

15.3 Nanotechnology in Algal Separation

Separation of algal cell from water phase is necessary for further application of biomass in a nondestructive way. Several types of membranes like polysulfone, polyvinylidene fluoride, and polyethersulfone have been used for the algal support as membrane has strong physical and chemical stability. But membrane fouling through hydrophobic mechanism leads to introduction of nanomaterials in the system to enhance hydrophilic, antifouling characteristics (Anjum et al. 2019). Nano-chitosan as flocculation agent shows better result than chitosan for the biomass recovery of microalgae *Nannochloropsis* sp. (Farid et al. 2013). Immobilization of living algal cells through the entrapment into nanomaterials within 3D inorganic, organic, or hybrid networks like electrospun nanofibers, laminar nanomaterials, and magnetic nanoparticle impregnated polymer shows much interest to the recent scientific community. These type nanomaterials supported algal cell give high impact for the removal of nitrate, phosphate, and other waste pollutants from wastewater.

Nanofiber production through electrospinning technique gives high surface-to-volume ratio with high porosity that leads to use as filter or membrane system for environmental purpose. Enormous potential of nanofibers has been used for integration of microorganism for filtration and purification purpose (San Keskin et al. 2015). There are several types of nanofibers like porous nanofibers, two-dimensional nanofibers/nets, bioinspired nanofibers, and cable-like nanofibers, and they can be fabricated through electrospinning method (Agarwal et al. 2019). *Chlamydomonas reinhardtii* microalgal cells are integrated in an electrospun polysulfone nanofibrous web for color textile industrial wastewater treatment. After 14-day interval time, about 73% decolorization of Reactive Black 5 is shown by algae-supported nanofiber where this percentage comes to ~68.8% and ~12.4% for only the use of microalgae cells and pristine polysulfone nanofibrous web, respectively. This type of decolorization technique also shows good reusability characteristics (San Keskin et al. 2015). Electrospun nanofiber mats have been used as the matrix by Eroglu and coworkers for immobilization of *Chlorella vulgaris* algal cells. The absence of microalgae in nontoxic, water-insoluble chitosan mat exhibits 32 ± 3% nitrate removal, where presence of microalgal cells gives 87 ± 4% removal of nitrate from water medium (Eroglu et al. 2012). Polyvinylidene fluoride (PVDF) hollow fiber membranes with nano-TiO_2 fabricate for *Chlorella vulgaris* cultivation. These algal membrane bioreactors have a capability to remove average 78% phosphorous and 34% nitrogen from wastewater (Hu et al. 2015). Algae-TiO_2/Ag bionano hybrid material shows high photocatalytic efficiency under visible light irradiation for the removal of Cr(VI) from wastewater. This system shows highest 91% Cr(VI) reduction at pH 2.0. Trapping experiment with different scavengers proved that the main reactive species is $O_2^{\bullet-}$ for photoreduction. In photoreduction mechanism, the photogenerated electron is captured by Ag and promoted photocatalytic activity of TiO_2 under visible light. During irradiation process, the dissolved organic substances of algal cells consume photo-excited holes and hydroxyl radical and enhance photoreduction of Cr(VI) on TiO_2 surface (Wang et al. 2017). Nano-Fe_3O_4 coated with polyethyleneimine (Fe_3O_4@PEI) has been used as magnetic seeds for the cultivation of *Chlorella pyrenoidosa* and *Scenedesmus obliquus* microalgae in real urban sewage. This type of novel flocculation technique gives high harvest efficiency with reusable characteristics and low operational cost (Liu et al. 2019) (Table 15.3).

Table 15.3 Nanofiber as algae support for the removal of pollutant from wastewater

Support	Algae	Pollutant	Removal%	Reference
Pristine polysulfone nanofibrous web	*Chlamydomonas reinhardtii*	Reactive black 5, reactive blue 221	72.97	San Keskin et al. (2015)
Electrospun nanofiber mats	*Chlorella vulgaris*	Nitrate	87	Eroglu et al. (2012)
Polyvinylidene fluoride hollow fiber membranes with nano-TiO_2	*Chlorella vulgaris*	Phosphorous	78	Hu et al. (2015)
Ultrafine TiO_2/Ag chitosan hybrid nanofiber mats	*Chlorella vulgaris*	Cr(VI)	91	Wang et al. (2017)

15.4 Future Prospective and Conclusion

Microalgae cultivation and nanotechnology show great efforts for the treatment of wastewater. "Bio-nanofactories" are very important for its ability to deliver eco-friendly green approach without the formation of secondary pollutant. Stable, non-toxic nanoparticle formation and its excellent photocatalytic activity make them highly effective for the photocatalytic wastewater treatment. Huge numbers of scientific studies are available on the nanoparticle-mediated catalytic transformation which deals with dye removal, degradation of organic pollutant, and adsorption of toxic pollutant to make the environment sustainable. The separation of microalgal cells by the nanofibers, magnetic nano-seeds, nano-chitosan, etc., enhance the activity and recyclability of the process. But this technique is still seeking for large-scale industrial application and real wastewater treatment. The characterization of nano-substance and its effect on environment, preparation cost for large-scale application, environmental policy, and optimum condition should be studied before its practical use. Microalgal cells also show high productivity for biofuel production and rapidly studied throughout the world. Hence, determination of suitable microalgae with design of plant for use in both wastewater treatment and biofuel production will open a new era for "nano bio-factories."

References

Aboelfetoh EF, El-Shenody RA, Ghobara MM (2017) Eco-friendly synthesis of silver nanoparticles using green algae (*Caulerpa serrulata*): reaction optimization, catalytic and antibacterial activities. Environ Monit Assess 189:349

Agarwal P, Gupta R, Agarwal N (2019) Advances in synthesis and applications of microalgal nanoparticles for wastewater treatment. J Nanotechnol 2019:1–10

Ahmad A, Senapati S, Khan MI et al (2003) Intracellular synthesis of gold nanoparticles by a novel alkalotolerant actinomycete, *Rhodococcus* species. Nanotechnology 14:824

Ali I (2012) New generation adsorbents for water treatment. Chem Rev 112:5073–5091. https://doi.org/10.1021/cr300133d

Anastas PT, Kirchhoff MM (2002) Origins, current status, and future challenges of green chemistry. Acc Chem Res 35:686–694

Anjum M, Miandad R, Waqas M et al (2019) Remediation of wastewater using various nanomaterials. Arab J Chem 12:4897–4919

Aziz N, Fatma T, Varma A, Prasad R (2014) Biogenic synthesis of silver nanoparticles using *Scenedesmus abundans* and evaluation of their antibacterial activity. Journal of Nanoparticles, Article ID 689419, 2014. https://doi.org/10.1155/2014/689419

Aziz N, Faraz M, Pandey R, Sakir M, Fatma T, Varma A, Barman I, Prasad R (2015) Facile algae-derived route to biogenic silver nanoparticles: Synthesis, antibacterial and photocatalytic properties. Langmuir 31:11605–11612. https://doi.org/10.1021/acs.langmuir.5b03081

Balaraman P, Balasubramanian B, Kaliannan D et al (2020) Phyco-synthesis of silver nanoparticles mediated from marine algae *Sargassum myriocystum* and its potential biological and environmental applications. Waste Biomass Valoriz 11:5255–5271

Barwal I, Ranjan P, Kateriya S, Yadav SC (2011) Cellular oxido-reductive proteins of *Chlamydomonas reinhardtii* control the biosynthesis of silver nanoparticles. J Nanobiotechnology 9:56

Beveridge TJ, Murray RG (1980) Sites of metal deposition in the cell wall of *Bacillus subtilis*. J Bacteriol 141:876–887

Bhattacharya P, Swarnakar S, Ghosh S et al (2019) Disinfection of drinking water via algae mediated green synthesized copper oxide nanoparticles and its toxicity evaluation. J Environ Chem Eng 7:102867

Bhattacharyya PN, Sonowal S, Bhattacharyya LH, Prasad R, Sarma H (2022) Bioremediation of cytostatic pharmaceutical and personal care products and emerging technologies. In: Sharma H, Dominguez DC, Lee W-Y (eds) Emerging Contaminants in the Environment. Elsevier 579–601

Borah D, Das N, Das N et al (2020) Alga-mediated facile green synthesis of silver nanoparticles: photophysical, catalytic and antibacterial activity. Appl Organomet Chem 34:e5597

Borase HP, Patil CD, Suryawanshi RK et al (2017) Mechanistic approach for fabrication of gold nanoparticles by *Nitzschia* diatom and their antibacterial activity. Bioprocess Biosyst Eng 40:1437–1446

Borowitzka MA (2013) High-value products from microalgae—their development and commercialisation. J Appl Phycol 25:743–756

Brayner R, Barberousse H, Hemadi M et al (2007) Cyanobacteria as bioreactors for the synthesis of Au, Ag, Pd, and Pt nanoparticles via an enzyme-mediated route. J Nanosci Nanotechnol 7:2696–2708

Brayner R, Yéprémian C, Djediat C et al (2009) Photosynthetic microorganism-mediated synthesis of akaganeite (β-FeOOH) nanorods. Langmuir 25:10062–10067

da Silva Ferreira V, ConzFerreira ME, Lima LMTR et al (2017) Green production of microalgae-based silver chloride nanoparticles with antimicrobial activity against pathogenic bacteria. Enzym Microb Technol 97:114–121

Dahoumane SA, Djediat C, Yéprémian C et al (2012a) Recycling and adaptation of *Klebsormidium flaccidum* microalgae for the sustained production of gold nanoparticles. Biotechnol Bioeng 109:284–288

Dahoumane SA, Djediat C, Yéprémian C et al (2012b) Species selection for the design of gold nanobioreactor by photosynthetic organisms. J Nanopart Res 14:883

Dahoumane SA, Wijesekera K, Filipe CDM, Brennan JD (2014) Stoichiometrically controlled production of bimetallic gold-silver alloy colloids using micro-alga cultures. J Colloid Interface Sci 416:67–72

Dahoumane SA, Mechouet M, Wijesekera K et al (2017) Algae-mediated biosynthesis of inorganic nanomaterials as a promising route in nanobiotechnology–a review. Green Chem 19:552–587

Darwesh OM, Matter IA, Eida MF et al (2019) Influence of nitrogen source and growth phase on extracellular biosynthesis of silver nanoparticles using cultural filtrates of *Scenedesmus obliquus*. Appl Sci 9:1465

Delanka-Pedige HMK, Munasinghe-Arachchige SP, Cornelius J et al (2019) Pathogen reduction in an algal-based wastewater treatment system employing *Galdieria sulphuraria*. Algal Res 39:101423

Dhingra R, Naidu S, Upreti G, Sawhney R (2010) Sustainable nanotechnology: through green methods and life-cycle thinking. Sustainability 2:3323–3338

Ebrahiminezhad A, Bagheri M, Taghizadeh S-M et al (2016) Biomimetic synthesis of silver nanoparticles using microalgal secretory carbohydrates as a novel anticancer and antimicrobial. Adv Nat Sci Nanosci Nanotechnol 7:15018

Edison TNJI, Atchudan R, Kamal C, Lee YR (2016) *Caulerpa racemosa*: a marine green alga for eco-friendly synthesis of silver nanoparticles and its catalytic degradation of methylene blue. Bioprocess Biosyst Eng 39:1401–1408

Eroglu E, Agarwal V, Bradshaw M et al (2012) Nitrate removal from liquid effluents using microalgae immobilized on chitosan nanofiber mats. Green Chem 14:2682–2685

Eroglu E, Chen X, Bradshaw M et al (2013a) Biogenic production of palladium nanocrystals using microalgae and their immobilization on chitosan nanofibers for catalytic applications. RSC Adv 3:1009–1012

Eroglu E, D'Alonzo NJ, Smith SM, Raston CL (2013b) Vortex fluidic entrapment of functional microalgal cells in a magnetic polymer matrix. Nanoscale 5:2627–2631

Farid MS, Shariati A, Badakhshan A, Anvaripour B (2013) Using nano-chitosan for harvesting microalga *Nannochloropsis* sp. Bioresour Technol 131:555–559

Ferstenberg LB, Stokes PM, Silverberg B (1975) An electron microscope study of copper in *Scenedesmus*. International conference on heavy metals in the environment, Toronto, Ontario, Canada, pp C298–C300

Feurtet-Mazel A, Mornet S, Charron L et al (2016) Biosynthesis of gold nanoparticles by the living freshwater diatom *Eolimna minima*, a species developed in river biofilms. Environ Sci Pollut Res 23:4334–4339

Gadd GM, Griffiths AJ (1977) Microorganisms and heavy metal toxicity. Microb Ecol 4:303–317

Gericke M, Pinches A (2006) Microbial production of gold nanoparticles. Gold Bull 39:22–28

Govarthanan M, Jeon C-H, Jeon Y-H et al (2020) Non-toxic nano approach for wastewater treatment using *Chlorella vulgaris* exopolysaccharides immobilized in iron-magnetic nanoparticles. Int J Biol Macromol 162:1241–1249

Govindaraju K, Basha SK, Kumar VG, Singaravelu G (2008) Silver, gold and bimetallic nanoparticles production using single-cell protein (*Spirulina platensis*) Geitler. J Mater Sci 43:5115–5122

Hall-Stoodley L, Costerton JW, Stoodley P (2004) Bacterial biofilms: from the natural environment to infectious diseases. Nat Rev Microbiol 2:95–108

Hassall K (1969) An asymmetric respiratory response occurring when fluoride and copper ions are applied jointly to *Chlorella vulgaris*. Physiol Plant 22:304–311

Hessel C, Allegre C, Maisseu M et al (2007) Guidelines and legislation for dye house effluents. J Environ Manag 83:171–180

Hu W, Yin J, Deng B, Hu Z (2015) Application of nano TiO_2 modified hollow fiber membranes in algal membrane bioreactors for high-density algae cultivation and wastewater polishing. Bioresour Technol 193:135–141

Hwang J-H, Church J, Lee S-J et al (2016) Use of microalgae for advanced wastewater treatment and sustainable bioenergy generation. Environ Eng Sci 33:882–897

Khalafi T, Buazar F, Ghanemi K (2019) Phycosynthesis and enhanced photocatalytic activity of zinc oxide nanoparticles toward organosulfur pollutants. Sci Rep 9:1–10

Khalid M (2020) Nanotechnology and chemical engineering as a tool to bioprocess microalgae for its applications in therapeutics and bioresource management. Crit Rev Biotechnol 40:46–63

Klaus T, Joerger R, Olsson E, Granqvist C-G (1999) Silver-based crystalline nanoparticles, microbially fabricated. Proc Natl Acad Sci U S A 96:13611–13614

Konishi Y, Tsukiyama T, Tachimi T et al (2007) Microbial deposition of gold nanoparticles by the metal-reducing bacterium *Shewanella* algae. Electrochim Acta 53:186–192

Kowshik M, Deshmukh N, Vogel W et al (2002) Microbial synthesis of semiconductor CdS nanoparticles, their characterization, and their use in the fabrication of an ideal diode. Biotechnol Bioeng 78:583–588

Leach RK, Blunt L, Chetwynd DG, Yacoot A (2002) Nanoscience advances in the UK in support of nanotechnology. Int J Nanosci 1:123–138

Lengke MF, Fleet ME, Southam G (2007) Biosynthesis of silver nanoparticles by filamentous cyanobacteria from a silver (I) nitrate complex. Langmuir 23:2694–2699

Li Y, Duan X, Qian Y et al (1999) Nanocrystalline silver particles: synthesis, agglomeration, and sputtering induced by electron beam. J Colloid Interface Sci 209:347–349

Liu Y, Rafailovich MH, Malal R et al (2009) Engineering of bio-hybrid materials by electrospinning polymer-microbe fibers. Proc Natl Acad Sci U S A 106:14201–14206

Liu Y, Jin W, Zhou X et al (2019) Efficient harvesting of *Chlorella pyrenoidosa* and *Scenedesmus obliquus* cultivated in urban sewage by magnetic flocculation using nano-Fe_3O_4 coated with polyethyleneimine. Bioresour Technol 290:121771

Luangpipat T, Beattie IR, Chisti Y, Haverkamp RG (2011) Gold nanoparticles produced in a microalga. J Nanopart Res 13:6439–6445

Mandal RP, Sekh S, Sarkar NS et al (2016) Algae mediated synthesis of cadmium sulphide nanoparticles and their application in bioremediation. Mater Res Express 3:55007

Mohseniazar M, Barin M, Zarredar H et al (2011) Potential of microalgae and lactobacilli in biosynthesis of silver nanoparticles. Bioimpacts 1:149–152

Patil MP, Kim G-D (2018) Marine microorganisms for synthesis of metallic nanoparticles and their biomedical applications. Colloids Surf B Biointerfaces 172:487–495

Prasad R (2016) Advances and Applications through Fungal Nanobiotechnology. Springer, International Publishing Switzerland (ISBN: 978-3-319-42989-2)

Prasad R (2017) Fungal Nanotechnology: Applications in Agriculture, Industry, and Medicine. Springer Nature Singapore Pte Ltd. (ISBN 978-3-319-68423-9)

Prasad R (2019a) Microbial Nanobionics: Basic Research and Applications. Springer International Publishing (ISBN 978-3-030-16534-5) https://www.springer.com/gp/book/9783030165338

Prasad R (2019b) Microbial Nanobionics: State of Art. Springer International Publishing (ISBN 978-3-030-16383-9) https://www.springer.com/gp/book/9783030163822

Prasad R, Pandey R, Barman I (2016) Engineering tailored nanoparticles with microbes: quo vadis. WIREs Nanomed Nanobiotechnol 8:316–330. https://doi.org/10.1002/wnan.1363

Princy KF, Gopinath A (2018) Optimization of physicochemical parameters in the biofabrication of gold nanoparticles using marine macroalgae *Padina tetrastromatica* and its catalytic efficacy in the degradation of organic dyes. J Nanostructure Chem 8:333–342

Pugazhendhi A, Prabhu R, Muruganantham K et al (2019) Anticancer, antimicrobial and photocatalytic activities of green synthesized magnesium oxide nanoparticles (MgONPs) using aqueous extract of *Sargassum wightii*. J Photochem Photobiol B Biol 190:86–97

Rabie AM, Abukhadra MR, Rady AM et al (2020) Instantaneous photocatalytic degradation of malachite green dye under visible light using novel green Co–ZnO/algae composites. Res Chem Intermed 46:1955–1973

Rahman A, Kumar S, Bafana A et al (2019) Individual and combined effects of extracellular polymeric substances and whole cell components of *Chlamydomonas reinhardtii* on silver nanoparticle synthesis and stability. Molecules 24:956

Rajaboopathi S, Thambidurai S (2017) Green synthesis of seaweed surfactant based CdO-ZnO nanoparticles for better thermal and photocatalytic activity. Curr Appl Phys 17:1622–1638

Ramakrishna M, Babu DR, Gengan RM et al (2016) Green synthesis of gold nanoparticles using marine algae and evaluation of their catalytic activity. J Nanostructure Chem 6:1–13

Rao MD, Pennathur G (2017) Green synthesis and characterization of cadmium sulphide nanoparticles from *Chlamydomonas reinhardtii* and their application as photocatalysts. Mater Res Bull 85:64–73

Salehi M, Biria D, Shariati M, Farhadian M (2019) Treatment of normal hydrocarbons contaminated water by combined microalgae–Photocatalytic nanoparticles system. J Environ Manag 243:116–126

San Keskin NO, Celebioglu A, Uyar T, Tekinay T (2015) Microalgae immobilized by nanofibrous web for removal of reactive dyes from wastewater. Ind Eng Chem Res 54:5802–5809

Sayadi MH, Salmani N, Heidari A, Rezaei MR (2018) Bio-synthesis of palladium nanoparticle using *Spirulina platensis* alga extract and its application as adsorbent. Surf Interfaces 10:136–143
Selvam GG, Sivakumar K (2015) Phycosynthesis of silver nanoparticles and photocatalytic degradation of methyl orange dye using silver (Ag) nanoparticles synthesized from *Hypnea musciformis* (Wulfen) JV Lamouroux. Appl Nanosci 5:617–622
Senapati S, Syed A, Moeez S et al (2012) Intracellular synthesis of gold nanoparticles using alga *Tetraselmis kochinensis*. Mater Lett 79:116–118
Sharma A, Sharma S, Sharma K et al (2016) Algae as crucial organisms in advancing nanotechnology: a systematic review. J Appl Phycol 28:1759–1774
Sharma M, Behl K, Nigam S, Joshi M (2018) TiO_2-GO nanocomposite for photocatalysis and environmental applications: a green synthesis approach. Vacuum 156:434–439
Sicard C, Brayner R, Margueritat J et al (2010) Nano-gold biosynthesis by silica-encapsulated micro-algae: a "living" bio-hybrid material. J Mater Chem 20:9342–9347
Singaravelu G, Arockiamary JS, Kumar VG, Govindaraju K (2007) A novel extracellular synthesis of monodisperse gold nanoparticles using marine alga, *Sargassum wightii* Greville. Colloids surf B Biointerfaces 57:97–101
Singh I, Singh S (2019) Study of algal mediated biosynthesis of nanoparticle: future of green nanotechnology. Curr Life Sci 5:7–14
Subramaniyam V, Subashchandrabose SR, Thavamani P et al (2015) *Chlorococcum* sp. MM11—a novel phyco-nanofactory for the synthesis of iron nanoparticles. J Appl Phycol 27:1861–1869
Vasistha S, Khanra A, Rai MP (2020) Influence of microalgae-ZnO nanoparticle association on sewage wastewater towards efficient nutrient removal and improved biodiesel application: an integrated approach. J Water Process Eng 39:101711
Wahid MH, Eroglu E, Chen X et al (2013) Functional multi-layer graphene–algae hybrid material formed using vortex fluidics. Green Chem 15:650–655
Wang L, Zhang C, Gao F et al (2017) Algae decorated TiO_2/Ag hybrid nanofiber membrane with enhanced photocatalytic activity for Cr (VI) removal under visible light. Chem Eng J 314:622–630
WHO (2014) Preventing diarrhoea through better water, sanitation and hygiene. World Health Organization, Geneva
Xu L, Guo C, Wang F et al (2011) A simple and rapid harvesting method for microalgae by in situ magnetic separation. Bioresour Technol 102:10047–10051
Zayadi RA, Bakar FA (2020) Comparative study on stability, antioxidant and catalytic activities of bio-stabilized colloidal gold nanoparticles using microalgae and cyanobacteria. J Environ Chem Eng 8:103843

Chapter 16
Surface-Functionalized Diatom Silica as a Bio-Absorbent for the Removal of Toxins from Water

Mahesh P. Bhat, U. T. Uthappa, and Mahaveer D. Kurkuri

Contents

M. P. Bhat
Agricultural Automation Research Center, Chonnam National University, Gwangju, Republic of Korea
e-mail: mahaveer.kurkuri@jainuniversity.ac.in

U. T. Uthappa
Department of Environment and Energy Engineering, Chonnam National University, Gwangju, Republic of Korea

M. D. Kurkuri
Centre for Nano and Material Sciences, Jain University, Bengaluru, Karnataka, India

D. Thangadurai et al. (eds.), *Bioprospecting Algae for Nanosized Materials*, Nanotechnology in the Life Sciences,
https://doi.org/10.1007/978-3-030-81557-8_16

16.1 Introduction

Water is an essential element of the universe for all living organisms, which is being used in various applications in industries and agriculture. The increased globalization resulted in the modernization of the industrial and agricultural sector. For all living creatures, water quality is very important for a healthy lifestyle and is used immensely in agriculture for the cultivation of plants. Water has now become polluted with many toxins because of the improper treatment of wastewater before releasing into the environment by factories such as leather, textiles, food, paper mills, rubber, pharmaceuticals, pesticides, cosmetics, pulp, paint, printing and plastics (Zhou et al. 2019). Furthermore, pollutants released from these industries are relatively carcinogenic. There are a huge number of small- and large-scale industries in developing countries that produce multiple waste materials as by-products that are significantly toxic to both humans and the environment. The scientific community is therefore continuously working on the removal of contaminants from the water.

In the smelting, paint, dye, petroleum processing, metal plating, printing, mining, electrochemical, pesticides and nuclear power industries, metal ions such as Cr^{3+}, Fe^{2+}, Fe^{3+}, Co^{2+}, Ni^{2+}, Cu^{2+}, Zn^{2+}, As^{3+}, As^{5+}, Cd^{2+}, Hg^{2+}, Pb^{2+}, Sb^{3+}, Th^{4+} and U^{6+} are usually used (Argun et al. 2007; Bhat et al. 2016; Padervand and Gholami 2013; Pan et al. 2010; Sarı et al. 2010; Sheng et al. 2012; Sprynskyy et al. 2015; Yusan et al. 2012; Zhu et al. 2011). Owing to unregulated industrializations where the wastewater containing heavy metals is released into the environment without being scientifically treated, these heavy metal ions become harmful pollutants. These metal ions are carcinogenic and non-biodegradable and can easily enter the human body, causing severe health problems, including human growth decline, cancer, nervous system damage and eventually death (Zawierucha et al. 2016). When the concentration of these toxic metal ions reaches above the permissible limit set by the World Health Organization (WHO), these metal ions are considered as contaminants. For the removal of metal ions from wastewater, many techniques such as coagulation, ozonation, electrochemical, adsorption, advanced oxidation method, microbial degradation, ion exchange and membrane filtration have been adapted (Ahmad et al. 2015; Barakat 2011; He et al. 2018; Yagub et al. 2014). However, due to its ease of operation, quick fast process, economical, feasible, easy regeneration and high performance, the adsorption process for wastewater treatment has gained popularity. Adsorption is a surface phenomenon that through electrostatic attraction, hydrogen bond interaction and π-π interaction, it binds adsorbate to the surface of adsorbent (Joseph et al. 2019; Sriram et al. 2020; Zheng et al. 2018). Due to the highly porous structure and high specific surface area, potential adsorbents such as activated carbon, xerogels, various silica materials such as MCM-41 and SBA-15 have been extensively studied and used for the removal of contaminants from wastewater (Cashin et al. 2018; Kyzas et al. 2019; Ptaszkowska-Koniarz et al. 2018). The synthetic methods for synthesizing these adsorbents, however, are time-consuming, laborious, not eco-friendly and not cost-effective. There is also a tremendous demand for replacing these synthetic absorbents with naturally available materials such as diatomaceous earth or diatomite (DE). DEs are inexpensive and can be easily adjusted to achieve optimum efficiency during the treatment of wastewater, being naturally available, and thus may replace conventional expensive adsorbents.

DEs are single-cell organized microalgae with attached silica shell walls called frustules, which are a remnant of dead algae and have porous structures arranged hierarchically (Fig. 16.1). Besides, the main composition of DE is in the form of amorphous silica, which is a non-toxic, three-dimensional hydrophilic shape, low density, porous in nature and abundantly available in marine areas, making it inexpensive. Therefore, much attention was paid to DE by the scientific community, and several scientific studies were performed to investigate them for various applications, including biosensors, gas sensors, chromatic separation, drug delivery, catalysts, dye removal, hydrogen storage, thermal energy storage, solar cells, batteries and supercapacitors (Aw et al. 2012; Bhat et al. 2019; Dolatabadi and de la Guardia 2011; Inchaurrondo et al. 2016; Kabiri et al. 2014; Leonardo et al. 2016; Losic et al. 2006; Sun et al. 2017a; Tian et al. 2016; Yang et al. 2011). Nature-gifted DE may

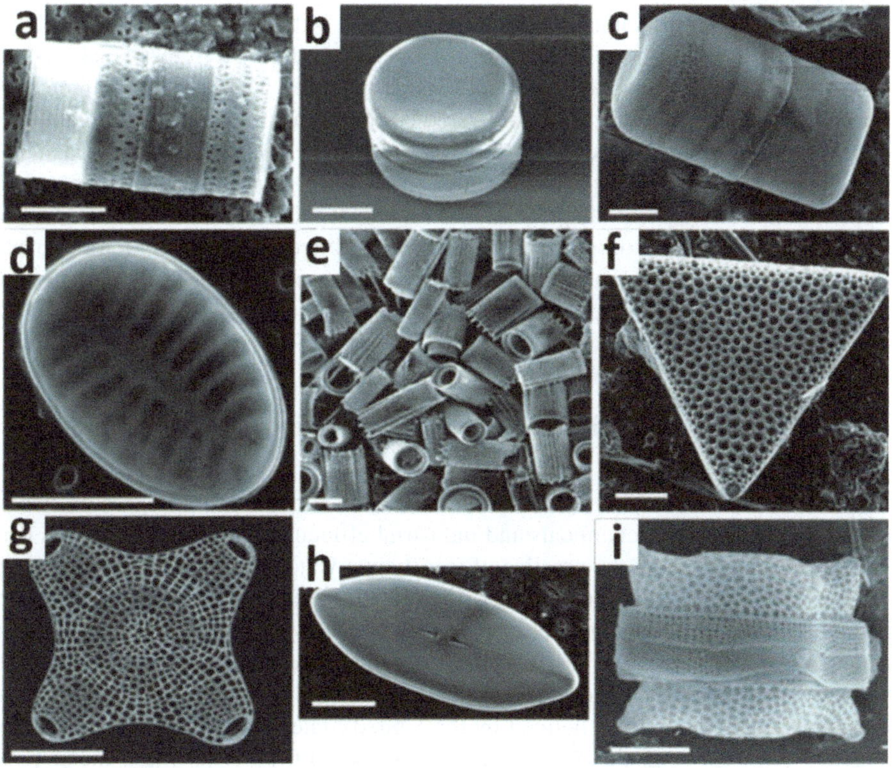

Fig. 16.1 Morphology of various diatom species. Scanning electron microscopy (SEM) images of (**a–d**, **f–i**) diatom frustules and (**e**) fossilized diatom biosilica (scale bar: 10 μm). (Copyright © Elsevier 2017, reprinted with permission from Tommasi et al. 2017)

also be one of the excellent materials for applications in water treatment. DE, however, has a few drawbacks, such as fewer reaction sites available, that is, hydroxyl groups, poor regeneration, the low surface-to-volume ratio and consumption of more adsorbents for contaminate trace removal. Thus to enhance the surface properties, it is very important to modify the surface of DE. To improve the properties of DE, the surface has been altered by the simple process and used for various applications such as drug delivery, supercapacitors, energy storage, dye degradation, heavy metal and dye removal (Bariana et al. 2013; Hua et al. 2012; Sriram et al. 2019; Uthappa et al. 2018a, b; Zhang et al. 2014). The advantages of these naturally available DE can be contrasted with other synthetic materials, that is, only dilute acids and water are needed for DE purification. Synthetic materials, however, require chemicals and more time for synthesis; through surface modifications with other active materials, the efficiency of DE can be improved or tuned by easy protocol. For synthetic materials, if the material efficiency is lower, it is important to make additional surface changes, which will be more expensive. DEs are extremely porous, measuring approximately 100–500 nm in pore diameter and three-dimensional (approximately 15–30 μm), which can make it more adaptable to the surface of the active material. The synthetic materials, however, are either less or extremely porous and cannot adapt to their surface with more active content. DEs are non-toxic, although the use of harmful chemicals during synthesis makes synthetic materials more toxic. DE and surface-modified DE are environmentally friendly candidates that are commonly studied for applications for water treatment. It clearly shows that the use of DE for water treatment applications is increasingly significant, which clearly shows that it is an excellent candidate for the specified applications. The current chapter focuses on recent modern advances in low-cost, naturally accessible adsorbent DE and surface-modified DE for metal ion removal. The content might be useful for large readers who use naturally available DE and surface-adjusted DE to work on water treatment.

16.2 Surface-Modified DE for the Ion Removal

Rapid urbanization and industrialization lead to the contamination of various water bodies with hazardous chemicals and industrial effluents. Water contamination has arisen from neglect and the practice of insufficient water treatment. The metal ions such as Cr^{3+}, Mn^{2+}, Fe^{2+}, Co^{2+}, Ni^{2+}, Cu^{2+}, Zn^{2+}, Hg^{2+} and Pb^{2+} are present in trace amounts in the environment (Patil et al. 2018). However, for the body, metal ions such as Fe^{2+}, Cu^{2+} and Zn^{2+} are necessary for a limited quantity. The ingestion of these ions in excess can lead to serious health risks. When their concentration reaches the allowable limit defined by the United States Environmental Protection Agency (USEPA) and the World Health Organization (WHO), these metal ions become contaminants (Table 16.1).

To date, various methods, such as ion exchange, reverse osmosis, membrane filtration, bioremediation, electrochemical treatment, precipitation, floatation and

Table 16.1 Permissible limits of various toxic metal ions

Toxic metal ions	Permissible limit	
	WHO	USEPA
Chromium	0.05	0.1
Iron	0.3	0.3
Cobalt	–	0.002
Nickel	0.07	–
Copper	2.0	1.0
Zinc	3.0	5.0
Arsenic	0.01	0.01
Cadmium	0.003	0.005
Mercury	0.006	0.002
Lead	0.01	0.015
Antimony	0.02	0.006
Uranium	0.015	0.03

adsorption techniques, have been used for water remediation. Of all, adsorption stands out as a result of low cost, functional applicability and reusability. Therefore, the removal of toxic ions using adsorption has attracted the attention of many researchers all over the world and is a hot field of research. In this regard, various biomaterials are used for the remediation of water. Diatoms stand out because of their biocompatibility and availability, which are unique. DE was used by Ibrahim et al. (2013) as a universal absorbent for multi-ion uptake such as Pb^{2+}, Cd^{2+}, Zn^{2+}, Cu^{2+} and Ni^{2+} at pH 4.7. Likewise, Shabani et al. (2017) ball-milled DE to develop nano-DE with a surface area of 119.5 m^2/g for the removal of Fe^{2+}, Cu^{2+}, Mn^{2+} and Cr^{3+} with q_m values of 111.1, 100.0, 142.8 and 33.2 mg/g, respectively. Further, ElSayed (2018) used DE for the adsorption of Al^{3+}, Ba^{2+}, Cd^{2+}, Cr^{3+}, Cu^{2+}, Fe^{2+}, Pb^{2+}, Mn^{2+}, Ni^{2+} and Zn^{2+} ions at pH 4 with 99% removal efficiency.

16.2.1 Chromium (Cr) Ion Removal

Chromium has been widely used in many industries for various applications such as wood preservation, dyeing, alloying, metallurgy, refractory and tannery. However, chromium is carcinogenic, mutagenic and teratogenic for living beings in excess concentrations. The metal typically occurs in oxidation states of Cr^{3+} and Cr^{6+} where Cr^{6+} is extremely toxic over Cr^{3+}. The removal of a high concentration of chromium is therefore necessary. In this regard, Guru et al. (2008) used column bed with different particle sizes of DE for Cr^{3+} adsorption which were tested at pH 6. DE with a particle size of 1.29 mm with a surface area of 20.4 m^2/g was optimized to display a q_m value of 28.1 mg/g. Li et al. (2009) updated DE for removal of Cr^{3+} at optimized pH 4 with MnO_2 and micro-emulsion. The study showed that with enhanced –OH, micro-emulsion-modified DE has greater removal efficiency than MnO_2-modified DE. The updated DE micro-emulsion showed a q_m value of 49.8 mg/g. Further, Gong et al. (2019) regenerated brewery spent DE (BRD) for Cr^{3+} removal at pH 7 by ammonified bacteria *Lysinibacillus fusiformis*. Material showed a removal efficiency of 71.7% with q_m value of 0.3 mg/g in comparison with bare BRD (47.1%). Although this technique is low cost and environmentally sustainable, the process of regeneration takes a long time (14 days). Further, Yuan et al. (2010) performed DE-modified magnetite nanoparticle by hydrosol method for Cr^{6+} adsorption at pH 2.5. With rapid adsorption, the study showed a q_m value of 69.1 mg/g. Electrostatic attraction accompanied by a redox phase was involved in the Cr^{6+} removal mechanism. Later, for the elimination of Cr^{6+} ions at pH 7, Du et al. (2014) modified DE with MnO_2 nanowires. A high q_m value of 197.6 mg/g was seen from the experiment (Fig. 16.2). Similarly, via the hydrothermal method for the removal of Cr^{6+} ions at pH 2 and 8, Du et al. (2015) synthesized flower, wire and sheet-like morphology nanoparticles in DE. However, the high surface area exhibited by wire-like morphology-modified DE was 144.2 m^2/g over flower and sheet morphologies with a q_m value of 101.4 mg/g. Abu-Zurayk et al. (2015) reported surfactant-modified DE for the removal of Ce^{3+} and Cr^{6+} at pH 4.5. The findings showed that the q_m values for Cr^{3+} and Cr^{6+} were 22.7 mg/g and 17.0 mg/g, respectively. The study showed less ability to adsorb, which may be attributed to material agglomeration due to elevated surface area. Sun et al. (2017b) modified DE surface with carbon by hydrothermal

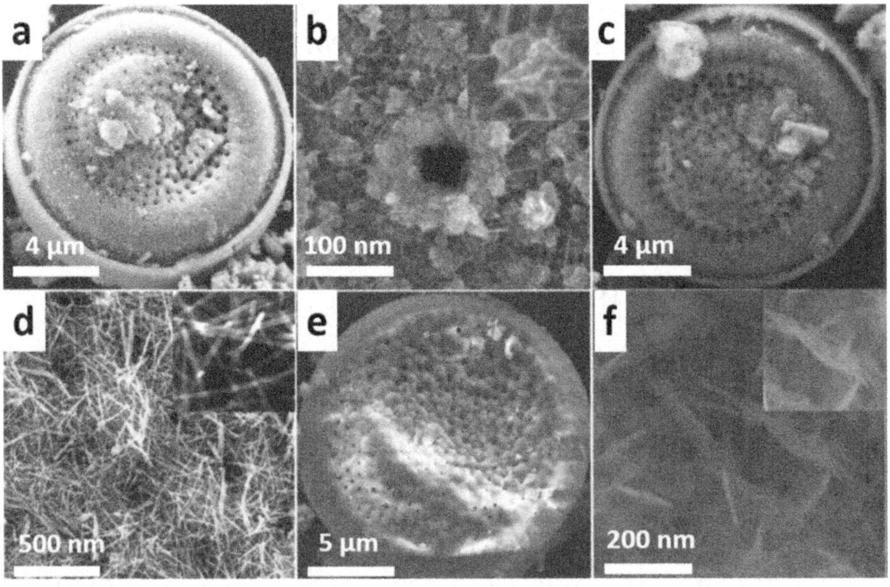

Fig. 16.2 SEM images of MnO_2/DE: (**a, b**) flower-like morphology, (**c, d**) wire-like morphology, and (**e, f**) sheet-like morphology. (Copyright © Elsevier 2014, reprinted with permission from Du et al. 2014)

method for the adsorption of Cr^{6+} at pH 2. The material showed a 19.6 mg/g q_m value in less than 30 min. Later, by surface modification with EDTA at pH 3, Sha et al. (2018) utilized industrial waste DE for the removal of Cr^{6+} ions. The substance was stated to have a surface area of 9.8 m^2/g with a 22.3 mg/g q_m value. Du et al. (2018) reported DE-modified Nb_2O_5 nanorods by hydrothermal method for Cr^{6+} removal from wastewater at pH 7. The composite showed a q_m value of 115 mg/g. Selim et al. (2018) modified DE for the removal of Cr^{6+} at pH 2 by hydrothermal method with MCM-41. The study showed a q_m value of 70.8 mg/g reached equilibrium within 30 min. In a study conducted by Abukhadra et al. (2019), they modified DE with Ni/NiO nanoparticles for the adsorption of Cr^{6+} which resulted in a q_m value of 338.4 mg/g. The study showed a prolonged toxin ion removal time of 11 h.

16.2.2 Cobalt (Co) Ion Removal

This metal is considered as heavy metal present in earth's crust which is odourless and hard. Cobalt is used in semiconductor applications, ceramics, hydrometers, nuclear plant, alloy and electroplate industries (Tofan et al. 2013). Sheng et al. (2012) used DE for the remediation of Co^{2+} ions at pH 7–8. Negatively charged DE surface possess a high tendency for Co^{2+} ions. In addition, the recycling studies showed good efficiency after six cycles for the removal of Co^{2+}. This study showed that without surface modification, DE could be a potential candidate to adsorb Co^{2+}. Surface-modified DE, however, may demonstrate even higher performance.

16.2.3 Nickel (Ni) Ion Removal

Nickel, known to be harmful to living organisms, is a heavy metal. It is, however, essential for the generation of red blood cells in trace amounts. The toxicity of nickel can lead to heart and liver disorders. Nickel, on the other hand, has been widely used in the food, fertilizer and steel industries (Dehghani et al. 2019). Dobor et al. (2015) used DE and cellulose mixture at different ratios for Ni^{2+} ion removal at pH 6 prior to carbonization using H_2SO_4. The optimized 15:5 ratio revealed a q_m value of 88.0 mg/g. Later, Sha et al. (2018) adapted the hydrothermal method for the surface modification of industrial DE with EDTA for the Ni^{2+} adsorption at pH 8. However, the observed q_m value was 9.4 mg/g, which could be attributed to the low surface area of the material. Further, Nefzi et al. (2018) developed palygorskite clay from DE for the removal of Ni^{2+} at pH 7 by precipitation technique. The study showed q_m value of 166.6 mg/g. This material may thus be an efficient candidate for the removal of Ni^{2+}. Similarly, Salih and Ghosh (2018b) modified DE with chitosan to obtain –OH and –NH_2 functional groups on the surface for the adsorption of Ni^{2+} ions. The composite showed a q_m value of 149.6 mg/g at pH 6.

16.2.4 Copper (Cu) Ion Removal

Copper is considered to be an important metal ion with anti-bacterial and anti-microorganism nature. High concentration copper waste from industrial effluents such as paper, textile, electroplating, tannery could put human health at risk (Rakati et al. 2019). Therefore, the need for the removal of high concentration of Cu^{2+} ions is important. In this regard, at different temperatures ranging from 400 to 1100 °C, Yuan et al. (2013) treated DE with APTES ((3-aminopropyl)triethoxysilane). The reaction at 800 °C resulted in 22.7 mg/g of good adsorption efficiency. Further, Nosrati et al. (2017) reported polyethyleneimine (PEI)-functionalized DE for the selective recovery of Cu^{2+} ions at pH 4. With a surface area of 61.1 m^2/g, adsorption reached equilibrium in 3 min. However, there were several steps involved in the synthesis protocol. In another study, DE modified with glutaraldehyde cross-linked PEI was recorded in the same community for adsorption of Cu^{2+} ions from acid mine drainage water at pH 4, resulting in a q_m value of 5.4 mg/g (Larsson et al. 2018). Later, by blending with DE and tungsten residues for Cu^{2+} remediation, Jing et al. (2018) reported a highly porous, high temperature and corrosion-resistant ceramsite. Increased adsorption potential with increased Cu^{2+} pH was observed in the results. The adsorption reached equilibrium with a q_m value of 14 mg/g in 5 h. Due to the lack of surface area, however, the material showed poor adsorption (Fig. 16.3). Nefzi et al. (2018) developed palygorskite clay from DE by precipitation method for Cu^{2+} adsorption at pH 4. The adsorption technique showed a 333.3 mg/g q_m value with equilibrium reached within 90 min. This strategy was cost-effective and followed a straightforward protocol with high potential for adsorption.

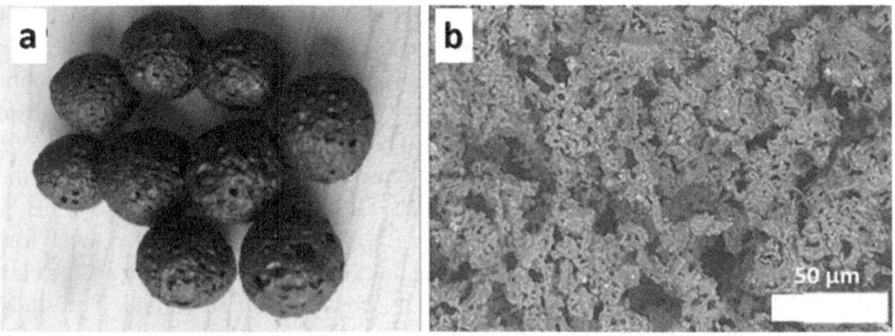

Fig. 16.3 (**a**) Image of ceramsites. (**b**) SEM image of surface morphology of ceramsite. (Copyright © Elsevier 2018, reprinted with permission from Jing et al. 2018)

16.2.5 Zinc (Zn) Ion Removal

For a healthy life, zinc is considered an important metal ion at the trace level. On the other hand, at higher concentrations, it can contribute to muscle desiccation, pancreatic diseases, stomach ache, nausea and anaemia (Bhowmik et al., 2010). In various industrial applications, zinc is dominant, such as steel, metal plating, batteries and mineral extraction (Salih and Ghosh 2018a). For the adsorption of Zn^{2+} ions at pH $>$ 7, Caliskan et al. (2011) modified DE with MnO_2. Negatively charged MnO_2 was highly attracted to positive charged Zn^{2+}. The material showed q_m and equilibrium values of 37.7 mg/g and 120 min, respectively. Salih and Ghosh (2018a) have surface-modified DE for the adsorption of Zn^{2+} ions with chitosan where they have reached a pH 6 q_m value of 127.4 mg/g. This technique, on the other hand, took 130 min to reach equilibrium. Sosa et al. (2019) used surface-modified DE with various ligands such as HEIDA (2-N-(hydroxyethyl)iminodiacetic acid) and EDTA (ethylenediaminetetraacetic acid) for the removal of Zn^{2+} ions at pH 5.5. The 10^{-2} M modified HEIDA and EDTA showed q_m value of 32 mg/g. The technique was tedious as the use of electrochemical workstation was involved in the estimation of ligand concentration. Therefore, chitosan acted as a good candidate for the Zn^{2+} remediation by modified DE (Fig. 16.4).

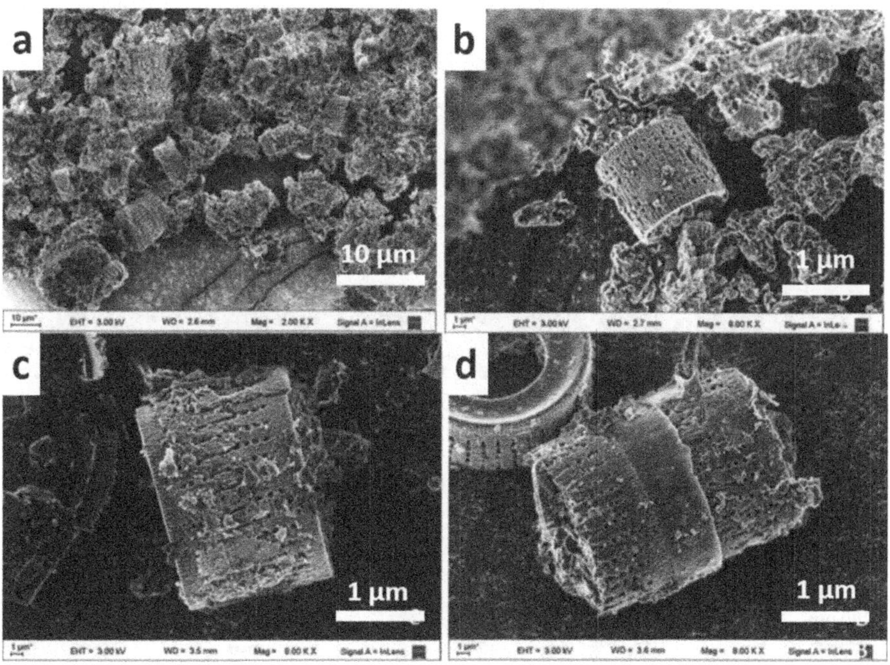

Fig. 16.4 SEM images of DE: (**a**, **b**) DE-HEID, (**c**) DE-EDTA and (**d**) single DE. (Copyright © Elsevier 2019, reprinted with permission from Sosa et al. 2019)

16.2.6 Arsenic (As) Ion Removal

Arsenic is one of the deadliest toxins which is carcinogenic and can lead to immunological and neurological disorders. Generally, arsenic will be present in two oxidation states in aqueous phases, such as As^{3+} (arsenite) and As^{5+} (arsenate). As^{3+} is considered as highly toxic over As^{5+} (Matschullat 2000). Thus, there is a high need for the removal of these contaminants from water. Bimetal oxide (Fe-Mn) incorporated into DE by the coating technique for As^{3+} adsorption at pH 7 was stated by Chang et al. (2009). A surface area of 15 m^2/g exhibited the produced material. The material's low surface area resulted in poor efficiency of adsorption. Later, for the As^{5+} remediation at pH 5.5, Pan et al. (2010) reported iron-oxide-coated DE with a surface area of 93 m^2/g. Although the technique was cost-effective, material efficiency was low. Further, Du et al. synthesized α-Fe_2O_3 nanowire deposited DE for the removal of As^{3+} and As^{5+} at pH 3.5. The material exhibits a surface area of 30 m^2/g with an adsorption capacity of 60.6 mg/g and 81.2 mg/g for As^{3+} and As^{5+}, respectively, which was observed within 15 min (Fig. 16.5). In addition, for the remediation of As^{3+} at pH 7.5–8, Knoerr et al. (2013) reported iron oxide altered DE by precipitation process. The material showed a 16 mg/g q_m value. For the adsorption of As^{3+} at pH 6, Pojananukij et al. (2017) developed nanoscale zero-valent overlays on DE. With a q_m value of 2.7 mg/g, the material was seen to be less effective. Similarly, Du et al. (2014) synthesized MnO_2 nanowires coated DE by hydrothermal method for As^{5+} removal at pH 7. The material resulted in a 178 m^2/g surface area with a 108.2 mg/g q_m value.

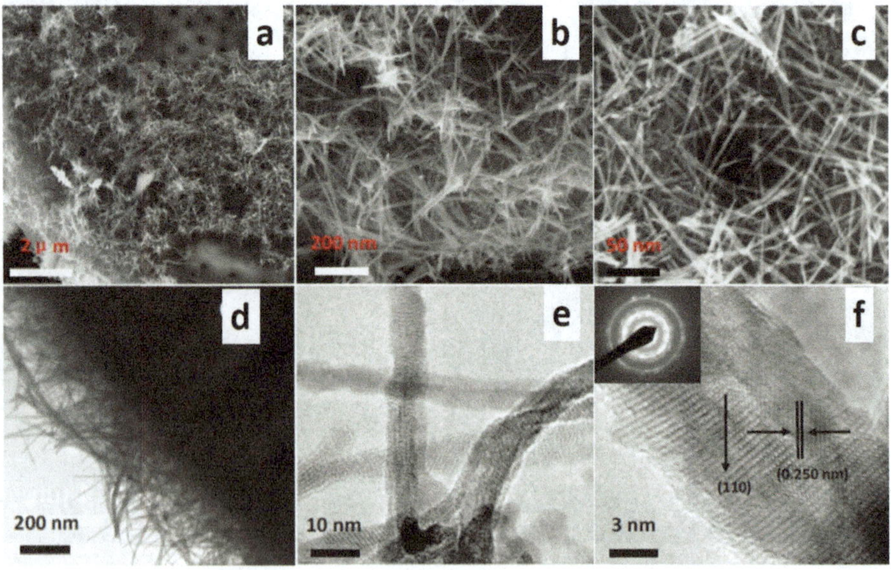

Fig. 16.5 (**a–c**) SEM, (**d**) TEM and (**e, f**) HRTEM images of α-Fe_2O_3 nanowires/DE. (Copyright © RSC 2013, reprinted with permission from Du et al. 2013)

16.2.7 Cadmium (Cd) Ion Removal

Cadmium has a harmful impact on human health and is a poisonous metal. Excessive ingestion of these metal ions can result in kidney, liver, cardiovascular and nervous system damages. Therefore, the removal of Cd^{2+} from potable water is necessary. In this view, Sun et al. (2019) synthesized Zn_2SiO_4 nanostructured silicates on DE through a hydrothermal process for Cd^{2+} adsorption at pH 5.5. With a q_m value of 31.0 mg/g, the material showed poor quality. In addition, Yu et al. (2016) synthesized mesoporous silica using DE as a silica source functionalized with $-NH_2$ for Cd^{2+} removal at pH 6. The study revealed a q_m value of 102.9 mg/g that needed 10 h to achieve equilibrium.

16.2.8 Antimony (Sb) Ion Removal

In general, antimony will be present in the oxidation state of +3. It has been used for a variety of industrial applications such as power grid, alloy, petroleum refineries, pewter, textile, latex, paint, ceramic, quartz and weaponries (Khalid et al. 2000). On the other hand, Sb^{3+} is health hazardous, causing accumulation of cholesterol, lower blood pressure, etc. Thus, to remove Sb^{3+} ions from water, Sari et al. (2010) used DE derived from nature to extract Sb^{3+} ions from water. DE was able to effectively adsorb Sb^{3+} at pH 6, showing a q_m value of 35.2 mg/g. This study showed that DE without any surface modification could be used as an Sb^{3+} absorbent. Thus, there is a room for further research for the effective removal of Sb+ using surface-modified DE.

16.2.9 Mercury (Hg) Ion Removal

Mercury is highly toxic, and it is hazardous to the health and environment. It is expected to cause damages to the cardiovascular system, kidney, bone, liver and neurons (Clarkson 1993). Geothermal phenomena, such as volcanic earthquakes, contribute to the environmental pollution of mercury. Hg^{2+} is known to be highly toxic and volatile and can accumulate easily in food. Industries such as oil refineries, pulp, chloralkali, batteries, electroplating and electronics are also the key cause of contamination with Hg^{2+} (Yardim et al. 2003). Thus, the need for the remediation of Hg^{2+} ions from water supplies is therefore strong. Therefore, Caner et al. (2015) synthesized DE surface-modified chitosan for the adsorption of Hg^{2+} from the water source at pH 5. Although the material has a lower surface area of 0.9 m^2/g, due to the presence of $-OH$ and $-NH_2$ functionalities, they have high adsorption potential. It was noticed that the q_m value was 116.2 mg/g. After 10 cycles, the substance maintained 61% efficiency. In addition, chitosan-functionalized DE for the adsorption of Hg^{2+} at pH 3 was stated by Fu et al. (2017). The substance showed a 15 m^2/g surface area with a q_m value of 195.7 mg/g. With 61% of retained efficiency after five cycles, adsorption reached equilibrium in 1 h. In another study, Fu et al. (2018) modified DE with chitosan for the removal of Hg^{2+} at pH 4. The material showed adsorption capacity of 102 mg/g. Later, for the visible identification and removal of Hg^{2+} from the aqueous medium, Patil et al. (2017) functionalized DE with rhodamine-derived chemosensor. The material displayed a 6.28 m^2/g surface area. In addition, the production of a cartridge was sufficiently effective to extract 30 mL of 1 ppm Hg^{2+} using 450 mg of material.

16.2.10 Lead (Pb) Ion Removal

Lead is a heavy metal ion that is considered harmful to living organisms and poisonous. Significant health damage such as the brain, liver, cardiovascular and kidney damages can be caused upon excess consumption of Pb^{2+} in water (Awual 2019). It has been extensively used in battery, paint and pigment industries, and effluents from these industries will contaminate the water bodies. Remediation of Pb^{2+} from water is therefore an important activity. Sheng et al. (2009) reported DE for the removal of Pb^{2+} at $<$ pH 7. Further, Irani et al. (2011) made a comparative study of DE, perlite and dolomite for the removal of Pb^{2+}. At pH 6, DE showed high q_m value of 25 mg/g, exhibiting high performance over other materials. Beheshti and Irani (2016) have developed nano-DE for the removal of Pb^{2+} ions at pH 5.5 by ball milling DE particles. The study showed a surface area of 119 m^2/g with a 103.1 mg/g q_m value. The procedure does not require any chemical treatment. In addition, for removal of Pb^{2+} at pH 5.5, Knoerr et al. (2011) used surface-modified DE with copper hydroxide. The resulting material displayed an efficiency of 130 mg/g for adsorption. Later, for Pb^{2+} adsorption at pH 6, Alijani et al. (2014) modified DE with carbon nanotubes through chemical vapour deposition (CVD). The material recycling studies showed 95% efficiency for up to six cycles. However, the method involves sophisticated synthesis and time-consuming instrumentation. Furthermore, Li et al. (2014) used surface-modified DE for Pb^{2+} remediation at pH 5.8 by redox reaction with MnO_2. The material registered a surface area of 24.1 m^2/g with a 93.5% removal efficiency (Fig. 16.6). Similarly, for the removal of Pb^{2+} at pH 6, Salih and Ghosh (2018b) reported chitosan-modified DE. The material possesses –OH, $-NH_2$ groups which were selective towards metal ion. The obtained surface area of material 14.4 m^2/g resulted in the q_m value of 175.2 mg/g. In another study, Dobor et al. (2015) reported the optimized ratio of 15:5 cellulose and DE mixtures using H_2SO_4 for the removal of Pb^{2+} at pH 5 by carbonization synthesized carbon-DE composite. The material showed a surface area of 14.5 m^2/g with a q_m value of 387.6 mg/g. However, the material could not show excellent results. Later, Abu-Zurayk et al. (2015) stated the removal of Pb^{2+} ions at pH 4.5 by surfactant-altered DE. The surfactant hexadecyltrimethylammonium bromide (HDTMABR)-modified DE showed a removal efficiency of 24.2 mg/g, which was higher than bare

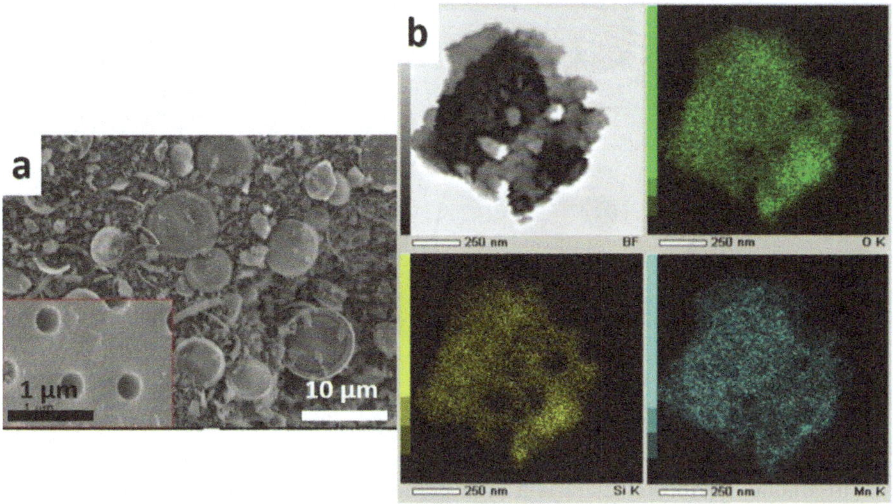

Fig. 16.6 (**a**) SEM image and (**b**) EDX element mapping of MnO_2-functionalized DE. (Copyright © Elsevier 2014, reprinted with permission from Li et al. 2014)

DE. Furthermore, Sun et al. (2019) used the hydrothermal process to synthesize Zn_2SiO_4 layer on DE with ethanediamine for the development of silicate nanostructures. The material displayed a surface area of 264 m^2/g with a 443 mg/g q_m value. At pH 5.5, the material demonstrated high removal efficiency.

16.2.11 Lanthanum (La) Ion Removal

It is a rare earth element used in alloys, glass, ceramics and for catalytic applications in industries (Wu et al. 2011). Zhou et al. (2016) used surface polymerized DE using polyacrylic acid for La^{3+} adsorption by UV exposure (Fig. 16.7). The material displayed an outstanding 90% efficiency of La^{3+} removal even after eight cycles with a q_m value of 139.5 mg/g.

16.2.12 Radioactive Metals such as Thorium and Uranium Ion Adsorption

Radioactive metals are considered to be highly dangerous, even at trace amounts. Thorium (Th^{4+}) and uranium (U^{6+}) are naturally available in earth crusts. They are not abundant in nature (Ilaiyaraja et al. 2013). The nuclear plants and ore mines are the major source of radioactive metal contamination (Salameh et al. 2017). The water bodies contaminated by radioactive metals are risky for health and the

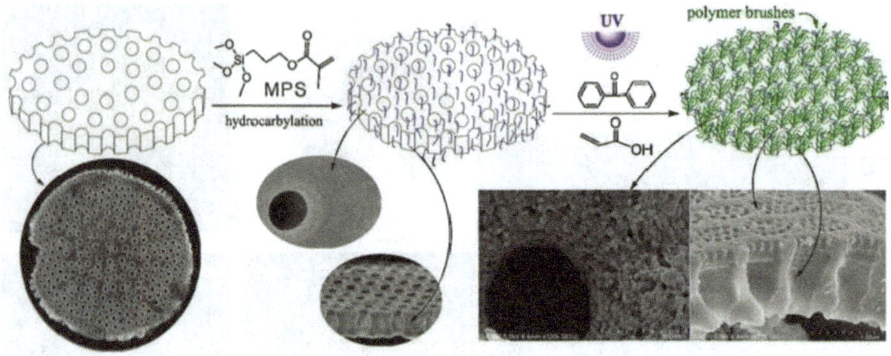

Fig. 16.7 Schematic representing the synthesis process of composite material for La^{3+} adsorption. (Copyright © Elsevier 2016, reprinted with permission from Zhou et al. 2016)

environment. Therefore, there is a high demand for the remediation of these ions from nuclear waste. Sheng et al. (2008) experimented to extract Th^{4+} from water sources using raw DE. At pH 3 DE, there was slow adsorption of Th^{4+} and the balance was reached in 150 min. Further, Yusan et al. (2012) reported calcinated and flux calcinated DE for Th^{4+} adsorption at pH 4. The experiment showed that the calcinated DE resulted in a high surface area of 21.9 m^2/g compared to a calcinated DE flux of 1.2 m^2/g. Both materials attained equilibrium in <20 min and expressed q_m value of 12.8 mg/g for calcinated and 11.4 mg/g for flux calcinated DE. Later, Salameh et al. (2017) used surface-modified DE with HDTMABR for the adsorption of U^{6+} and Th^{4+} at pH 4 and 3, respectively, from aqueous medium. The material exhibits equilibrium within 30 min with q_m values of 38.4 mg/g and 46.0 mg/g for U^{6+} and Th^{4+}, respectively. Further, Sprynskyy et al. (2010) used surface-modified DE with HDTMABR for the removal of U^{6+} at pH 6.2 with q_m value of 158.8 mg/g. Later, same group has used DE modified with ionic liquid for the adsorption of U^{6+} at pH 4.3 (Sprynskyy et al. 2015). The adsorption attained equilibrium within 5 min. The material showed good adsorption efficiency of 86.7 mg/g. Similarly, Lu et al. (2017) modified DE with magnesium silicate by hydrothermal method for the adsorption of U^{6+} at pH 5. The modified material showed surface area of 12 m^2/g followed by equilibrium attained in 3 h with q_m value of 31.5 mg/g. Senol et al. (2019) developed polyacrylamide-functionalized DE for the removal of uranyl at neutral pH. The material showed surface area of 51.8 m^2/g with good removal efficiency.

16.3 Conclusion

A detailed survey of the use of DE in the removal of metal ions from water is given in this chapter. Metal ions are a class of emerging pollutants which have been extensively studied for removal by DE. Due to the limited functionalities on bare DE, they show lesser removal efficiencies. Bare DE shows a maximum adsorption capacity of 25 mg/g for Pb^{2+}, 28.1 mg/g for Cr^{3+} and 35.2 mg/g for Sb^{3+}. Nevertheless, the adsorption potential substantially increased to 443 mg/g for Pb^{2+} after surface modifications with Zn_2SiO_4, and subsequently, Ni-modified DE shows 338.4 mg/g for Cr^{6+} adsorption. Chitosan was another excellent surface modification material since the presence of several functional groups could promote higher adsorption efficiency, although it has a comparatively smaller surface area. To increase the surface area and adsorption capacities, the alteration of the DE surface by MnO_2 also gave rise to excellent composite materials. With low-cost synthesis protocols, palygorskite prepared via DE has shown outstanding adsorption properties. Surface-modified DE shows outstanding adsorption properties for chitosan, MnO_2, Nb_2O_5, Ni, cellulose and Zn_2SiO_4. There were no major changes in the surface area and adsorption of metal ions when changing the surface of DE with iron oxide. There were also no major changes in removal efficiency for chemical modifications with HDTMABR, EDTA, GA-PEI, micro-emulsion and HEIDA. Chemical modifications thus did not lead to a greater increase in adsorption relative to other approaches. In order to avoid sources of secondary contaminants, it is necessary to consider the reusability of adsorbents, but the majority of the studies did not carry out reusability experiments. More effort must be made in the future to increase the reusability efficiency of modified DE for the removal of metal ions. An analysis of the removal efficiency in the presence of other contaminants has to be assessed for real-time applications. In conclusion, the use of DE and improvement of the efficiency of adsorption by surface modifications offer a lower cost and greener solution to water remediation technologies.

Acknowledgment The authors acknowledge the financial support of DST, India (DST/BDTD/EAG/2019), DST/TDT/DDP-31/2021 and Jain University. Prof. Kyeong Hwan Lee honors Chonnam National University for the support.

References

Abukhadra MR, Sayed MA, Rabie AM, Ahmed SA (2019) Surface decoration of diatomite by Ni/NiO nanoparticles as hybrid composite of enhanced adsorption properties for malachite green dye and hexavalent chromium. Colloids Surf A Physicochem Eng Asp 577:583–593

Abu-Zurayk RA, Al Bakain RZ, Hamadneh I, Al-Dujaili AH (2015) Adsorption of Pb (II), Cr (III) and Cr (VI) from aqueous solution by surfactant-modified diatomaceous earth: equilibrium, kinetic and thermodynamic modeling studies. Int J Miner Process 140:79–87

Ahmad A, Mohd-Setapar SH, Chuong CS, Khatoon A, Wani WA, Kumar R, Rafatullah M (2015) Recent advances in new generation dye removal technologies: novel search for approaches to reprocess wastewater. RSC Adv 5:30801–30818

Alijani H, Beyki MH, Shariatinia Z, Bayat M, Shemirani F (2014) A new approach for one step synthesis of magnetic carbon nanotubes/diatomite earth composite by chemical vapor deposition method: application for removal of lead ions. Chem Eng J 253:456–463

Argun ME, Dursun S, Ozdemir C, Karatas M (2007) Heavy metal adsorption by modified oak sawdust: thermodynamics and kinetics. J Hazard Mater 141:77–85

Aw MS, Simovic S, Yu Y, Addai-Mensah J, Losic D (2012) Porous silica microshells from diatoms as biocarrier for drug delivery applications. Powder Technol 223:52–58
Awual MR (2019) Novel conjugated hybrid material for efficient lead (II) capturing from contaminated wastewater. Mater Sci Eng C 101:686–695
Barakat M (2011) New trends in removing heavy metals from industrial wastewater. Arab J Chem 4:361–377
Bariana M, Aw MS, Kurkuri M, Losic D (2013) Tuning drug loading and release properties of diatom silica microparticles by surface modifications. Int J Pharm 443:230–241
Beheshti H, Irani M (2016) Removal of lead (II) ions from aqueous solutions using diatomite nanoparticles. Desalin Water Treat 57:18799–18805
Bhat MP, Patil P, Nataraj S, Altalhi T, Jung H-Y, Losic D, Kurkuri MD (2016) Turmeric, naturally available colorimetric receptor for quantitative detection of fluoride and iron. Chem Eng J 303:14–21
Bhat MP, Kigga M, Govindappa H, Patil P, Jung H-Y, Yu J, Kurkuri M (2019) A reversible fluoride chemosensor for the development of multi-input molecular logic gates. New J Chem 43:12734–12743
Bhowmik D, Chiranjib K, Kumar S (2010) A potential medicinal importance of zinc in human health and chronic. Int J Pharm 1:05–11
Caliskan N, Kul AR, Alkan S, Sogut EG, Alacabey I (2011) Adsorption of Zinc (II) on diatomite and manganese-oxide-modified diatomite: a kinetic and equilibrium study. J Hazard Mater 193:27–36
Caner N, Sarı A, Tüzen M (2015) Adsorption characteristics of mercury (II) ions from aqueous solution onto chitosan-coated diatomite. Ind Eng Chem Res 54:7524–7533
Cashin VB, Eldridge DS, Yu A, Zhao D (2018) Surface functionalization and manipulation of mesoporous silica adsorbents for improved removal of pollutants: a review. Environ Sci Water Res Technol 4:110–128
Chang F, Qu J, Liu H, Liu R, Zhao X (2009) Fe–Mn binary oxide incorporated into diatomite as an adsorbent for arsenite removal: preparation and evaluation. J Colloid Interface Sci 338:353–358
Clarkson TW (1993) Mercury: major issues in environmental health. Environ Health Perspect 100:31–38
Dehghani MH, Sarmadi M, Alipour MR, Sanaei D, Abdolmaleki H, Agarwal S, Gupta VK (2019) Investigating the equilibrium and adsorption kinetics for the removal of Ni (II) ions from aqueous solutions using adsorbents prepared from the modified waste newspapers: a low-cost and available adsorbent. Microchem J 146:1043–1053
Dobor J, Perényi K, Varga I, Varga M (2015) A new carbon–diatomite earth composite adsorbent for removal of heavy metals from aqueous solutions and a novel application idea. Microporous Mesoporous Mater 217:63–70
Dolatabadi JEN, de la Guardia M (2011) Applications of diatoms and silica nanotechnology in biosensing, drug and gene delivery, and formation of complex metal nanostructures. TrAC Trends Analyt Chem 30:1538–1548
Du Y, Zheng G, Wang J, Wang L, Wu J, Dai H (2014) MnO_2 nanowires in situ grown on diatomite: highly efficient absorbents for the removal of Cr (VI) and As (V). Microporous Mesoporous Mater 200:27–34
Du Y, Wang L, Wang J, Zheng G, Wu J, Dai H (2015) Flower-, wire-, and sheet-like MnO_2-deposited diatomites: highly efficient absorbents for the removal of Cr (VI). J Environ Sci 29:71–81
Du Y, Wang X, Wu J, Qi C, Li Y (2018) Adsorption and photoreduction of Cr (VI) via diatomite modified by Nb_2O_5 nanorods. Particuology 40:123–130
ElSayed EE (2018) Natural diatomite as an effective adsorbent for heavy metals in water and wastewater treatment (a batch study). Water Sci 32:32–43
Fu Y, Xu X, Huang Y, Hu J, Chen Q, Wu Y (2017) Preparation of new diatomite–chitosan composite materials and their adsorption properties and mechanism of Hg (II). R Soc Open Sci 4:170829

Fu Y, Huang Y, Hu J, Zhang Z (2018) Preparation of chitosan/amine modified diatomite composites and adsorption properties of Hg (II) ions. Water Sci Technol 77:1363–1371
Gong X, Tian W, Wang L, Bai J, Qiao K, Zhao J (2019) Biological regeneration of brewery spent diatomite and its reuse in basic dye and chromium (III) ions removal. Process Saf Environ Prot 128:353–361
Gürü M, Venedik D, Murathan A (2008) Removal of trivalent chromium from water using low-cost natural diatomite. J Hazard Mater 160:318–323
He Y, Jiang DB, Chen J, Zhang YX (2018) Evaluation of MnO_2-templated iron oxide-coated diatomites for their catalytic performance in heterogeneous photo Fenton-like system. J Hazard Mater 344:230–240
Hua M, Zhang S, Pan B, Zhang W, Lv L, Zhang Q (2012) Heavy metal removal from water/wastewater by nanosized metal oxides: a review. J Hazard Mater 211:317–331
Ibrahim SS, Ibrahim HS, Ammar NS, Abdel Ghafar HH, Jamil TS, Farahat M (2013) Applicability of Egyptian diatomite for uptake of heavy metals. Desalin Water Treat 51:2343–2350
Ilaiyaraja P, Deb AKS, Sivasubramanian K, Ponraju D, Venkatraman B (2013) Removal of thorium from aqueous solution by adsorption using PAMAM dendron-functionalized styrene divinyl benzene. J Radioanal Nucl Chem 297:59–69
Inchaurrondo N, Font J, Ramos CP, Haure P (2016) Natural diatomites: efficient green catalyst for Fenton-like oxidation of Orange II. Appl Catal B 181:481–494
Irani M, Amjadi M, Mousavian MA (2011) Comparative study of lead sorption onto natural perlite, dolomite and diatomite. Chem Eng Technol 178:317–323
Jing Q-X, Wang Y-Y, Chai L-Y, Tang C-J, Huang X-D, Huan G, Wei W, Wei Y (2018) Adsorption of copper ions on porous ceramsite prepared by diatomite and tungsten residue. T Nonferr Metal Soc 28:1053–1060
Joseph L, Jun BM, Jang M, Park CM, Muñoz-Senmache JC, Hernández-Maldonado AJ, Heyden A, Yu M, Yoon Y (2019) Removal of contaminants of emerging concern by metal-organic framework nanoadsorbents: a review. Chem Eng J 369:928–946
Kabiri S, Kurkuri MD, Kumeria T, Losic D (2014) Frit-free PDMS microfluidic device for chromatographic separation and on-chip detection. RSC Adv 4:15276–15280
Khalid N, Ahmad S, Toheed A, Ahmed J (2000) Potential of rice husks for antimony removal. Appl Radiat Isot 52:31–38
Knoerr R, Brendlé J, Lebeau B, Demais H (2011) Elaboration of copper hydroxide phase modified diatomite and their application in lead ions immobilization. New J Chem 35:461–468
Knoerr R, Brendlé J, Lebeau B, Demais H (2013) Preparation of ferric oxide modified diatomite and its application in the remediation of As (III) species from solution. Microporous Mesoporous Mater 169:185–191
Kyzas GZ, Bomis G, Kosheleva RI, Efthimiadou EK, Favvas EP, Kostoglou M, Mitropoulos AC (2019) Nanobubbles effect on heavy metal ions adsorption by activated carbon. Chem Eng J 356:91–97
Larsson M, Nosrati A, Kaur S, Wagner J, Baus U, Nydén M (2018) Copper removal from acid mine drainage-polluted water using glutaraldehyde-polyethyleneimine modified diatomaceous earth particles. Heliyon 4:e00520
Leonardo S, Prieto-Simón B, Campàs M (2016) Past, present and future of diatoms in biosensing. Trends Analyt Chem 79:276–285
Li E, Zeng X, Fan Y (2009) Removal of chromium ion (III) from aqueous solution by manganese oxide and microemulsion modified diatomite. Desalination 238:158–165
Li S, Li D, Su F, Ren Y, Qin G (2014) Uniform surface modification of diatomaceous earth with amorphous manganese oxide and its adsorption characteristics for lead ions. Appl Surf Sci 317:724–729
Losic D, Rosengarten G, Mitchell JG, Voelcker NH (2006) Pore architecture of diatom frustules: potential nanostructured membranes for molecular and particle separations. J Nanosci Nanotechnol 6:982–989

Lu S, Hu J, Chen C, Chen X, Gong Y, Sun Y, Tan X (2017) Spectroscopic and modeling investigation of efficient removal of U (VI) on a novel magnesium silicate/diatomite. Sep Purif Technol 174:425–431

Matschullat J (2000) Arsenic in the geosphere—a review. Sci Total Environ 249:297–312

Nefzi H, Abderrabba M, Ayadi S, Labidi J (2018) Formation of palygorskite clay from treated diatomite and its application for the removal of heavy metals from aqueous solution. Water 10:1257

Nosrati A, Larsson M, Lindén JB, Zihao Z, Addai-Mensah J, Nydén M (2017) Polyethyleneimine functionalized mesoporous diatomite particles for selective copper recovery from aqueous media. Int J Miner Process 166:29–36

Padervand M, Gholami MR (2013) Removal of toxic heavy metal ions from waste water by functionalized magnetic core–zeolitic shell nanocomposites as adsorbents. Environ Sci Pollut Res 20:3900–3909

Pan Y-F, Chiou CT, Lin T-F (2010) Adsorption of arsenic (V) by iron-oxide-coated diatomite (IOCD). Environ Sci Pollut Res 17:1401–1410

Patil P, Bhat MP, Gatti MG, Kabiri S, Altalhi T, Jung H-Y, Losic D, Kurkuri M (2017) Chemodosimeter functionalized diatomaceous earth particles for visual detection and removal of trace mercury ions from water. Chem Eng J 327:725–733

Patil P, Ajeya KV, Bhat MP, Sriram G, Yu J, Jung H-Y, Altalhi T, Kigga M, Kurkuri MD (2018) Real-time probe for the efficient sensing of inorganic fluoride and copper ions in aqueous media. ChemistrySelect 3:11593–11600

Pojananukij N, Wantala K, Neramittagapong S, Lin C, Tanangteerpong D, Neramittagapong A (2017) Improvement of As (III) removal with diatomite overlay nanoscale zero-valent iron (nZVI-D): adsorption isotherm and adsorption kinetic studies. Water Sci Technol Water Supply 17:212–220

Ptaszkowska-Koniarz M, Goscianska J, Pietrzak R (2018) Removal of rhodamine B from water by modified carbon xerogels. Colloids Surf A Physicochem Eng Asp 543:109–117

Rakati KK, Mirzaei M, Maghsoodi S, Shahbazi A (2019) Preparation and characterization of poly aniline modified chitosan embedded with ZnO-Fe_3O_4 for Cu (II) removal from aqueous solution. Int J Biol Macromol 130:1025–1045

Salameh SI, Khalili FI, Al-Dujaili AH (2017) Removal of U (VI) and Th (IV) from aqueous solutions by organically modified diatomaceous earth: evaluation of equilibrium, kinetic and thermodynamic data. Int J Miner Process 168:9–18

Salih SS, Ghosh TK (2018a) Adsorption of Zn (II) ions by chitosan coated diatomaceous earth. Int J Biol Macromol 106:602–610

Salih SS, Ghosh TK (2018b) Highly efficient competitive removal of Pb (II) and Ni (II) by chitosan/diatomaceous earth composite. J Environ Chem Eng 6:435–443

Sarı A, Çıtak D, Tuzen M (2010) Equilibrium, thermodynamic and kinetic studies on adsorption of Sb (III) from aqueous solution using low-cost natural diatomite. Chem Eng J 162:521–527

Selim AQ, Mohamed EA, Mobarak M, Zayed AM, Seliem MK, Komarneni S (2018) Cr (VI) uptake by a composite of processed diatomite with MCM-41: isotherm, kinetic and thermodynamic studies. Microporous Mesoporous Mater 260:84–92

Şenol ZM, Arslan DŞ, Şimşek S (2019) Preparation and characterization of a novel diatomite-based composite and investigation of its adsorption properties for uranyl ions. J Radioanal Nucl Chem 321:791–803

Sha H, Wu Y, Fan Y (2018) Utilization of industrial waste as a novel adsorbent: mono/competitive adsorption of chromium (VI) and nickel (II) using diatomite waste modified by EDTA. Appl Organomet Chem 32:e3977

Shabani KS, Ardejani FD, Badii K, Olya ME (2017) Preparation and characterization of novel nano-mineral for the removal of several heavy metals from aqueous solution: batch and continuous systems. Arab J Chem 10:S3108–S3127

Sheng G, Hu J, Wang X (2008) Sorption properties of Th (IV) on the raw diatomite—effects of contact time, pH, ionic strength and temperature. Appl Radiat Isot 66:1313–1320

Sheng G, Wang S, Hu J, Lu Y, Li J, Dong Y, Wang X (2009) Adsorption of Pb (II) on diatomite as affected via aqueous solution chemistry and temperature. Colloids Surf A Physicochem Eng Asp 339:159–166
Sheng G, Dong H, Li Y (2012) Characterization of diatomite and its application for the retention of radiocobalt: role of environmental parameters. J Environ Radioact 113:108–115
Sosa GL, Morantes CF, Flores FM, Sánchez RMT, Zalts A, Ramirez SA (2019) Characterization of diatomaceous earth modified by organic ligands for enhanced zinc adsorption. J Environ Chem Eng 7:103197
Sprynskyy M, Kovalchuk I, Buszewski B (2010) The separation of uranium ions by natural and modified diatomite from aqueous solution. J Hazard Mater 181:700–707
Sprynskyy M, Kowalkowski T, Tutu H, Cukrowska EM, Buszewski B (2015) Ionic liquid modified diatomite as a new effective adsorbent for uranium ions removal from aqueous solution. Colloids Surf A Physicochem Eng Asp 465:159–167
Sriram G, Bhat MP, Kigga M, Uthappa U, Jung H-Y, Kumeria T, Kurkuri MD (2019) Amine activated diatom xerogel hybrid material for efficient removal of hazardous dye. Mater Chem Phys 235:121738
Sriram G, Uthappa U, Losic D, Kigga M, Jung H-Y, Kurkuri MD (2020) Mg–Al-layered double hydroxide (LDH) modified diatoms for highly efficient removal of Congo red from aqueous solution. Appl Sci 10:2285
Sun XW, Zhang YX, Losic D (2017a) Diatom silica, an emerging biomaterial for energy conversion and storage. J Mater Chem A 5:8847–8859
Sun Z, Yao G, Xue Y, Sun W, Zheng S (2017b) In situ synthesis of carbon @ diatomite nanocomposite adsorbent and its enhanced adsorption capability. Particul Sci Technol 35:379–386
Sun L, Wang J, Wu J, Wang T, Du Y, Li Y, Li H (2019) Constructing nanostructured silicates on diatomite for Pb (II) and Cd (II) removal. J Mater Sci 54:6882–6894
Tommasi ED, Gielis J, Rogato A (2017) Diatom frustule morphogenesis and function: a multidisciplinary survey. Marine Genomics 35:1–18
Tian L, Zhang J, Shi H, Li N, Ping Q (2016) Adsorption of malachite green by diatomite: equilibrium isotherms and kinetic studies. J Dispers Sci Technol 37:1059–1066
Tofan L, Teodosiu C, Paduraru C, Wenkert R (2013) Cobalt (II) removal from aqueous solutions by natural hemp fibers: batch and fixed-bed column studies. Appl Surf Sci 285:33–39
Uthappa U, Brahmkhatri V, Sriram G, Jung H-Y, Yu J, Kurkuri N, Aminabhavi TM, Altalhi T, Neelgund GM, Kurkuri MD (2018a) Nature engineered diatom biosilica as drug delivery systems. J Control Release 281:70–83
Uthappa U, Sriram G, Brahmkhatri V, Kigga M, Jung H-Y, Altalhi T, Neelgund GM, Kurkuri MD (2018b) Xerogel modified diatomaceous earth microparticles for controlled drug release studies. New J Chem 42:11964–11971
Uthappa U, Sriram G, Arvind O, Kumar S, Neelgund GM, Losic D, Kurkuri MD (2020) Engineering MIL-100 (Fe) on 3D porous natural diatoms as a versatile high performing platform for controlled isoniazid drug release, Fenton's catalysis for malachite green dye degradation and environmental adsorbents for Pb^{2+} removal and dyes. Appl Surf Sci 528:146974
Wu D, Zhang L, Wang L, Zhu B, Fan L (2011) Adsorption of lanthanum by magnetic alginate-chitosan gel beads. J Chem Technol Biotechnol 86:345–352
Yagub MT, Sen TK, Afroze S, Ang HM (2014) Dye and its removal from aqueous solution by adsorption: a review. Adv Colloid Interf Sci 209:172–184
Yan S, Huo W, Yang J, Zhang X, Wang Q, Wang L, Pan Y, Huang Y (2018) Green synthesis and influence of calcined temperature on the formation of novel porous diatomite microspheres for efficient adsorption of dyes. Powder Technol 329:260–269
Yang W, Lopez PJ, Rosengarten G (2011) Diatoms: self assembled silica nanostructures, and templates for bio/chemical sensors and biomimetic membranes. Analyst 136:42–53
Yardim MF, Budinova T, Ekinci E, Petrov N, Razvigorova M, Minkova V (2003) Removal of mercury (II) from aqueous solution by activated carbon obtained from furfural. Chemosphere 52:835–841

Yu Z-H, Zhang Y-F, Zhai S-R, Wang Y, Pan Y-Z, Meng C-G (2016) Amino-modified mesoporous sorbents for efficient cd (II) adsorption prepared using non-chemical diatomite as precursor. J Sol-Gel Sci Technol 78:110–119

Yuan P, Liu D, Fan M, Yang D, Zhu R, Ge F, Zhu J, He H (2010) Removal of hexavalent chromium [Cr(VI)] from aqueous solutions by the diatomite-supported/unsupported magnetite nanoparticles. J Hazard Mater 173:614–621

Yuan P, Liu D, Tan D-Y, Liu K-K, Yu H-G, Zhong Y-H, Yuan A-H, Yu W-B, He H-P (2013) Surface silylation of mesoporous/macroporous diatomite (diatomaceous earth) and its function in Cu (II) adsorption: the effects of heating pretreatment. Microporous Mesoporous Mater 170:9–19

Yusan S, Gok C, Erenturk S, Aytas S (2012) Adsorptive removal of thorium (IV) using calcined and flux calcined diatomite from Turkey: evaluation of equilibrium, kinetic and thermodynamic data. Appl Clay Sci 67:106–116

Zawierucha I, Kozlowski C, Malina G (2016) Immobilized materials for removal of toxic metal ions from surface/groundwaters and aqueous waste streams. Environ. Sci: Process Impacts 18:429–444

Zhang YX, Li F, Huang M, Xing Y, Gao X, Li B, Guo ZY, Guan YM (2014) Hierarchical NiO moss decorated diatomites via facile and templated method for high performance supercapacitors. Mater Lett 120:263–266

Zheng N-C, Wang Z, Long J-Y, Kong L-J, Chen D-Y, Liu Z-Q (2018) Shape-dependent adsorption of CeO_2 nanostructures for superior organic dye removal. J Colloid Interface Sci 525:225–233

Zhou Q, Yang H, Yan C, Luo W, Li X, Zhao J (2016) Synthesis of carboxylic acid functionalized diatomite with a micro-villous surface via UV-induced graft polymerization and its adsorption properties for Lanthanum (III) ions. Colloids Surf A Physicochem Eng Asp 501:9–16

Zhou Y, Lu J, Zhou Y, Liu Y (2019) Recent advances for dyes removal using novel adsorbents: a review. Environ Pollut 252:352–365

Zhu H, Jiang R, Fu Y-Q, Jiang J-H, Xiao L, Zeng G-M (2011) Preparation, characterization and dye adsorption properties of γ-Fe_2O_3/SiO_2/chitosan composite. Appl Surf Sci 258:1337–1344

Chapter 17
Microalgal Nanotechnology for the Remediation of Environmental Pollutants

Sonam Dwivedi and Iffat Zareen Ahmad

Contents

S. Dwivedi · I. Z. Ahmad (✉)
Natural Products Laboratory, Department of Bioengineering and Biosciences, Integral University, Lucknow, Uttar Pradesh, India
e-mail: iffat@iul.ac.in

D. Thangadurai et al. (eds.), *Bioprospecting Algae for Nanosized Materials*, Nanotechnology in the Life Sciences,
https://doi.org/10.1007/978-3-030-81557-8_17

17.1 Introduction

The bioremediation process involves degrading or detoxifying toxic, mutagenic, and carcinogenic compounds through the incorporation of microorganisms like microalge, fungi, bacteria, and plants (Prasad and Aranda 2018). Bioremediation is cost-effective and convenient method for changing contaminants to nontoxic compounds (Mandal et al. 2012). Bioremediation incompetence relies on the microorganism used and environmental conditions. As reported previously, bioremediation of wastewater uses fungi (Mishra and Malik 2013; Kaushik and Malik 2015), bacteria (Wasi et al. 2013), and microalgae (Prajapati et al. 2013). Phycoremediation is a process used to eradicate the excess toxics from the industrial waste. Microalgae have the ability to phycoremediate and detoxify the industrial waste stream. Microalgae can grow easily, therefore, enhances bioremediation efficiency and provides a new insight for this problem (Hammed et al. 2016).

Microalgae have been reported to use pesticides, hydrocarbons, endocrine disruptors, and cyanides as carbon and nitrogen sources. Moreover, microalgal cell wall consists of carbohydrate, therefore, able to remove toxic agents from the wastewater by the process of bioabsorption and potentially reduced the level of nitrogen, phosphorous, and heavy metals from the wastewater. Microalgae can remediate the industrial pollutants present in the wastewater discharges, livestock and poultry pollutants, atmospheric deposition originated widely from air pollutants, surface runoff discharged from agricultural lands to the water bodies, and anthropogenic sources (Hammed et al. 2016).

Environmental contamination is the serious issue globally, and it is expanding day by day. Air pollutants are CO, chlorofluorocarbons, volatile organic compounds, hydrocarbons, and nitrogen oxides. Water and soil polluting substances are organic and inorganic compounds, sewage water, industrial effluents, pesticides, fertilizers, and oil spills (Tripathy et al. 2021). The expansion of nanotechnology improved the monitoring systems and cleanup process. Nanomaterials act as good adsorbents, catalysts, and sensor (Mohamed 2017).

Nanoparticles are widely used in many sectors of economy. Nowadays, with the help of green technology, nanoparticles are also implemented in the environmental remediation with reduced toxicity. Green nanotechnology uses the biological sources which are nontoxic and eco-friendly (Prasad et al. 2018; Sarma et al. 2021). Algae are commonly used for the biosynthesis of various nanoparticles because they grow rapidly and are easy to handle. Bioactive compounds present in microalgae have several roles in the formation of nanoparticles, they act as reducing, capping, and stabilizing agents (Aziz et al. 2014, 2015). Algal nanomaterials are classified into four types, that is, nanometal, nano-oxide, carbon nanomaterials, and quantum dots (Wang et al. 2019). This chapter highlights the recent updates on microalgae used in the remediation of environmental pollutants by using different nanotechnological approaches.

17.2 Microalgal Nanotechnologies for Remediation

It is essential that the materials used for the remediation treatment process are not the pollutant itself. Therefore, biodegradable materials are fascinating sources for this field of application. Biodegradable materials not only provide the safe and eco-friendly option for the customer, but also there is no production of waste materials. Furthermore, target-specific detoxification of contaminants has enhanced the efficiency of remediation techniques. New techniques have developed that combine the surface-modified nanomaterials to overcome the problems involved in the removal process of pollutants (Kamat and Meisel 2003). Environmental pollutants removed by microalgae through different methods are shown in Fig. 17.1 and Table 17.1.

17.2.1 Water Pollutant Remediation

Increased production of nanostructures and nanocomposites for water remediation has gained attention because of its cost-effectiveness, being nontoxic in nature, and eco-friendly. Algae have the capability to reduce metal ions, and make them a suitable candidate for the synthesis of nanoparticles. For the microalgal nanoparticles synthesis, various parameters have been taken such as the effects of environmental condition, kinetics, and mechanism for nanoparticles formation, that is, pH of the reaction mixture, selection of appropriate strains, light intensity, ionic strength, and biological transition of metal cations (Agarwal et al. 2019).

Various methods have been developed to synthesize metallic microalgal nanoparticles having superior quality nanocrystals (Dahoumane et al. 2017). The biomolecules directly taken from the disrupted microalgal cells can be used (Patel et al. 2015; El-Sheekh and El-Kassas 2016); for the synthesis of silver nanoparticles, cell-free culture supernatants of cyanobacteria and chlorophyta species were used (Patel et al. 2015). Various microalgal species have been used for the synthesis of nanoparticles, such as chlorophyta, haptophyta, and ochrophyta (Merin et al. 2010; Mohseniazar et al. 2011; Dahoumane et al. 2014). Silver nanoparticles from *Spirulina platensis* and *Lyngbya majuscula* have been reported previously (Rosken et al. 2014; El-Sheekh and El-Kassas 2016). Gold nanoparticles synthesis using *Chlorella vulgaris* have been reported (Mahdieh et al. 2012). Also, cadmium and platinum nanoparticles synthesis using microalgae have been reported (https://www.climatecolab.org/contests/2015/wastemanagement/c/proposal/1320133).

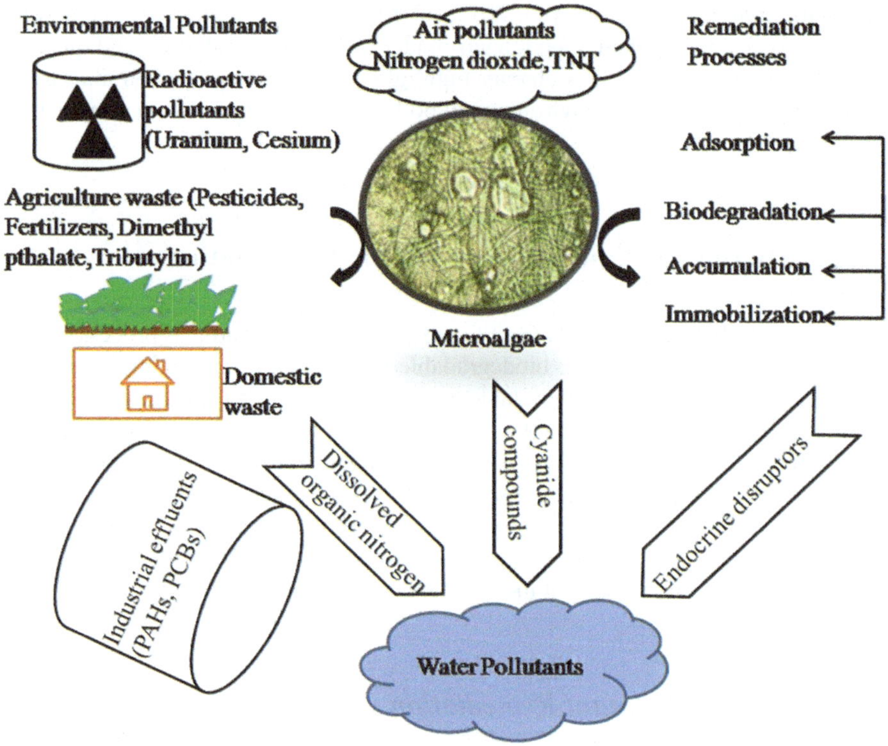

Fig. 17.1 Removal of environmental pollutants by microalgae

Nanofibers possess unique physical, chemical, and optical properties and they are used in a wide variety of applications, like air filtrations, template self-assembly in layered materials, and microsensors (Ramakrishna et al. 2006; Wang et al. 2016; Anjum et al. 2019). Nanofibers also can be hybridized with biomacromolecles and thereby can be used for wastewater treatment (Yoon et al. 2008). Algae combined with the TiO_2/Ag nanomaterial are used to remove the Cr(IV) under UV irradiation (Wang et al. 2017). Polysulfone nanofiber hybrid with microalgae *Chlamydomonas reinhardtii* is used for the removal of reactive dyes from water (San Keskin et al. 2015). Immobilized microalgal cells removed the nitrate from the liquid effluents, therefore, help to eradicate the organic waste and heavy metals from wastewater and play a crucial role in environmental industry (Eroglu et al. 2012). Water pollution is mainly caused by nanoplastics and bioplastics, and extensive preventive measures have been taken to eradicate the problems. An extracellular polymeric substance (EPS) produced by freshwater *Cyanothece* sp. has high bioflocculant capacity and can be used to replace the synthetic flocculants used in wastewater treatment (Cunha et al. 2020). Microalgal nanoparticles have the capability to reduce environmental pollutants and prevent human health from endangerment. Microalgae can be used in wastewater treatment for the removal of nitrogen, phosphorus, and heavy metals.

Table 17.1 Removal of a wide range of environmental pollutant exudes from different sources of contamination by microalgae

Group of pollutants	Type of pollutant	Source of contamination	Microalgae	References
Dissolved organic nitrogen (DON)	Urea, amino acids, amino sugars, proteins, nucleic acids, fulvic acids, and humic acids	Rivers, lakes, and marine ecosystem	*Selenastrum capricornutum*	Urgun-Demirtas et al. (2008), Sattayatewa et al. (2009)
Cyanide compounds	Hydrogen cyanide, sodium cyanide, potassium cyanide, and thiocyanates	Metal finishing, gold ore processing, plastic manufacturing, food processing, and steel production	*Nitzschia pellucida*	Vanelslander et al. (2011)
Endocrine disruptors	Alkylphenol and bisphenol A	Wastewater bodies	*Chlorella fusca*	Ji et al. (2014), Hom-Diaz et al. (2015)
Nitrogen pollutant	Nitrogen oxide (NOx), ammonia nitrogen, and TNT	Inorganic nitrogenous fertilizers and the production of wastes from human and animal populations	*Acrosiphonia coalita*, *Porphyra yezoensis*, and *Portieria hornemannii*	Abdel-Raouf et al. (2012), Mondal et al. (2019)
Heavy metal	Copper, zinc, nickel, cobalt, lead, chromium, and cadmium	Industrial waste	*Cladophora glomerata*	Abdel-Raouf et al. (2012)
Radioactive pollutant	Radium, thorium, and uranium	Industrial waste	*Cystoseira indica* and *Desmodesmus armatus*	Mondal et al. (2019), Kim et al. (2020)
Polycyclic aromatic hydrocarbon	Phenanthrene, fluoranthene, and pyrene	Industrial waste	*Selenastrum capricornutum*	Tam et al. (2010)
Polychlorinated biphenyls		Industrial waste	*Caepidium antarcticum* and *Desmarestia* sp.	Chun et al. (2013)
Phenolics	2,4-Dimethylphenol and 2-chlorophenol	Industrial waste	*Chlorella* sp., *Scenedesmus obliquus*, and *Spirulina maxima*	Klekner and Kosaric (1992), Kruefu et al. (2012)

(continued)

Table 17.1 (continued)

Group of pollutants	Type of pollutant	Source of contamination	Microalgae	References
Oil, petroleum, and hydrocarbon	n-Alkanes and polycyclic hydrocarbons	Industrial waste	*Scenedesmus obliquus* and *Nitzschia linearis*	Tang et al. (2010)
	Pesticides	Agricultural waste	*Gracilaria verrucosa*, *Scenedesmus*, *Chlamydomonas*, *Stichococcus*, and *Chlorella*	Taghizade et al. (2018), Nie et al. (2020)
	Fertilizers	Agricultural waste	*Chlorella* sp.	Savci (2012)
	Dimethyl phthalate	Agricultural waste	*Closterium lunula*, *Synechocystis* sp. PCC6803, *Synechococcus* sp. PCC7942, and *Cyanothece* sp. PCC7822	Mondal et al. (2019)
	Tributyltin	Agricultural waste	*Ankistrodesmus falcatus*, *Skeletonema costatum*, *Chlorella vulgaris*, and *Chlorella* sp.	Mondal et al. (2019)
Polyfluorinated compounds	Polyfluorinated compounds (PFCs), perfluorosulfonates (PFSAs), and perfluorocarboxylic acids (PFCAs)	PFCs in dust and air (forensic nozzles with filter housings for dust collection and polyurethane foam)	*Chlorella pyrenoidosa* and *Selenastrum capricornutum*	Mondal et al. (2019)
Pharmaceutical active compounds	Antibiotics, tranquilizers, diuretics, and psychiatric drugs	Wastewater treatment plants	*Chlamydomonas reinhardtii* and *Chlorella sorokiniana*	Mondal et al. (2019)

Hence, used for the purification of wastewater (Agarwal et al. 2019). Water scarcity is one of the major global problems along with the deprived energy resources and environmental degradation. Microalgal cultivation is one of the promising ways to lower the environmental impacts by removing (80–100%) concomitant pollutants and increasing carbon fixation rate. The microalgal biomass contains omega-3-fatty acids, pigments, amino acids, and high sugar content, thereby making them a suitable candidate for the production of biofuels and highly valuable products like carotenoids and protein-rich supplements (Shahid et al. 2020; Aziz et al. 2017; Rawat et al. 2021).

17.2.2 Remediation of Dissolved Organic Nitrogen

Dissolved organic nitrogen (DON) causes pollution of surface water. Bioavailable DON (ABDON) is a portion of DON utilized by algal species (Gobler et al. 2002; Glibert et al. 2007). DON contains metabolite urea, free and bounds amino acids, cell amino sugars, proteins, nucleic acids (RNA, DNA), fulvic acids, and humic acids (as given in Table 17.1), and large proportion of total dissolved nitrogen is present in rivers, lakes, and marine ecosystem. Surface water contains higher amount of DON than the members of N pool (Berman and Bronk 2003). DON was initially thought to be unavailable to microalgae and microorganisms. However, studies have shown that microalgae play a direct or indirect role in DON removal (Hammed et al. 2016). Under optimum conditions DON became available to living organisms in water ecosystems (Urgun-Demirtas et al. 2008; Sattayatewa et al. 2009; Simsek et al. 2012). Photochemical decomposition of DON released low-molecular-weight compounds, such as urea, amino acids, and ammonium which can stimulate the growth of algae (Vähätalo and Zepp 2005; Vähätalo and Järvinen 2007). Bioavailable DON (ABDON) for the algae can be determined by the removal of ABDON in *Selenastrum capricornutum* (Urgun-Demirtas et al. 2008; Sattayatewa et al. 2009; Simsek et al. 2012).

17.2.3 Remediation of Cyanide Compounds

Cyanides are toxic compounds containing carbon-nitrogen radicals (CN^-) (e.g., hydrogen cyanide, sodium cyanide, potassium cyanide, and thiocyanates) and they originate from manufacturing companies such as metal finishing, gold ore processing, plastic manufacturing, food processing, and steel production (Gould et al. 2012). Microalgae are used for the elimination of tightly bonded cyanide compounds from wastewater (Gurbuz et al. 2009). Although growth of microalgae could be initially retarded by cyanide, later, some microalgae withstanding the higher concentration (up to 400 mg L^{-1}) of cyanide become acclimatized (Hammed et al. 2016). Microalgae utilized cyanide through bioconversion of simple organic or inorganic molecules. Diatom *Nitzschia pellucida* produced two metabolites bromine and iodine cyanide which showed allelopathic activity that contribute significantly to the atmospheric halocarbon (Vanelslander et al. 2012). Various pathways involved in the cyanide metabolism occur in the plants, fungi, and bacteria, such as hydrolytic pathways, oxidative pathways, reductive pathways, transfer pathways, and synthesis pathways, but not reported in microalgae (Gupta et al. 2010; Gould et al. 2012). However, microalgae showed metabolized cyanide from wastewater. Therefore, more research is needed to investigate the role of cyanide metabolism in microalgae.

17.2.4 Remediation of Endocrine Disruptors

Endocrine disruptors are endocrine-disrupting compounds that are known as water pollutants such as synthetic and natural estrogens. They caused adverse impacts on human health (Hom-Diaz et al. 2015). Endocrine disruptors imposed risk to reproductive health, change in morphology, interference in sex differentiation, increased proliferation of breast cancer cells, and mutated the cellosaurus cell lines (Gattullo et al. 2012).

Microalgae have the capability to remove endocrine disruptors from wastewater (Nadal et al. 2006; Della et al. 2008; Li et al. 2009; Shi et al. 2010; Gattullo et al. 2012; Zhou et al. 2013; Hom-Diaz et al. 2015). *Chlorella fusca* was able to remove most of bisphenol A in the presence of light (Hirooka et al. 2005). The endocrine disruptors are removed by the methods of biodegradation and photolytic degradation (Shi et al. 2010). Rate of remediation depends on the molecular weight of the endocrine disruptors and the type of algal species. Immobilized microalgae can also be used for the removal process (Ji et al. 2014; Hom-Diaz et al. 2015). Further studies are needed to address the high rate of bioremediation by microalgae.

17.3 Air Pollutants Remediation

The pollutants found in the air are different types of toxic gases (nitrogen oxides, sulfur oxides, carbon oxides, ozone, etc.), suspended particulate matter, and volatile organic compounds (benzene, toluene, ethylbenzene, xylene, formaldehyde). Nanomaterials have some important properties such as large specific surface areas, size distribution, molecular structure, hydrophilicity/lipophilicity, and high reactivity. All these physicochemical characteristics make nanomaterials suitable for a wide range of environmental applications such as adsorbents, catalysts, and sensors (Neto et al. 2019).

Metal-based nanomaterials removed the various pollutants from wastewater like heavy metals and chlorinated organic compounds. Metal and metal oxide nanomaterials possess the high adsorption capacity and fast kinetics (Santhosh et al. 2016). TiO_2 NPs have been widely studied for the properties of catalysis, cleaning and purification, semiconducting, and gas sensing and is also cost-effective (Adesina 2004; Li et al. 2008).

Recently algae-based bioreactors have been used to clean up the skies of densely packed cities. It used the carbon dioxide and transferred the clean air to the environment. This system is capable of removing much of carbon as equivalent to 400 trees. These algal bioreactors work through HAVAC system. The HAVAC stands for heating, ventilation, and air conditioning. In this system, carbon dioxide is absorbed through tubes and chambers. The biomaterial feed on carbondioxide allowing the bioreactor to produce clean air in the environment (https://www.theburnin.com/technology/hypergiant-algae-bioreactor-cleans-carbon-based-air-pollution-2019-9/).

Greenhouse gases, especially CO_2 emission, are about 68% of total emissions. The transportation system and power plants are the major sources of CO_2 emissions. Greenhouse gases cause global warming which has high impact on human life and the climate. Several research and development projects have been performed to reduce the level of CO_2 in the environment. Microalgae are an efficient biological approach for CO_2 fixation (Xie et al. 2014; Jalilian et al. 2020). CO_2 can be used to grow microalgae by injecting flue gas from power plants into the microalgal culture by cultivating into the raceway pond to minimize atmospheric CO_2 (Arata et al. 2013). An algal lamp provides environmental benefits by removing excess carbon dioxide. An algal lamp removes 1 ton of carbon dioxide from the atmosphere per year, which is 10 times the approximately 0.1 tons of carbon dioxide stored and sequestered by an evergreen ash tree in a year. It has also been suggested that when these algal lamps die, they can be used as biofuels (http://sustainable-nano.com/2015/03/03/algae-lamps/).

17.3.1 Nitrogen Pollutants Remediation

Nitrogen pollution is of great concern, with nitrogen oxide (NOx) in the air and ammonia nitrogen in water being two major pollutants (Table 17.1). Culturing microalgae in NOx or wastewater with high concentrations of ammonia nitrogen would both reduce environmental pollution and provide a source of nitrogen for microalgal culture. Strategies were developed to use microalgae for the bioremediation of N pollution, especially NOx and ammonia N, in air and water. Microalgal utilization of NOx and ammonia N varies tremendously among algal strains and species. Large-scale and long-term production will help to optimize the cultivation and improve the economic viability of microalgal bioremediation of environmental pollutants. Efforts have been made to develop algae-based environmental applications to increase the basic research into microalgal biology, to screen the microalgal strains for high tolerance to pollutants, and for broad adaptability (Chang and Wang 2020). Increased use of inorganic nitrogenous fertilizers and the production of wastes from human and animal populations, due to the accumulation of nitrogen (N) in the environment, are the major concerns for health and have been the causative factor of methemoglobinemia (Abdel-Raouf et al. 2012).

Trinitrotoluene (TNT) is a widespread explosive at military sites. Various literatures reported that microalgae have the ability for the degradation of TNTs and other nitroaromatic compounds. Microalgae have the enzyme nitrate reductase which can be used in explosives as a source of nitrogen. *Acrosiphonia coalita*, *Porphyra yezoensis*, and *Portieria hornemannii* (Table 17.1) showed removal of TNT by reducing its nitrogen (Mondal et al. 2019).

17.4 Industrial Pollutants Remediation

17.4.1 Inorganic Pollutants

17.4.1.1 Heavy Metals

Heavy metal remediation by algae implicated the biosorption, change in cellular structure, modification, and immobilization. To achieve implementation of algal biosorption technology in industrial and environmental remediation requires biomass concentration, initial pH, time, and temperature (Zeraatkar et al. 2016). Cu, Zn, Ni, Co, Pb, Cr, and Cd are the major industrial pollutants as shown in Table 17.1 (Venkateswarlu et al. 1994; Kaushik et al. 1997). Bioaccumulation of metals by algae is the best method to remediate wastewater from industrial pollutants. Algae are also grown in pond with nutrition and maintenance. Marine and freshwater algae uptake heavy metals such as Co, Mo, Ca, Mg, Cu, Zn, Cr, Pb, and Se by the biological transport and accumulate these metals within the cells (Afkar et al. 2010; Kumar and Gaur 2011; Chen et al. 2012). Due to this property, algae can be used in heavy metal remediation to obtain clean water (Yee et al. 2004; Han et al. 2007; Ajjabi and Chouba 2009; Tuzen and Sari 2010). *Cladophora glomerata* accumulated Zn^{2+}, Cu^{2+}, Pb^{2+}, and Cr^{3+} as well as Ni^{2+}, Cd^{2+}, Co^{2+}, Fe^{2+}, and Mn^{2+} to clean the artificial freshwater (Table 17.1). Algae are also good accumulators of organochlorides and tributyl tin (Abdel-Raouf et al. 2012).

17.4.1.2 Radioactive Pollutants

Microalgae have the capability to remove radionuclides like radium, thorium, and uranium by the method of biosorption. *Cystoseira indica* brown algae remove uranium (Table 17.1) (Mondal et al. 2019). Photoheterotrophic microalga *Desmodesmus armatus* removes cesium (Cs^+) by using magnetic nanoparticles. *Desmodesmus armatus* grown in potassium (K^+)-starved condition improves the accumulation of Cs^+ and with the addition of volatile fatty acids enhanced removal of Cs^+ (Kim et al. 2020).

17.4.2 Organic Pollutants

The organic industrial pollutants degraded by algae are phenolics, pesticides, polyaromatic hydrocarbons (PAHs), and polychlorinated biphenyls (PCBs). Microalgae *Ankistrodesmus braunii* and *Scenedesmus quadricauda* degraded different phenols such as catechol, tyrosol, hydroxytyrosol, p-hydroxy benzoic acid, ferulic acid, p-coumaric acid, synaptic acid, caffeic acid, and vanillic acid

(Subashchandrabose et al. 2013). The *Chlorella* sp. dechlorinated 2-chlorophenol, while *Scenedesmus* sp. degraded 2,4-dinitrophenol (2,4-DNP). Oil and petroleum produced the major pollutants due to the widespread usage of automobiles and motors in industries. *Scenedesmus obliquus* and *Nitzschia linearis* grew in a medium containing 0.1% crude oil and removed n-alkanes and PAHs (Gamila et al. 2004). Organic pollutants degraded by algae and cyanobacteria are as follows:

17.4.2.1 Phenolics

Chlorella sp., *Scenedesmus obliquus*, and *Spirulina maxima* degraded the phenolic compounds found in industrial pollutants. 2,4-dimethylphenol and 2-chlorophenol were degraded by *Chlorella* and 2,4-dinitrophenol was degraded by *Scenedesmus* (Klekner and Kosaric 1992) (Table 17.1). Niobium-loaded zinc oxide nanoparticles (Nb-loaded ZnO NPs) in the range of 20 and 40 nm were synthesized by flame spray pyrolysis (FSP) technique which degrade phenol (Kruefu et al. 2012). Manganese ferrite nanopowder is prepared for thermal decomposition and used as a catalyst for the oxidative degradation of phenol in aqueous solutions in the presence of potassium peroxydisulfate as oxidant (Stonia et al. 2017).

17.4.2.2 Oil, Petroleum, and Hydrocarbon

Two microalgal species (*Chlorella vulgaris* and *Dunaliella tertiolecta*) along with ZnO nanoparticles remove the hydrocarbon from oily water samples. The toxicities of nanoparticles and paraffin hydrocarbons and their combination with the microalgae were evaluated in terms of cell growth and chlorophyll content (Salehi et al. 2019). Microalgae (*Chlorella sorokiniana*) together with bacteria (*Pseudomonas migulae*) have biodegraded 200–500 mg/L of phenanthrene under photosynthetic conditions (Muñoz et al. 2003). *Prototheca zopfii* has been reported to remove hydrocarbon through cell-free system and immobilized algal cells (Suzuki and Yamaya 2005; Ueno et al. 2006; de-Bashan and Bashan 2010). The immobilized thermotolerant strains selectively degraded aliphatic hydrocarbons (Ueno et al. 2006). *Scenedesmus obliquus* degrades the polycyclic hydrocarbons rather than n-alkanes, similarly *Nitzschia linearis* degrades the n-alkanes instead of polycyclic hydrocarbons. Microalgae play a significant role in the remediation of petroleum-contaminated water by degrading the light extractable petroleum hydrocarbons at the site (Jacques and McMartin 2009). Several microalgal-bacterial consortia have been found for remediation of hydrocarbons like *S. obliquus* that degrades the organic hydrocarbons from crude oil (Tang et al. 2010).

17.4.2.3 Polyaromatic Hydrocarbons (PAHs)

Persistent pollutants cause adverse effects to human health and environment. Most polychlorinated biphenyls (PCBs) and polycyclic aromatic hydrocarbons (PAHs) are toxic and stable in the environment (Borji et al. 2020). PAHs are toxic, carcinogenic organic compounds, and freshwater green microalga *Selenastrum capricornutum* immobilized in alginate beads removes mixed PAHs. These microalgal beads were effective in removing and biodegrading mixed PAHs at its dissolved concentrations, namely, phenanthrene, fluoranthene, and pyrene (Tam et al. 2010) (Table 17.1).

17.4.2.4 Polychlorinated Biphenyls (PCBs)

These are very hazardous pollutants because of their toxicity, carcinogenicity, wide distribution, and slow biodegradation in the environment (Meagher 2002; Perelo 2010). Some monohydroxylated PCBs are potent endocrine disrupters and some PCB metabolites are involved in neurotoxicity with a hydroxy group in the meta or para position. PCBs were used extensively in a variety of industrial applications due to its thermal stability and persist in aquatic sediments (Dhankher et al. 2012). Benthic microalgae have been used for the first time for in situ remediation (Yamamoto et al. 2008). Brown algae *Caepidium antarcticum* and *Desmarestia* sp. have the ability to associate its exudates with hydrophobic pollutants such as polychlorinated biphenyls (PCBs). Lack of effective electron donors and acceptors to promote in situ PCB biodegradation as well as the application of an electric potential to soil/sediment could be an effective means of providing electron-donors/acceptors to PCB dechlorination and degrading microorganisms, suggesting that this approach could be a cost-effective, environmentally sustainable strategy to remediate PCBs in situ (Chun et al. 2013).

17.5 Agricultural Pollutants Remediation

Wide application of agrochemicals like pesticides and fertilizers is for increased food production because of global rise in population. But excessive use of these chemicals is hazardous to the environment (Handojo et al. 2020). Various nanomaterials, including metal and metal-based silica, carbonaceous, and polymeric nanomaterials, for remediation of agricultural waste have been used (Handojo et al. 2020).

The widespread use of pesticides to boost agricultural productivity has detrimental effect on environment. Nanotechnology offers advantage in the area of pollution sensing and prevention. Nanosensors are capable of detecting toxic pollutants at very minute levels. Organic pesticides and industrial pollutants can be degraded into harmless components through photocatalysis using metal oxide semiconductor nanostructures (Baruah and Dutta 2009). The unicellular green photosynthetic algae *Chlamydomonas reinhardtii* was used to develop paper-based algal biosensor for the optical detection of nanoencapsulated atrazine (a herbicide). This is totally an environment-friendly device and cost-effective. Nanoencapsulated atrazine was detected through fluorescence spectrophotometry at wavelength ranging between 0.5 and 200 nm (Scognamiglio et al. 2019). Recently the use of green algae and cyanobacteria as fertilizer in agriculture is feasible strategy with environmental benefits. Algal biofertilizers also possess several advantages such as biocontrolling plant pathogens, reducing the use of chemicals, decline in greenhouse gas emissions, and improve the soil organic carbon, which cannot be achieved by chemical fertilizers. Cyanobacterial inoculums save the cost of fertilizers and reduce the environmental pollution by preventing the leakage of biofertilizer nutrients from farmland (Guo et al. 2020). Industrialization increases the problem of effluent disposal and maintaining the quality of natural ecosystem. Replacing the conventional approach of disposal or dilution, new technologies of bioremediation, detoxification, accumulation, and disinfection of the system are being developed. Exploitation of cyanobacteria in bioremediation is one of the recent trends. Cyanobacteria *Tolypothrix tenuis* was treated with fertilizer industry effluent to find out its nutrient removal efficiency (PO_4, SO_4, and NO_2 uptake) and changes in the metabolism (growth-chlorophyll-a, carotenoids, proteins, carbohydrate, phenols, and amino acids).

17.5.1 Pesticides

Pesticides have become a serious environmental and health hazard. Therefore, an efficient technology is essential for complete mineralization of pesticides to nontoxic forms. Nanotechnology offers many potential benefits to improve existing environmental technologies using new materials with effective performance, resulting in less consumption of energy and materials (Taghizade et al. 2018). Pesticide pollution is mainly due to dichlorodiphenyltrichloroethane (DDT). It is a major problem that has attracted great attention from researchers and the public. Organochlorine pesticides (OCPs) because of its poor degradability, toxicity, and long durability have been banned in many countries. Atrazine, glyphosate, simazine, imidacloprid, mecoprop, and isoproturon are commonly used pesticides. The major reason of pesticide contamination is surface runoff from agricultural land, condensation of volatile pesticides, improper transportation and storage of pesticides, improper spraying of pesticides, and inappropriate disposal of wastewater from pesticide industries. Recent research using microalgae for pesticide removal are mostly done on a laboratory scale, but there is need to establish methods for the large-scale treatment of pesticides using microalgae (Nie et al. 2020).

17.5.2 Fertilizers

Artificial or organic fertilizers cause serious environmental problems and they contaminate freshwater as well as marine ecosystems. Fertilizer pollution is nonpoint source pollution and it is harder to regulate (https://homeguides.sfgate.com/fertilizers-pollutants-78452.html). Maximum efficiency and highest quality product in food agriculture are needed for consumer fulfillment. Rates of nutrients in the soil affect the quality of yield. Therefore, fertilization is a priority at all times. Excessive fertilizers are the matter of concern for environmental health and reported adverse effects. Excessive fertilization, heavy metal accumulation, water eutrophication, and accumulation of nitrate can cause air pollution, which can lead to problems such as the greenhouse effect (Savci 2012). Cyanobacteria are involved in the biogeochemical cycle of carbon, nitrogen, and oxygen, so they can survive in wet soils and significantly affect the nutritional status, structural stability, and crop productivity. The microalgae *Chlorella* sp. are used in the production of biofuel, as fertilizer, and also in pollution control. The constraint of culturing of *Chlorella* sp. for large-scale production is suitable in f/2 medium. Three concentrations of commercial fertilizer with N/P/K ratio of 21:21:21 (0.07 g/L, 0.70 g/L, and 1.33 g/L) were used as alternative culture media to grow *Chlorella*. Cyanobacteria as a biofertilizer reduced global warming by decreasing CO_2 gas. Cyanobacteria are used for improving the quality of food products, physicochemical properties of soil, controlling soil-borne diseases, as nutraceuticals, and also in pharmaceuticals. Cyanobacterial biofertilizers are economical and environment-friendly (Chittora et al. 2020).

17.5.3 Dimethyl Phthalate

Dimethyl phthalate (DMP) is a phthalate ester and majorly used as plasticizers, insect repellent, in cosmetic products, paints, and automobiles. Microalgae used in the remediation of DMP like *Closterium lunula* degrades the DMP in the presence of inorganic carbon. The *Synechocystis* sp. PCC6803, *Synechococcus* sp. PCC7942, and *Cyanothece* sp. PCC7822 are also involved in the biodegradation of DMP (Mondal et al. 2019).

17.5.4 Tributyltin

Tributyltin or TBT has multiple uses, it is used as an antifouling agent in paints, antifungal agent in textiles, and also for making biocides. Several studies reported that microalgae are used in the remediation of TBT. In the process of biodegradation, TBT is converted to dibutyltin (DBT) and monobutyltin (MBT), respectively, by *Ankistrodesmus falcatus*, *Skeletonema costatum*, *C. vulgaris*, and *Chlorella* sp. (Mondal et al. 2019).

17.6 Remediation of Emerging Pollutants

Emerging pollutants (EPs) detected in the environment are very toxic and face difficulties in the degradation process. Due to lack of technology, these compounds were not detected earlier, hence, they are named EPs (Mondal et al. 2019). Emerging pollutants include a variety of compounds like antibiotics, drugs, steroids, endocrine disruptors, hormones, industrial additives, chemicals, and also microbeads and microplastics (https://www.unenvironment.org/events/un-environment-event/emerging-pollutants-wastewater-increasing-threat-0). Recently, new compound come into scene like perfluorocarbons, UV-filters, synthetic musks, parabens, siloxanes, neonicotinoids, and drug residues (Salthammer 2020). These pollutants are determined in air, water, soil, sediments, and ecological receptors even at a very lower concentration (Gavrilescu et al. 2015). Microalgae can bioremediate EPs through three processes, that is, bioadsorption, biouptake, and biodegradation. Microalgal bioadsorption occurs when EPs are either adsorbed to cell wall components or onto organic substances excreted by the cells. The biouptake involves the active transport of the contaminant into the cell, where it binds to intracellular proteins and other compounds. Microalgal biodegradation process transforms the complex compounds into simpler molecules through catalytic metabolic degradation. Biodegradation provides one of the best methods for the remediation of EPs (Sutherland and Ralph 2019).

17.6.1 Polyfluorinated Compounds

Polyfluorinated compounds (PFCs) include perfluorosulfonates (PFSAs) and perfluorocarboxylic acids (PFCAs). They are incorporated as polymer in the textile industry, as surface treatment material in furniture and carpet factories, as well as in nonstick cookware. *Chlorella pyrenoidosa* and *Selenastrum capricornutum* have been used in the bioremediation of PFCs (Mondal et al. 2019).

17.6.2 Pharmaceutical Active Compounds

Pharmaceutical active compounds (PhACs) are biologically active compounds and hazardous to the environment. PhACs are antibiotics, tranquilizers, diuretics, and psychiatric drugs. PhACs are also emerging contaminants because they remain unregulated and impact the environment, water supply, and also human health. PHACs enter the environment through diffuse sources, agricultural waste, and untreated waste coming from waste water treatment plants. Many studies have reported that PhACs have been removed using a microalgal bioremediation process. For example, a study on waste water containing antibiotics has been removed through microalgal technology. High removal rates were reported in microalgae

C. reinhardtii and *C. sorokiniana*, which grew rapidly on antibiotic sulfapyridine. Further research is needed to explore the remediation of emerging pollutants using microalgae (Mondal et al. 2019).

17.7 Strategic Approaches for the Remediation of Environmental Pollutants by Microalgae

17.7.1 Immobilized Microalgae for Removing Pollutants

An immobilized microalga is defined as a living cell that artificially or naturally prevented moving from its original location to all parts of the aqueous system. This technique is used for removing the pollutant (De-Bashan and Bashan 2010). Six different types of immobilization have been used: covalent coupling, affinity immobilization, adsorption, confinement in liquid–liquid emulsion, capture behind semipermeable membrane, and entrapment in polymers. All these types of immobilization are collectively called passive and active immobilization consisting flocculent agents, chemical attachment, and gel encapsulation (De-Bashan and Bashan 2010). The most commonly used method of immobilization is entrapment and encapsulation in polymers. In polymeric immobilization systems, there is physical separation between the microorganisms and the treated wastewater. The microorganisms are immobilized alive within the polymer because its pore is smaller than the microorganisms, while the fluid flows through it and sustains its metabolism and eventual growth. This method is eco-friendly and produces no health hazards as well as no further pollution, and moreover, their end-products can be converted to additional by-products like fertilizers or biofuel that may further reduce costs. This system produces less sludge and is simpler to maintain. Microalgal immobilization technology can treat several different contaminants simultaneously. There are still many technical aspects to be developed, such as need to improve polymers themselves, to create alloys of organic polymers by mixing different polymers, to improve diffusion of effluents, development of simple immobilization procedures, as well as proper selection of polymer (De-Bashan and Bashan 2010).

17.7.2 Adsorption

Adsorption is the process of removing pollutants by using an absorbent. The absorbent may be a mineral or of biological origin (Bai and Acharya 2017). Carbon-based nanomaterials can be used as absorbent for the removal of antibiotics. Adsorption process is widely used for the removal of industrial waste (Ahmed et al. 2015; Zhang et al. 2016; Huang et al. 2017; Zhou et al. 2017; Xiong et al. 2018a, b). The adsorption by microalgae can be achieved by the assembly of functional groups and polymer on the algal cell walls (Davis et al. 2003; Xiong et al. 2018a, b). The adsorption process depends on various factors like physiological state of the organism, age of the cells, availability of micronutrients, and the environmental conditions such as pH, temperature, and the presence of particular ions. *Chlamydomonas* sp., *Chlorella miniata*, *Scenedesmus platydiscus*, *Scenedesmus quadricauda*, *Selenastrum capricornutum*, and *Synechocystis* sp. exhibit biosorption and biodegradation processes in polyaromatic hydrocarbon. *Selenastrum capricornutum* has been cultivated for the biosorption of fluoranthene, pyrene, and benzo-α-pyrene (Pathak et al. 2018). Biosorption is also used for the removal of metals, phenols, and dyes from wastewater (Pathak et al. 2018).

17.7.3 Biodegradation

Biodegradation is described as breakdown of organic substance by the use of microalgae or living organisms. It occurs within or outside of the algal cells, resulting in the production of some end products which are further consumed by the algal cells in their metabolic process. Phenanthrene and fluoranthene are biodegraded and accumulated by the two diatoms *Skeletonema costatum* and *Nitzschia* sp. simultaneously. Microalgae also accumulated and degraded the polyaromatic hydrocarbons. The capability of *Nitzschia* sp. was higher than *S. costatum* (Pathak et al. 2018). Algae degrade the antibiotic by two methods: cometabolic and metabolic degradation. In cometabolic degradation, nonspecific enzymes are used with additional carbon and energy source, while in the metabolic process sole carbon and energy sources with specific enzymes are used. For example, levofloxacin biodegraded by *C. vulgaris* (Leng et al. 2020).

17.7.4 Accumulation

Accumulation is an intracellular process to remove pollutant from water, soil, and plants by acting as filters (Davis et al. 2003; Bai and Acharya 2017). Algae accumulation has been reported for the removal of antibiotics such as trimethoprim, sulfamethoxazole, and doxycycline (Bai and Acharya 2017; Prata et al. 2018). Accumulation of antibiotics produces reactive oxygen species which are essential to maintain the cellular level which in excess cause death, so it is necessary to maintain normal concentration at cellular level (Xiong et al. 2018a, b). Accumulation in combination with biodegradation in algal cells completes the assimilation of some antibiotics. The accumulation of sulfamethazine in *C. pyrenoidosa* was observed and then followed by biodegradation (Sun et al. 2017).

17.8 Advantages of Microalgal Nanotechnology Remediation

1. It does not emit anything harmful to the environment.
2. It has become popular as consumers of the technology are becoming more environment conscious. This will give benefits to investors at long run in certain areas.
3. It requires less cost for maintenance. This reduces operating cost and hence overall cost on the long run.
4. It uses renewable natural resources and hence will never run out of vital resources such as water and electricity.
5. It will slowdown effects of global warming due to reduction in CO_2 emissions.

17.9 Disadvantages of Microalgal Nanotechnology Remediation

1. Initial investment or implementation cost is very high.
2. People are still not familiar with the technology and hence will take time to adopt it for larger population.
3. The technology is still evolving and many of the products are at R&D stage. Hence, people are unaware of performance results.
4. Lack of skilled human resources to install or implement the green technology-based products or systems.
5. In most of the countries, policies have not been finalized for the green technology-based systems.

17.10 Scope of the Remediation Marketing

Latest reports on the global environmental remediation market found that this industry showed significant growth over last few years and anticipated a continuous growth within the forecast period of 2020–2026 (https://www.giiresearch.com/report/vmr935438-global-environmental-remediation-market-research.html). The global bioremediation market report showed significant CAGR of 7.72% during the forecast period from 2019 to 2028 (https://www.asdreports.com/market-research-report-515382/global-bioremediation-market). The global bioremediation market segmented to component, process, medium, and region. North America dominated the bioremediation market in the year 2018 with 36.22% and 32.40% rise in the upcoming year till 2028. North America has large number of remediation companies besides new products launch over there. Asia-Pacific contributes to the second largest position in the bioremediation market because of increased industrialization and urbanization over there. The global bioremediation market is based on product, process, and region. The components of bioremediation market are microorganisms, instrument, accessories, consumables, and services. Processes which have been used from the remediation are phytoremediation, bioventing, bioleaching, bioslurping, biostimulation, and bioaugmentation. Medium for source of remediation are soil, water, radioactive, etc. Application determines the environmental pollutants stream like agriculture, wastewater, aquaculture industrial waste, etc. Region determines where the bioremediation takes place like in North America, Europe, Asia-Pacific, and rest of the world. Demands for bioremediation increased to clean the environment globally. In situ remediation process developed rapidly in 2018 with 23.09% CAGR, biostimulation grew with 16.91%. Moreover, soil remediation hike in the year 2018 due to increased contaminants found in soil and various techniques are extensively used to eradicate the pollutants completely. Industrial pollutants remediation industries also contribute to the market in 2018 with 27.09% CAGR besides agriculture and aquaculture shares (21.18% in 2018).

17.10.1 Companies in the Remediation Market Globally

These companies include Organica Biotech Pvt. Ltd, Amalgan Biotech, Inc., Sarva Bio Remed, LLC, EOS Remediation LLC, Biofuture Ltd, Biodyne USA, LLC, Newterra Ltd, Novozymes, Thermo Fisher Scientific, Inc., Drylet, LLC, Microgen Biotech Limited, Probiosphere, QM Environmental Services Nederland, and Shandong Sukahan Bio-Technology Co. Ltd.

17.11 Future Projections

17.11.1 Boost Research and Analysis to Evaluate the Effects of Remediation

More research will be needed to explore the role of remediation. Moreover, remediation process is advantageous, economic, feasible, has higher public acceptance, and also has a high rate of contaminant reduction (Tangahu et al. 2011). Bioinformatics and biostatistics analyses will help to identify and evaluate the efficient metabolic processes, enzymes, genes, or operons capable of bioremediation specific pollutants. However, genomics, proteomics, and metabolomics concerned with bioremediation help to explore in devising possible solutions targeting specific pollutants and identify and compare the gene and protein sequences which are capable of efficient removal of contaminants. Diversity and phylogenetic studies are important bioremediation processes. Knowledge of these factors will help predict the behaviors of organisms under different conditions of bioremediation (Jafari et al. 2013). However, existing research has determined sequences and structural information of cellular components involved in bioremediation, but extensive research is still lacking in its diversity and modification for higher efficiency products. Aside from the removal of pollutants, the end-products or by-products are also important in the remediation system (Jonsson and Haller 2014; Tripathi et al. 2016). The future of bioremediation as a technology depends on biotic and abiotic factors that influence the natural selection of organisms capable of removing the pollutants. Moreover, genomes and proteasomes information will also contribute to significant designing of efficient bioremediation systems. Sequence information can help to determine the different variants of important genes and proteins existing in nature as well as molecular pathways can also influence bioremediation process.

17.11.2 Improve Engineering Applications Using Nanotechnology

There is a need to develop "smarter" and sophisticated nanomaterials which have the ability to perform multiple functions such as one nanomaterial remediate multiple pollutant (hydrophobic and hydrophilic). We can make the nanoparticles that destroy a wide spectrum of pollutants and improve delivery systems. All these engineering improvements can increase the ability to remediate most of the hazardous waste sites and minimizing potential harm.

17.11.2.1 Bioelectrochemical Systems (BES) for Water Pollution

Every year billions of money was used to treat trillions of liters of wastewater to consume limited amount of energy. Moreover, this water contains organic pollutants so can be used to produce electricity, hydrogen, and high-value chemicals like caustic soda. Novel systems for groundwater and wastewater remediation will be needed to develop. This approach will help to remove pollutants like TPHs, PAHs, antibiotics, metalloids, and pesticides. A mathematical model will also be developed. Hybrid systems in combination with BES reactors have been developed for soil remediation. Endogenous microorganism activity combined with the plants upgrade the remediation process.

17.11.2.2 Development of Novel Technology for Metal Removal

Depending on the metallic contaminants different bacterial strains can be used. Microbes contain various metal ions at different concentrations, which can be determined through physical-chemical techniques or electron microscope.

17.11.2.3 Improvement of Biostimulation/Bioaugmentation Technologies

In biostimulation process, changes occur in the local environment in order to stimulate endogenous bacteria for remediation. This can be done by the addition of phosphorus, nitrogen, oxygen, or carbon (in the form of molasses). In bioaugmentation, addition of bacterial cultures is required to scale up the process of remediation of a contaminant. Both these techniques are employed to degrade the TPHs and PAHs in the presence of metals and metalloids. Moreover, microalgae as biofertilizing agents as well as with the addition of alternate sources such organic amendments, nontoxic synthetic chelators, and biosurfactants can be designed.

17.11.2.4 Develop Ecopile Technology (Phytoremediation in Combination with Biostimulation/Bioaugmentation)

The ecopile process involves mitigation of hydrocarbon degraders with the known pollutant which effects the root growth and results in stimulation of microbial activity in the rhizosphere. A number of ecopiles are developed by using different pollutants such as TPH, PAH, potentially toxic metals, and metalloids in EU and China.

References

Abdel-Raouf N, Al-Homaidan A, Ibraheem IBM (2012) Microalgae and wastewater treatment. Saudi J Biol Sci 19(3):257–275

Adesina AA (2004) Industrial exploitation of photocatalysis: progress, perspectives and prospects. Catal Surv Asia 8(4):265–273

Afkar E, Ababna H, Fathi AA (2010) Toxicological response of the green alga *Chlorella vulgaris*, to some heavy metals. Am J Environ Sci 6(3):230

Agarwal P, Gupta R, Agarwal N (2019) Advances in synthesis and applications of microalgal nanoparticles for wastewater treatment. J Nanotechnol 2019:1–9

Ahmed MB, Zhou JL, Ngo HH, Guo W (2015) Adsorptive removal of antibiotics from water and wastewater: progress and challenges. Sci Total Environ 532:112–126

Ajjabi LC, Chouba L (2009) Biosorption of Cu^{2+} and Zn^{2+} from aqueous solutions by dried marine green macroalga *Chaetomorpha linum*. J Environ Manag 90(11):3485–3489

Anjum M, Miandad R, Waqas M, Gehany F, Barakat MA (2019) Remediation of wastewater using various nano-materials. Arab J Chem 12(8):4897–4919

Arata S, Strazza C, Lodi A, Del Borghi A (2013) *Spirulina platensis* culture with flue gas feeding as a cyanobacteria-based carbon sequestration option. Chem Eng Technol 36(1):91–97

Aziz N, Fatma T, Varma A, Prasad R (2014) Biogenic synthesis of silver nanoparticles using *Scenedesmus abundans* and evaluation of their antibacterial activity. Journal of Nanoparticles, Article ID 689419, 2014. https://doi.org/10.1155/2014/689419

Aziz N, Faraz M, Pandey R, Sakir M, Fatma T, Varma A, Barman I, Prasad R (2015) Facile algae-derived route to biogenic silver nanoparticles: Synthesis, antibacterial and photocatalytic properties. Langmuir 31:11605–11612. https://doi.org/10.1021/acs.langmuir.5b03081

Aziz N, Prasad R, Ibrahim AIM, Ahmed AIS (2017) Promising applications for the production of biofuels through algae. In: Microbial Biotechnology (Patra JK, Vishnuprasad CN, Das G), Springer Nature Singapore Pte Ltd. 81–103

Bai X, Acharya K (2017) Algae-mediated removal of selected pharmaceutical and personal care products (PPCPs) from Lake Mead water. Sci Total Environ 581:734–740

Baruah S, Dutta J (2009) Nanotechnology applications in pollution sensing and degradation in agriculture: a review. Environ Chem Lett 7(3):191–204

Berman T, Bronk DA (2003) Dissolved organic nitrogen: a dynamic participant in aquatic ecosystems. Aquat Microb Ecol 31(3):279–305

Borji H, Ayoub GM, Al-Hindi et al (2020) Nanotechnology to remove polychlorinated biphenyls and polycyclic aromatic hydrocarbons from water: a review. Environ Chem Lett 11:1–18

Chen H, Wang Q (2020) Microalgae-based nitrogen bioremediation. Algal Res 46:101775

Chen CY, Chang HW, Kao PC et al (2012) Biosorption of cadmium by CO_2-fixing microalga *Scenedesmus obliquus* CNW-N. Bioresour Technol 105:74–80

Chittora D, Meena M, Barupal T, Swapnil P (2020) Cyanobacteria as a source of biofertilizers for sustainable agriculture. Biochem Biophys Rep 22:100737

Chun CL, Payne RB, Sowers KR, May HD (2013) Electrical stimulation of microbial PCB degradation in sediments. Water Res 47(1):141–152

Cunha C, Silva L, Paulo J et al (2020) Microalgal-based biopolymer for nano- and microplastic removal: a possible biosolution for wastewater treatment. Environ Pollut 263:114385

Dahoumane SA, Wijesekera K, Filipe CD, Brennan JD (2014) Stoichiometrically controlled production of bimetallic gold-silver alloy colloids using micro-alga cultures. J Colloid Interface Sci 416:67–72

Dahoumane SA, Mechouet M, Wijesekera K et al (2017) Algae-mediated biosynthesis of inorganic nanomaterials as a promising route in nanobiotechnology – a review. Green Chem 19(3):552–587

Davis TA, Volesky B, Mucci A (2003) A review of the biochemistry of heavy metal biosorption by brown algae. Water Res 37(18):4311–4330

De-Bashan LE, Bashan Y (2010) Immobilized microalgae for removing pollutants: review of practical aspects. Biotech 101(6):1611–1627

Della GM, Pinto G, Pistillo P et al (2008) Biotransformation of ethinylestradiol by microalgae. Chemosphere 70(11):2047–2053

Dhankher OP, Pilon-Smites EA, Meagher RB, Doty S (2012) Biotechnological approaches for phytoremediation. In: A. Arie, MH Paul (eds), *Plant Biotechnology and Agriculture* (pp 309–329). Academic Press

El-Sheekh MM, El-Kassas HY (2016) Algal production of nano-silver and gold: their antimicrobial and cytotoxic activities: a review. J Genet Eng Biotechnol 14(2):299–310

Eroglu E, Agarwal V, Bradshaw M, Chen X et al (2012) Nitrate removal from liquid effluents using microalgae immobilized on chitosan nanofiber mats. Green Chem 14(10):2682–2685

Gamila HA, Ibrahim MBM, Sakamoto W et al (2004) Algal bioassay for evaluating the role of algae in bioremediation of crude oil: I-isolated strains. Bull Environ Contam Toxicol 73(5):883–889

Gattullo CE, Bährs H, Steinberg CE, Loffredo E (2012) Removal of bisphenol A by the freshwater green alga *Monoraphidium braunii* and the role of natural organic matter. Sci Total Environ 416:501–506

Gavrilescu M, Demnerová K, Aamand J et al (2015) Emerging pollutants in the environment: present and future challenges in biomonitoring, ecological risks and bioremediation. New Biotechnol 32(1):147–156

Glibert PM, Wazniak CE, Hall MR, Sturgis B (2007) Seasonal and interannual trends in nitrogen and brown tide in Maryland's coastal bays. Ecol Appl 17(sp5):S79–S87

Gobler CJ, Renaghan MJ, Buck NJ (2002) Impacts of nutrients and grazing mortality on the abundance of *Aureococcus anophagefferens* during a New York brown tide bloom. Limnol Oceanogr 47(1):129–141

Gould WD, King M, Mohapatra BR, Cameron RA et al (2012) A critical review on destruction of thiocyanate in mining effluents. Miner Eng 34:38–47

Guo S, Wang P, Wang X, Zou M et al (2020) Microalgae biotechnology for food, health and high value products. In: Asraful A, Xu JL, Wang Z (eds) Microalgae as biofertilizer in modern agriculture. Springer, Singapore, pp 397–411

Gupta N, Balomajumder C, Agarwal VK (2010) Enzymatic mechanism and biochemistry for cyanide degradation: a review. J Hazard Mater 176(1–3):1–13

Gurbuz F, Ciftci H, Akcil A (2009) Biodegradation of cyanide containing effluents by *Scenedesmus obliquus*. J Hazard Mater 162(1):74–79

Hammed AM, Prajapati SK, Simsek S et al (2016) Growth regime and environmental remediation of microalgae. Algae 31(3):189–204

Han X, Wong YS, Wong MH, Tam NFY (2007) Biosorption and bioreduction of Cr (VI) by a microalgal isolate *Chlorella miniata*. J Hazard Mater 146(1–2):65–72

Handojo L, Ikhsan NA, Mukti RR, Indarto A (2020) Agrochemicals detection, treatment and remediation. In: Prasad MNV (ed) Nanotechnology for remediations of agrochemicals, 1st edn. Elsevier, Butterworth-Heinemann, pp 535–567

Hirooka T, Nagase H, Uchida K et al (2005) Biodegradation of bisphenol A and disappearance of its estrogenic activity by the green alga *Chlorella fusca* var. *vacuolata*. Environ Toxicol Chem 24(8):1896–1901

Hom-Diaz A, Llorca M, Rodríguez-Mozaz S, Vicent T et al (2015) Microalgae cultivation on wastewater digestate: β-estradiol and 17α-ethynylestradiol degradation and transformation products identification. J Environ Manag 155:106–113

Huang D, Wang X, Zhang C et al (2017) Sorptive removal of ionizable antibiotic sulfamethazine from aqueous solution by graphene oxide-coated biochar nanocomposites: influencing factors and mechanism. Chemosphere 186:414–421

Jacques NR, McMartin DW (2009) Evaluation of algal phytoremediation of light extractable petroleum hydrocarbons in subarctic climates. Remediat J 20(1):119–132

Jafari M, Danesh YR, Ghoosta Y (2013) Molecular techniques in fungal bioremediation. In: Goltapeh EM, Danesh YR, Varma A (eds) Fungi as Bioremediators, 1st edn. Springer, Berlin, Heidelberg, pp 453–465

Jalilian N, Najafpour GD, Khajouei M (2020) Macro and micro algae in pollution control and biofuel production – a review. Chem Bio Eng Rev 7(1):18–33

Ji MK, Kabra AN, Choi J et al (2014) Biodegradation of bisphenol A by the freshwater microalgae *Chlamydomonas mexicana* and *Chlorella vulgaris*. Ecol Eng 73:260–269

Jonsson A, Haller H (2014) Sustainability aspects of in-situ bioremediation of polluted soil in developing countries and remote regions. In: Maria CHS (ed) Environmental risk assessment of soil contamination, pp 57–86

Kamat PV, Meisel D (2003) Nanoscience opportunities in environmental remediation. C R Chim 6(8–10):999–1007

Kaushik P, Malik A (2015) Mycoremediation of synthetic dyes: an insight into the mechanism, process optimization and reactor design. In: Singh SN (ed) Microbial degradation of synthetic dyes in wastewaters. Springer, Cham, pp 1–25

Kaushik S, Sahu BK, Lawania RK, Tiwari RK (1997) Occurrence of heavy metals in lentic water of Gwalior region. J Postgrad Med 3(4):137–140

Kim I, Choi GG, Nam SW et al (2020) Enhanced removal of cesium by potassium-starved microalga, *Desmodesmus armatus* SCK, under photoheterotrophic condition with magnetic separation. Chemosphere 252:126482

Klekner V, Kosaric N (1992) Degradation of phenols by algae. Environ Technol 13(5):493–501

Kruefu V, Ninsonti H, Wetchakun N et al (2012) Photocatalytic degradation of phenol using Nb-loaded ZnO nanoparticles. Eng J 16(3):91–100

Kumar D, Gaur JP (2011) Metal biosorption by two cyanobacterial mats in relation to pH, biomass concentration, pretreatment and reuse. Biotech 102(3):2529–2535

Leng L, Wei L, Xiong Q et al (2020) Use of microalgae based technology for the removal of antibiotics from wastewater: a review. Chemosphere 238:124680

Li Q, Mahendra S, Lyon DY et al (2008) Antimicrobial nanomaterials for water disinfection and microbial control: potential applications and implications. Water Res 42(18):4591–4602

Li R, Chen GZ, Tam NFY, Luan TG et al (2009) Toxicity of bisphenol A and its bioaccumulation and removal by a marine microalga *Stephanodiscus hantzschii*. Ecotoxicol Environ Saf 72(2):321–328

Mahdieh M, Zolanvari A, Azimee AS (2012) Green biosynthesis of silver nanoparticles by *Spirulina platensis*. Sci Iran 19(3):926–929

Mandal AK, Sarma PM, Singh B et al (2012) Bioremediation: an environment friendly sustainable biotechnological solution for remediation of petroleum hydrocarbon contaminated waste. ARPN J Sci Tech 2(8):1–12

Meagher RB (2002) Phytoremediation of toxic elemental and organic pollutant. Curr Opin Plant Biol 3(2):153–162

Merin DD, Prakash S, Bhimba BV (2010) Antibacterial screening of silver nanoparticles synthesized by marine micro algae. Asian Pac J Trop Med 3(10):797–799

Mishra A, Malik A (2013) Recent advances in microbial metal bioaccumulation. Crit Rev Environ Sci Technol 43(11):1162–1222

Mohamed EF (2017) Nanotechnology: future of environmental air pollution control. Environ Manag Sustain Dev 6(2):429–454

Mohseniazar M, Barin M, Zarredar H et al (2011) Potential of microalgae and lactobacilli in biosynthesis of silver nanoparticles. Bioimpacts 1(3):149

Mondal M, Halder G, Oinam G et al (2019) Bioremediation of organic and inorganic pollutants using microalgae. In: Singh JS (ed) New and future developments in microbial biotechnology and bioengineering. Elsevier, pp 223–235

Munoz R, Guieysse B, Mattiasson B (2003) Phenanthrene biodegradation by an algal-bacterial consortium in two-phase partitioning bioreactors. Appl Microbiol Biotechnol 61(3):261–267

Nadal M, Wargent JJ, Jones KC et al (2006) Influence of UV-B radiation and temperature on photodegradation of PAHs: preliminary results. J Atmos Chem 55(3):241–252

Nie J, Sun Y, Zhou Y, Kumar M et al (2020) Bioremediation of water containing pesticides by microalgae: mechanisms, methods, and prospects for future research. Sci Total Environ 707:136080

Neto VDOS, Freire PDTC, do Nascimento RF (2019) Removal of pesticides and volatile organic pollutants with nanoparticles. In: Nascimento RD, Ferreira RD, Paula AD, Neto VS (eds) Nanomaterials applications for environmental matrices, 1st edn. Elsevier, pp 405–426

Patel V, Berthold D, Puranik P, Gantar M (2015) Screening of cyanobacteria and microalgae for their ability to synthesize silver nanoparticles with antibacterial activity. Biotechnol Rep 5:112–119

Pathak B, Gupta S, Verma R (2018) Biosorption and biodegradation of polycyclic aromatic hydrocarbons (PAHs) by microalgae. In: Grégorio C, Eric L (eds) Green adsorbents for pollutant removal. Springer, Cham, pp 215–247

Perelo LW (2010) In situ and bioremediation of organic pollutants in aquatic sediments. J Hazard Mater 177(1-3):81–89

Prajapati SK, Kaushik P, Malik A, Vijay VK (2013) Phycoremediation and biogas potential of native algal isolates from soil and wastewater. Bioresour Technol 135:232–238

Prasad R, Aranda E (2018) Approaches in Bioremediation: The New Era of Environmental Microbiology and Nanobiotechnology. Springer International Publishing (ISBN: 978-3-030-02369-0) https://www.springer.com/gp/book/9783030023683

Prasad R, Jha A, Prasad K (2018) Exploring the Realms of Nature for Nanosynthesis. Springer International Publishing (ISBN 978-3-319-99570-0) https://www.springer.com/978-3-319-99570-0

Prata JC, Lavorante BR, Maria da Conceição BSM, Guilhermino L (2018) Influence of microplastics on the toxicity of the pharmaceuticals procainamide and doxycycline on the marine microalgae *Tetraselmis chuii*. Aquat Toxicol 197:143–152

Ramakrishna S, Fujihara K, Teo WE et al (2006) Electrospun nanofibers: solving global issues. Mater Today 9(3):40–50

Rawat J, Gupta PK, Pandit S, Prasad R, Pande V (2021) Current perspectives on integrated approaches to enhance lipid accumulation in microalgae. 3 Biotech 11, 303. https://doi.org/10.1007/s13205-021-02851-3

Rösken LM, Körsten S, Fischer CB et al (2014) Time-dependent growth of crystalline Au^0-nanoparticles in cyanobacteria as self-reproducing bioreactors: *Anabaena* sp. J Nanopart Res 16(4):2370

Salehi M, Biria D, Shariati M, Farhadian M (2019) Treatment of normal hydrocarbons contaminated water by combined microalgae – Photocatalytic nanoparticles system. J Environ Manag 243:116–126

Salthammer T (2020) Emerging indoor pollutants. Int J Hyg Environ Health 224:113423

San Keskin NO, Celebioglu A, Uyar T, Tekinay T (2015) Microalgae immobilized by nanofibrous web for removal of reactive dyes from wastewater. Ind Eng Chem Res 54(21):5802–5809

Santhosh C, Velmurugan V, Jacob G (2016) Role of nanomaterials in water treatment applications: a review. Chem Eng J 306:1116–1137

Sarma H, Joshi S, Prasad R, Jampilek J (2021) Biobased Nanotechnology for Green Applications. Springer International Publishing (ISBN 978-3-030-61985-5) https://www.springer.com/gp/book/9783030619848

Sattayatewa C, Pagilla K, Pitt P et al (2009) Organic nitrogen transformations in a 4-stage Bardenpho nitrogen removal plant and bioavailability/biodegradability of effluent DON. Water Res 43(18):4507–4516

Savci S (2012) An agricultural pollutant: chemical fertilizer. Int J Environ Sci Technol 3(1):73

Scognamiglio V, Antonacci A, Arduini F et al (2019) An eco-designed paper-based algal biosensor for nanoformulated herbicide optical detection. J Hazard Mater 373:483–492

Shahid A, Malik S, Zhu H et al (2020) Cultivating microalgae in wastewater for biomass production, pollutant removal, and atmospheric carbon mitigation; a review. Sci Total Environ 704:135303

Shi W, Wang L, Rousseau DP, Lens PN (2010) Removal of estrone, 17α-ethinylestradiol, and 17ß-estradiol in algae and duckweed-based wastewater treatment systems. Environ Sci Pollut Res Int 17(4):824–833

Simsek H, Kasi M, Wadhawan T et al (2012) Fate of dissolved organic nitrogen in two stage trickling filter process. Water Res 46(16):5115–5126

Stoia M, Muntean C, Militaru B (2017) $MnFe_2O_4$ nanoparticles as new catalyst for oxidative degradation of phenol by peroxydisulfate. J Environ Sci 53:269–277

Subashchandrabose SR, Ramakrishnan B, Megharaj M, Venkateswarlu K, Naidu R (2013) Mixotrophic cyanobacteria and microalgae as distinctive biological agents for organic pollutant degradation. Environ Int 51:59–72

Sun M, Lin H, Guo W et al (2017) Bioaccumulation and biodegradation of sulfamethazine in *Chlorella pyrenoidosa*. J Ocean Univ China 16(6):1167–1174

Sutherland DL, Ralph PJ (2019) Microalgal bioremediation of emerging contaminants – opportunities and challenges. Water Res 164:114921

Suzuki T, Yamaya S (2005) Removal of hydrocarbons in a rotating biological contactor with biodrum. Process Biochem 40(11):3429–3433

Taghizade FT, Mehrdadi N, Baghdadi M, Nabi Bidhendi GR (2018) Application of nanotechnology in pesticides removal from aqueous solutions – a review. J Nanosci Nanotechnol 14(1):43–56

Tam NFY, Chan MN, Wong YS et al (2010) Removal and biodegradation of polycyclic aromatic hydrocarbons by immobilized microalgal beads. WIT Trans Ecol Environ 140:391–402

Tang X, He LY, Tao XQ et al (2010) Construction of an artificial microalgal-bacterial consortium that efficiently degrades crude oil. J Hazard Mater 181(1–3):1158–1162

Tangahu BV, Sheikh Abdullah SR, Basri H et al (2011) A review on heavy metals (As, Pb, and Hg) uptake by plants through phytoremediation. Int J Chem Eng 2011:939161

Tripathi V, Edrisi SA, O'Donovan A et al (2016) Bioremediation for fueling the biobased economy. Trends Biotechnol 34(10):775–777

Tripathy A, More RD, Gupta S, Samuel J, Singh J, Prasad R (2021) Present and future prospect of algae: A potential candidate for sustainable pollution mitigation. The Open Biotechnology Journal 15:142–156. https://doi.org/10.2174/1874070702115010142

Tuzen M, Sarı A (2010) Biosorption of selenium from aqueous solution by green algae (*Cladophora hutchinsiae*) biomass: equilibrium, thermodynamic and kinetic studies. Chem Eng J 158(2):200–206

Ueno R, Wada S, Urano N (2006) Synergetic effects of cell immobilization in polyurethane foam and use of thermotolerant strain on degradation of mixed hydrocarbon substrate by *Prototheca zopfii*. Fish Sci 72(5):1027–1033

Urgun-Demirtas M, Sattayatewa C, Pagilla K (2008) Bioavailability of dissolved organic nitrogen in treated effluents. Water Environ Res 80(5):397–406

Vähätalo AV, Järvinen M (2007) Photochemically produced bioavailable nitrogen from biologically recalcitrant dissolved organic matter stimulates production of a nitrogen-limited microbial food web in the Baltic Sea. Limnol Oceanogr 52(1):132–143

Vähätalo AV, Zepp RG (2005) Photochemical mineralization of dissolved organic nitrogen to ammonium in the Baltic Sea. Environ Sci Technol 39(18):6985–6992

Vanelslander B, Paul C, Grueneberg J et al (2012) Daily bursts of biogenic cyanogen bromide (BrCN) control biofilm formation around a marine benthic diatom. Proc Natl Acad Sci U S A 109(7):2412–2417

Vanelslander B, Pohnert G, Sabbe K, Vyverman W (2011) Chemical warfare between microalgae: biogenetic bromine cyanide (BrCN) controls biofilm formation around a marine benthic diatom. In: J. Mees, J. Seys (eds), VLIZ Young Marine Scientists' Day. Vlaams Instituut Voor de Zee(VLIZ), Oostende, 102

Venkateswarlu V, Reddy PM, Kumar B (1994) Heavy-metal pollution in the rivers of Andhra Pradesh, India. J Environ Biol 15(4):275–282

Wang X, Yu J, Sun G, Ding B (2016) Electrospun nanofibrous materials: a versatile medium for effective oil/water separation. Mater Today 19(7):403–414

Wang L, Zhang C, Gao F et al (2017) Algae decorated TiO_2/Ag hybrid nanofiber membrane with enhanced photocatalytic activity for Cr (VI) removal under visible light. Chem Eng J 314:622–630

Wang F, Guan W, Xu L et al (2019) Effects of nanoparticles on algae: adsorption, distribution, ecotoxicity and fate. Appl Sci 9(8):1534

Wasi S, Tabrez S, Ahmad M (2013) Use of *Pseudomonas* spp. for the bioremediation of environmental pollutants: a review. Environ Monit Assess 185(10):8147–8155

Xie YP, Ho SH, Chen CY et al (2014) Simultaneous enhancement of CO_2 fixation and lutein production with thermo-tolerant *Desmodesmus* sp. F51 using a repeated fed-batch cultivation strategy. Biochem Eng J 86:33–40

Xiong W, Zeng G, Yang Z et al (2018a) Adsorption of tetracycline antibiotics from aqueous solutions on nanocomposite multi-walled carbon nanotube functionalized MIL-53 (Fe) as new adsorbent. Sci Total Environ 627:235–244

Xiong JQ, Kurade MB, Jeon BH (2018b) Can microalgae remove pharmaceutical contaminants from water? Trends Biotechnol 36(1):30–44

Yamamoto T, Goto I, Kawaguchi O, Minagawa K, Ariyoshi E, Matuda O (2008) Phytoremediation of shallow organically enriched marine sediments using benthic microalgae. Mar Pollut Bull 57(1–5):108–115

Yee N, Benning LG, Phoenix VR, Ferris FG (2004) Characterization of metal-cyanobacteria sorption reactions: a combined macroscopic and infrared spectroscopic investigation. Environ Sci Technol 38(3):775–782

Yoon K, Hsiao BS, Chu B (2008) Functional nanofibers for environmental applications. J Mater Chem 18(44):5326–5334

Zeraatkar AK, Ahmadzadeh H, Talebi AF et al (2016) Potential use of algae for heavy metal bioremediation, a critical review. J Environ Manag 181:817–831

Zhang C, Lai C, Zeng G et al (2016) Efficacy of carbonaceous nanocomposites for sorbing ionizable antibiotic sulfamethazine from aqueous solution. Water Res 95:103–112

Zhou GJ, Peng FQ, Yang B, Ying G (2013) Cellular responses and bioremoval of nonylphenol and octylphenol in the freshwater green microalga *Scenedesmus obliquus*. Ecotoxicol Environ Saf 87:10–16

Zhou W, Wang J, Chen P et al (2017) Bio-mitigation of carbon dioxide using microalgal systems: advances and perspectives. Renew Sust Energ Rev 76:1163–1175

Index

D. Thangadurai et al. (eds.), *Bioprospecting Algae for Nanosized Materials*, Nanotechnology in the Life Sciences,
https://doi.org/10.1007/978-3-030-81557-8

D

T

The manufacturer's authorised representative in the EU is Springer Nature Customer Service Centre GmbH, Europaplatz 3, 69115 Heidelberg, Germany. If you have any concerns regarding our products, please contact ProductSafety@springernature.com

Printed and bound by CPI Group (UK) Ltd, Croydon, CR0 4YY
22/09/2026
02220954-0006